Study Guide

to Accompany

Seeley, Stephens, Tate

Anatomy and Physiology

Philip Tate
Phoenix College

James Kennedy
Phoenix College

Rodney Seeley
Idaho State University

Original Illustrations by D. Michael Dick

Times Mirror/Mosby
College Publishing

ST. LOUIS . TORONTO . BOSTON . LOS ALTOS

Editor: David Kendric Brake
Developmental Editor: Jean Babrick
Project Manager: Patricia Gayle May
Production Editor: Mary Drone
Designer: John Rokusek
Camera Ready Production: Pepe Productions

Preface

To the Student

This study guide is designed to accompany Anatomy and Physiology by Seeley, Stephens, and Tate. Each chapter in the study guide, and the order of topics within the chapter, corresponds to a chapter in the text. This makes it possible for you to study systematically and also makes it easier for you to find or to review information. Read the entire chapter in the text before you use this study guide. It is designed to help you understand and master the subject of anatomy and physiology.

FEATURES

Focus

Each chapter begins with a focus statement, briefly reviewing some of the main points of the text chapter. This is not a chapter summary; you will find that in the text. The chapter summary is useful. Reading it should become a routine part of your study habits. In the study guide, the focus statement sets the stage by reminding you of the major concepts you should have learned by reading the textbook.

Word Parts

Many of the words used in anatomy and physiology come from Latin, Greek, or other languages. In their original language the words are descriptive; and learning the origin of words can help you remember what the word means. Many of these words are made up of a word root containing the meaning plus a prefix or a suffix that modifies the meaning. Take, for example, "cyte", the word root that means cell. If the prefix "chondro-" (meaning cartilage), is added to the word root, the word chondrocyte is formed. A chondrocyte is a cartilage cell. If the suffix "-ology" (meaning study of), is added to the word root, the word cytology is formed. Cytology is the study of cells. Knowing the fundamental parts of a word makes its meaning clearer and makes it easier to remember the word.

The word parts exercise lists the important word parts of the vocabulary in the chapter and gives their meaning. You should find and write down an example of a word for each word part and relate the meaning of the word part to the definition of the word. The glossary and the list of word roots, prefixes, suffixes, and combining forms on the inside cover of the textbook provide additional information. If you can't find an example, check the answer key.

Content Learning Activity

This section of the study guide contains a variety of exercises including matching, completion, ordering, and labeling activities, arranged by order of topics in the text. Each part begins with a quotation from the text or a statement that identifies the subject to be covered. Occasionally, you will find a "bulletin" statement describing important information. Just because that information is not a question does not mean it is unimportant. Quite the contrary. The "bulletin" statements are added to the learning manual because they will help you to understand the material, so pay attention to them.

The content learning activity is not a test; it is a strategy to help you learn. Don't guess! If you learn something incorrectly it is difficult to relearn it correctly. Use the textbook or your lecture notes for help whenever you are not sure of an answer. The emphasis here is on learning the content, hence the name of this section. The content questions cover the material in the same sequence as it is presented in the text. Learning the material in this order makes it easier to relate pieces of information to each other, and makes it easier to remember the information.

After completing the exercises check your answers against the answer key. If you missed a question, check the text to make sure you now understand the correct answer. Before going on to the next section of the learning manual, review this section to be sure you understand and remember the content. Cover the answers you have written with a piece of paper and mentally answer each exercise once more as you review.

Quick Recall

The quick recall section asks you to list, name, or briefly describe some aspect of the chapter's content. Although this section can be completed rapidly, do not confine yourself to quickly writing down the answers. As you complete each quick recall question, use it to trigger more information in your mind. For example, if the quick recall question asks you to name the two major regions of the body, do that, then think of their definition, what their various subparts are, visualize them, and so on. This section should be enjoyable and satisfying because it will demonstrate that you have learned the basic information about the material. Verify your answers against the answer key.

Mastery Learning Activity

The mastery learning activity lets you see what you have learned and if you can use that information. It consists of multiple choice questions that are similar to the questions on the exams you will take for a grade, so it is really a "practice" test and should be taken as a test. However, don't guess. This "practice" test is also a learning tool. If you don't know the answer for sure, admit it and then find out what the correct answer is. Some of the questions require recall of information. Others may state the information somewhat differently than the way it appeared in either the text or study guide. This is entirely fair, since in real life you must be able to recognize the information no matter how it is reworded, and you should even be able to express the information in your own words. Another goal of this section is to make you think about the relationship between different bits of information or concepts, so some of the questions are more complex than those requiring only recall. Finally, some questions in this section ask you to use what you have learned to solve new problems.

After you have answered these questions, check the answer key. In addition to the answers, there is a detailed explanation of why a particular answer was correct. Sometimes an explanation of why a choice is incorrect is also given. These explanations are provided because this section is more difficult than the preceding sections. Make sure you understand why each answer is correct. Check the textbook, ask another student, talk with your instructor, but make sure you know. The mastery learning activity will show you the areas that you need to concentrate on further. Use it to improve your understanding of anatomy and physiology.

The format of this section allows you to write the answers to the questions beside each question. If you cover the answers, you can retake the test. Don't be satisfied until you get at least 90% of the questions correct.

Final Challenges

This section of the study guide corresponds to the concept questions at the end of each chapter in the text. These questions challenge you to apply information to new situations, analyze data and come to conclusions, synthesize solutions, and evaluate problems. Some of the problem solving questions in the mastery learning section have given you practice for the questions in this section. In addition, explanations are provided to help you see how to go about solving questions of this type. Even though explanations are given, write down your answers to the questions on a separate sheet of paper. Writing is a good way to organize your thoughts and most of us can benefit from practice in writing. A good way to see if you have communicated your thoughts effectively is to have another student read your answers and see if they make sense.

The questions contain useful information, but they are not designed primarily to help you learn specific information. Rather, they emphasize the thought processes necessary to solve problems. If all you do is read the question and quickly look up the answer, you have defeated the purpose of this section. Think about the questions and develop your reasoning skills. Long after you have forgotten a particular bit of information, these skills will be useful, not only for anatomy and physiology related problems, but for many other aspects of your life as well. We hope that you not only see the benefit of possessing problem solving skills, but will come to appreciate that solving problems is fun!

A Final Thought

Good luck with all aspects of the anatomy and physiology course you are about to begin. For most students this is a challenging course and we hope that the study guide makes things a little easier and a little clearer. We are confident that when you have completed the course you will be proud of what you have accomplished. Just remember to enjoy the learning process as you go along.

Philip Tate
James Kennedy
Rodney Seeley

ACKNOWLEDGMENTS

The development and production of this study guide involved much more than the work of the authors and we gratefully acknowledge the assistance of the many other individuals involved. We wish to thank our families for their support, for their encouragement and understanding from the beginning to the end of the project helped to sustain our efforts. A special note of appreciation must be extended to Erica Michaels - for her input was invaluable during the development of the format of the study guide. We also wish to acknowledge David Brake and Jean Babrick at Times Mirror/Mosby College Publishing for their assistance in making the study guide a unique and valuable asset for the student. Our thanks to everyone involved with the artwork taken from the textbook, and in particular, our thanks to D. Michael Dick for the illustrations drawn especially for the study guide. We also wish to recognize the contribution of Sue Pepe to the design, layout, and production of the study guide. To the reviewers listed below, our gratitude for their thorough and thoughtful critiques, which resulted in a significant improvement of the study guide. Thank you.

Anthony Chee
Houston Community College

James Fawcett
University of Nebraska/Omaha

Dale Fishbeck
Youngstown State University

Bonnie Gordon
Memphis State University

James Hall
Central Piedmont Community College

Kenneth Kaloustian
Quinnipiac College

Harvey Liftin
Broward Community College

Betty Orr
Sinclair Community College

Clarence Wolfe
Northern Virginia Community College

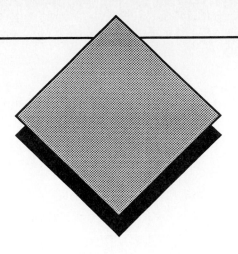

Contents

The Human Organism

FOCUS: The human organism is often examined at seven structural levels: chemical, organelles, cells, tissues, organs, organ systems, and the organism. Anatomy examines the structure of the human organism, and physiology investigates its processes. Structure and process interact to maintain homeostasis through negative feedback mechanisms. Human anatomy can be examined by sectioning the body into three planes: sagittal, transverse, and frontal. The trunk contains three cavities: thoracic, abdominal, and pelvic. These cavities and the organs they contain are lined with serous membranes.

WORD PARTS

Give an example of a new vocabulary word that contains each word part.

WORD PART	MEANING	EXAMPLE
cyto-	cell	1. _____
-logy	study of; thought	2. _____
homeo-	the same; steady	3. _____
-stasis	standing; staying	4. _____
viscer-	organs of the body cavity	5. _____
sagitt-	an arrow	6. _____
pleur-	the side; a rib	7. _____
retro-	behind; back of	8. _____
pariet-	wall	9. _____
cardio-	the heart	10. _____

Anatomy and Physiology

66 *Anatomy is the scientific discipline that investigates the body's structure.* *99*

Match these terms with the correct statement or definition:

Anatomic imaging
Cytology
Histology

Regional anatomy
Surface anatomy
Systemic anatomy

o __Cytology__ 1. Study of the structural features of cells.

o __Histology__ 2. Study of tissues.

o __systemic anatomy__ 3. Study of the body by systems (a group of structures that have one or more common functions.)

o __surface anatomy__ 4. Use of external landmarks such as bony projections to locate deeper structures.

☞ Physiology is the scientific discipline that deals with the vital processes of living things.

Structural and Functional Organization

66 The body can be conceptually considered at seven structural levels. *99*

A. Match these terms with the correct statement or definition:

Cell
Chemical level
Organ
Organism

Organelle
Organ system
Tissue

o __Organelle__ 1. Structure within a cell that performs one or more specific functions.

o __Organ__ 2. Two or more tissue types that perform one or more common functions.

o __Cell__ 3. Basic living subunit of all plants and animals.

o __Organ System__ 4. Group of organs classified as a unit because of a common set of functions.

o __tissue__ 5. Group of cells with similar structure and function, plus the extracellular substances located between the cells.

B. Using the terms from part A, arrange the structural levels of the body in order, from smallest to largest.

o 1. __Chemical level__   o 5. __Organ__

o 2. __Cell__   o 6. __Organ system__

o 3. __Organelle__   o 7. __Organism__

o 4. __Tissue__

SYSTEMS	COMPONENTS	FUNCTIONS
Cardiovascular	Brain, spinal cord, nerves	Exchanges gases between blood and the air
Endocrine	Mouth, esophagus, stomach, intestines	Removes waste products from circulatory system
Integumentary	Bones, cartilage	
Lymphatic		Allows body movement, maintains posture
Reproductive		

C. Complete the table by selecting the correct system, component or function from the lists above.

ORGAN SYSTEMS OF THE BODY

SYSTEM	MAJOR COMPONENTS	FUNCTIONS
(1)	Skin, hair, nails	Protects, prevents water loss
Skeletal	(2)	Protects, supports, produces blood cells
Muscular	Muscles attached to the skeleton	(3)
Nervous	(4)	Detects sensation, controls movement
(5)	Glands such as thyroid, pituitary, and adrenals	Regulation of metabolism and reproduction
(6)	Heart, blood vessels, and blood	Transports nutrients, wastes, gases
(7)	Lymph vessels and nodes	Removes foreign substances from blood, maintains tissue fluid
Respiratory	Lungs, respiratory passages	(8)
Digestive	(9)	Process of digestion, absorption of nutrients
Urinary	Kidneys, urinary bladder	(10)
(11)	Gonads, accessory structures and genitals	Process of reproduction, sexual function

1. _Integumentary_

2. _bones_
   cartilage

3. _allows body movmnt maintains posture_

4. _brain spinal cord, nerves_

5. _Endocrine_

6. _Cardiovascular_

7. _Lymphatic_

8. _exchanges gases between blood & air_

9. _stomach intestine mouth, esophagus_

10. _removes waste prod. from circ. sys_

11. _Reproductive_

Homeostasis

"Homeostasis is the existence of a relatively constant environment within the body."

Match these terms with the correct statement or definition:

Negative feedback
Positive feedback

_____ 1. Maintains homeostasis by resisting or reducing any deviation from an ideal normal value.

_____ 2. Medical therapy is often designed to help this type of feedback.

_____ 3. When a deviation from a normal value occurs, the response is to increase the deviation.

Directional Terms

"Directional terms always refer to the body in the anatomical position."

Match these terms with the correct statement or definition:

Anterior	Lateral
Caudal	Medial
Cephalic	Posterior
Deep	Proximal
Distal	Superficial
Inferior	Superior
Dorsal	Ventral

inferior or caudal 1. "Lower than" or "toward the tail" (two terms).

superior or cephalic 2. "Higher than" or "toward the head" (two terms).

anterior or ventral 3. "Toward the front" or "toward the belly" (two terms).

posterior / dorsal 4. "Toward the back (of the body)" (two terms).

distal 5. "Farther than another structure from the point of attachment to the trunk".

lateral 6. "Away from the midline".

deep ✓ medial 7. "Away from the surface".

Planes

"A plane is an imaginary flat surface passing through the body or an organ."

A. Match these terms with the correct statement or definition:

Frontal (coronal) plane	Sagittal plane
Longitudinal	Transverse plane
Oblique	

transvervse 1. Divides the body into superior and inferior portions and runs parallel to the surface of the ground.

_____coronal_____ 2. Runs vertically through the body and divides the body into anterior and posterior portions.

_____longitudinal_____ 3. A cut through the long axis of an organ.

_____oblique_____ 4. A cut across the long axis of an organ at any angle other than a right angle.

 A midsagittal section divides the body into equal right and left halves. A parasagittal section divides the body into right and left parts to one side of the midline.

B. Match these terms with the correct planes labeled in Figure 1-1:

Frontal (coronal) plane
Midsagittal plane
Transverse plane

1. ___coronal/frontal___
2. ___transverse___
3. ___midsagittal___

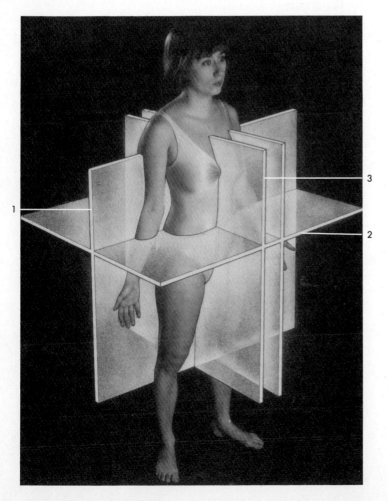

Figure 1-1

C. Match these terms with the
correct section in Figure 1-2:

Longitudinal section
Oblique section
Transverse section

1. _longitudinal_
2. _transverse_
3. _oblique_

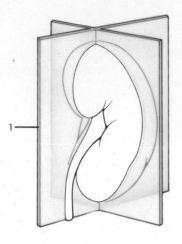
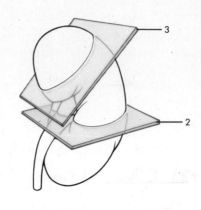

Figure 1-2

Body Regions

"The body is commonly divided into several regions.**"**

Using the list of body regions provided, complete the following statements:

Abdomen   Pectoral
Arm   Pelvic
Leg   Thigh
Limbs   Trunk
Neck   Wrist

The appendicular region consists of the _(1)_ and their associated
girdles. The axial region includes the head, _(2)_, and _(3)_. The
upper limb is attached to the body by the _(4)_ girdle, whereas the
lower limb is attached by the _(5)_ girdle. The upper limb is divided
into the _(6)_, forearm, _(7)_, and hand. The lower limb is divided into
the _(8)_, _(9)_, ankle and foot. The trunk is divided into the thorax,
_(10)_, and pelvis.

1. _limbs_
2. _Trunk_
3. _neck_
4. _pectoral_
5. _pelvic_
6. _Arm_
7. _wrist_
8. _thigh_
9. _leg_
10. _abdomen_

The abdominal region can be subdivided into four quadrants by imaginary lines - one horizontal and one vertical - that intersect at the umbilicus (navel).

Body Cavities

" The body contains several large trunk cavities that do not open to the exterior of the body. "

A. Match these terms with the correct statement or definition:

Abdominal cavity
Pelvic cavity
Thoracic cavity

Thoracic cavity 1. Cavity surrounded by the rib cage, divided by the mediastinum, and bounded inferiorly by the diaphragm.

pelvic cavity 2. Cavity bounded by the abdominal muscles and the superior bones of the pelvis.

abdominal cav. 3. Cavity containing the stomach and kidneys.

pelvic cavity 4. Small space enclosed by the bones of the pelvis.

There is no physical separation between the abdominal and pelvic cavities, which are sometimes called the abdominopelvic cavity.

B. Match these terms with the correct statement or definition:

Mesentery
Parietal serous membrane
Pericardial membrane
Peritoneal membrane

Pleural membrane
Retroperitoneal
Visceral serous membrane

visceral serous memb. 1. Portion of a serous membrane in contact with an organ.

Pleural memb. 2. Serous membrane that surrounds the lungs and lines the thoracic cavity.

Peritoneal memb. 3. Serous membrane that lines the abdominal and pelvic cavities and their organs.

Mesentery 4. Double-layered serous membrane that anchors some abdominal organs to the body wall.

Retroperitoneal 5. Location of organs covered only by parietal peritoneum.

 A potential space or cavity is located between the visceral and parietal serous membranes. The cavity is filled with serous fluid that reduces friction between the visceral and parietal serous membranes.

C. Match these terms with the correct cavity or structure labeled in Figure 1-3:

Abdominal cavity
Diaphragm
Mediastinum
Pelvic cavity
Pericardial cavity
Peritoneal cavity
Pleural cavity
Thoracic cavity

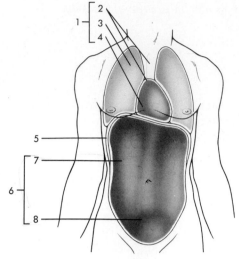

Figure 1-3

1. _Thoracic cavity_
2. _mediastinum_
3. _pleural_
4. _pericardial_
5. _peritoneal_
6. _abdominal_
7. _diaphram_
8. _pelvic_

diaphram
peritoneal
abdominal

QUICK RECALL

1. **List the four primary tissue types.**

   Epithelial, Connective, Muscle & Nervous

2. **List the two kinds of feedback mechanisms found in living things.**

   + & –

3. **Describe the anatomical position.**

   erect & hands facing forward arms by sides

4. **List the three major planes used to section the human body.**

   Sagittal, transverse, frontal (Coronal)

5. List the three major planes used to section an organ of the human body.

longittudinal, oblique, transverse (cross)

6. List the three trunk cavities of the human body.

thoracic, abdominal, pelvic

7. Name the three serous membranes lining the trunk cavities and their organs.

pericardial - heart, pleural - lungs

8. List four retroperitoneal organs.

kidneys,

& peritoneal adrenal glands, pancreas, portions of intestine & urinary bladder

MASTERY LEARNING ACTIVITY

Place the letter corresponding to the correct answer in the space provided.

<u>a</u> 1. Physiology
   a. deals with the vital processes or functions of living organisms.
   b. includes the subdiscipline of cytology.
   c. is defined as the study of tissues.
   d. all of the above

<u>c</u> 2. An organ is
   a. a specialized structure within a cell that carries out a specific function.
   b. at a lower level of organization than a cell.
   c. two or more tissues that perform a specific function.
   d. a group of cells that perform a specific function.
   e. a and b

<u>d</u> 3. The systems that are the most important in the regulation or control of the other systems of the body are the
   a. circulatory and muscular systems.
   b. circulatory and endocrine systems.
   c. nervous and circulatory systems.
   d. nervous and endocrine systems.

____ 4. Negative feedback mechanisms
   a. make deviations from normal smaller.
   b. maintain homeostasis.
   c. are responsible for an increased rate of sweating when air temperature is higher than body temperature.
   d. a and b
   e. all of the above

<u>a</u> 5. Which of the following terms do NOT mean the same thing when referring to a human in the anatomical position?
   a. superior and dorsal
   b. anterior and ventral
   c. inferior and caudal
   d. posterior and dorsal

<u>d</u> 6. A term that means "toward the attached end of a limb" is
   a. medial.
   b. lateral.
   c. distal.
   d. proximal.
   e. superficial.

ich of the following directional
ns are paired most appropriately as
osites?
   a. superficial and deep
   b. medial and proximal
   c. tal and lateral
   d. superior and posterior
   e. anterior and inferior

___d___ 8. The chin is _____ to the umbilicus
   (belly button).
   a. caudad
   b. anterior
   c. posterior
   d. superior
   e. lateral

___a___ 9. A section that divides the human body
   into anterior and posterior parts is a
   _____ section.
   a. frontal
   b. sagittal
   c. transverse

___b___ 10. Which of the following terms is
   correctly defined?
   a. The arm is that part of the upper
      limb between the shoulder and
      wrist.
   b. The leg is that part of the lower limb
      between the knee and ankle.
   c. The appendicular region includes the
      head, neck, and trunk.
   d. The pectoral girdle is where the
      lower limbs attach.

___a___ 11. The pelvic cavity contains the
   a. urinary bladder.
   b. liver.
   c. stomach.
   d. spleen.
   e. kidneys.

___b___ 12. The thoracic cavity is separated from
   the abdominal cavity by the
   a. sternum.
   b. diaphragm.
   c. septum.
   d. mediastinum.
   e. mesentery.

___c___ 13. The lungs are found within the
   a. mediastinum.
   b. pericardial cavity.
   c. thoracic cavity.
   d. all of the above

___c___ 14. Given the following characteristics:
   1. reduces friction between organs
   2. functions in secretion and lubrication
   3. lines cavities that open to the exterior
   4. lines cavities that do NOT open
      to the exterior

   Which of the above are characteristic of
   serous membranes?
   a. 1, 2
   b. 1, 2, 3
   c. 1, 2, 4
   d. 2, 3, 4
   e. 1, 2, 3, 4

___c___ 15. Given the following cavities:
   1. abdominal
   2. pelvic
   3. oral
   4. pericardial

   Which cavities are lined by serous
   membranes?
   a. 1, 2
   b. 1, 2, 3
   c. 1, 2, 4
   d. 2, 3, 4
   e. 1, 2, 3, 4

___e___ 16. Given the following organ and space
   combinations:
   1. heart and pericardial space
   2. lungs and pleural space
   3. stomach and peritoneal space
   4. kidney and peritoneal space

   kidneys are retroperitoneal

   Which of the organs is correctly paired
   with a space that surrounds that organ?
   a. 1, 2
   b. 1, 2, 3
   c. 1, 2, 4
   d. 2, 3, 4
   e. 1, 2, 3, 4

10

_e_ 17. Given the following body cavity and
membrane combinations:
1. abdominal cavity and peritoneum
2. thoracic cavity and pleural
membrane
3. pericardial cavity and pericardial
membrane
4. pelvic cavity and peritoneum

Which of the body cavities are correctly
paired with a membrane lining that
body cavity?
a. 1, 2
b. 2, 3
c. 3, 4
d. 1, 2, 3
e. 1, 2, 3, 4

_a_ 18. Which of the following membrane
combinations line portions of the
diaphragm?
a. parietal pleura - parietal peritoneum
b. parietal pleura - visceral peritoneum
c. visceral pleura - parietal peritoneum
d. visceral pleura - visceral peritoneum

_e_ 19. Which of the following organs are
retroperitoneal?
a. kidneys
b. pancreas
c. adrenal gland
d. urinary bladder
e. all of the above

_X_ 20. Blood pH levels were monitored in an
experiment. On the graph below at
point A, the subject ate a large amount
of antacid (Tums). As a result, blood
pH increased to point B.

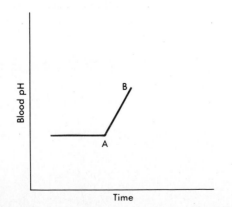

Graphed below are three possible
responses (C) to the increase in
blood pH.

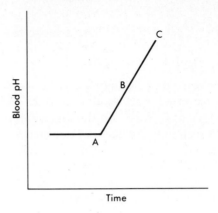

Response 1

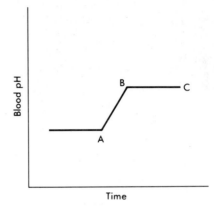

Response 2

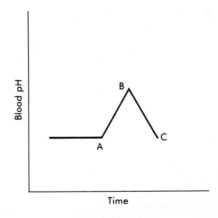

Response 3

Which of the responses graphed above
represents a positive feedback
mechanism?
a. Response 1
b. Response 2
c. Response 3

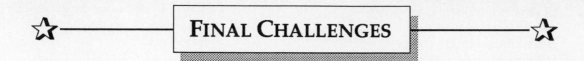
Use a separate sheet of paper to complete this section.

1. The anatomical position of a cat refers to the animal standing erect on all four limbs and facing forward. Based on the origin (etymology) of the directional terms, what two terms would indicate movement toward the tail? What two terms would mean movement toward the animal's belly? Compare these terms with those referring to a human in the anatomical position.

2. Complete the following statements, using the correct directional term for a human being.
   a. The knee is _____ to the ankle.
   b. The foot is _____ to the ankle.
   c. The ear is _____ to the nose.
   d. The nose is _____ to the lips.
   e. The lips are _____ to the teeth.
   f. The heart is _____ to the sternum (breastbone).

3. When blood sugar levels decrease, the hunger center in the brain is stimulated. Is this part of a negative or positive feedback system? Explain.

4. When food enters the stomach, it stimulates the stomach to secrete acid and enzymes. The more food within the stomach, the greater the amount of secretions. The secretions convert the food into a partially digested, semiliquid material called chyme, which passes into the small intestine, where the digestive process is completed. Is this an example of a positive or negative feedback system? Explain.

5. A man has been shot in the abdomen. The bullet passed through the large intestine and lodged in the kidney. Name, in order, the serous membranes through which the bullet passed.

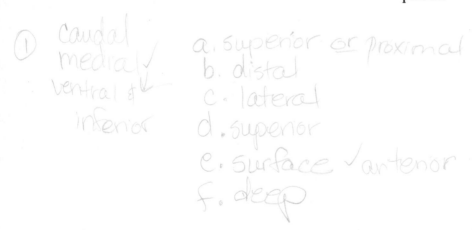

① caudal
medial ✓
ventral ✓
inferior

a. superior or proximal
b. distal
c. lateral
d. superior
e. surface ✓ anterior
f. deep

ANSWERS TO CHAPTER 1

WORD PARTS

1. cytology
2. cytology; histology; physiology; neurophysiology
3. homeostasis
4. homeostasis
5. viscera; visceral
6. sagittal, parasagittal
7. pleura; pleural; pleurisy
8. retroperitoneal
9. parietal
10. cardiovascular; pericardium; pericardial

CONTENT LEARNING ACTIVITY

Anatomy and Physiology

1. Cytology
2. Histology
3. Systemic anatomy
4. Surface anatomy

Structural and Functional Organization

A.
1. Organelle
2. Organ
3. Cell
4. Organ system
5. Tissue

B.
1. Chemical level
2. Organelle
3. Cell
4. Tissue
5. Organ
6. Organ system
7. Organism

C.
1. Integumentary
2. Bones, cartilage
3. Allows body movement, maintains posture
4. Brain, spinal cord, nerves
5. Endocrine
6. Cardiovascular
7. Lymphatic
8. Exchanges gases between blood and air
9. Mouth, esophagus, stomach, intestines
10. Removes waste products from circulatory system
11. Reproductive

Homeostasis

1. Negative feedback
2. Negative feedback
3. Positive feedback

Directional Terms

1. Inferior; Caudal
2. Superior; Cephalic
3. Anterior; Ventral
4. Posterior; Dorsal
5. Distal
6. Lateral
7. Deep

Planes

A. 1. Transverse plane
   2. Frontal (coronal) plane
   3. Longitudinal
   4. Oblique

B. 1. Frontal (coronal) plane
   2. Transverse plane
   3. Midsagittal plane

C. 1. Longitudinal section
   2. Transverse section
   3. Oblique section

Body Regions

1. Limbs
2. Neck
3. Trunk
4. Pectoral
5. Pelvic
6. Arm
7. Wrist
8. Thigh
9. Leg
10. Abdomen

Body Cavities

A. 1. Thoracic cavity
   2. Abdominal cavity
   3. Abdominal cavity
   4. Pelvic cavity

B. 1. Visceral serous membrane
   2. Pleural membrane
   3. Peritoneal membrane
   4. Mesentery
   5. Retroperitoneal

C. 1. Thoracic cavity
   2. Mediastinum
   3. Pleural cavity
   4. Pericardial cavity
   5. Diaphragm
   6. Peritoneal cavity
   7. Abdominal cavity
   8. Pelvic cavity

QUICK RECALL

1. Epithelial, connective, muscle, and nervous tissues
2. Negative and positive
3. A person standing erect with feet facing forward, arms hanging to the sides, and palms of the hands facing forward with the thumbs to the outside
4. Sagittal, transverse (horizontal) and frontal (coronal)
5. Longitudinal, cross (transverse), and oblique
6. Thoracic, abdominal, and pelvic cavities
7. Pleural, pericardial, and peritoneal serous membranes
8. Kidneys, adrenal glands, pancreas, portions of the intestine, and the urinary bladder

MASTERY LEARNING ACTIVITY

1. A. Physiology deals with the vital processes or functions of living organisms. Anatomy includes the subdiscipline of cytology (the study of cells). Histology is the study of tissues.

2. C. An organ is two or more tissues that perform a specific function. An organelle is a specialized structure within a cell and is at a lower level of organization than a cell. A tissue is a group of cells that perform a specific function.

14

3. D. The nervous and endocrine systems are the most important regulatory systems of the body. The circulatory system transports gases, nutrients, and waste products. The muscular system is responsible for movement.

4. E. Negative feedback mechanisms maintain homeostasis by making deviations from normal smaller. When air temperature is greater than body temperature, body temperature tends to increase. Sweating decreases body temperature (reduces the deviation from normal) and helps to maintain homeostasis. It is a negative feedback mechanism.

5. A. For a human in the anatomical position, superior (cranial, cephalad) is toward the head. Dorsal (posterior) refers to the back.

6. D. Proximal means toward the attached end, and distal means away from the attached end of a limb. Medial is toward the midline, and lateral is away from the midline (toward the side). Superficial is toward or on the body's surface.

7. A. The correct answer is based on knowing the definitions of the terms listed and making a judgment about which pair of terms is the best set of opposites.

8. D. The chin is superior to (higher than) the umbilicus.

9. A. A frontal section divides the human body into anterior and posterior sections, a sagittal section into left and right parts, and a transverse section into superior and inferior parts.

10. B. The lower limbs are the thigh (hip to knee), leg (knee to ankle), ankle, and foot. The upper limbs are the arm (shoulder to elbow), forearm (elbow to wrist), wrist, and hand. The axial region includes the head, neck, and trunk, whereas the appendicular region consists of the limbs and their girdles. The pectoral girdle is the point of attachment of the upper limbs, and the pelvic girdle is the point of attachment of the lower limbs.

11. A. The pelvic cavity contains the urinary bladder, the internal reproductive organs, and the lower part of the digestive system (rectum).

12. B. The diaphragm divides the ventral body cavity into the thoracic cavity and the abdominopelvic cavity.

13. D. The heart is surrounded by the pericardial cavity, which is contained within the mediastinum. The mediastinum is a wall of organs that divides the thoracic cavity into two parts.

14. C. Serous membranes line the trunk cavities, produce serous fluid, protect organs against friction, and do not open to the exterior.

15. C. Serous membranes line the trunk cavities and related organs. This includes the peritoneal, pleural, and pericardial membranes. Mucous membranes line the surface of organs and cavities that open to the exterior of the body. The oral cavity is lined by a mucous membrane.

16. B. The kidneys are retroperitoneal; therefore they are NOT surrounded by the peritoneal space.

17. E. The abdominal and pelvic cavities are lined by the peritoneum, the thoracic cavity by the pleural membrane, and the pericardial cavity by the pericardial membrane.

18. A. The thoracic cavity, including the superior surface of the diaphragm, is lined by the parietal pleura. The abdominal cavity, including the inferior surface of the diaphragm, is lined by the parietal peritoneum.

19. E. Retroperitoneal refers to organs that are located behind the peritoneum, i.e., they are between the body wall and the parietal peritoneum. All of the organs listed are retroperitoneal.

20. A. First, one must be able to interpret the graphs. For response 1, pH increased still further from the normal value, for response 2 there was no further increase, and for response 3 pH returned to normal. Next, the definitions of positive and negative feedback must be applied to the graphs. Because positive feed back mechanisms increase the difference between a value and its normal value (homeostasis), response 1 represents a positive feedback mechanism. Negative feed back mechanisms resist further change (response 2) or return the values to normal (response 3).

1. In the cat, caudal (tail) and posterior (following) are toward the tail; ventral (belly) and inferior (lower) are toward the belly. In humans, caudal and inferior are toward where the tail would be; ventral and anterior (before) are toward the belly.

2. A. proximal
   B. distal
   C. lateral
   D. superior
   E. anterior (ventral)
   F. deep

3. It is part of a negative feedback system. Stimulation of the hunger center can result in eating, and the ingested food causes blood sugar levels to increase (return to homeostasis).

4. It is negative feedback because the secretions result in partial digestion of the food and the emptying of the stomach. Once the stomach is empty, stimulation of secretion decreases.

5. After passing through the abdominal wall, the parietal peritoneum is the first membrane pierced. In passing through the large intestine, the visceral peritoneum, the large intestine, and then the visceral peritoneum are penetrated. To enter the kidney, which is retroperitoneal, the bullet passes through the parietal peritoneum.

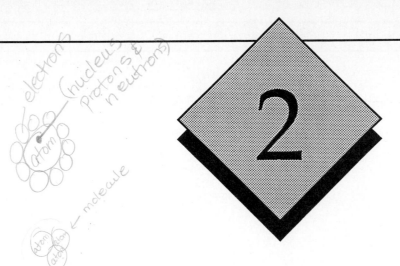

Chemical Principles
Of Life

FOCUS: Chemistry is the study of the composition and structure of matter and the reactions that matter can undergo. Matter is composed of atoms which consist of a nucleus (protons and neutrons) surrounded by electrons. Atoms can combine to form molecules. The chemical bonds between the atoms of molecules include ionic, covalent, and hydrogen bonds. Organic molecules contain carbon (and usually hydrogen) atoms, and inorganic molecules include all other molecules (except for carbon dioxide). Acids are defined as proton (H+) donors, and bases are proton acceptors. The pH scale measures the hydrogen concentration of a solution. Carbohydrates are built from monosaccharides, lipids from glycerol and fatty acids, proteins from amino acids, and nucleic acids (DNA and RNA) from nucleotides.

WORD PARTS

Give an example of a new vocabulary word that contains each word part.

WORD PART	MEANING	EXAMPLE
end-	within; inside	1. _____
ex-	out; from	2. _____
erg-	work; power	3. _____
syn-	together	4. _____
-thesis	an arranging	5. _____
carbo-	coal; carbon-containing	6. _____

WORD PART	MEANING	EXAMPLE
hydr-	water	7. _____
poly-	many	8. _____
mono-	one	9. _____
facchar-	sugar	10. _____

CONTENT LEARNING ACTIVITY

Basic Chemistry

" *Chemistry is the scientific discipline that deals with the composition and structure of* **"**
substances and with the reactions they undergo.

A. Match these terms with the correct statement or definition:

Atom   Nucleus
Electron   Neutron
Element   Proton
Energy shell

_____ 1. Smallest particle into which an element can be divided using conventional chemical techniques.

_____ 2. Matter composed of atoms of only one kind.

_____ 3. Subatomic particles with no electrical charge.

_____ 4. Subatomic particles which revolve around the nucleus, and with very little mass.

_____ 5. Energy level of an electron as it orbits the nucleus.

☞ Energy is the ability to do work.

B. Match these terms with the correct parts of an atom labeled in Figure 2-1:

Electron
Nucleus
Neutron

Orbital
Proton

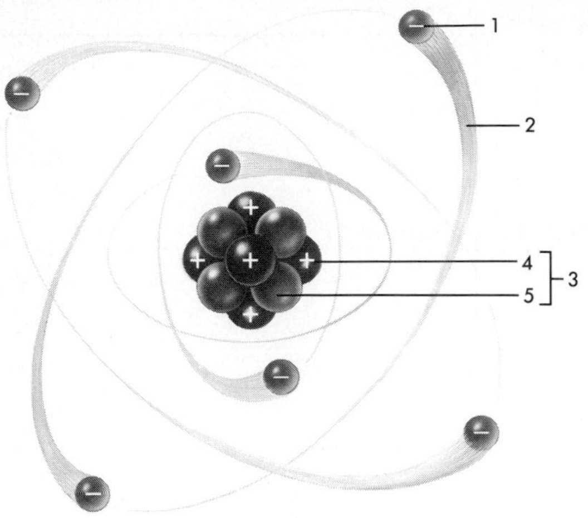

Figure 2-1

1. _____

2. _____

3. _____

4. _____

5. _____

C. Match these terms with the correct statement or definition:

Atomic number
Atomic weight
Avogadro's number

Isotopes
Mass number

_____ 1. Number of protons in an atom.

_____ 2. Number of protons and neutrons in an atom.

_____ 3. Weight in grams of Avogadro's number of atoms of an element.

_____ 4. Number of atoms in one mole of an element.

_____ 5. Elements that have the same atomic number, but different mass numbers, e.g., hydrogen, deuterium, and tritium.

Electrons and Chemical Bonds

66 *Chemical bonds are formed when the outermost electrons are transferred or shared between atoms.* **99**

Using the terms provided, complete the following statements:

Covalent
Electrolytes
Electrons
Gained

Hydrogen
Ionic
Ions
Lost

1. _____

2. _____

3. _____

4. _____

5. _____

The chemical behavior of an atom is largely determined by the (1) in its outermost energy shell. (2) bonds result when one atom loses an electron and another accepts that electron. Atoms that have donated or accepted electrons are called (3) and may be positively or negatively charged. A cation is an ion that has (4) one or more electrons, and an anion is an ion that has (5) one or more electrons. Cations and anions are sometimes called _(6)_ because they can conduct electrical current when they are in solution. _(7)_ bonds result when two atoms complete their energy shells by sharing electrons. Hydrogen atoms covalently bound to oxygen or nitrogen have a slight positive charge. The attraction of the hydrogen atom to the small negative charge associated with oxygen or nitrogen atoms of other molecules is called a _(8)_ bond.

6. _____

7. _____

8. _____

☞ Polar covalent bonds occur when electrons are not shared equally by atoms.

Chemical Reactions

❝A chemical reaction is the process by which atoms or molecules interact to either form or break chemical bonds.**❞**

Using the terms provided, complete the following statements:

Anabolism	Exchange
Catabolism	Oxidation-reduction
Decomposition	Synthesis

When two or more atoms, ions, or molecules combine to form a new and larger molecule, the process is called a _(1)_ reaction; all these reactions in the body are collectively referred to as _(2)_. When larger molecules are broken down to form smaller molecules, ions, or atoms, the process is called a _(3)_ reaction. Collectively, all these breakdown reactions in the body are called _(4)_. _(5)_ reactions occur when one compound is broken down and a new compound is synthesized, such as when phosphate is transferred from ATP to glucose. _(6)_ reactions involve the transfer of electrons in which one atom donates an electron to another atom.

1. _____

2. _____

3. _____

4. _____

5. _____

6. _____

☞ The substances that combine in a chemical reaction are called the reactants, and the substances that are formed are the products.

Energy Relationships

❝A specific amount of energy is associated with chemical bonds, and as chemical bonds are broken**❞** or formed, energy relationships change.

A. Match these terms with the correct statement or definition:

Potential energy
Kinetic energy

_____ 1. Kind of energy that exists in chemical bonds, i.e., stored energy.

_____ 2. Energy that does work, e.g., heat.

B. Match these terms with the
   correct statement or definition:

             Activation energy
             Endergonic
             Exergonic

_____ 1. Reaction that gives off energy.

_____ 2. Type of reaction that is typical of catabolism.

_____ 3. Type of reaction that is typical of anabolism.

_____ 4. Amount of energy required to start a chemical reaction.

Reversible Reactions

66_Some reactions can proceed from reactants to products and from products to reactants spontaneously._**99**

Match these terms with the
correct statement or definition:

             Equilibrium
             Reversible reaction

_____ 1. Chemical reaction that can proceed from reactants to products and from products to reactants spontaneously.

_____ 2. At this point the ratio of products and reactants in a chemical reaction is constant.

Rate of Chemical Reactions

66_The rate at which a chemical reaction proceeds is influenced by several factors...._**99**

Using the terms provided, complete the following statements:

Catalyst   Enzymes
Concentration   Increases
Decreases

Chemical reactions are influenced by several factors, including how easily the substances react with one another. If the concentration of reactants increases, the rate of the chemical reaction _(1)_. When the temperature decreases, the speed of chemical reactions _(2)_. A _(3)_ is a substance that increases the rate at which a chemical reaction proceeds by reducing the activation energy of the reaction without itself being permanently changed or depleted. Protein molecules in the body that act as catalysts are called _(4)_. Regulation of chemical events in the cell is due primarily to mechanisms that control either the _(5)_ or the activity of enzymes.

1. _____

2. _____

3. _____

4. _____

5. _____

Water

"Water plays an important role in many chemical reactions.**"**

Match these terms with the
correct statement or definition:

Condensation   Hydrogen bonding
Dissociation   Hydrolysis

_____ 1. Major force that holds water molecules together as a liquid.

_____ 2. Process that occurs when ionic substances dissolve in water.

_____ 3. Synthesis reaction in which water is a product.

Solutions and Concentrations

"Any liquid that contains dissolved substances is called a solution. The concentration of solute**"** particles dissolved in solvents can be expressed in several ways.

Match these terms with the
correct statement or definition:

Molality   Percent by weight
Molarity   Solute
Osmolality   Solvent

_____ 1. Substance dissolved in the solution.

_____ 2. Method of expressing concentration that divides the weight of the solute by the weight of the solvent.

_____ 3. One mole of solute in 1 kg of water.

_____ 4. The molality of a solution times the number of particles into which the solute dissociates.

Acids and Bases

"Many molecules are classified as acids or bases.**"**

A. Match these terms with the
correct statement or definition:

Acids   Buffers
Bases   Salts

_____ 1. Substances that are proton ($H+$) donors.

_____ 2. Substances that accept hydrogen ions.

_____ 3. Molecules consisting of a cation other than H^+ and an anion other than OH^-.

_____ 4. Chemicals that resist changes in the pH of a solution when acids or bases are added.

☞ Strong acids or bases dissociate completely when dissolved in water, but weak acids or bases only partly dissociate.

B. Match these terms with the correct statement or definition:

Acidic solution
Basic (alkaline) solution

Neutral solution

_____ 1. pH of 7, e.g., pure water

_____ 2. pH less than 7

_____ 3. Greater concentration of hydroxide ions than hydrogen ions.

Oxygen and Carbon Dioxide

"Oxygen and carbon dioxide are two important inorganic molecules.**"**

Match these terms with the correct statement or definition:

Carbon dioxide
Oxygen

_____ 1. Comprises about 21% of the gas in the atmosphere.

_____ 2. Used in the final step of the oxidation-reduction reactions that extract energy from food molecules.

_____ 3. Produced when organic molecules are metabolized.

Carbohydrates

"Carbohydrates are composed of oxygen, hydrogen and carbon, with the general**"** molecular formula (CH_2O).

A. Match these terms with the correct statement or definition:

Monosaccharides
Disaccharides

Polysaccharides

_____ 1. Simple sugars, (e.g., glucose) that are the building blocks for other carbohydrates.

_____ 2. Sucrose, lactose, and other double sugars.

_____ 3. Many glucose molecules bonded together.

☞ Isomers such as glucose and fructose are molecules that have the same number and type of atoms but differ in their three-dimensional arrangements.

B. Match these terms with the correct statement or definition:

Cellulose
Glucose

Glycogen
Starch

_____ 1. The carbohydrate most immediately available for energy.

_____ 2. Energy storage molecule in muscle and the liver.

_____ 3. Structural component of plant cell walls that is undigestible by humans.

Lipids

“Lipids contain a lower ratio of oxygen to carbon than do carbohydrates, and dissolve in nonpolar organic solvents such as acetone or alcohol.”

A. Match these terms with the correct statement or definition:

Carboxyl group   Glycerol
Fatty acid   Saturated

_____ 1. These two combine to form fat.

_____ 2. Straight hydrocarbon chain with a carboxyl group attached.

_____ 3. Double-bonded oxygen and a hydroxyl group attached to a carbon atom.

_____ 4. Fatty acid that has only single covalent bonds between carbon atoms.

B. Match these terms with the correct statement or definition:

Fats   Prostaglandins
Fat-soluble vitamins Steroids
Phospholipids

_____ 1. Monoglycerides, diglycerides, and triglycerides.

_____ 2. Components of the cell membrane that have polar and nonpolar ends and form a double layer.

_____ 3. Important in regulating tissue inflammation and repair.

_____ 4. Cholesterol, bile acids, estrogens, progesterone, and testosterone.

Proteins

“Proteins are often large molecules that are involved in the regulation of chemical reactions or function as structural components within the body.”

Using the terms provided, complete the following statements:

Active site   Primary
Amino acid   Quaternary
Cofactor   Secondary
Enzymes   Tertiary
Essential   Vitamins
Peptide

1. _____

2. _____

3. _____

4. _____

5. _____

6. _____

The building blocks for proteins are 20 basic types of _(1)_ molecules. Humans can synthesize 12 of these from simple organic molecules, but the remaining eight are called _(2)_ and must be included in the diet. Covalent bonds formed between amino acid molecules during protein synthesis are called _(3)_ bonds. The _(4)_ structure of a protein is determined by the sequence of amino acids bound to each other, whereas the _(5)_ structure of a protein is determined by the hydrogen bonds between amino acids that cause the protein to fold or coil into helices or pleated sheets. Folding of helices or pleated sheets by formation of covalent bonds between sulfur atoms of two amino acids and attraction or repulsion of part of the protein molecule to water produces the _(6)_ structure of proteins. The

shape produced as two or more proteins combine to form a
functional unit is the _(7)_ structure. _(8)_ are protein molecules that
act as catalysts for chemical reactions. These molecules are highly
specific because of their three-dimensional shape, which
determines the structure of the enzyme's _(9)_. Some enzymes
require additional, nonprotein substances to be functional. This
_(10)_ can be an ion or a complex of organic molecules. Some _(11)_
serve as cofactors for certain enzymes.

7. _____

8. _____

9. _____

10. _____

11. _____

Nucleic Acids and ATP

66 *Nucleic acids direct the activities of the cell and are involved in protein synthesis. ATP stores* **99**
and provides energy for the cell.

Match these terms with the
correct statement or definition:

Adenosine triphosphate (ATP)   Nucleotide
DNA   RNA

_____ 1. Building block for nucleic acids, consisting of a monosaccharide, an
organic base, and a phosphate group.

_____ 2. Long, double strand of nucleotides arranged in a helix.

_____ 3. Its organic bases include uracil; a single strand of nucleotides.

_____ 4. Energy currency of the cell.

Molecular Diagrams

Match these terms with the correct
molecular diagram in Figure 2-2:

Amino acid   Monosaccharide
ATP   Nucleotide
Fat   Polypeptide
Fatty acid   Polysaccharide
Glycerol

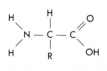

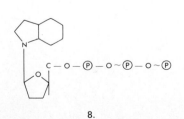

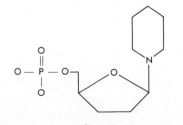

1. _____

2. _____

3. _____

4. _____

5. _____

6. _____

7. _____

8. _____

9. _____

Figure 2-2

1. List three subatomic particles and give their charge.

2. List three types of bonds between atoms.

3. Name four types of chemical reactions.

4. List four factors that affect the rate of chemical reactions.

5. List the properties of water that make it well-suited for living organisms.

6. Name the four types of organic molecules found in living things. For each type of organic molecule, list its building block.

7. List three kinds of carbohydrates.

8. List two important functions of carbohydrates in the human body.

9. Name three functions of lipids in the human body.

10. List three important functions of protein in the human body.

11. List the organic bases for DNA, RNA, and ATP.

Place the letter corresponding to the correct answer in the space provided.

_____ 1. The smallest unit of an element that still retains the properties of that element is a (an)
   a. electron.
   b. molecule.
   c. neutron.
   d. proton.
   e. atom.

_____ 2. The number of electrons in an atom is equal to
   a. Avogadro's number.
   b. the mass number.
   c. the atomic number.
   d. the number of neutrons.

_____ 3. Carbon forms chemical bonds with other atoms by achieving a stable or complete outermost energy shell. This is accomplished when carbon
   a. loses two electrons.
   b. gains two electrons.
   c. shares two electrons.
   d. loses four electrons.
   e. shares four electrons.

_____ 4. A polar covalent bond between atoms occurs when
   a. one of the atoms has a greater affinity for electrons than the other atom.
   b. the atoms attract the electrons equally.
   c. an electron from one of the atoms is completely transferred to the other atom.
   d. the molecule becomes ionized.
   e. a hydrogen atom is shared between two different atoms.

_____ 5. A cation is a (an)
   a. uncharged atom.
   b. positively charged atom.
   c. negatively charged atom.
   d. atom that has gained an electron.
   e. c and d

_____ 6. In a decomposition reaction
   a. the reactants gain electrons.
   b. atoms are transferred from one molecule to another.
   c. large molecules are broken down to form small molecules.
   d. all of the above

_____ 7. The energy required to start a chemical reaction is called
   a. equilibrium energy.
   b. nonequilibrium energy.
   c. endergonic energy.
   d. exergonic energy.
   e. activation energy.

_____ 8. In freely reversible reactions
   a. there is a small activation energy.
   b. reactants form products and products form reactants.
   c. the ratio of products to reactants remains constant.
   d. there is little potential energy difference between reactants and products.
   e. all of the above

_____ 9. The rate of chemical reactions is influenced by
   a. the concentration of the reactants.
   b. temperature.
   c. enzymes.
   d. all of the above

_____ 10. Water
   a. is composed of two oxygen atoms and one hydrogen atom.
   b. has a low specific heat.
   c. is composed of polar molecules into which ionic substances dissociate.
   d. is produced in a hydrolysis reaction.

_____ 11. When sugar is dissolved in water, the water is called the
   a. solute.
   b. solvent.
   c. solution.

_____ 12. A 1-osmolal solution of glucose and a 1-osmolal solution of sodium chloride
    a. have the same number of water molecules.
    b. have the same number of osmotic particles.
    c. have the same number of molecules.
    d. a and b
    e. a and c

_____ 13. A solution with a pH of 5 is a (an) _____ and contains more _____ hydrogen ions than a neutral solution.
    a. base, more
    b. base, less
    c. acid, more
    d. acid, less

_____ 14. A buffer
    a. slows down chemical reactions.
    b. speeds up chemical reactions.
    c. increases the pH of solutions.
    d. maintains a relatively constant pH.
    e. b and c

_____ 15. Oxygen
    a. is an inorganic molecule.
    b. is released from glucose when glucose is broken down for energy.
    c. combines with carbon dioxide and is released from the body.
    d. all of the above

_____ 16. Which of the following is an example of a carbohydrate?
    a. glycogen
    b. wax
    c. steroid
    d. DNA
    e. none of the above

_____ 17. The basic units or building blocks of fats are
    a. simple sugars (monosaccharides).
    b. double sugars (disaccharides).
    c. amino acids.
    d. glycerol and fatty acids.
    e. nucleotides.

_____ 18. A peptide bond joins together
    a. amino acids.
    b. fatty acids.
    c. disaccharides.
    d. nucleotides.
    e. none of the above

_____ 19. DNA
    a. is the genetic material.
    b. is a single strand of nucleotides.
    c. contains the nucleotide uracil.
    d. all of the above

_____ 20. ATP
    a. is formed by the addition of a phosphate group to ADP.
    b. is formed with energy released during catabolism reactions.
    c. provides the energy for anabolism reactions.
    d. contains three phosphate groups.
    e. all of the above

Use a separate sheet of paper to complete this section.

1. The atomic number of nitrogen is 7 and the mass number is 14. What is the number of protons, neutrons, and electrons in an atom of nitrogen?

2. Nitrogen atoms have five electrons in their outermost shell. How many hydrogen atoms, which have one electron in their outermost shell, will combine with nitrogen to form a noncharged molecule? What kind of chemical bond is formed?

3. Two substances, A and B, can combine to form substance C:

$$A + B \longrightarrow C$$

Substance A and B each dissolve in water to form a colorless solution, whereas substance C forms a red solution. Using this information, explain the following two experiments.

A. When solutions A and B are combined, no color change takes place. However, when the combined solution is heated, it turns red.

B. When solution A and B are combined, no color change takes place. However, when substance D is added to the combined solution, it turns red. Later, the exact amount of substance D that was added is recovered from the solution.

4. When placed in a solution, potassium chloride (KCl) dissociated to form potassium ions (K^+) and chloride ions (Cl^-). On the other hand, fructose (a sugar) does not dissociate in water. Suppose you have a 1-osmolal solution of potassium chloride and a 1-osmolal solution of fructose. Which solution has the larger number of molecules. Compared to the solution with the smaller number of molecules, how many more molecules are in the solution with the larger number of molecules. Give the actual number of molecules.

5. Given that blood is buffered by the following reactions:

$$CO_2 + H_2O \longleftrightarrow H_2CO_3 \longleftrightarrow H^+ + HCO_3^-$$

what will happen to blood pH if a person holds his breath?

6. Suppose you have two substances and you know that one is a carbohydrate and one is a lipid. How could you tell which one was the carbohydrate and which one was the lipid, if you had available the materials found in a bathroom (hint: medicine cabinet, make-up kit).

ANSWER TO CHAPTER 2

1. endergonic
2. exergonic
3. endergonic; exergonic; energy
4. synthesis
5. synthesis
6. carbohydrate; carboxyl

7. carbohydrate; hydrolysis; hydrogen; hydroxide; dehydration
8. polysaccharide; polypeptide; polymer
9. monosaccharide; monoglyceride; monomer
10. disaccharide; diglyceride

CONTENT LEARNING ACTIVITY

Basic Chemistry

A.
1. Atom
2. Element
3. Neutron
4. Electron
5. Energy shell

B.
1. Electron
2. Orbital
3. Nucleus

4. Proton
5. Neutron

C.
1. Atomic number
2. Mass number
3. Atomic weight
4. Avogadro's number
5. Isotopes

Electrons and Chemical Bonds

1. Electrons
2. Ionic
3. Ions
4. Lost

5. Gained
6. Electrolytes
7. Covalent
8. Hydrogen

Chemical Reactions

1. Synthesis
2. Anabolism
3. Decomposition

4. Catabolism
5. Exchange
6. Oxidation-reduction

Energy Relationships

A.
1. Potential energy
2. Kinetic energy

B.
1. Exergonic
2. Exergonic
3. Endergonic
4. Activation energy

Reversible Reactions

1. Reversible reaction   2. Equilibrium

Rate of Chemical Reactions

1. Increases   4. Enzymes
2. Decreases   5. Concentration
3. Catalyst

Water

1. Hydrogen bonding   3. Condensation
2. Dissociation

Solutions and Concentrations

1. Solute   3. Molality
2. Percent by weight   4. Osmolality

Acids and Bases

A. 1. Acids   B. 1. Neutral solution
    2. Bases   2. Acidic solution
    3. Salts   3. Basic (alkaline) solution
    4. Buffers

Oxygen and Carbon Dioxide

1. Oxygen   3. Carbon dioxide
2. Oxygen

Carbohydrates

A. 1. Monosaccharides   B. 1. Glucose
    2. Disaccharides   2. Glycogen
    3. Polysaccharides   3. Cellulose

Lipids

A. 1. Glycerol; fatty acid   B. 1. Fats
    2. Fatty acid   2. Phospholipids
    3. Carboxyl group   3. Prostaglandins
    4. Saturated   4. Steroids

Proteins

1. Amino acid
2. Essential
3. Peptide
4. Primary
5. Secondary

6. Tertiary
7. Quaternary
8. Enzymes
9. Active site
10. Cofactor
11. Vitamins

Nucleic Acids and ATP

1. Nucleotide
2. DNA

3. RNA
4. ATP

Molecular Diagrams

1. Monosaccharide
2. Polysaccharide
3. Glycerol
4. Fatty acid
5. Fat

6. Amino acid
7. Polypeptide
8. ATP
9. Nucleotide

QUICK RECALL

1. Protons- positive charge; neutrons- no charge; and electrons- negative charge
2. Ionic, covalent, and hydrogen bonds
3. Synthesis, decomposition, exchange, and oxidation-reduction
4. How easily substances react, concentration of reactants, temperature of reactants, enzymes present, and whether a reaction is endergonic or exergonic
5. Liquid at a temperature range common to life, has a high specific heat, many substances dissolve in it, and it is an effective lubricant

6. Carbohydrates: monosaccharides; Fats: glycerol and fatty acids; Proteins: amino acids; Nucleic acids: nucleotides
7. Monosaccharides, disaccharides, and polysaccharides
8. Energy and structural functions
9. Energy, structural, protection, insulation, regulation, vitamins
10. Regulation, transport, protection, contraction, structure, energy
11. DNA: adenine, thymine, guanine, cytosine
RNA: adenine, uracil, guanine, cytosine
ATP: adenine

MASTERY LEARNING ACTIVITY

1. E. An element consists of atoms of only one kind, and the atoms retain the properties of the element. A molecule is composed of two or more atoms joined together. Neutrons, protons, and electrons are parts of atoms.

2. C. The atomic number is the number of protons in an atom. Since the number of protons is equal to the number of electrons, it is also the number of electrons. The mass number is the number of protons and neutrons in an atom, and Avogadro's number is the number of atoms in a mole.

3. E. When a carbon atom shares four electrons, it increases the number of electrons in its outer most shell to eight, the number necessary for a stable outermost energy shell.

4. A. A polar covalent bond is due to unequal sharing of electrons, with one end of the molecule being more negative than the other end. Nonpolar covalent bonds share electrons equally between atoms, whereas ionic bonds involve complete transfer of electrons.

5. B. A cation is positively charged due to the loss of an electron.

6. C. Decomposition reactions involve the break down of molecules. Electrons are transferred in oxidation-reduction reactions, and atoms or molecules are transferred in exchange reactions.

7. E. Activation energy is required to start chemical reactions. Exergonic reactions give off energy, and endergonic reactions require the addition of energy.

8. E. In freely reversible reactions there is little potential energy difference between reactants and products: therefore little activation energy is required for a reaction to proceed. The reactants form products and vice versa until an equilibrium is reached in which the ratio of products to reactants remains constant.

9. D. The rate of chemical reactions increases with temperature, the concentration of the reactants, and with enzymes that catalyze the reaction in question.

10. C. The polar nature of water molecules makes it possible for polar molecules and ions to dissolve in the water. Water has one oxygen and two hydrogen ions, has a high specific heat, and is used up in hydrolysis reactions (produced in condensation reactions).

11. B. A solution is a solvent (water) into which a solute (sugar) is dissolved.

12. C. A 1-osmolal solution of one substance by definition has the same number of solute particles (but not necessarily the same number of molecules) and water molecules as a 1-osmolal solution of another substance. Since the sodium chloride dissociates into sodium and chloride ions but the glucose does not dissociate, the 1-osmolal solution of sodium chloride has one half the number of molecules as the 1-osmolal glucose solution.

13. C. Solutions with a pH of less than seven are acidic (above seven they are basic). Acid solutions contain more hydrogen ions than neutral solutions.

14. D. Buffers resist a change in pH. Enzymes speed up chemical reactions they catalyze.

15. A. Oxygen is an inorganic molecule (does not contain carbon). Carbon dioxide is released from glucose as glucose is broken down for energy.

16. A. Glycogen is composed of long chains of glucose. Waxes and steroids are lipids, and DNA is a nucleic acid.

17. D. The basic building blocks of fats are glycerol and fatty acids, the basic building blocks of carbohydrates are monosaccharides, and the basic building blocks of proteins are amino acids.

18. A. Peptide bonds are covalent bonds formed between amino acids during condensation reactions.

19. A. DNA, the genetic material, is a double strand of nucleotides. RNA is a single strand of nucleotides and contains the base uracil (DNA has the base thymine).

20. E. Energy from catabolism reactions is used to add a phosphate group to ADP (which has 2 phosphate groups) and form ATP (which has 3 phosphate groups). When ATP is broken down to ADP, energy is released that may be used in anabolism reactions.

1. The atomic number is equal to the number of protons (7), and the number of protons equals the number of electrons (7). The number of neutrons (7) is equal to the mass number minus the atomic number (14 - 7).

2. Three hydrogen atoms can each form a covalent bond with a nitrogen atom, so that the outermost shells of the nitrogen and hydrogen atoms are stable or complete. Thus the nitrogen atom has eight electrons in its outermost shell (its five original electrons plus three electrons shared with the hydrogen atoms), and each hydrogen atom has two electrons in its outermost shell (its original electron plus one electron shared with nitrogen). The molecule formed is ammonia.

3. A. Substances A and B require activation energy for the reaction to proceed, and this activation energy is provided by the heat.
B. Substance D is a catalyst that lowers the activation energy, but is not used up or altered by the reaction.

4. The osmolality of a solution is equal to the molality of the solution times the number of particles into which the solute dissociates. A 1-osmolal solution of KCl contains one half a mole of KCl (1/2 molal x 2 = 1 osmolal) and a 1-osmolal solution of fructose contains a mole of fructose (1 molal x 1 = 1 osmolal). Therefore fructose has one half a mole more molecules than KCl. Since a mole contains 6.023×10^{23} molecules (Avogadro's number), the fructose solution has 3.012×10^{23} more molecules. Note, however, that both solutions contain the same number of particles; i.e., the number of K^+ and Cl^- ions combined is equal to the number of fructose molecules.

5. Holding one's breath would cause an increase in blood carbon dioxide levels because carbon dioxide is not eliminated by the respiratory system. As a result, more carbonic acid (H_2CO_3) would form and then dissociate into hydrogen ions (H^+) and bicarbonate ions (HCO_3^-). The increased number of hydrogen ions would cause a decrease in blood pH.

6. Carbohydrates are usually polar molecules that dissolve in water, whereas lipids usually do not dissolve in water but will dissolve in organic solvents such as alcohol or acetone. One could try to dissolve each substance in water, alcohol (medicinal supply), or acetone (fingernail polish remover). Since many carbohydrates are sugars, one could also taste each substance.

Structure And Function Of The Cell

FOCUS: The basic unit of the human body is the cell, which consists of organelles that are specialized to perform a variety of functions. The plasma membrane regulates the movement of materials into and out of the cell by diffusion, facilitated diffusion, active transport and other mechanisms. Chromosomes in the nucleus contain DNA, which controls the activities of the cell through the production of mRNA (transcription), which leaves the nucleus and at ribosomes causes the production of enzymes (translation). The chemical reactions that occur within the cell such as the production of ATPs in mitochondria are regulated by the enzymes. Also produced at ribosomes are proteins (e.g.,microtubules and microfilaments) that provide structural support to the cell. Some of these proteins, the spindle fibers, are involved with the separation of chromosomes during cell division. Cells divide by mitosis during growth or repair of tissues and divide by meiosis to produce gametes.

WORD PARTS

Give an example of a new vocabulary word that contains each word part listed.

WORD PART	MEANING	EXAMPLE
inter-	between	1. _intercellular_
intra-	within	2. _intracellular_
extra-	outside	3. _extracellular_
iso-	equal	4. _____
hypo-	under; less than	5. _hypodermis_

WORD PART	MEANING	EXAMPLE
hyper-	over; above	6. _____
-ton-	tension	7. _____
lys-	cut; loosen; dissolve	8. _____
-some	body	9. *chromosome*
-kin-	to move	10. _____

CONTENT LEARNING ACTIVITY

Cell Structures and Their Functions

"*Cells are highly organized units.***"**

Match these terms with the
correct statement or definition:

Cytoplasm   Organelles
Nucleus   Protoplasm

Protoplasm _____

Nucleus _____

Cytoplasm _____

Organelles _____

1. Living matter, of which all cells are composed.

2. Location of genetic material of the cell.

3. Protoplasm surrounding the nucleus of the cell.

4. Structures that perform specific cellular functions.

Structure of the Plasma Membrane

"*The plasma, or cell membrane is the outermost component of a cell.***"**

A. Match these terms with the
correct statement or definition:

Extracellular   Intracellular
Intercellular   Plasma (cell) membrane

_____

1. These two terms refer to substances outside the cell membrane.

_____

2. The functions of this structure are to enclose and support the cell and to determine what moves into and out of the cell.

B. Using the list of terms provided, complete the following statements:

1. _____

2. _____

Hydrophilic   Phospholipid
Hydrophobic   Protein
Lipid bilayer

The central layer of the plasma membrane is composed of a double layer of molecules called a (1) . (2) molecules, which have a polar end and a nonpolar end, predominate in this layer.

The nonpolar ends of the phospholipids, which repel water, are called (3) and face each other in the interior of the plasma membrane. The polar ends are exposed to water inside and outside the cell and are called (4) . (5) molecules "float" on both the inner and outer surfaces of the lipid bilayer and give the membrane the appearance that it consists of three layers.

3. _____

4. _____

5. _____

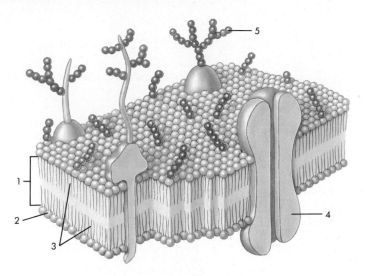

Figure 3-1

C. Match these terms with the correct cell membrane parts labeled in Figure 3-1:

Carbohydrate molecule
Hydrophilic region of
  phospholipid
Hydrophobic region of
  phospholipid
Lipid bilayer
Protein

1. _____

2. _____

3. _____

4. _____

5. _____

Movement Through the Plasma Membrane

❝*The cell membrane separates the extracellular from the intracellular material and is selectively* **❞** *permeable, allowing some substances to pass through it, but not others.*

Match these terms with the correct statement or definition:

Carrier molecules
Cell membrane channels
Lipid bilayer

_____ 1. Molecules that are soluble in lipid dissolve in this layer; acts as a barrier to most polar substances.

_____ 2. Allow molecules of only a certain size range to pass through.

_____ 3. Positively charged, this structure causes positive ions to pass through less readily than neutral or negatively charged ions.

_____ 4. Large polar molecules cannot pass through the cell membrane in significant amounts unless transported by these.

Diffusion

“*Diffusion of molecules is an important means by which substances move through the***”**
extracellular and intracellular fluids in the body.

A. Match these terms with the
   correct statement or definition:

   Diffusion   Solvent
   Solute

_____ 1. Predominant liquid or gas in a solution.

_____ 2. Product of the constant random motion of all atoms, molecules, or ions
in a solution.

_____ 3. Tendency for solute molecules to move from an area of high
concentration to an area of low concentration in a solution.

☞ The concentration gradient is the concentration difference between two points,
divided by the distance between those two points. Viscosity is a measure of the
resistance to flow of a liquid.

B. Match these terms with the
   correct statement or definition:

   Decrease
   Increase

_____ 1. Change in the rate of diffusion when there is an increase in the
concentration gradient or an increase in temperature.

_____ 2. Change in the rate of diffusion when there is a decrease in the size of
diffusing molecules.

_____ 3. Change in the rate of diffusion when there is an increase in viscosity.

Osmosis

“*Osmosis is important to cells because large volume changes caused by water***”**
movement disrupt normal cell function.

Match these terms with the
correct statement or definition:

Crenation   Isosmotic
Hyposmotic   Isotonic
Hypotonic   Lysis
Hyperosmotic   Osmotic pressure
Hypertonic

_____ 1. Force required to prevent the movement of water across a selectively
permeable membrane.

_____ 2. Solution with a lower osmotic pressure and fewer solute particles than
another solution.

_____ 3. When a cell is placed in this type of solution, the cell, by definition,
neither swells nor shrinks.

_____ 4. When a cell is placed in this type of solution, the cell, by definition,
shrinks.

_____ 5. Rupture of the cell, which may occur from placing a cell into a
hypotonic solution.

Mediated Transport Mechanisms

66_Mediated transport mechanisms involve carrier molecules within the cell membrane that function to_**99** _move large water soluble molecules or electrically charged molecules across the cell membrane._

A. Match these terms with the correct statement or definition:

Active site
Competition

Saturation
Specificity

_____ 1. Each carrier molecule binds to and transports only a single type of molecule.

_____ 2. Part of the carrier molecule that binds to and transports the molecules.

_____ 3. Rate of transport of molecules across the membrane is limited by the number of available carrier molecules.

_____ 4. Result of similar molecules binding to the carrier molecule.

B. Match these terms with the correct statement or definition:

Active transport
Facilitated diffusion

_____ 1. Does not require metabolic energy (ATPs).

_____ 2. Process that moves substances into or out of cells from a high to a low concentration.

_____ 3. Can move substances against a concentration gradient.

Endocytosis and Exocytosis

66Endocytosis and exocytosis involve bulk movement of material into and out of the cell.**99**

Match these terms with the correct statement or definition:

Endocytosis
Exocytosis

Phagocytosis
Pinocytosis

_____ 1. Includes both phagocytosis and pinocytosis and refers to bulk uptake of material by the formation of a vesicle.

_____ 2. Means "cell eating" and is the ingestion of solid particles.

_____ 3. Movement of secretory vesicles to the cell membrane, where the membrane of the vesicle fuses with the cell membrane and the content of the vesicle is eliminated from the cell.

Nucleus

"The information contained in the nucleus determines most of the chemical events that occur in the cell.**"**

Using the terms provided, complete the following statements:

Chromatin   Nuclear envelope
Chromosomes   Nuclear pores
Histones   Nucleoli
Messenger RNA   Nucleus

The _(1)_ is a large, membrane-bound organelle usually located near the center of the cell. It is surrounded by a _(2)_ composed of two membranes separated by a space. At many points the inner and outer membranes come together to form holes called the _(3)_. The 23 pairs of _(4)_ characteristic of human cells are located within the nucleus. Except during cell division, DNA and associated protein are dispersed through out the nucleus as thin strands called _(5)_. The proteins associated with DNA include _(6)_ and some acidic proteins that play a role in the regulation of DNA function. In the nucleus, DNA determines the structure of _(7)_, which moves out of the nucleus through nuclear pores into the cytoplasm. Within the cytoplasm mRNA is involved in the synthesis of proteins. _(8)_ are somewhat rounded, dense, well-defined nuclear bodies with no surrounding membrane that number from one to four per nucleus depending on the cell.

1. _____

2. _____

3. _____

4. _____

5. _____

6. _____

7. _____

8. _____

Ribosomes and Endoplasmic Reticulum

"The functional ribosome, composed of ribosomal RNA and protein, consists of a large**"** subunit and a smaller subunit joined together; the endoplasmic reticulum forms broad flattened sacs and tubules that interconnect.

Match these terms with the correct statement or definition:

Cisternae   Rough endoplasmic reticulum
Ribosome   Smooth endoplasmic reticulum

_____ 1. Site where mRNA and tRNA come together and assemble amino acids to form proteins.

_____ 2. A polysome is a cluster of these structures.

_____ 3. Tubules and flattened sacs with many ribosomes attached.

_____ 4. Abundant amounts of this structure are found in cells that secrete proteins.

_____ 5. Found in the cell when lipid synthesis, detoxification processes, or storage of calcium ions occurs.

☞ The outer membrane of the nuclear envelope is continuous with the endoplasmic reticulum.

Golgi Apparatus and Secretory Vesicles

❝*The Golgi apparatus is specialized smooth endoplasmic reticulum and secretory vesicles are*❞ *membrane-bound structures that pinch off from the Golgi apparatus.*

Match these terms with the correct statement or definition:

Glycoproteins
Golgi apparatus

Lipoproteins
Secretory vesicles

_____ 1. Concentrates and chemically modifies proteins to form glycoproteins and lipoproteins.

_____ 2. Carbohydrates attached to proteins.

_____ 3. Structures that pinch off from Golgi apparatuses and release their contents to the exterior of the cell by exocytosis.

Lysosomes, Peroxisomes, and Cytoplasmic Inclusions

❝*There are several types of membrane-bound vesicles in the cell.*❞

Match these terms with the correct statement or definition:

Autophagia
Cytoplasmic inclusions

Lysosomes
Peroxisomes

_____ 1. Membrane-bound vesicles containing a variety of hydrolytic enzymes that function as intracellular digestive systems.

_____ 2. Process of digesting organelles of the cell that are no longer functional.

_____ 3. Membrane-bound bodies that contain a variety of oxidative enzymes that either decompose or synthesize hydrogen peroxide.

_____ 4. Granules of several types found in the cytoplasm of cells; some contain lipochrome pigments that tend to increase in amount with age.

☞ Phagocytic vesicles fuse with the lysosomes to expose the phagocytized particles to hydrolytic enzymes.

Mitochondria

❝*Mitochondria are small, spherical, rod-shaped or thin filamentous*❞ *structures found throughout the cytoplasm.*

Using the terms provided, complete the following statements:

ATP
Cristae
Electron-transport chain

Matrix
Oxidative metabolism

Mitochondria are the major sites of _(1)_ production within cells. Mitochondria have inner and outer membranes separated by a space. The inner membranes have numerous infoldings called _(2)_ that project like shelves into the interior of the mitochondria. A complex series of mitochondrial enzymes forms two major enzyme systems, which are responsible for _(3)_ and most ATP synthesis. The enzymes of the citric acid (Kreb's) cycle comprise the _(4)_, and the enzymes of the _(5)_ are embedded with the inner membrane.

1. _____

2. _____

3. _____

4. _____

5. _____

Microtubules, Centrioles, Spindle Fibers, Cilia, and Flagella

"Microtubules form essential components of certain cell organelles such as**"** centrioles, spindle fibers, cilia and flagella.

Match these terms with the
correct statement or definition:

Basal body	Cilia
Centriole	Flagella
Centrosome	Tubulin

_____ 1. Protein units found in microtubules.

_____ 2. Specialized zone of cytoplasm close to the nucleus containing
two centrioles.

_____ 3. Two small cylindrical organelles oriented perpendicular to each other,
each containing nine triplets of microtubules.

_____ 4. Short appendages that project from the surface of the cell and are
capable of moving; they vary in number from one to thousands per cell.

_____ 5. A modified centriole located in the cytoplasm at the base of each cilium
and flagellum.

Microfilaments, Microvilli, Stereocilia, and Intermediate Fibers

"Microfilaments provide structure to the cytoplasm and mechanical support for microvilli and stereocilia.**"**

Match these terms with the
correct statement or definition:

Intermediate filaments	Microvilli
Microfilaments	Stereocilia

_____ 1. Small fibrils that form bundles, sheets, or networks in the cytoplasm
of cells.

_____ 2. Cylindrically shaped extensions of the cell membrane supported by
microfilaments; increase cell surface area; commonly found in
the intestine and kidney.

_____ 3. Elongated microvilli that do not move and are commonly found in
parts of the male reproductive tract.

_____ 4. Protein fibers intermediate in size between microtubules
and microfilaments.

Cell Diagram

Match these terms with the
cell parts labeled on Figure 3-2:

Centrioles	Plasma membrane
Golgi apparatus	Ribosomes
Lysosome	Rough endoplasmic reticulum
Microtubule	Secretory vesicle
Mitochondrion	Smooth endoplasmic reticulum
Nucleolus	

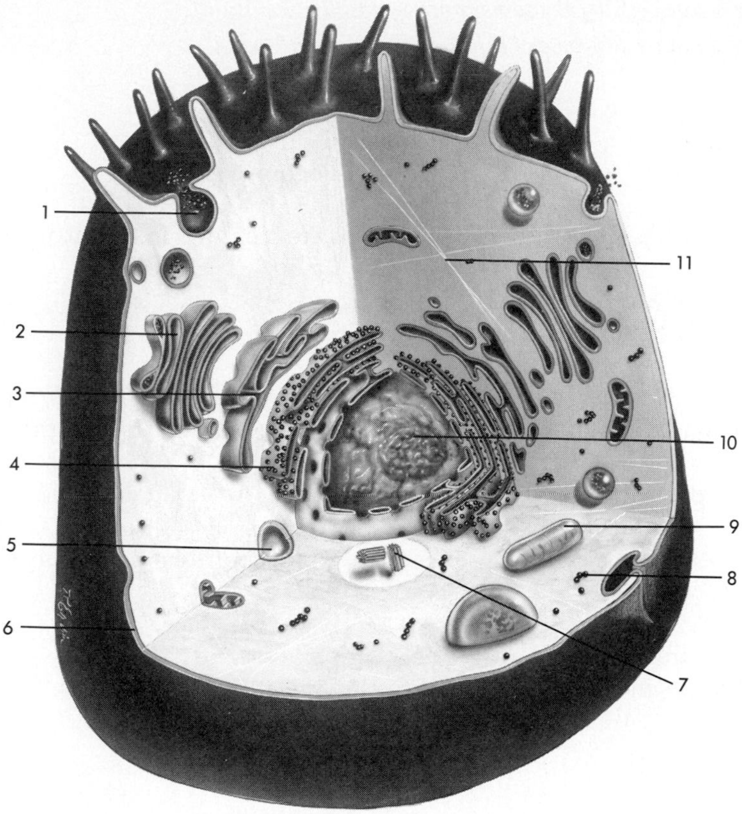

Figure 3-2

1. _____

2. _____

3. _____

4. _____

5. _____

6. _____

7. _____

8. _____

9. _____

10. _____

11. _____

Protein Synthesis

"Events that lead to protein synthesis begin in the nucleus and end in the cytoplasm.**"**

A. Match these terms with the correct statement or definition:

mRNA   Transcription
rRNA   Translation
tRNA

_____ 1. This process occurs when double strands of a DNA segment separate and RNA nucleotides pair with DNA nucleotides.

_____ 2. This type of RNA carries information in groups of three nucleotides, called codons, and each codon codes for a specific amino acid.

_____ 3. This type of RNA has an anticodon and binds to a specific amino acid.

_____ 4. This process involves the synthesis of polypeptide chains at the ribosome in response to information contained in mRNA molecules.

☞ The proteins produced in a cell function either as enzymes or structural components inside and outside the cell.

Mitosis

"Nearly all cell divisions in the body occur by the same process, and the resultant "daughter" cells have the same amount and type of DNA as the "parent" cells.

A. Match these terms with the correct statement or definition:

Autosome
Cytokinesis
Homologous

Interphase
Mitosis
Sex chromosomes

_____ 1. In human cells, 22 pairs of chromosomes are included in this group.

_____ 2. Members of each pair of autosomal chromosomes, which look structurally alike, are described by this term.

_____ 3. Following duplication of the genetic material within the nucleus, the genetic material is distributed into two new nuclei through this process.

_____ 4. Division of the cell's cytoplasm to produce two new daughter cells.

_____ 5. Period between cell divisions when DNA and centrioles are replicated.

☞ During replication of DNA, each strand of DNA serves as a template for the production of a new strand of DNA. This results in the production of two identical DNA molecules.

B. Match these terms with the correct statement or definition:

Astral fibers
Centromere

Chromatids
Spindle fibers

_____ 1. Identical pieces of DNA that are joined together at one point by the centromere.

_____ 2. Specialized region that joins two chromatids together.

_____ 3. Microtubules that radiate from the centrioles and end blindly.

_____ 4. Microtubules that project toward the equator and may attach to the centromeres of the chromosomes.

C. Match these terms with the correct statement or definition as they relate to the events in mitosis:

Anaphase
Metaphase

Prophase
Telophase

_____ 1. Chromosomes become visible.

_____ 2. Chromosomes (each with two chromatids) align along the equator with spindle fibers attached to their centromeres.

_____ 3. Separation of chromatids occurs.

_____ 4. Migration of each set of chromosomes to the centrioles is completed.

D. Match these terms with the phases of mitosis and the cell parts involved in mitosis labeled in Figure 3-3:

Anaphase
Astral fiber
Centriole
Centromere

Metaphase
Prophase
Spindle fiber
Telophase

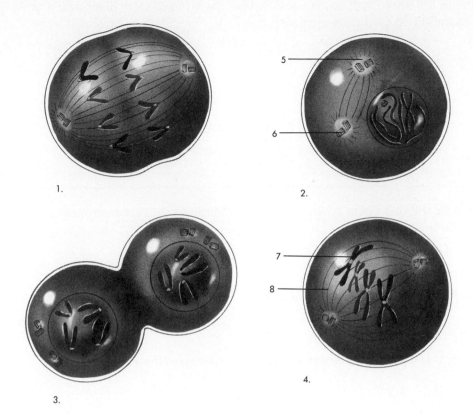

Figure 3-3

1. _____ 4. _____ 7. _____

2. _____ 5. _____ 8. _____

3. _____ 6. _____

Meiosis

66*Meiosis is the special process in which the nucleus undergoes two divisions to produce gametes,*99 *each containing half the number of chromosomes in the parent cell.*

Match these terms with the
correct statement or definition:

Crossing over   Haploid
Diploid   Interkinesis
Gamete   Tetrad

_____ 1. Reproductive cells; a spermatozoon in a male or an oocyte in a female.

_____ 2. Refers to the complement of chromosomes contained in a gamete.

_____ 3. In prophase I, the four chromatids of a homologous pair of chromosomes join together, or synapse, to form this structure.

_____ 4. Short period of time between the formation of daughter cells from the first division of meiosis and the second meiotic division.

_____ 5. When tetrads are formed, some of the chromosomes break apart, and part of one chromatid is exchanged for part of another chromatid.

45

Genetic diversity occurs for two reasons. First, crossing over produces chromatids with different DNA content; and second, there is a random distribution of the genetic material received from each parent.

QUICK RECALL

1. List the two major types of molecules found in the cell membrane and give their functions.

2. List the factors that affect the rate and direction of diffusion in a solution.

3. List six types of movement of materials across cell membranes.

4. List three characteristics of mediated transport mechanisms.

5. List three terms used to describe the tendency of cells to shrink or swell when placed in a solution.

Complete the chart below by writing in the organelle described by the structures and functions listed.

Organelle	Structure	Function
6. _____	Membrane-bound vesicles pinched off from the Golgi apparatus.	Contents released to the exterior of the cell by exocytosis.
7. _____	Broad, flattened sacs and tubules that interconnect, no ribosomes attached.	Lipid synthesis and detoxification.
8. _____	Surrounded by double-layered envelope with pores.	Contains DNA in the form of chromatin (chromosomes) which produces RNA.

9. _____ Composed primarily of protein units called tubulin. | Support for the cytoplasm of the cell; involved in cell division; essential component of centrioles, spindle fibers, cilia, and flagella.

10. _____ Cylindrically shaped extensions of the cell membrane supported by microfilaments. | Increase cell surface area for absorption.

11. _____ Closely packed stacks of curved cisternae composed of smooth endoplasmic reticulum. | Concentrates and packages materials for secretion from the cell.

12. _____ Two subunits composed of ribosomal RNA and protein. | Site where mRNA and tRNA come together to assemble amino acids into proteins.

13. _____ Membrane-bound vesicles that contain hydrolytic enzymes. | Breakdown of phagocytized particles; autophagia.

14. _____ Small, spherical, rod-shaped or filamentous, double membrane with infoldings of the inner membrane called cristae. | Most ATP synthesis in the cell.

15. _____ Nine evenly-spaced, longitudinally oriented, parallel units; each unit consisting of three parallel microtubules joined together. | Move to each side of the nucleus during cell division; special microtubules called spindle fibers develop from the surrounding region.

16. _____ Broad, flattened sacs and tubules that interconnect; ribosomes attached. | Synthesis of proteins.

17. _____ One to four rounded, dense, well-defined nuclear bodies. | Production and assembly of ribosomes.

18. _____ Granules in the cytoplasm. | Metabolically inert, include lipid-containing pigments and glycogen.

19. _____	Appendages from the surface of the cell with two centrally located microtubules and nine peripheral pairs of microtubules joined together.	Movement of materials over the surface of the cells.
20. _____	Membrane-bound vesicles that contain a variety of oxidative enzymes such as catalases.	Contain enzymes for oxidation-reduction reactions in liver cells and steroid-producing cells.
21. _____	Small protein fibrils that form bundles, sheets or networks.	Provide structure to cytoplasm and mechanical support to microvilli and stereocilia.
22. _____	Protein fibrils, intermediate in size between microfilaments and microtubules.	Provide structure to cytoplasm; part of cytoskeleton.

23. List three types of RNA.

24. Name the four phases of mitosis.

MASTERY LEARNING ACTIVITY

Place the letter corresponding to the correct answer in the space provided.

_____ 1. Which of the following are functions of the proteins found in cell membranes?
a. form pores
b. act as carrier molecules
c. act as enzymes
d. all of the above

_____ 2. Lipid-soluble molecules diffuse through the _____; water-soluble molecules diffuse through the _____.
a. membrane pores; membrane pores
b. membrane pores; cell membranes
c. cell membranes; membrane pores
d. cell membranes; cell membranes

3. The end result of diffusion is
   a. all net movement of molecules ceases.
   b. some molecules become more concentrated.
   c. determined by the amount of energy expended in the diffusion process.
   d. determined by the rate at which diffusion took place.

4. Which of the following statements is true about osmosis?
   a. Always involves a membrane that is impermeable to at least one kind of solute molecule.
   b. The number of solute molecules present, not the kind, determines the osmotic pressure.
   c. The greater the difference in solute concentrations across the semipermeable membrane, the greater the osmotic pressure difference across the semipermeable membrane.
   d. Osmosis always involves difference in solvent concentration.
   e. all of the above

5. Cells placed in a hypertonic solution will
   a. swell.
   b. shrink.
   c. neither swell nor shrink.
   d. this term does not describe the tendency of cells to shrink or swell.

6. Container A contains a 10% salt solution and container B a 20% salt solution. If the two solutions were connected, the net movement of water by diffusion would be from _____ to _____, and the net movement of the salt by diffusion would be from _____ to _____.
   a. A,B,A,B,
   b. A,B,B,A
   c. B,A,A,B
   d. B,A,B,A

7. Suppose that a woman was running a long distance race. During the race she lost a large amount of hyposmotic sweat. You would expect her cells to
   a. shrink.
   b. swell.
   c. remain the same.

8. Suppose that a man is doing heavy exercise in the hot summer sun. He sweats profusely. He then drinks a large amount of distilled water. You would expect his tissue cells to
   a. shrink.
   b. swell.
   c. remain the same.

9. The basic principal of the artificial kidney is to pass blood through very small blood channels made up of thin membranes. On the other side of the membrane is dialyzing fluid. Substances are exchanged between the blood and the dialyzing fluid by diffusion. The membrane is porous enough to allow all materials in the blood, except the plasma proteins and blood cells, to diffuse through. It is known that normal blood plasma has the following concentrations (the exact units are unimportant. Just realize that the larger the number, the greater the concentration):

Sodium	Potassium	Urea	Glucose
142	5	26	100

   The intent of the artificial kidney is to remove waste products such as urea while maintaining important constituents of the blood plasma such as sodium, potassium, and glucose as close as possible to their normal levels. Which of the proposed dialyzing fluids would accomplish this task?

	Sodium	Potassium	Urea	Glucose
a)	0	0	0	0
b)	142	5	0	125
c)	0	0	26	100
d)	148	10	0	90
e)	152	10	0	125

10. Which of the following statements is (are) true about facilitated diffusion?
   a. Net movement is with the concentration gradient.
   b. It requires the expenditure of energy.
   c. It does not require a carrier molecule.
   d. It moves materials through membrane pores.
   e. a and b

11. The concentration of fluids in the human intestine tends to be isosmotic with the blood plasma. The intestinal membrane is essentially impermeable to magnesium salts. If a person ingested a large amount of magnesium salts for treatment of constipation, you would expect his tissue cells to
   a. shrink.
   b. swell.
   c. remain the same.

12. A patient is suffering from acute cerebral edema, a very dangerous condition because too much pressure in the cranial vault obstructs the flow of blood to the brain. A doctor suggests that an appropriate treatment to reduce the edema would be to inject the patient with a hypertonic solution of mannitol. Which of the following statements is essential to the doctor's thesis that a hypertonic mannitol solution will relieve the edema?
   a. Mannitol reduces the osmotic concentration (pressure) of the blood.
   b. Mannitol reduces the permeability of cells to water.
   c. Mannitol stimulates the kidneys to excrete large quantities of salts.
   d. Mannitol is unable to enter cells.

13. In diabetes mellitus, due to insufficient insulin production, glucose cannot enter cells. Instead it accumulates in the blood plasma. Which of the following statements would be true under these conditions?
   a. The osmolality of the plasma would decrease.
   b. The osmotic pressure of the plasma would remain the same.
   c. Water would move out of the tissues into the plasma.
   d. Tissue cells would swell.

14. Given the following characteristics:
   1. specificity
   2. competition
   3. saturation

   Which are characteristics of active transport?
   a. 1
   b. 1, 2
   c. 1, 3
   d. 2, 3
   e. 1, 2, 3

15. A process that requires energy and moves large molecules in solution (i.e., not particulate matter) into cells is called
   a. diffusion.
   b. facilitated diffusion.
   c. pinocytosis.
   d. phagocytosis.
   e. exocytosis.

16. An experimenter made up a series of solutions of different concentrations of a substance. Cells were then placed in each solution and, after 1 hour, the amount of the substance that had moved into the cells was determined. The results are graphed below:

Concerning the part of the graph from A to B:
   a. the transport process appears to require membrane pores.
   b. the substance moving into the cells must be lipid soluble.
   c. movement is against the concentration gradient.
   d. concentration of the substance within the cell fails to increase.
   e. carriers are unsaturated.

17. In an experiment the rate of transfer of an amino acid into a cell was carefully monitored. Part way through the experiment, at time A, a metabolic inhibitor was introduced. The results of this experiment are graphed below.

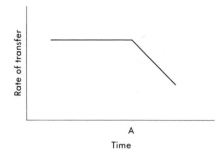

On the basis of these data it can be concluded that the mechanism responsible for the movement of the amino acid was
a. diffusion.
b. facilitated diffusion.
c. active transport.
d. there is insufficient information to reach a conclusion.

18. Cells that initially contained NO intracellular glucose were placed in a series of glucose solutions of different concentrations. After 5 minutes, the amount of intracellular glucose was determined. The results of the experiment are graphed below.

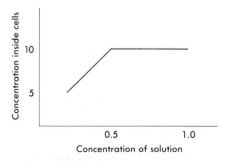

Which transport mechanism is responsible for moving glucose into the cell?
a. diffusion
b. facilitated diffusion
c. active transport
d. inadequate data to support a conclusion

19. Two different molecules, A and B, can both move easily across cell membranes when only one of them is present. However, when both types are present, the movement of A is greatly reduced, whereas the movement of B is only slightly reduced. This information suggests that the movement of molecule A is by _____, and the movement of molecule B is by _____.
a. diffusion, diffusion
b. diffusion, active transport
c. active transport, diffusion
d. active transport, active transport

20. Cytoplasm is found
a. in the nucleus.
b. between the plasma membrane and nuclear membrane.
c. inside the plasma membrane.
d. throughout the cell.

21. A large structure normally visible in the nucleus of a cell?
a. endoplasmic reticulum
b. mitochondria
c. nucleolus
d. lysosome
e. none of the above

22. The cell organelle that is involved in the storage, modification, and packaging of secretory materials?
a. nucleolus
b. ribosome
c. mitochondria
d. Golgi apparatus
e. chromosomes

23. Which of the following organelles produces large amounts of ATP?
a. nucleus
b. mitochondria
c. ribosomes
d. endoplasmic reticulum
e. lysosomes

_____ 24. Which of the following organelles would one expect to be present in large amounts in a cell responsible for secreting a lipid?
a. rough endoplasmic reticulum
b. lysosomes
c. smooth endoplasmic reticulum
d. ribosomes
e. a and c

_____ 25. Which of the following functions are red blood cells incapable of performing?
a. synthesize ATP
b. consume oxygen
c. synthesize new protein
d. use glucose as a nutrient
e. all of the above

_____ 26. If you observed the following characteristics in an electron micrograph of a cell:
1. microvilli
2. many mitochondria
3. the cell lines a cavity
4. no lysosome
5. little smooth endoplasmic reticulum
6. few vacuoles

On the basis of these observations alone, which of the following is likely to be a major function of that cell?
a. active transport
b. secretion of protein
c. intracellular digestion
d. secretion of a lipid

_____ 27. A portion of an mRNA molecule that determines one amino acid in a polypeptide chain is called a
a. nucleotide.
b. gene.
c. codon.
d. nucleotide sequence.
e. nucleoside.

_____ 28. In which of the following organelles is mRNA synthesized?
a. nucleus
b. ribosome
c. endoplasmic reticulum
d. nuclear envelope
e. peroxisome

_____ 29. The organelle of the cell that serves as the site of protein synthesis?
a. ribosome
b. lysosome
c. Golgi apparatus
d. nucleolus
e. vacuole

_____ 30. Transfer RNA
a. specifies the chemical structure of enzymes.
b. duplicates itself at mitosis.
c. provides a template for mRNA.
d. a and c
e. none of the above

_____ 31. Choose the consequence that most specifically predicts the response of a cell to a substance that inhibits messenger RNA synthesis.
a. inhibits protein synthesis
b. inhibits DNA synthesis
c. inhibits fat synthesis
d. stimulates protein synthesis
e. stimulates fat synthesis

_____ 32. Given the following activities:
1. repair
2. growth
3. gamete production
4. differentiation

Which of the activities are the result of mitosis?
a. 2
b. 3
c. 1, 2
d. 3, 4
e. 1, 2, 4

_____ 33. Chromosomes disperse to become chromatin during
a. anaphase.
b. metaphase.
c. prophase.
d. telophase.

34. The major function of meiosis is to ensure that each of the resultant daughter cells
   a. has the same number and kind of chromosomes as the mother cell.
   b. has one half the number of chromosomes as the mother cell.
   c. has one half the number and one of each pair of chromosomes from the mother cell.
   d. none of the above

35. The sex of an individual is determined by their sex chromosomes. An individual is male if he has chromosomes
   a. MM.
   b. MF.
   c. XX.
   d. XY.

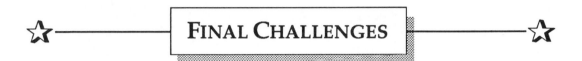

FINAL CHALLENGES

Use a separate sheet of paper to complete this section.

1. Given that you can make solutions of different concentrations (using a solute that does not cross the cell membrane), design an experiment that could be used to determine the osmolality of the cytoplasm of red blood cells.

2. Sometimes isotonic glucose solutions are administered intravenously to patients who can not otherwise take adequate amounts of nourishment. A nursing student hypothesized that such a treatment would result in the patient producing a large amount of dilute urine. Explain why you agree or disagree with this hypothesis.

3. Two different molecules, A and B, can both easily cross cell membranes when only one of them is present. It has also been found that even when both molecules are present, the rate of movement of each molecule is unaffected by the presence of the other molecule.

   A. Given that the method of transport of molecules A and B was either diffusion or facilitated diffusion, what additional information or experiments would be necessary to determine the method of transport?

   B. Given that the method of transport of molecules A and B was either facilitated diffusion or active transport, what additional information or experiments would be necessary to determine the method of transport?

4. A suspension of 10 million cells per milliliter was prepared from two different types of tissue, A and B. After incubating the cells for a period of time, the oxygen consumption (milliliters of oxygen used per minute) was determined with the following results:

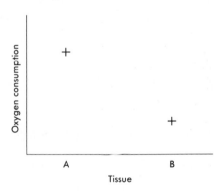

On the basis of these data can it be concluded that tissue A has the greatest number of mitochondria? Why or why not?

5. Suppose that you have available for use a light microscope and an electron microscope. Design an experiment that would adequately test the hypothesis that cells synthesize enzymes for the purpose of secretion.

6. Suppose you have the following results from an experiment: (1) a stimulus caused a cell to secrete a lipid, and (2) treatment of the cell with an inhibitor of mRNA did not affect the response of the cell to the stimulus. Describe the most likely sequence of events that lead to lipid secretion in this cell in response to a stimulus.

ANSWERS TO CHAPTER 3

WORD PARTS

1. Intercellular; intermediate
2. Intracellular
3. Extracellular
4. Isosmotic; isotonic
5. Hyposmotic; hypotonic

6. Hyperosmotic; hypertonic
7. Isotonic; hypotonic; hypertonic
8. Lysosome
9. Lysosome; peroxisome; centrosome
10. Cytokinesis; interkinesis

CONTENT LEARNING ACTIVITY

Cell Structures and Their Functions

1. Protoplasm
2. Nucleus

3. Cytoplasm
4. Organelles

Structure of the Plasma Membrane

A. 1. Extracellular; intercellular
    2. Plasma (cell) membrane

B. 1. Lipid bilayer
    2. Phospholipid
    3. Hydrophobic
    4. Hydrophilic
    5. Protein

C. 1. Lipid bilayer
    2. Hydrophilic region of phospholipid
    3. Hydrophobic region of phospholipid
    4. Protein
    5. Carbohydrate molecule

Movement Through the Plasma Membrane

1. Lipid bilayer
2. Cell membrane channels

3. Cell membrane channels
4. Carrier molecules

Diffusion

A. 1. Solvent
    2. Diffusion
    3. Diffusion

B. 1. Increase
    2. Increase
    3. Decrease

Osmosis

1. Osmotic pressure
2. Hyposmotic
3. Isotonic

4. Hypertonic
5. Lysis

Mediated Transport Mechanisms

A. 1. Specificity
   2. Active site
   3. Saturation
   4. Competition

B. 1. Facilitated diffusion
   2. Facilitated diffusion
   3. Active transport

Endocytosis and Exocytosis

1. Endocytosis
2. Phagocytosis

3. Exocytosis

Nucleus

1. Nucleus
2. Nuclear envelope
3. Nuclear pores
4. Chromosomes

5. Chromatin
6. Histones
7. Messenger RNA
8. Nucleoli

Ribosomes and Endoplasmic Reticulum

1. Ribosome
2. Ribosome
3. Rough endoplasmic reticulum

4. Rough endoplasmic reticulum
5. Smooth endoplasmic reticulum

Golgi Apparatus and Secretory Vesicles

1. Golgi apparatus
2. Glycoproteins

3. Secretory vesicles

Lysosomes, Peroxisomes, and Cytoplasmic Inclusions

1. Lysosomes
2. Autophagia

3. Peroxisomes
4. Cytoplasmic inclusions

Mitochondria

1. ATP
2. Cristae
3. Oxidative metabolism

4. Matrix
5. Electron transport chain

Microtubules, Centrioles, Spindle Fibers, Cilia, and Flagella

1. Tubulin
2. Centrosome
3. Centriole

4. Cilia
5. Basal body

Microfilaments, Microvilli, Stereocilia, and Intermediate Fibers

1. Microfilaments
2. Microvilli

3. Stereocilia
4. Intermediate filaments

Cell Diagram

1. Secretory vesicle
2. Golgi apparatus
3. Smooth endoplasmic reticulum
4. Rough endoplasmic reticulum
5. Lysosome
6. Plasma membrane

7. Centrioles
8. Ribosomes
9. Mitochondrion
10. Nucleolus
11. Microtubule

Protein Synthesis

1. Transcription
2. mRNA

3. tRNA
4. Translation

Mitosis

A. 1. Autosome
2. Homologous
3. Mitosis
4. Cytokinesis
5. Interphase

B. 1. Chromatids
2. Centromere
3. Astral fibers
4. Spindle fibers

C. 1. Prophase
2. Metaphase
3. Anaphase
4. Telophase

D. 1. Anaphase
2. Prophase
3. Telophase
4. Metaphase
5. Astral fiber
6. Centriole
7. Centromere
8. Spindle fiber

Meiosis

1. Gamete
2. Haploid
3. Tetrad

4. Interkinesis
5. Crossing over

QUICK RECALL

1. Phospholipids: form the lipid bilayer that separates the inside of the cell from the outside; proteins: structural supports, enzymes, membrane channels, carrier molecules, and receptor molecules

2. Magnitude of concentration gradient, temperature of solution, size of diffusing molecules, and viscosity of the solvent

3. Diffusion, osmosis, facilitated diffusion, active transport, phagocytosis, pinocytosis, and exocytosis

4. Specificity, saturation, and competition
5. Isotonic, hypotonic, and hypertonic
6. Secretory vesicles
7. Smooth endoplasmic reticulum
8. Nucleus
9. Microtubules
10. Microvilli
11. Golgi apparatus
12. Ribosomes
13. Lysosomes
14. Mitochondria
15. Centrioles
16. Rough endoplasmic reticulum
17. Nucleoli
18. Cytoplasmic inclusions
19. Cilia
20. Peroxisomes
21. Microfilaments
22. Intermediate filaments
23. Messenger RNA, transfer RNA, and ribosomal RNA
24. Prophase, metaphase, anaphase, and telophase

MASTERY LEARNING ACTIVITY

1. D. Proteins act as carrier molecules, function as enzymes, and line the pores of cell membranes.

2. C. Water-soluble molecules cannot diffuse through the membrane because of its lipid nature. If small enough, water-soluble molecules diffuse through membrane pores. Lipid-soluble molecules can diffuse through the membrane proper.

3. A. When net movement of molecules ceases, an equilibrium condition of a uniform distribution of molecules is achieved. This is the end result of diffusion.

   b is incorrect, since with a uniform distribution molecules will not be more concentrated.

   c is incorrect. Diffusion is a passive transport mechanism requiring no expenditure of energy.

   d is incorrect, since the rate of diffusion has no effect on the equilibrium condition. A uniform distribution may be achieved quickly or slowly, but the end result is the same.

4. E. When a membrane does not permit the passage of at least one solute molecule, the result is an unequal concentration of solvent that diffuses through the membrane (osmosis). The number of solute molecules on each side of the membrane determines the osmotic pressure difference.

5. B. By definition, cells placed in a hypertonic solution will shrink. If placed in a hypotonic solution, they will swell, and in an isotonic solution they will neither swell nor shrink.

6. B. Since A has 10% salt, it has 90% water, B has 20% salt, and thus 80% water. Substances diffuse from areas of high concentration to low. Thus water will diffuse from A (90%) to B (80%), and salt will diffuse from B (20%) to A (10%).

7. A. Sweat is less concentrated than blood plasma. Thus, compared to blood plasma, more water is lost than salts. This increases the osmotic concentration of the blood plasma. Water then moves from the tissues into the plasma, resulting in shrinkage of the cells.

8. B. Replacing the lost water and salts with distilled water results in a more dilute blood plasma. Relatively speaking, then, the blood plasma has more water than previously, and thus water moves from the blood plasma into the tissues, causing them to swell.

9. B. With this choice there will be no net movement of sodium or potassium, since the concentrations of these substances are the same in both the dialyzing fluid and the blood plasma. Urea definitely moves from the blood plasma (26) into the dialyzing fluid (0). There is a slight movement of glucose from the dialyzing fluid into the blood plasma.

   The basic principal involved is that substances diffuse from areas of high concentration to areas of low concentration. With this is mind, a comparison of each proposed dialyzing fluid with normal blood plasma must be made. The object is to find one that will maintain sodium, potassium, and glucose levels as close to the normal levels as possible, yet still remove urea.

a is incorrect. All of the substances involved are removed from the blood plasma.

c is incorrect. There is a loss of sodium and potassium, but no net exchange of urea.

d is incorrect. There is a gain of sodium and potassium and a loss of glucose.

e is incorrect. There is a gain of sodium, potassium, and glucose.

10. A. Facilitated diffusion moves materials with the concentration gradient, does not require the expenditure of energy, and does use a carrier molecule. Materials move through membrane pores by simple diffusion.

11. A. Since the magnesium salts tend to remain in the intestine, they cause the intestinal contents to be more concentrated (hyperosmotic) with respect to the blood plasma. Relatively speaking, then, the gut would contain less water than the blood plasma. Water moves from the blood plasma into the intestine. By a chain reaction, as water leaves the blood plasma, the concentration of the plasma relative to the tissues changes. As a result, water moves from the tissues (causing them to shrink) into the blood plasma.

12. D. Since mannitol cannot enter cells, injection of a hyperosmotic mannitol solution raises the osmolality (osmotic concentration) of the blood. Thus water moves out of the tissues (brain) into the blood, reducing the edema. Eventually, the mannitol and water is excreted by the kidneys.

a is incorrect. Adding a hyperosmotic mannitol solution raises the osmotic concentration (pressure) of the blood.

b is incorrect. Reducing the permeability of cells to water makes it even harder to relieve the edema (remove the water). Actually, mannitol has no effect on cell permeability to water.

c is incorrect. Excretion of large amounts of salts by the kidneys lowers the osmotic concentration (pressure) of the blood. Thus water moves from the blood into tissues. Of course, this makes the edema condition worse. Actually, mannitol has no effect on salt excretion by the kidneys.

13. C. When glucose concentrations increase, the blood plasma contains proportionately less water. Thus water moves from the tissues into the blood plasma.

a is incorrect. As glucose concentrations increase, osmolality of the plasma increases.

b is incorrect. As osmolality (concentration) increases, the osmotic pressure of the blood plasma increases.

d is incorrect. As water moves from the tissues (see answer C) into the blood plasma, the tissue cells shrink.

14. C. Diffusion and facilitated diffusion do not require energy. Phagocytosis moves particulate matter into cells, pinocytosis moves large molecules in solution into cells, and exocytosis moves substances out of cells.

15. E. Active transport has specificity because molecules are transported by means of a carrier molecule. Because a carrier is involved, similar molecules may compete for the carrier. Also, once all the carrier molecules are in use, saturation occurs.

16. E. Since the amount transferred after point B levels off, despite increasing concentrations, it can be concluded that carrier-mediated transport is involved. In section A - B of the graph the carrier system is not saturated: thus increasing concentrations result in increased transport of the substance as more and more carrier molecules are used. From these data there is no way to tell if the carrier-mediated process is facilitated diffusion or active transport.

17. C. Since the metabolic inhibitor greatly reduced the rate of transfer of the amino acid, it can be concluded that the transport mechanism involved requires energy. Of the three transport mechanisms listed, only active transport uses energy.

18. C. Since the curve levels off (exhibits saturation), simple diffusion can be ruled out. Note that the cell started with NO intracellular glucose. After 5 minutes, however, the intracellular concentration of glucose was always greater than the extracellular concentration. Clearly this could occur only if active transport was involved.

19. D. The data suggest that both types of molecules are competing for the same carrier molecule. Apparently molecule B is a much better competitor, since its rate of transport is hardly affected by the presence of molecule A. Diffusion does not require carrier molecules, so both molecules must be moved by active transport.

20. B. The nucleus is defined as everything inside the nuclear membrane. Everything between the nuclear membrane and the plasma membrane is cytoplasm.

21. C. Normally the only large structure visible in the nucleus is the nucleolus. It is made up of RNA. The chromosomal material is dispersed, so the chromosomes are not visible except during cell division.

22. D. The Golgi apparatus is an extension of the endoplasmic reticulum (ER). Rough ER (with ribosomes) produces proteins that are transferred to the Golgi apparatus. Smooth ER (no ribosomes) produces lipids that are handled in a similar fashion.

23. B. Mitochondria are sometimes referred to as the powerhouse of the cell. The enzymes of the citric acid cycle (Kreb's cycle) and the cytochrome oxidase system (electron transport system) are located within the mitochondria. Therefore it is the site of oxidative metabolism and the major site of ATP synthesis within the cell. To answer this question and similar questions about the cellular organelles that could be asked, one must memorize their functions.

24. C. Smooth endoplasmic reticulum produces lipids that pass into the Golgi apparatus, where they are packaged into vesicles and then secreted. Rough endoplasmic reticulum (ER with ribosomes) produces proteins, which are then packaged by the Golgi apparatus. Lysosomes contain digestive enzymes.

25. C. The red blood cell has no nucleus and therefore cannot synthesize mRNA, which is required for new protein synthesis. Red blood cells can synthesize ATP, consume oxygen, and use glucose because of enzymes that were produced in the red blood cell before it lost its nucleus.

26. A. The cell lines a cavity, indicating that substances entering or leaving the cavity must be transported or must diffuse across that cellular barrier. The microvilli increase the surface area of the cell exposed to the lumen of the cavity, and numerous mitochondria suggest that energy, in the form of ATP, is used by the cell. Thus the data suggest that the cell is involved in active transport.

    Lipid secretion is unlikely with the small amount of smooth endoplasmic reticulum and few vacuoles. Also, without secretory vacuoles, protein secretion is unlikely.

27. C. A codon refers to a sequence of three organic bases (nucleotides) in a mRNA molecule. At the ribosome the codon determines which tRNA can combine with the mRNA. The particular tRNA that combines with the organic bases of mRNA is attached to a specific amino acid. Therefore a codon determines one amino acid in a polypeptide chain.

28. A. mRNA is synthesized within the nucleus. The sequence of organic bases that make up mRNA is determined by DNA within the nucleus. That is, DNA acts as a template for mRNA. The sequence of organic bases in mRNA subsequently determines the amino acid sequence of a protein.

29. A. At the ribosomes, mRNA and tRNA interact to form proteins.

30. E. Transfer RNA has as a component a sequence of three organic bases. These organic bases can pair with a unique sequence of three organic bases on mRNA, which was determined by the sequence of organic bases in the nuclear DNA. There are in excess of 20 different types of tRNA. Each type of tRNA is bound to a specific amino acid. Therefore the function of tRNA is to combine with a specific sequence of organic bases on mRNA and participate in determining the sequence of amino acids in a polypeptide.

31. A. Messenger RNA carries "instructions" for protein synthesis from the chromosomes (DNA) to the ribosomes. A substance that prevented the production of mRNA synthesis would prevent protein synthesis.

32. E. All of the characteristics listed except gamete production occur by mitosis. Gamete production is the result of meiosis.

33. D. During telophase the cell returns to the interphase condition. The chromosomes unravel to form chromatin, the nuclear membrane reforms, and the nucleolus reappears.

34. C. Meiosis results in the daughter cells having one half of the number of chromosomes as the mother cell. It also results in the daughter cells having one of each homologous pair of chromosomes (one of each type of chromosome found in the mother cell).

35. D. XY combination produces a male, and XX produces a female.

 FINAL CHALLENGES

1. One approach is to make up different solutions of known osmolalities. Red blood cells could be placed in each solution and observed to see if they change shape. The solution in which the red blood cells neither swell nor shrink is an isotonic solution. Since the solute molecules do not cross the cell membrane, this would also be an isosmotic solution. The osmolality of the cytoplasm of the red blood cell would be equal to the concentration of this solution.

2. Agree. Initially, since the glucose solution is isotonic, there is only an increase in blood volume. However, as the glucose is used up (metabolized), the osmolality of the blood decreases. This means that the blood is more dilute than usual; the kidneys compensate by producing large quantities of dilute urine. This removes the excess water from the blood, restoring it to its normal volume and osmolality (homeostasis).

3. A. Diffusion of molecules A and B could be distinguished from facilitated diffusion in two ways:

   1. If the process is facilitated diffusion, there could be competition for a carrier molecule. The fact that neither molecule affects the rate of transport of the other could mean that the process is not facilitated diffusion or that two different carrier molecules are involved. Introduction of other molecules similar to A or B could resolve this issue.

   2. If the process is facilitated diffusion, then increasing the concentration gradient for molecules A and B should result in saturation of the carrier molecule and a leveling off of the rate of transport. If the process is diffusion, increasing the concentration gradient results in increasing rates of transport.

   B. Active transport of molecules A and B could be distinguished from facilitated diffusion (or diffusion) in two ways:

   1. Determine the concentration gradient and, if movement is against the concentration gradient, the transport mechanism is active transport.

   2. Introduce a metabolic inhibitor. Since active transport requires energy, this stops the movement of molecules A and B.

4. It is logical to conclude that tissue A has the greatest number of mitochondria (which use oxygen to produce ATPs). However, it is not necessary to make that conclusion. Other possibilities are that tissue A has larger mitochondria or mitochondria that have a greater enzymatic activity that tissue B.

5. Enzymes are proteins. The presence of well-developed rough endoplasmic reticulum, Golgi apparatuses, and secretory vacuoles is strong evidence for the secretion of a protein.

6. The fact that an inhibitor of mRNA synthesis does not inhibit the secretion of lipid suggests that the response of the cell involves the release of stored lipids or the activation of enzymes that already exist. Neither of these processes require protein 0 synthesis, which is blocked by a mRNA inhibitor.

Histology:
Study Of Tissues

FOCUS: The cells of the body are specialized to form four basic types of tissues. Epithelial tissue consists of a single or a multiple layer of cells with little intracellular material between the cells. It covers the free surfaces of the body, providing protection, regulating the movement of materials, and producing secretions (glands). Connective tissue holds structures together (i.e., tendons and ligaments), is a site of fat storage and blood cell production, and supports other tissues (i.e., cartilage and bone). The extracellular matrix produced by connective tissue cells is responsible for many of the functional characteristics of connective tissue. Muscle tissue has the ability to contract and is responsible for body movement (skeletal muscle), blood movement (cardiac muscle), and movement of materials through hollow organs (smooth muscle). Nervous tissue is specialized to conduct electrical impulses and functions to control and coordinate the activities of other tissues. The four tissue types develop from the primary germ layers (endoderm, mesoderm, and ectoderm) in the embryo. In some locations epithelial and connective tissues form serous or mucous membranes. Inflammation is a process that isolates and destroys injurious agents. Some tissues recover from injury by replacing lost cells with cells of the same type (regeneration), but other tissues can only repair by the addition of cells of a different type (replacement), leading to scar formation and loss of function.

WORD PARTS

Give an example of a new vocabulary word that contains each word part.

WORD PART	MEANING	EXAMPLE
epi-	upon; over	1. _____
meso-	middle	2. _____
desmo-	band; ligament	3. _____

WORD PART	MEANING	EXAMPLE
apo-	away from; off	4. _____
acin-	grape; sac	5. _____
crin-	to separate	6. _____
mero-	a part	7. _____
holo-	entire	8. _____
-blast	formative	9. _____
-clast	broken	10. _____

CONTENT LEARNING ACTIVITY

Classification of Epithelium

66_Epithelia are classified according to the number of cell layers and the shape of the cells._99

A. Match these terms with the correct statement or definition:

Pseudostratified epithelium   Stratified columnar epithelium
Simple columnar epithelium   Stratified cuboidal epithelium
Simple cuboidal epithelium   Stratified squamous epithelium
Simple squamous epithelium   Transitional epithelium

_____ 1. Single layer of cube-shaped cells.

_____ 2. Multiple layers of tall, thin cells.

_____ 3. Layers of cells that appear cubelike when an organ is relaxed and flattened when the organ is distended by fluid.

_____ 4. Single layer of flat, often hexagonal cells.

_____ 5. Single layer of cells; some cells are tall and thin and reach the free surface, and others do not.

_____ 6. Multiple layers of cells in which the basal layer is cuboidal and becomes flattened at the surface. It may be moist or keratinized.

☞ The basement membrane is a specialized type of extracellular material that is secreted by the epithelial cells on the side opposite their free surface.

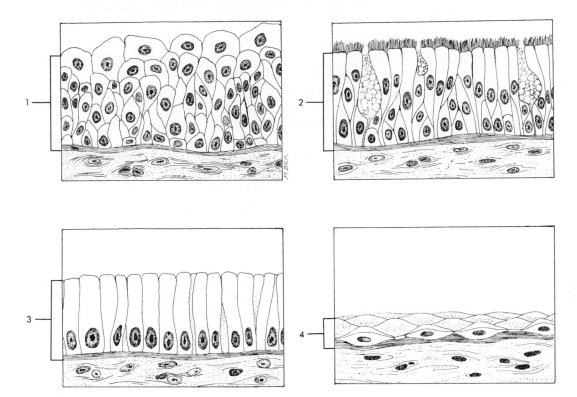

Figure 4-1

B. Match these terms with the types of tissue in Figure 4-1:

Pseudostratified epithelium
Simple columnar epithelium
Simple squamous epithelium
Transitional epithelium

1. _____

2. _____

3. _____

4. _____

Functional Characteristics

"*Epithelial tissues perform many functions.***"**

A. Match these terms with the correct statement or definition:

Simple epithelium
Stratified epithelium

_____ 1. Found in areas where protection is a major function.

_____ 2. Found in organs where principal functions are diffusion, filtration, secretion, or absorption.

_____ 3. Found in areas such as the mouth, skin, throat, anus, and vagina.

B. Match these terms with the correct statement or definition:

Cuboidal or columnar
Squamous

_____ 1. Epithelial cells that are involved with secretion or absorption.

_____ 2. Epithelial cells involved with diffusion or filtration.

C. Match these terms with the correct statement or definition:

Ciliated   Smooth
Folded   With microvilli

_____ 1. Cell surface that reduces friction.

_____ 2. Propel materials along the cell surface.

_____ 3. Cell surface that has rigid sections alternating with flexible sections.

_____ 4. Cell surface that greatly increases surface area.

D. Using the terms provided, complete the following statements:

Basement membrane   Zonula adherens
Desmosomes   Zonula occludens
Gap junctions

Epithelial cells secrete glycoproteins that attach the cells to the basement membrane and to each other. This relatively weak binding is reinforced by _(1)_, disk-shaped structures with especially adhesive glycoproteins that bind cells to each other. Hemidesmosomes, similar to one half of a desmosome, attach epithelial cells to the _(2)_. _(3)_ form permeability barriers by completely surrounding the cell and preventing the passage of materials in between cells. On the other hand, _(4)_ are small channels that allow the passage of ions and small molecules between cells.

1. _____

2. _____

3. _____

4. _____

Glands

"_Glands are secretory organs._**"**

A. Match these terms with the correct statement or definition:

Endocrine
Exocrine

_____ 1. Glands with a duct.

_____ 2. Glands with no duct that secrete hormones.

B. Match these terms with the correct statement or definition:

Acinar or alveolar   Straight
Coiled   Tubular
Compound   Unicellular
Simple

_____ 1. Exocrine glands composed of one cell; e. g., goblet cells.

_____ 2. Exocrine glands with ducts that branch repeatedly.

_____ 3. Exocrine glands that end in saclike structures.

_____ 4. Tubular exocrine glands with no coiling.

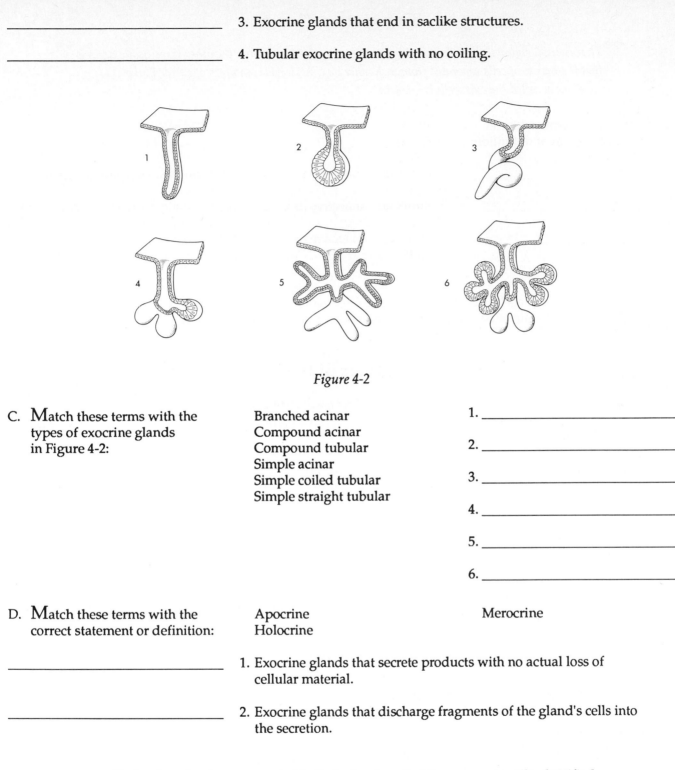

Figure 4-2

C. Match these terms with the types of exocrine glands in Figure 4-2:

Branched acinar
Compound acinar
Compound tubular
Simple acinar
Simple coiled tubular
Simple straight tubular

1. _____

2. _____

3. _____

4. _____

5. _____

6. _____

D. Match these terms with the correct statement or definition:

Apocrine   Merocrine
Holocrine

_____ 1. Exocrine glands that secrete products with no actual loss of cellular material.

_____ 2. Exocrine glands that discharge fragments of the gland's cells into the secretion.

☞ Endocrine glands are so varied in their structure that they are not easily classified.

Connective Tissue

66 *The essential characteristic that distinguishes connective tissue from the other three tissue types is* 99 *that it consists of cells separated from each other by considerable amounts of extracellular substances called the extracellular matrix.*

A. Match these terms with the correct statement or definition:

Blasts
Clasts

Cytes

_____ 1. Suffix for connective tissue cells that form the extracellular matrix.

_____ 2. Suffix for connective tissue cells that maintain the extracellular matrix.

_____ 3. Suffix for connective tissue cells that break down the extracellular matrix.

☞ The extracellular matrix has three components: (1) protein fibers, (2) ground substance consisting of nonfibrous protein and other molecules, and (3) fluid.

Protein Fibers of the Matrix

66 *Three types of protein fiber are in connective tissue.* 99

Match these terms with the correct statement or definition:

Collagen
Elastin

Reticular fibers

_____ 1. These protein fibers are the most common protein in the body and are strong and flexible, but inelastic.

_____ 2. These protein molecules are very short, thin collagen fibers that branch to form a network.

_____ 3. This protein gives the tissue in which it is found an elastic or rubbery characteristic. The structure of this molecule is similar to a coiled metal spring.

Nonprotein Matrix Molecules

66 *There are two types of large, nonprotein molecules of the intracellular matrix.* 99

Match these terms with the correct statement or definition:

Ground substance
Hyaluronic acid

Proteoglycan

_____ 1. Shapeless background material of the extracellular matrix.

_____ 2. Constitute part of the ground substance, and are composed of repeating disaccharides; good lubricant for joint cavities.

_____ 3. Large and composed of proteins and polysaccharides.

Connective Tissue Matrix with Fibers as the Primary Feature

66*Connective tissue that has a matrix with fibers as a primary feature has two subtypes, fibrous and special.*99

A. Match these terms with the correct statement or definition:

Dense   Loose (areolar)
Fibroblasts   Macrophages

_____ 1. Fibrous connective tissue in which protein fibers fill nearly all the extra cellular space.

_____ 2. Fibrous connective tissue in which protein fibers form a lacy network with numerous fluid-filled spaces.

_____ 3. Cells that produce the fibers of connective tissue.

_____ 4. Cells that engulf bacteria and cell debris within connective tissue.

B. Match these terms with the correct statement or definition:

Irregular dense
Regular dense

_____ 1. Dense connective tissue that contains protein fibers predominantly oriented in the same direction; strong in one direction.

_____ 2. Dense connective tissue that contains protein fibers that can be arranged as a network of randomly oriented fibers; strong in many directions.

C. Match these terms with the correct statement or definition:

Dense irregular collagenous   Dense regular collagenous
Dense irregular elastic   Dense regular elastic

_____ 1. Connective tissue found in tendons and ligaments; dense collagen fibers oriented in the same direction.

_____ 2. Connective tissue found in special ligaments such as the nuchal ligament; dense elastin fibers oriented in the same direction.

_____ 3. Dense connective tissue with randomly oriented fibers of elastin; found in elastic arteries.

_____ 4. Dense connective tissue with randomly oriented collagen fibers; found in the skin (dermis).

☞ The nuchal ligament is along the posterior of the neck and helps to hold the head upright.

D. Match these terms with the correct statement or definition:

Adipose   Reticular
Dendritic cells   Yellow bone marrow
Red bone marrow

_____ 1. Consists of fat cells that contain large amounts of lipid, and can be either brown or yellow (white) in color.

_____ 2. Forms the framework of lymphatic tissue, bone marrow, and the liver.

_____ 3. Cells that occur between reticular fibers and look very much like reticular cells, but are part of the immune system.

_____ 4. Found in the bone and contains adipose tissue.

Matrix With Both Protein Fibers and Ground Substance

" *There are two types of connective tissue that have a matrix with both protein fibers and ground substance.* **"**

A. Match these terms with the correct statement or definition:

Chondrocytes   Hyaline cartilage
Elastic cartilage   Lacunae
Fibrocartilage

_____ 1. Cartilage cells.

_____ 2. Spaces in which cartilage cells are located.

_____ 3. Very smooth tissue with a glassy, translucent matrix. It is found in areas where strong support and some flexibility are needed (e.g., rib cage and rings of bronchi and trachea)

_____ 4. Connective tissue with thick bundles of collagen dispersed through the matrix. It is found in areas that must withstand a great deal of pressure such as the knee.

B. Match these terms with the correct statement or definition:

Cancellous bone   Lacunae
Compact bone   Osteocytes
Hydroxyapatite   Trabeculae

_____ 1. Complex salt crystal that constitutes the mineral (inorganic) portion of bone.

_____ 2. Bone cells.

_____ 3. Spaces occupied by bone cells.

_____ 4. Plates of bone.

_____ 5. Essentially solid bone.

☞ Blood is unique among the connective tissues because the matrix between the cells is liquid.

Identification of Connective Tissue

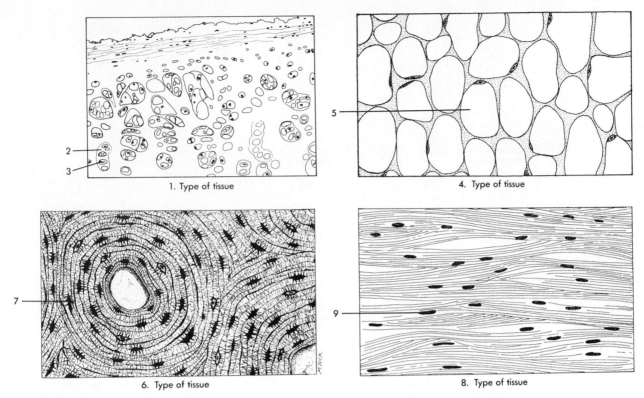

1. Type of tissue

4. Type of tissue

6. Type of tissue

8. Type of tissue

Figure 4-3

Match these terms with the correct tissue type or structure labeled in Figure 4-3:

Adipose
Bone
Cartilage
Chondrocyte
Dense regular connective tissue

Fat droplet
Fibroblast
Lacuna
Osteocyte

1. _____

2. _____

3. _____

4. _____

5. _____

6. _____

7. _____

8. _____

9. _____

Muscle Tissue

❝*The main characteristics of muscle tissue are that it is contractile and it is responsible for movement.*❞

A. Match these terms with the correct statement or definition:

Cardiac
Skeletal

Smooth

_____ 1. Striated, voluntary muscle cells.

_____ 2. Striated, involuntary muscle cells.

_____ 3. Cylindrical cells that branch, with a single, centrally located nucleus.

_____ 4. Spindle-shaped cells with a single, centrally located nucleus.

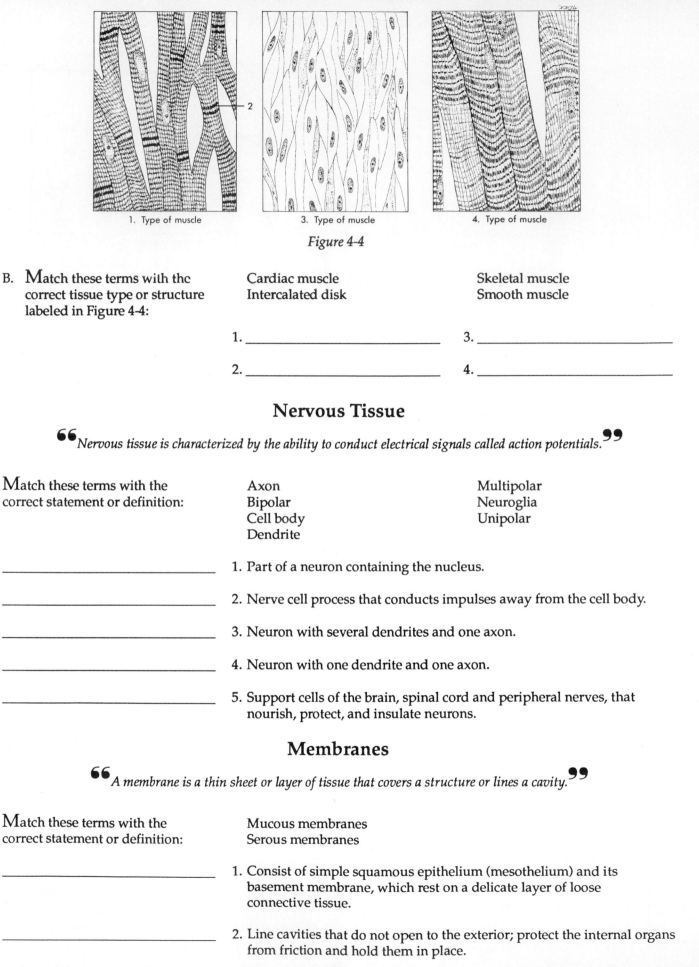

1. Type of muscle   3. Type of muscle   4. Type of muscle

Figure 4-4

B. Match these terms with the correct tissue type or structure labeled in Figure 4-4:

Cardiac muscle   Skeletal muscle
Intercalated disk   Smooth muscle

1. _____   3. _____

2. _____   4. _____

Nervous Tissue

"*Nervous tissue is characterized by the ability to conduct electrical signals called action potentials.***"**

Match these terms with the correct statement or definition:

Axon   Multipolar
Bipolar   Neuroglia
Cell body   Unipolar
Dendrite

_____ 1. Part of a neuron containing the nucleus.

_____ 2. Nerve cell process that conducts impulses away from the cell body.

_____ 3. Neuron with several dendrites and one axon.

_____ 4. Neuron with one dendrite and one axon.

_____ 5. Support cells of the brain, spinal cord and peripheral nerves, that nourish, protect, and insulate neurons.

Membranes

"*A membrane is a thin sheet or layer of tissue that covers a structure or lines a cavity.***"**

Match these terms with the correct statement or definition:

Mucous membranes
Serous membranes

_____ 1. Consist of simple squamous epithelium (mesothelium) and its basement membrane, which rest on a delicate layer of loose connective tissue.

_____ 2. Line cavities that do not open to the exterior; protect the internal organs from friction and hold them in place.

70

_____ 3. Consist of epithelial cells and their basement membrane, which rest on a thick layer of loose connective tissue called the lamina propria.

_____ 4. Line cavities that open to the outside of the body; functions include protection, absorption, and secretion.

Inflammation

"*The inflammatory response mobilizes the body's defenses, isolates and destroys microorganisms and other* **"** *infectious agents, and removes foreign materials and damaged cells so that tissue repair can proceed.*

Match these terms with the correct statement or definition:

Disturbance of function   Pain
Edema   Vasodilation
Mediators of inflammation

_____ 1. Chemical substances that are released or activated in tissues and adjacent blood vessels after a person is injured.

_____ 2. Expansion of blood vessels, which produces symptoms of redness and heat.

_____ 3. Swelling of a tissue because of fluid accumulation.

_____ 4. Clotting of blood and other proteins, which "walls off" the site of the injury from the rest of the body.

_____ 5. Result of edema and some mediators stimulating nerves.

_____ 6. Result of pain, limitation of movement resulting from edema, and tissue destruction.

Tissue Repair

"*Tissue repair is the substitution of viable cells for dead cells, and it can occur by regeneration or replacement.***"**

Using the terms provided, complete the following statements:

Labile   Replacement
Permanent   Secondary union
Primary union   Stable
Regeneration

In (1) , the new cells are of the same type as those that were destroyed, whereas in (2) , a new type of tissue develops that eventually causes the loss of some tissue function. (3) cells continue to divide throughout life. Damage to these cells can be completely repaired by regeneration. (4) cells do not actively replicate after growth ceases, but they do retain the ability to divide, if necessary and are capable of regeneration. (5) cells can not replicate, and, if killed, they are replaced by a different type of cell. If the edges of a wound are close together such as in a surgical incision, the wound heals by a process called (6) . If the edges of the wound are not close together or if there has been extensive loss of tissue, the process is called (7) .

1. _____

2. _____

3. _____

4. _____

5. _____

6 _____

7. _____

1. List eight kinds of epithelium based on numbers of cell layers and shape of cells.

2. List four functions performed by epithelial cells.

3. List the four types of free cell surfaces of epithelial cells.

4. Name five types of cell connections.

5. List three types of exocrine glands based on how products leave the cell.

6. Name three types of protein fibers found in connective tissue.

7. List four types of fibrous tissue found in the human body.

8. Name three types of cartilage found in the human body.

9. Name two types of bone found in the human body.

10. List three types of special connective tissue.

11. List three types of muscle cells found in the human body.

12. List the three primary germ layers and one derivative from each.

13. **List the two major categories of membranes found in the human body.**

14. **List the five major symptoms of inflammation.**

15. **List the three categories into which cells can be classified according to their regenerative ability.**

MASTERY LEARNING ACTIVITY

Place the letter corresponding to the correct answer in the space provided.

_____ 1. Given the following characteristics:
   1. capable of contraction
   2. covers all free body surfaces
   3. lacks blood vessels
   4. comprises various glands
   5. anchored to connective tissue by a basement membrane

   Which of the above are characteristic of epithelial tissue?
   a. 1, 2, 3
   b. 2, 3, 5
   c. 3, 4, 5
   d. 1, 2, 3, 4
   e. 2, 3, 4, 5

_____ 2. A tissue that covers a surface, is one cell layer thick, and is composed of flat cells is
   a. simple squamous epithelium.
   b. simple cuboidal epithelium.
   c. simple columnar epithelium.
   d. stratified squamous epithelium.
   e. transitional epithelium.

_____ 3. Epithelium composed of two or more layers of cells with only the deepest layer in contact with the basement membrane is known as
   a. stratified epithelium.
   b. simple epithelium.
   c. pseudostratified epithelium.
   d. columnar epithelium.

_____ 4. Stratified epithelium is usually found in areas of the body where the principal activity is
   a. filtration.
   b. protection.
   c. absorption.
   d. diffusion.
   e. none of the above

_____ 5. An epithelial cell with microvilli would most likely be found
   a. lining blood vessels.
   b. lining the lungs.
   c. in serous membranes.
   d. lining the digestive tract.

_____ 6. A tissue that contains cells with the following characteristics:
1. covers a surface
2. one layer of cells
3. cells are flat

Performs which of the functions listed below?
a. phagocytosis
b. active transport
c. secretion of many complex lipids and proteins
d. is adapted to allow certain substances to diffuse across it
e. a and b

_____ 7. A type of junction between epithelial cells whose ONLY function is to prevent the cells from coming apart (provides mechanical strength) is the
a. desmosome.
b. gap junction.
c. dermatome.
d. tight junction.

_____ 8. Pseudostratified ciliated epithelium can be found lining the
a. digestive tract.
b. trachea.
c. thyroid gland.
d. kidney tubules.
e. a and b

_____ 9. Cuboidal epithelium can be found in the
a. mesothelium.
b. endothelium.
c. vagina.
d. oviducts.
e. thyroid gland.

_____ 10. In parts of the body such as the urinary bladder, where considerable distension occurs, one can expect to find the which of the following type of cells?
a. cuboidal
b. pseudostratified
c. transitional
d. squamous

_____ 11. A histologist observed the characteristics listed below while viewing a tissue with a microscope:
1. consists of more than one layer of cells
2. surface cells are alive
3. cells near the free surface are flat
4. cells in the basal layers appear to be undergoing mitosis
5. there appear to be many cell-to-cell attachments in all layers

Choose the MINIMUM number of characteristics that would allow one to classify the tissue observed as moist, stratified squamous epithelium.
a. 1
b. 1, 2
c. 1, 2, 3
d. 1, 2, 3, 4
e. 1, 2, 3, 4, 5

_____ 12. Those glands that lose their connection with the epithelium during embryonic development and thus empty their secretions directly into the blood are called
a. exocrine glands.
b. apocrine glands.
c. endocrine glands.
d. merocrine glands.
e. holocrine glands.

_____ 13. Glands that accumulate secretions and release them only when the individual secretory cells rupture and die are
a. merocrine glands.
b. holocrine glands.
c. apocrine glands.

_____ 14. The fibers in connective tissue are formed by
a. fibroblasts.
b. adipocytes.
c. osteoblasts.
d. macrophages.

_____ 15. Extremely delicate fibers that make up the framework for organs such as the liver, spleen, and lymph nodes are
a. collagen fibers.
b. elastic fibers.
c. reticular fibers.
d. microvilli.
e. cilia.

16. A tissue that contains a large amount of extracellular collagen organized as parallel fibers would probably be found in
    a. a muscle.
    b. a tendon.
    c. adipose tissue.
    d. bone.
    e. cartilage.

17. Which of the following is true of adipose tissue?
    a. site of energy storage
    b. a type of connective tissue
    c. acts as a protective cushion
    d. functions as a heat insulator
    e. all of the above

18. Fibrocartilage is found in the
    a. cartilage rings of the trachea.
    b. costal cartilages.
    c. intervertebral disks.
    d. a and b
    e. all of the above

19. A tissue in which cells are located in lacunae surrounded by a hard matrix of hydroxyapatite is
    a. hyaline cartilage.
    b. bone.
    c. nervous tissue.
    d. dense fibrous connective tissue.
    e. fibrocartilage.

20. Blood is an example of
    a. epithelial tissue.
    b. connective tissue.
    c. muscle tissue.
    d. nervous tissue.
    e. none of the above

21. Which of the following are characteristic of smooth muscle?
    a. under voluntary control
    b. striated
    c. capable of spontaneous contraction
    d. a and b
    e. all of the above

22. Which of the following statements about nervous tissue is NOT true?
    a. Neurons have cytoplasmic extensions called axons.
    b. Electrical signals (action potentials) are conducted along axons.
    c. Bipolar neurons have two axons.
    d. Neurons are protected and nourished by neuroglia.

23. A bullet wound that passes through one's upper arm without hitting bone could contact which of the following tissue types?
    a. nervous
    b. muscle
    c. connective
    d. epithelial
    e. all of the above

24. Concerning the germ layers:
    a. there are four germ layers.
    b. the germ layers give rise to the four tissue types.
    c. all the germ layers are first formed in the middle layer of the embryo.
    d. all of the above

25. Linings of the digestive, respiratory, urinary and reproductive tracts are composed of
    a. serous membranes.
    b. mesothelium.
    c. mucous membranes.
    d. endothelium.
    e. a and b

26. The lamina propria
    a. is a layer of smooth muscle.
    b. is located under (deep to) mucous membranes.
    c. binds the skin (integument) to underlying tissues.
    d. all of the above

27. Chemical mediators of inflammation
    a. stimulate nerve endings to produce the symptom of pain.
    b. increase the permeability of blood vessels.
    c. cause vasodilation (expansion) of blood vessels.
    d. are released into or activated in tissues following injury.
    e. all of the above

_____ 28. Which of the following tissues are capable of mitosis throughout life and thus have the ability to undergo repair by producing new cells to replace damaged cells?
a. nervous tissue (neurons)
b. connective tissue
c. epithelial tissue
d. a and b
e. all of the above

_____ 29. Permanent cells
a. divide and replace damaged cells in replacement tissue repair.
b. form granulation tissue.
c. are responsible for removing scar tissue.
d. are replaced by a different cell type if they are destroyed.
e. are replaced during regeneration tissue repair.

_____ 30. Which of the following is true about granulation tissue?
a. produced by fibroblasts
b. a type of connective tissue
c. may turn into scar tissue
d. a and b
e. all of the above

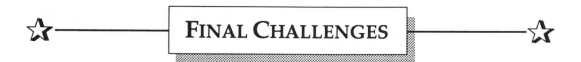

FINAL CHALLENGES

Use a separate sheet of paper to complete this section.

1. On a histology exam, Slide Mann was asked to identify the types of epithelial tissue lining the surface of an organ. He identified the first tissue as stratified squamous epithelium and the second tissue as stratified cuboidal. In both cases he was wrong. Given than the tissues both came from the same organ, what was the epithelial type?

2. Ciliated cells move materials over their surfaces. Explain why ciliated epithelial cells are found in the tube system of the respiratory system, but are not found in the duct system of the urinary system.

3. Some tendons are formed into a broad sheet of tissue called an aponeurosis. If it were necessary to surgically cut through an aponeurosis to reach a deeper structure, would it be best to make a longitudinal or a transverse section through the aponeurosis? Explain.

4. Slide Mann was examining a ligament under the microscope. Slide knew that ligaments attached bones to bones, so he was surprised to observe a large number of elastic fibers in the ligament. Why did it seem inappropriate for a ligament to have elastic fibers?

5. When Slide Mann asked his instructor about the ligament with elastic fibers, the instructor responded with a question for Slide, "Do you think the ligament joined the bones of the spine (vertebrae) to each other, or did it join the thigh bone (femur) to the hip bone (coxa)?" Explain why this question should make everything clear to Slide.

6. A muscle is striated, has a single nucleus, is under involuntary control, and is capable of spontaneous contractions. List all the pairs of traits (e.g. striated, involuntary control) from this list that would identify the muscle as cardiac muscle.

ANSWERS TO CHAPTER 4

1. epidermis; epithelium
2. mesothelium, mesoderm
3. desmosome; hemidesmosome
4. apocrine
5. acinar

6. endocrine; exocrine; apocrine; merocrine; holocrine
7. merocrine
8. holocrine
9. osteoblast; fibroblast
10. osteoclast

CONTENT LEARNING ACTIVITY

Classification of Epithelium

A.
1. Simple cuboidal epithelium
2. Stratified columnar epithelium
3. Transitional epithelium
4. Simple squamous epithelium
5. Pseudostratified epithelium
6. Stratified squamous epithelium

B.
1. Transitional epithelium
2. Pseudostratified columnar epithelium
3. Simple columnar epithelium
4. Simple squamous epithelium

Functional Characteristics

A.
1. Stratified epithelium
2. Simple epithelium
3. Stratified epithelium

B.
1. Cuboidal or columnar
2. Squamous

C.
1. Smooth
2. Ciliated
3. Folded
4. With microvilli

D.
1. Desmosomes
2. Basement membrane
3. Zonula occludens
4. Gap junctions

Glands

A.
1. Exocrine
2. Endocrine

B.
1. Unicellular
2. Compound
3. Acinar or alveolar
4. Straight

C.
1. Simple straight tubular
2. Simple acinar
3. Simple coiled tubular
4. Branched acinar
5. Compound tubular
6. Compound acinar

D.
1. Merocrine
2. Apocrine

Connective Tissue

1. Blasts
2. Cytes

3. Clasts

Protein Fibers of the Matrix

1. Collagen
2. Reticular fibers

3. Elastin

Nonprotein Matrix Molecules

1. Ground substance
2. Hyaluronic acid

3. Proteoglycan

Connective Tissue Matrix With Fibers as the Primary Feature

A. 1. Dense
    2. Loose (areolar)
    3. Fibroblasts
    4. Macrophages

B. 1. Regular dense
    2. Irregular dense

C. 1. Dense regular collagenous
    2. Dense regular elastic
    3. Dense irregular elastic
    4. Dense irregular collagenous

D. 1. Adipose
    2. Reticular
    3. Dendritic cells
    4. Yellow bone marrow

Matrix With Both Protein Fibers and Ground Substance

A. 1. Chondrocytes
    2. Lacunae
    3. Hyaline cartilage
    4. Fibrocartilage

B. 1. Hydroxyapatite
    2. Osteocytes
    3. Lacunae
    4. Trabeculae
    5. Compact bone

Identification of Connective Tissue

1. Cartilage
2. Lacuna
3. Chondrocyte
4. Adipose tissue
5. Fat droplet

6. Bone
7. Osteocyte
8. Dense regular connective tissue
9. Fibroblast

Muscle Tissue

A. 1. Skeletal
   2. Cardiac
   3. Cardiac
   4. Smooth

B. 1. Cardiac muscle
   2. Intercalated disk
   3. Smooth muscle
   4. Skeletal muscle

Nervous Tissue

1. Cell body
2. Axon
3. Multipolar

4. Bipolar
5. Neuroglia

Membranes

1. Serous membranes
2. Serous membranes

3. Mucous membranes
4. Mucous membranes

Inflammation

1. Mediators of inflammation
2. Vasodilation
3. Edema

4. Coagulation
5. Pain
6. Disturbance of function

Tissue Repair

1. Regeneration
2. Replacement
3. Labile
4. Stable

5. Permanent
6. Primary union
7. Secondary union

QUICK RECALL

1. Simple squamous, simple cuboidal, simple columnar, stratified squamous, stratified cuboidal, stratified columnar, pseudostratified, and transitional epithelium
2. Protection, secretion, transportation, absorption, filtration, and diffusion
3. Smooth, ciliated, folded, and with microvilli
4. Desmosomes, hemidesmosomes, zonula adherens, zonula occludens, and gap junctions
5. Merocrine, apocrine, and holocrine
6. Collagen, reticular fibers, and elastin
7. Dense regular collagenous, dense irregular collagenous, dense regular elastic, and dense irregular elastic
8. Hyaline, fibrocartilage, and elastic cartilage
9. Compact and cancellous
10. Reticular, adipose, and bone marrow
11. Skeletal, cardiac, and smooth muscle
12. Ectoderm: brain, spinal cord, peripheral nerves; endoderm: lining of digestive system, trachea, bronchi, lungs, liver, thyroid; mesoderm: bone, cartilage, tendons, muscle, and blood
13. Serous and mucous
14. Redness, heat, swelling, pain, and disturbance of function
15. Labile, stable, and permanent

1. E. Muscle tissue, not epithelial tissue, is capable of contraction. The other statements are true for epithelial tissue.

2. A. The tissue is simple (one cell layer thick), squamous (flat cells) epithelium (covers a surface).

3. A. Since the description specified two or more layers, only stratified and pseudostratified epithelium are possible candidates for a correct answer. Of these two choices, stratified epithelium has only the deepest layer in contact with the basement membrane. Pseudostratified epithelium has all cells contacting the basement membrane.

4. B. Stratified epithelium (many layers of cells) is protective in function and is found in areas subjected to friction (mouth, pharynx, esophagus, anus, vagina, and skin). Simple squamous epithelium would be associated with filtration and diffusion. Secretion is typically a function of cuboidal or columnar shaped cells.

5. D. Microvilli greatly increase the surface area of epithelial cells. This allows greater absorption at the surface of the cell, an ideal trait for cells that must absorb materials from the intestine.

6. D. One layer of flat cells covering a surface describes simple squamous epithelium. This tissue forms a thin sheet through which diffusion can occur. Examples are found in blood vessels (endothelium), air sacs (alveoli) in the lungs, and in the kidneys (Bowman's capsule). Secretion, phagocytosis, active transport, and absorption would be expected of larger cells (cuboidal or columnar).

7. A. Desmosomes provide mechanical strength. Gap junctions provide a way for materials to be exchanged between cells. Tight junctions prevent extracellular materials from passing between cells.

8. B. Pseudostratified ciliated epithelium is found in the nasal cavity (upper respiratory tract) and in the trachea (lower respiratory tract). The cilia move mucous with entrapped particulate matter and microorganisms to the back of the throat, where it is swallowed. This helps to keep the respiratory tract clear. The digestive tract has simple columnar epithelium; it is not ciliated. The kidney tubules and the thyroid gland have cuboidal epithelium.

9. E. Cuboidal epithelial cells are often involved in secretion. They are found in the thyroid gland, salivary glands, sweat glands, and meninges. They also are found in the ducts of glands, the kidney tubules, and the covering of the ovaries.

10. C. Transitional epithelium is composed of cells that can flatten and slide over each other. This allows stretching of the tissue and makes transitional epithelium an ideal lining for the urinary bladder.

11. C. To identify a tissue as being stratified squamous epithelium one must know that the tissue consists of more than a single layer (1), that the tissue lines a surface (2 and 3), and that the cells near the surface are flat (3).

    Next, one must distinguish between moist, stratified squamous epithelium (lining of the mouth, esophagus, and vagina) and keratinized, stratified squamous epithelium (outer layer of the skin). In moist, stratified squamous epithelium the surface cells are alive (2). Keratinized squamous epithelium has surface cells that do not have nuclei and are dead.

    The first three characteristics are the MINIMUM necessary to identify the tissue as moist, stratified squamous epithelium.

12. C. Endocrine glands empty their secretions directly into the blood. Exocrine glands (apocrine, merocrine, and holocrine glands) empty their secretions by means of ducts.

13. B. Holocrine glanis accumulate secretions, then release the secretions by rupturing. This kills the cell. Apocrine glands also accumulate secretions, but release the secretions by pinching off a small portion of the cell. The remainder of the cell lives and continues to accumulate secretions. Merocrine glands form and discharge their secretions in a cyclic fashion. There is no cell destruction, as occurs in holocrine and apocrine glands, during this process.

14. A. Fibroblasts secrete the ground substance and fibers of connective tissue. Adipocytes are cells within loose connective tissue that store fat droplets within their cytoplasm. Osteoblasts secrete the ground substance, fibers, and mineral salts that compose bone. Macrophages are cells that are capable of phagocytosis.

15. C. The internal framework of some organs and many glands consists of reticular fibers. Collagen fibers are typically found in tendons and ligaments. Elastic fibers are found in the external ear (auricle), elastic cartilage, elastic ligaments, and the walls of arteries.

16. B. A list of characteristics is presented. The characteristics are consistent with dense, regularly arranged, collagenic connective tissue; to answer the question correctly, one must be aware of that relationship. Among the choices presented only B (tendon) is composed of dense, regularly arranged, collagenic connective tissue.

17. E. Adipose tissue is a type of connective tissue. It acts as a protective cushion (around the kidneys, for example) and functions as a heat insulator (under the skin).

18. C. There are three types of cartilage: hyaline cartilage (embryonic skeleton, costal cartilage, cartilage rings of the trachea, and articular cartilage on the ends of bones), elastic cartilage (external ear, epiglottis, and auditory tubes), and fibrocartilage (intervertebral disks and articular disks).

19. B. Bone is formed of a mineralized matrix called hydroxyapatite. Cells within the matrix are surrounded by small spaces called lacunae.

20. B. You simply had to know that blood is an example of connective tissue.

21. C. correct. Smooth muscle is capable of spontaneous contractions, is under involuntary control, and is not striated.

22. C. Bipolar neurons have one axon and one dendrite.

23. E. The question requires that one must know what tissue types are components of the upper arm. Epithelial (part of the skin), connective tissue (part of the skin and material around muscle), muscle, and nervous tissue (especially if pain sensations are a result) would be penetrated.

24. B. The germ layers are so named because they give rise to the four tissue types. There are three germ layers: the ectoderm (outer layer), mesoderm (middle layer), and endoderm (inner layer).

25. C. Cavities or organs that open to the exterior are lined by mucous membranes. Serous membranes (mesothelium) line the ventral body cavity. Endothelium lines blood vessels.

26. B. The lamina propria is a layer of loose connective tissue that lies under the epithelial cells of mucous membranes.

27. E. Chemical mediators such as histamine, kinins, prostoglandins, and leukotrienes, are released or activated in tissue following injury to the tissue. The mediators cause vasodilation, increase vascular permeability, and stimulate neurons.

28. C. Essentially an individual is born with all the nerve cells they will ever have. Epithelial and connective tissue cells are capable of mitosis throughout life.

29. D. Permanent cells cannot replicate and are replaced by other types of cells if destroyed (replacement tissue repair). Labile and stable cells can replicate (regeneration tissue repair). Granulation tissue is produced by fibroblasts and may remain as a scar.

30. E. Following tissue damage, a blood clot forms that temporarily binds the damaged tissue. This eventually is invaded by fibroblasts and blood vessels. The fibroblasts produce fibers that replace the clot, and the blood vessels provide revascularization of the area. The fibroblasts and blood vessels form granulation tissue. The granulation tissue will form a scar if it is not reabsorbed and replaced by the original tissue at the damage site.

1. The tissue is transitional epithelium, a stratified epithelium that lines organs such as the urinary bladder and ureters. When the organ is stretched, the cells of transitional epithelium become squamouslike; and when the organ is not stretched, the cells are roughly cuboidal in shape.

2. One possible factor is the amount of material moved. Cilia can move a small amount of mucus each day out of the lungs, but are not effective in moving the large amount of urine excreted each day. In the urinary system movement of urine is due to contraction of smooth muscle and gravity. Movement of mucus by contraction of the tube system of the lungs could interfere with breathing, and gravity does not move mucus in a superior direction. In addition, the tube system of the lungs is always open, creating a space that allows the cilia to beat and move materials.

3. Aponeuroses, like tendons, consist of dense, regular, collagenous connective tissue, with the the collagen fibers arranged parallel to each other and to the long axis of the aponeurosis. A cross section would sever most of the fibers, causing considerable tissue damage and making tissue repair more difficult. A longitudinal section would damage fewer fibers.

4. Ligaments join bones to bones and usually do not stretch. This holds the bones in proper relationship to each other. One would not expect elastic fibers because it would allow the ligaments to stretch, possibly causing misalignment of the bones.

5. The elastic fibers would be expected in the ligaments connecting the bones of the back (vertebrae). These ligaments, called the ligamentum flavae, are similar to the ligamentum nuchae that help to hold the head in place. They allow the spine to stretch and then return to its original position. Thus they allow flexibility and help to maintain the position of the vertebrae by pulling them back into the correct position. The vertebrae have a number of bony processes that help to prevent misalignment of the vertebrae (see Chapter 7).

6. There are three pairs of traits that would identify the muscle as cardiac muscle:
   1. striated, involuntary control
   2. striated, spontaneous contraction
   3. striated, single nucleus

   The other three possible pairs of traits describe cardiac muscle or smooth muscle:
   4. single nucleus, involuntary control
   5. single nucleus, spontaneous contraction
   6. involuntary control, spontaneous contraction

Integumentary System

FOCUS: The integumentary system consists of the skin, hair, nails, and a variety of glands. The epidermis of skin provides protection against abrasion, ultraviolet light, and water loss and produces vitamin D. The dermis provides structural strength and contains blood vessels involved in temperature regulation. The skin is attached to underlying tissue by the hypodermis, which is a major site of fat storage. Hair and nails consists of dead, keratinized epithelial cells.

WORD PARTS

Give an example of a new vocabulary word that contains each word part.

WORD PART	MEANING	EXAMPLE
fasci-	band	1. _____
stria-	channel	2. _____
papill-	nipple	3. _____
a-	not; without	4. _____
vas-	vessel	5. _____
kerat-	horn	6. _____
melan-	black	7. _____
strat-	to spread; layer	8. _____
luc-	light; clear	9. _____
corn-	horn	10. _____

CONTENT LEARNING ACTIVITY

Introduction

❝*The integumentary system, consisting of the skin and accessory structures such as hair, nails, and a variety of glands, is the largest organ system in the body.*❞

Match these terms with the
correct statement or definition:

Dermis   Hypodermis
Epidermis

_____ 1. Dense, irregular connective tissue.

_____ 2. Layer of epithelial tissue that covers the dermis.

_____ 3. Layer of loose connective tissue that attaches skin to underlying tissue.

Hypodermis

❝*The skin is attached to underlying bone or muscle by the hypodermis.*❞

Match these terms with the
correct statement or definition:

Fibroblasts, fat cells and macrophages
Loose connective tissue

_____ 1. The type of tissue found in the hypodermis.

_____ 2. The main types of cells within the hypodermis.

☞ The hypodermis is sometimes called subcutaneous tissue or superficial fascia.

Dermis

❝*The dermis is responsible for most of the structural strength of the skin.*❞

Match these terms with the
correct statement or definition:

Papillae   Reticular layer
Papillary layer   Striae

_____ 1. Deep layer of dermis; a fibrous layer that blends into the hypodermis.

_____ 2. Projections from the dermis into the epidermis.

_____ 3. Lines visible through the epidermis produced by the rupture of the dermis.

Epidermis

"The epidermis is a stratified squamous epithelium separated from the dermis by a basement membrane."

A. Match these terms with the correct statement or definition:

Desquamate   Langerhans cells
Keratinization   Melanocytes
Keratinocytes   Strata

_____ 1. Cells that produce a protective protein called keratin; the most abundant epidermal cell.

_____ 2. Cells in the epidermis that are part of the immune system.

_____ 3. To slough or be lost from the surface of the epidermis.

_____ 4. A process that occurs in epidermal cells during their movement from deeper epidermal layers to the surface.

_____ 5. Layers or regions of the epidermis.

B. Match these terms with the correct statement or definition:

Keratohyalin   Stratum germinativum
Lamellar bodies   Stratum granulosum
Stratum basale   Stratum lucidum
Stratum corneum   Stratum spinosum

_____ 1. Deepest portion of the epidermis; a single layer of cells; the site of production of most epidermal cells.

_____ 2. Epidermal layer superficial to the stratum basale, consisting of eight to ten layers of many-sided cells.

_____ 3. Name for the stratum of the epidermis that includes both the stratum basale and the stratum spinosum.

_____ 4. Derives its name from protein granules contained in the cells and is superficial to the stratum spinosum.

_____ 5. Nonmembrane bound protein granules found in the cells of the stratum granulosum.

_____ 6. Structures that move to the cell membrane and release their lipid contents into the intercellular space; the lipids are responsible for the permeability characteristics of the epidermis.

_____ 7. Clear, thin zone above the stratum granulosum; absent in most skin.

_____ 8. Most superficial stratum of the epidermis; dead cells surrounded by a hard protein envelope and filled with keratin; the keratin provides structural strength.

C. Match these terms with the parts labeled in Figure 5-1:

Dermis
Epidermis
Stratum basale
Stratum corneum

Stratum germinativum
Stratum granulosum
Stratum lucidum
Stratum spinosum

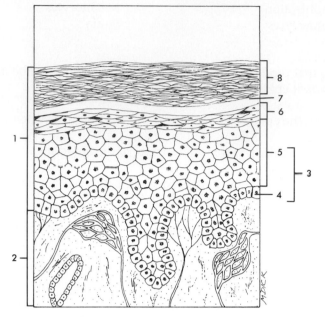

Figure 5-1

1. _____

2. _____

3. _____

4. _____

5. _____

6. _____

7. _____

8. _____

Thick Skin and Thin Skin

66*Skin is classified as thick or thin based on the structure of the epidermis.*99

Match these terms with the correct statement or definition:

Callus
Corn

Thick skin
Thin skin

_____ 1. In this type of skin the stratum lucidum is usually absent; hair is found in this type of skin.

_____ 2. The papillae of the dermis of this type of skin produce curving ridges that produce fingerprints and footprints.

_____ 3. Thickened area of thin or thick skin due to greatly increased number of layers of stratum corneum.

_____ 4. Cone-shaped structure that develops in thin or thick skin over a bony prominence.

Skin Color

66*Skin color is determined by pigments in the skin, by blood circulating through the skin,*99
and by the thickness of the stratum corneum.

Using the list of terms provided, complete the following statements:

Albinism
Carotene
Cyanosis
Melanocytes

Melanin
Melanosomes
Vitiligo

_(1)_ , a brown to black pigment, is responsible for most skin color.It is produced by _(2)_ , irregularly shaped cells with many long processes that extend between the keratinocytes of the stratum basale and stratum spinosum. Melanin is packaged into vacuoles called _(3)_ , which are released from the cell processes by exocytosis. A single mutation can prevent the manufacture of melanin, resulting in _(4)_ . _(5)_ , the development of patches of white skin, occurs because the melanocytes in the affected area are destroyed, apparently by an autoimmune response. _(6)_ is a yellow pigment found in plants such as carrots. When large amounts of this pigment are consumed, the excess accumulates in the stratum corneum and fat cells of the dermis and hypodermis, causing the skin to develop a yellowish tint. A decrease in blood oxygen content produces _(7)_ , a bluish skin color, whereas an abundant supply of oxygenated blood produces a reddish hue.

1. _____

2. _____

3. _____

4. _____

5. _____

6. _____

7. _____

Hair

❝_The presence of hair is one of the characteristics common to all mammals._**❞**

A. Match these terms with the correct statement or definition:

Lanugo   Vellus hairs
Terminal hairs

_____

_____

_____

_____

1. Delicate unpigmented hair that covers the fetus.

2. Long, coarse, pigmented hairs that replace lanugo on the scalp, eyebrows, and eyelids.

3. Short, fine, unpigmented hairs that replace lanugo over most of the body.

4. The type of hair that replaces vellus hairs at puberty.

B. Match these terms with the correct statement or definition:

Cortex   Medulla
Cuticle   Root
Hair bulb   Shaft

_____

_____

_____

_____

_____

1. Portion of hair protruding above the skin.

2. Central axis of the hair that consists of two or three layers of cells containing soft keratin.

3. Concentric layer that forms the bulk of the hair and consists of cells containing hard keratin.

4. Outermost layer of the hair shaft and root, composed of a single overlapping layer of cells containing hard keratin.

5. An expanded knob at the base of the hair root.

C. **Match these terms with the correct statement or definition:**

Arrector pili
Dermal root sheath

Epithelial root sheath
Matrix

_____ 1. Layers of epithelial cells immediately surrounding the root of the hair.

_____ 2. Mass of undifferentiated epithelial cells inside the hair bulb that produces the hair and internal epithelial root sheath.

_____ 3. Smooth muscle cells that attach to the hair follicle dermal root sheath and the papillary layer of the dermis.

☞ Hair is produced in cycles that involve a growth stage alternating with a resting stage.

D. **Match these terms with the parts of a hair follicle labeled in Figure 5-2:**

Arrector pili
Dermal root sheath
Hair bulb
Hair root
Hair shaft
Matrix
Papilla

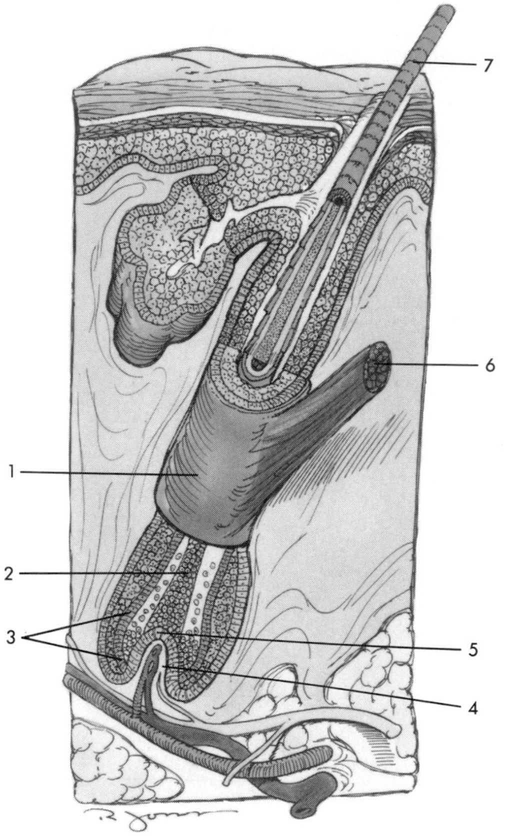

1. _____

2. _____

3. _____

4. _____

5. _____

6. _____

7. _____

Figure 5-2

Glands

❝_The major glands of the skin . . . . consist of epidermal tissue that extends into the dermis_❞ _or through the dermis into the hypodermis._

A. **Match these terms with the correct statement or definition:**

Apocrine sweat glands
Ceruminous glands
Merocrine sweat glands

Sebaceous glands
Sebum

_____ 1. Oily white substance rich in lipids.

_____ 2. Holocrine gland that produces sebum, which oils the hair and skin surface, prevents drying, and provides protection against bacteria.

_____ 3. Simple, coiled, tubular glands that open to the surface of the skin and secrete an isotonic fluid that is mostly water.

_____ 4. Compound coiled tubular glands that secrete a complex, viscous organic substance that is metabolized by bacteria to produce body odor.

B. Match these terms with the types of glands labeled in Figure 5-3:

Apocrine sweat gland
Merocrine sweat gland
Sebaceous gland

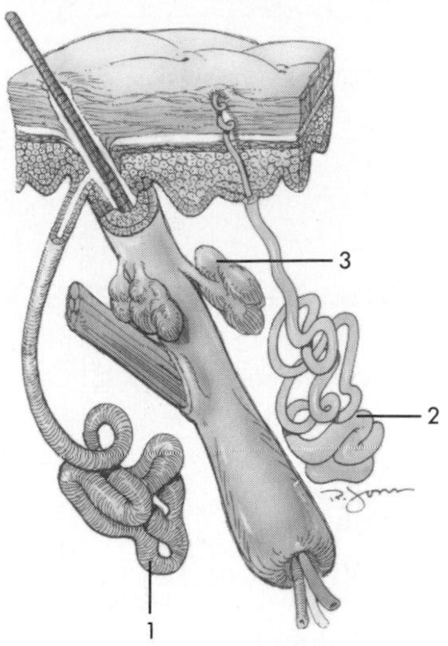

Figure 5-3

1. _____

2. _____

3. _____

Nails

"_Nails protect the ends of the digits, aid in manipulation and grasping of small objects, and are used for scratching._**"**

A. Match these terms with the correct statement or definition:

Eponychium   Nail fold
Hyponychium   Nail groove
Lunula   Nail matrix
Nail bed   Nail root
Nail body

_____ 1. Proximal portion of the nail that is covered by skin.

_____ 2. Portion of the skin that covers the lateral and proximal edges of the nail.

_____ 3. Holds the edges of the nail in place.

_____ 4. Cuticle; the stratum corneum of the nail fold that grows onto the nail body.

_____ 5. Nail root and nail body attach to this structure.

_____ 6. Proximal portion of the nail bed that produces most of the nail.

_____ 7. Whitish, crescent- shaped area at the base of the nail.

B. Match these terms with the
parts of the nail labeled
in Figure 5-4:

Body
Eponychium   *Figure 5-4*
Lunula
Nail root

1. _____

2. _____

3. _____

4. _____

Functions of the Integumentary System

❝The integumentary system has many functions in the body.**❞**

Match these terms with the   Excretion   Temperature regulation
correct statement or definition: Protection   Vitamin D production
                                   Sensation

_____   1. Accomplished by the skin as a physical barrier, as a permeability
                               barrier, as a barrier against ultraviolet light, and as a barrier
                               against abrasion.

_____   2. Carried out by producing sweat and increasing or decreasing blood
                               vessel diameter.

_____   3. Begins when a sterol in the skin is exposed to ultraviolet light and is
                               converted to provitamin D.

_____   4. Detection of touch, temperature, and pain.

_____   5. Occurs to a very slight degree with sweat production when some urea,
                               uric acid, and ammonia are lost.

The Effects of Aging on the Integumentary System

❝As the body ages, many changes occur in the integumentary system.**❞**

Using the terms provided, complete the following statements:

Decrease(s)   Increase(s)

As the body ages, blood flow to the skin _(1)_ , and the thickness
of skin _(2)_ . Elastic fibers in the dermis _(3)_ , and the skin tends
to sag. A _(4)_ in the activity of sebaceous and sweat glands results
in dry skin and a _(5)_ in thermoregulatory ability. The _(6)_ in
ability to sweat can contribute to death from heat prostration in
elderly individuals. The number of functioning melanocytes _(7)_ ,
but in some localized areas, especially the hands and face,
melanocytes _(8)_ to produce age spots. White or gray hairs also
occur because of a _(9)_ in melanin production.

1. _____

2. _____

3. _____

4. _____

5. _____

6. _____

7. _____

8. _____

9. _____

Integumentary Disorders

The integumentary system is directly affected by many disorders.

A. Match these terms with the correct statement or definition:

Acne
Decubitis ulcers
Ringworm and athlete's foot
Warts

_____ 1. Disorder of the hair follicles and sebaceous glands that involves testosterone and bacteria.

_____ 2. Viral infection of the epidermis.

_____ 3. Fungal infections that affect the keratinized portion of the skin.

_____ 4. Disorder caused by ischemia and necrosis of the hypodermis.

B. Match these terms with the correct statement or definition:

Basal cell carcinoma
Malignant melanoma
Squamous cell carcinoma

_____ 1. Cancer that begins in the stratum basale and extends into the dermis to produce an open ulcer; the most frequent type of skin cancer.

_____ 2. Cancer that typically produces a nodular, keratinized tumor confined to the epidermis.

_____ 3. Rare form of skin cancer that usually arises from a preexisting mole; the skin cancer that is most often fatal if not diagnosed and treated early.

QUICK RECALL

1. Name the two layers of the dermis.

2. List the five strata of the epidermis from the deepest to the most superficial.

3. List three types of hair found at different stages of development in humans.

4. Name the two stages in the hair production cycle.

5. List four types of glands associated with the skin.

6. List four protective functions of the integumentary system.

7. List two ways the integumentary system functions to regulate the temperature of the body.

8. List four effects that aging has on the integumentary system.

MASTERY LEARNING ACTIVITY

Place the letter corresponding to the correct answer in the space provided.

_____ 1. The hypodermis
   a. is the layer of skin where hair is produced.
   b. is the layer of skin where nails are produced.
   c. connects the dermis and the epidermis.
   d. produces the basement membrane.
   e. none of the above

_____ 2. Fingerprint and footprint patterns are a result of the development of the
   a. stratum corneum.
   b. dermis.
   c. hypodermis.
   d. stratum germinativum.

_____ 3. The layer of the skin that replaces (the place where mitosis is actively occurring) cells lost from the outer layer of the epidermis is the
   a. stratum corneum.
   b. stratum germinativum.
   c. reticular layer of the dermis.
   d. hypodermis.

_____ 4. If a splinter penetrated the skin of the sole of the foot to the second epidermal layer from the surface, the last layer to be damaged would be the
   a. stratum granulosum.
   b. stratum basale.
   c. stratum corneum.
   d. stratum lucidum.

_____ 5. In what area of the body would you expect to find an especially thick stratum corneum?
   a. back of the hand
   b. abdomen
   c. over the shin
   d. bridge of the nose
   e. heel of the foot

_____ 6. The function of melanin in the skin is
   a. lubrication of the skin.
   b. prevention of skin infections.
   c. protection from ultraviolet light.
   d. to reduce water loss.
   e. to help regulate body temperature.

7. Concerning skin color, which of the following statements is NOT correctly matched?
   a. skin appears yellow - carotene present
   b. no skin pigmentation (albinism) - genetic disorder
   c. skin tans - increased melanin production
   d. skin appears blue (cyanosis) - oxygenated blood
   e. negroes darker than caucasians - more melanin in negroes

8. Hair
   a. is produced by the stratum germinativum.
   b. consists of dead epithelial cells.
   c. is colored by melanin.
   d. is the same thing as fur.
   e. all of the above

9. A hair follicle is
   a. an extension of the epidermis deep into the dermis.
   b. an extension of the dermis deep into the epidermis.
   c. often associated with sweat glands.
   d. a and c
   e. b and c

10. Smooth muscles that produce "goose flesh" when they contract and are attached to hair follicles are called
    a. external root sheaths.
    b. arrector pili.
    c. dermal papillae.
    d. internal root sheaths.

11. Sebum
    a. oils the hair.
    b. is produced by sweat glands.
    c. consists of dead cells from hair follicle.
    d. all of the above

12. A congenital lack of sweat glands would primarily affect one's ability to
    a. secrete waste products.
    b. flush out secretions that accumulate in the hair follicle.
    c. oil the skin.
    d. prevent bacteria from growing on the skin.
    e. control one's body temperature in warm environments.

13. An experimenter wishes to determine the relationship between body temperature and water loss in an animal. The animal is placed in a chamber where the temperature of the air can be controlled. At each air temperature, the amount of water lost and the body temperature of the animal were measured. The results are graphed below:

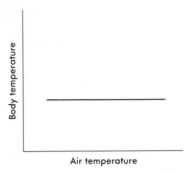

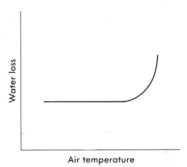

Given the following statements:
1. The animal has sweat glands.
2. The animal conserves water at low air temperatures.
3. The skin of the animal is resistant to water loss.
4. The animal uses water loss to maintain body temperature at a constant level at all air temperatures.

Using the experimental results, which of the statements can you conclude are true?
a. 2, 3
b. 1, 2, 3
c. 1, 3, 4
d. 2, 3, 4
e. 1, 2, 3, 4

_____ 14. Nails
   a. appear pink due to a special pigment within nail cells.
   b. are an outgrowth of the dermal papillae.
   c. contain lots of hard keratin.
   d. all of the above

_____ 15. While building the patio deck to his house, an anatomy and physiology instructor hit his finger with a hammer. He responded to this stimulus by saying, "Gee, I hope I didn't irreversibly damage the _____, because if I did, my fingernail will never grow back."
   a. nail bed
   b. stratum corneum
   c. nail matrix
   d. eponychium
   e. a and b

_____ 16. To increase heat loss from the body, one would expect
   a. dilation of dermal capillaries.
   b. constriction of dermal capillaries.
   c. increased sweating.
   d. a and c
   e. b and c

_____ 17. Body odor
   a. is due to special scent glands located under the arm pits.
   b. occurs when bacteria break down the organic secretions of apocrine glands.
   c. is due to a chemical reaction of sweat with the air.
   d. is due to sebum.
   e. is due to keratin.

_____ 18. Skin aids in maintaining the calcium and phosphate levels of the body at the optimum levels by participating in the production of
   a. Vitamin A.
   b. Vitamin B.
   c. Vitamin D.
   d. melanin.
   e. keratin.

_____ 19. On a sunny spring day a student decided to initiate her annual tanning ritual. However, in doing so she fell asleep while sunbathing. After awakening she noticed that the skin on her back was burned. She experienced redness, blisters, edema and significant pain. The burn was nearly healed about 10 days later. The burn was best classified as a
   a. first degree burn.
   b. second degree burn.
   c. third degree burn.

_____ 20. A burn patient is admitted into the hospital while you are present in the emergency room. Someone asks for an estimate of the percentage of surface area burned. Remembering the formula for assessing the extent of burns, you take a shot at it. If the patient was burned on the entire left upper and lower limbs and over three quarters of the anterior and posterior trunk, what would be the estimate?
   a. 54%
   b. 63%
   c. 45%
   d. 72%
   e. none of the above

Use a separate sheet of paper to complete this section.

1. The rate of water loss from the skin of the hand was measured. Following the measurement the hand was soaked in alcohol for 15 minutes. After all the alcohol was removed from the hand, the rate of water loss was again determined. Compared to the rate of water loss before soaking the hand in alcohol, what difference, if any, would you expect in the rate of water loss after soaking the hand in alcohol?

2. It has been several weeks since Goodboy Player has competed in a tennis match. After the match he discovers that a blister has formed beneath an old callus on his foot and the callus has fallen off. When he examines the callus he discovers that it appears yellow. Can you explain why?

3. Why is it difficult to surgically remove a large tattoo without causing scar tissue to form? (hint: why do tattoos appear bluish in color?)

4. Consider the following statement: Dark-skinned children are more susceptible to rickets (insufficient calcium in the bones) than fair-skinned children. Defend or refute this statement.

5. Given what you know about the cause of body odor, propose some ways to prevent the condition.

6. Pulling on hair can be quite painful, yet cutting hair is not painful. Explain.

7. A man is using gasoline to clean automobile parts. There is an explosion, and he is burned over the entire anterior surface of the trunk and the entire anterior surface of both lower limbs. Approximately how much of the man's surface area was damaged?

8. Dandy Chef has been burned on the arm. The doctor, using a forceps, pulls on a hair within the area that was burned. The hair easily pulls out. What degree of burn did the patient have and how do you know?

ANSWERS TO CHAPTER 5

1. fascicle; fasciculi
2. striae
3. papillae; papillary
4. avascular
5. avascular; vascular

6. keratin; keratinocyte
7. melanocyte; melanin; melanosomes
8. strata; stratified; stratum
9. stratum lucidum
10. cornified; stratum corneum; corn

CONTENT LEARNING ACTIVITY

Introduction

1. Dermis
2. Epidermis

3. Hypodermis

Hypodermis

1. Loose connective tissue

2. Fibroblasts, fat cells, and macrophages

Dermis

1. Reticular layer
2. Papillae

3. Striae

Epidermis

A. 1. Keratinocytes
   2. Langerhans cells
   3. Desquamate
   4. Keratinization
   5. Strata

B. 1. Stratum basale
   2. Stratum spinosum
   3. Stratum germinativum
   4. Stratum granulosum
   5. Keratohyalin
   6. Lamellar bodies

   7. Stratum lucidum
   8. Stratum corneum

C. 1. Epidermis
   2. Dermis
   3. Stratum germinativum
   4. Stratum basale
   5. Stratum spinosum
   6. Stratum granulosum
   7. Stratum lucidum
   8. Stratum corneum

Thick Skin and Thin Skin

1. Thin skin
2. Thick skin

3. Callus
4. Corn

Skin Color

1. Melanin
2. Melanocytes
3. Melanosomes
4. Albinism

5. Vitiligo
6. Carotene
7. Cyanosis

Hair

A. 1. Lanugo
    2. Terminal hairs
    3. Vellus hairs
    4. Terminal hairs

B. 1. Shaft
    2. Medulla
    3. Cortex
    4. Cuticle
    5. Hair bulb

C. 1. Epithelial root sheath
    2. Matrix
    3. Arrector pili

D. 1. Dermal root sheath
    2. Hair root
    3. Hair bulb
    4. Papilla
    5. Matrix
    6. Arrector pili
    7. Hair shaft

Glands

A. 1. Sebum
    2. Sebaceous glands
    3. Merocrine sweat glands
    4. Apocrine sweat glands

B. 1. Apocrine sweat gland
    2. Merocrine sweat gland
    3. Sebaceous gland

Nails

A. 1. Nail root
    2. Nail fold
    3. Nail groove
    4. Eponychium
    5. Nail bed
    6. Nail matrix
    7. Lunula

B. 1. Body
    2. Lunula
    3. Eponychium
    4. Nail root

Functions of the Integumentary System

1. Protection
2. Temperature regulation
3. Vitamin D production

4. Sensation
5. Excretion

The Effects of Aging on the Integumentary System

1. Decreases
2. Decreases
3. Decrease
4. Decrease
5. Decrease

6. Decrease
7. Decreases
8. Increase
9. Decrease

Integumentary Disorders

A. 1. Acne
    2. Warts
    3. Ringworm and athlete's foot
    4. Decubitis ulcers

B. 1. Basal cell carcinoma
    2. Squamous cell carcinoma
    3. Malignant melanoma

QUICK RECALL

1. Reticular and papillary layer
2. Stratum basale, stratum spinosum, stratum granulosum, stratum lucidum, and stratum corneum
3. Lanugo, vellus hairs, and terminal hairs
4. Growth stage and resting stage
5. Sebaceous glands, apocrine sweat glands, merocrine sweat glands, and ceruminous glands
6. Prevents water loss, physical barrier to microorganisms, protection against mechanical damage, protection against ultraviolet light, protection against damage to eyes and digits
7. Vasodilation/vasoconstriction and sweat production
8. Decreased blood flow to skin, skin thinner, decreased elastic fibers, decreased activity of sweat and sebaceous glands, decrease melanocyte activity, and age spots in some areas

MASTERY LEARNING ACTIVITY

1. E. The hypodermis, which is a layer of loose connective tissue (areolar tissue), connects the dermis to underlying tissue such as bone or muscle. It is not really part of the skin. The hypodermis is sometimes called subcutaneous tissue and is the site for some injections. The basement membrane connects the epidermis to the dermis. The stratum germinativum of the epidermis produces the basement membrane, hair, and nails.

2. B. The dermis consists of two layers, the papillary layer, and the reticular layer. The papillary layer fits closely to the basement membrane of the epidermis and has projections (papillae) that extend into the epidermis. These contain capillaries and sensory receptors. On the hand and feet they cause the overlying epidermis to form fingerprint and footprint patterns. The reticular layer of the dermis contains connective tissue fibers that are continuous with the hypodermis.

3. B. The epidermis is divided into four layers. Only the innermost of these layers, the stratum germinativum, is capable of mitosis. The other layers consist of dying or dead cells. This occurs because only the stratum germinativum is close enough to blood vessels (found in the dermis) to receive adequate oxygen and nutrients. The dermis and the hypodermis do not contribute to the cells that eventually reach the surface of the skin.

4. D. In thick skin (sole of foot) from the outside in, the layers of the epidermis are the stratum corneum, stratum lucidum, stratum granulosum, stratum spinosum and stratum basale.

5. E. The stratum corneum consists of many rows of dead, cornified, squamous epithelium. Areas of the body subjected to abrasion such as the heel of the foot or calluses have especially thick stratum corneums. This provides lots of protection for the underlying tissues. As cells of the stratum corneum are worn off, they are replaced by cells from deeper layers of the epidermis.

6. C. Melanin, produced by the melanocytes in the epidermis, is a dark pigment that prevents the penetration of ultraviolet light. Keratin, found in the stratum corneum of the epidermis, helps to reduce water loss. Lubrication of the skin and prevention of skin infections is due to secretions of skin glands. Body temperature regulation occurs by vasodilation and vasoconstriction of the vessels within the dermis of the skin.

7. D. The skin appears blue due to deoxygenated blood. This is a useful diagnostic characteristic that could indicate respiratory and/or circulatory dysfunction resulting in inadequate oxygen delivery to tissues. Another more normal cause of cyanosis is cold. In this case blood flow to the surface of the body is reduced (vasoconstriction) to conserve heat, and the skin appears pale or bluish.

8. E. The stratum germinativum of the epidermis produces the hair. New cells formed at the base of the hair quickly die, so the hair shaft is composed of dead epithelial cells. Just as skin is colored due to melanin produced by melanocytes, varying amounts of melanin account for hair color. Fur is just lots of hair.

9. A. A hair follicle is an extension of the epidermis deep into the dermis. It consists of two layers: the stratum germinativum of the epidermis, which gives rise to the hair; and an outer layer of connective tissue from the dermis. Typical sweat glands open directly onto the surface of the skin. Sebaceous glands are associated with hair follicles.

10. B. The arrector pili muscle is attached to the hair follicle and the papillary layer of the dermis. When it contracts, the pull simultaneously causes the hair to stand up and a small section of skin to bulge, producing a "goose flesh". This same response is seen in many mammals in response to cold. Causing the hair to stand up produces a thicker layer of fur, which provides greater insulation.

11. A. Sebum is an oily substance rich in lipids that is produced by sebaceous glands. With few exceptions, sebaceous glands are associated with hair follicles. The sebum oils the hair and surrounding skin, preventing drying.

12. E. The primary role of sweat glands is in the regulation of body temperature in warm temperatures. Although it is true that a small amount of waste products are excreted in sweat, lack of sweat glands would not be harmful in this regard. The kidneys really are responsible for waste product excretion. Typical sweat glands are not associated with hair follicles but open directly onto the skin surface. Sebaceous glands are associated with hair follicles and produce sebum. Sebum oils hair and the skin and prevents the growth of some bacteria.

13. A. The animal maintained a constant body temperature (homeostasis) for all air temperatures used in the experiment. This was not due to water loss alone, since the rate of water loss remained constant at several of the lower air temperatures. Undoubtedly other mechanisms such as vasodilation or vasoconstriction were involved in maintaining body temperature. Once air temperature increased enough, however, these other mechanisms could not prevent a rise in body temperature. At this point water loss increased dramatically and was responsible for keeping body temperature from increasing.

The fact that water loss was constant at the lower air temperatures suggests that the animal conserves water at these temperatures. This is accomplished because the skin is resistant to water loss. When the need arises, however, large amounts of water can be lost. There are two reasonable ways this can happen: from sweat glands (as in humans) or by panting (as in dogs). The data presented do not allow one to conclude which mechanism was used.

14. C. The nail is made up of two epidermal layers, the stratum corneum and the stratum lucidum. The hardness of the nail is due to large amounts of keratin (a normal component of skin). The nail appears pink because the nail is semitransparent and the underlying vascular tissue can be seen. Nails grow from the nail matrix at the proximal end of the nail.

15. C. The nail bed upon which the nail rests is formed by the stratum germinativum of the epidermis. However, the nail does not form here. Instead, it is produced by the stratum germinativum in the nail matrix at the proximal end of the nail. Part of the nail matrix can be seen as a whitish area at the base of the nail, the lunula. The eponychium (cuticle) does not affect nail growth.

16. D. Increased sweating cools the body as the sweat evaporated from the skin surface. Dilation of dermal capillaries produces increased blood flow through the skin, moving heat through warm blood from the body core to the surface and then to the environment.

17. B. Located in the axilla, around the anus, the labia majora of females and the scrotum of males are modified sweat glands (apocrine glands) that produce secretions rich in organic matter. When these secretions are decomposed by bacteria, which use them as a source of energy, body odor develops. Because normal sweat does not contain the organic matter, sweat in other areas of the body does not smell. Try sniffing a sweaty forearm sometime. These aprocrine sweat glands do not become active until puberty, so body odor is usually not a problem in children.

18. C. When exposed to sunlight (ultraviolet light), the skin produces vitamin D, which is absorbed into the blood. The vitamin D is metabolically transformed, first in the liver, then in the kidneys, into a compound that assists in the absorption of calcium and phosphate from the small intestine.

19. B. Usually sunburns are first degree burns. In this case, because of the blistering, it is a second degree burn.

20. A. The estimate is:

Left upper limb:	9%
Left lower limb:	18%
3/4 of trunk:	27%
Total:	54%

 ## FINAL CHALLENGES

1. Alcohol is an organic solvent and removes the lipids from skin, especially in the stratum corneum. Since the lipids normally prevent water loss, after soaking the hand in alcohol, the rate of water loss can be expected to increase.

2. Carotene, a yellow pigment from ingested plants, accumulates in lipids. The stratum corneum of a callus has more layers of cells than other noncallused parts of the skin and the cells in each layer are surrounded by lipids. The carotene in the lipids make the callus appear yellow.

3. The tattoo is located in the dermis; to remove the tattoo, the overlying epidermis and most of the dermis must be removed. The wound produced is much like a second degree burn. After removing a small tattoo the edges of the wound could be sutured together, producing a small scar (see primary union in Chapter 4). After removing a large tattoo considerable scar tissue would form, because most of the epithelial tissue regeneration must occur from the edge of the wound (see secondary union in Chapter 2).

4. Rickets is a children's disease due to a lack of vitamin D. Without vitamin D there is insufficient absorption of calcium from the small intestine, resulting in soft bones. If adequate vitamin D is ingested, rickets is prevented, whether one is dark or fair skinned. However, if dietary vitamin D is lacking, when the skin is exposed to ultraviolet light, it can produce a provitamin that can be transformed into vitamin D. Dark-skinned children are more susceptible to rickets because the additional melanin in their skins screens out the ultraviolet light and they produce less vitamin D.

5. Since body odor is due to the breakdown of the organic secretions of apocrine sweat glands, one possibility is to remove the secretions by washing. Another method is to kill the bacteria. The aluminum salts in some antiperspirants do this. Antiperspirants also reduce the watery secretions of merocrine sweat glands, but these secretions are not the cause of body odor. Deodorants mask body odor with another scent but do not prevent it.

6. The hair follicle (but not the hair) is well supplied with nerve endings that can detect movement or pulling of the hair. The hair itself is dead, keratinized epithelium so cutting the hair is not painful.

7. About 36% was damaged: 18% from the trunk and 18% from the lower limbs.

8. Since the hair follicle pulled out easily, the hair follicle has been destroyed. Therefore repair can only occur from the edge of the wound and this is a third degree burn.

Skeletal System: Histology And Development

FOCUS: The skeletal system consists of ligaments, tendons, cartilage, and bones. Ligaments attach bones to bones, and tendons attach muscles to bones. Cartilage forms the articular surfaces of bone, is the site of bone growth in length, and provides flexible support within the ear, nose, and ribs. Bone protects internal organs and provides a system of levers that makes possible body movements. Bone can form by intramembranous (within a membrane) or endochondral (within cartilage) ossification, grows in length at the epiphyseal plate, and increases in width beneath the periosteum. Bone is a dynamic tissue constantly undergoing remodeling to adjust to stress and is capable of repair.

WORD PARTS

Give an example of a new vocabulary word that contains each word part.

WORD PART	MEANING	EXAMPLE
peri-	around	1. _____
lacun-	space, hollow	2. _____
lamell-	leaf, layer	3. _____
cancel-	crossbar, lattice	4. _____
chondr-	cartilage	5. _____
oste-	bone	6. _____

WORD PART	MEANING	EXAMPLE
pro-	first, earliest	7. _____
gen-	cause, birth	8. _____
troph-	to nourish	9. _____
malac-	soft	10. _____

CONTENT LEARNING ACTIVITY

Introduction

“_The skeletal system consists of bones and their associated connective tissues,_**”** _including cartilage, tendons, and ligaments._

Match these terms with the correct statement or definition:

Bone   Ligament
Cartilage   Tendon

_____ 1. Very rigid tissue that functions to maintain body shape, protect internal organs, provide a system of levers upon which muscles act to produce body movement, and store minerals.

_____ 2. Somewhat rigid tissue that forms a smooth surface at bone articulations, provides flexible support, and provides a model for most adult bones.

_____ 3. Strong band of fibrous connective tissue that attaches muscle to bone.

General Histology

“_Connective tissue consists of cells separated from each other by an extracellular matrix._**”**

Match these terms with the correct statement or definition:

Collagen   Proteoglycan monomers
Proteoglycan aggregates

_____ 1. Most abundant protein in the human body.

_____ 2. Molecule with a protein core that has numerous polysaccharides attached.

_____ 3. Proteoglycan monomers plus hyaluronic acid.

Tendons and Ligaments

" *Tendons and ligaments are composed of dense, regular connective tissue.* **"**

Using the terms provided, complete the following statements:

Appositional   Fibrocyte
Endotendineum   Fascicles
Epitendineum   Interstitial
Fibroblast   Peritendineum

The collagen fibers of tendons are surrounded by (1) ,
comprised of loose connective tissue. The collagen fibers
combine to form (2) surrounded by a fibrous (3) . Several
fascicles form a tendon covered by a fibrous (4) . The cells of
developing tendons and ligaments are spindle-shaped (5) .
Once a fibroblast becomes completely surrounded by matrix, it is
a (6) . Tendons and ligaments grow by (7) growth when
surface fibroblasts divide to produce additional fibroblasts and
secrete matrix outside of existing fiber. (8) growth occurs when
fibroblasts proliferate and secrete matrix inside the tissue.

1. _____
2. _____
3. _____
4. _____
5. _____
6. _____
7. _____
8. _____

Hyaline Cartilage

" *Hyaline cartilage is the type of cartilage most intimately associated with bone function and development.* **"**

Match these terms with the
correct statement or definition:

Chondroblasts   Lacunae
Chondrocytes   Perichondrium

_____ 1. Cells that produce new cartilage matrix on the outside of more
mature cartilage.

_____ 2. Cartilage cells that are surrounded by matrix.

_____ 3. Double-layered connective tissue sheath that surrounds cartilage.

Bone Matrix

" *Bone matrix is composed of both organic and inorganic material.* **"**

Match these terms with the
correct statement or definition:

Canaliculi   Lamellae
Collagen   Osteoblast
Hydroxyapatite   Osteoclast
Lacunae   Osteocyte

_____ 1. Major organic component of bone.

_____ 2. Bone cells that produce, but are not surrounded by, bone matrix.

_____ 3. Large cells with several nuclei that break down bone matrix.

_____ 4. Thin sheets or layers of bone matrix.

_____ 5. Spaces occupied by long, thin cell processes extending between lacunae.

Bone Anatomy

❝About half the bones of the body are 'long bones' and the remaining half have some other shape.**❞**

A. Match these terms with the correct statement or definition:

Diaphysis	Marrow
Endosteum	Medullary cavity
Epiphyseal plate	Perforating (Sharpey's) fibers
Epiphyseal line	Periosteum
Epiphysis	

_____ 1. Shaft of a long bone.

_____ 2. Area of growth between the diaphysis and epiphysis.

_____ 3. Ossified epiphyseal plate.

_____ 4. Large cavity within the diaphysis.

_____ 5. Substance that fills bone cavities.

_____ 6. Connective tissue that connects the periosteum to bone.

_____ 7. Membrane that lines the medullary cavity.

B. Match these terms with the correct parts of the long bone labeled in Figure 6-1:

Articular cartilage
Cancellous bone
Compact bone
Diaphysis
Epiphyseal line
Epiphysis
Medullary cavity
Periosteum

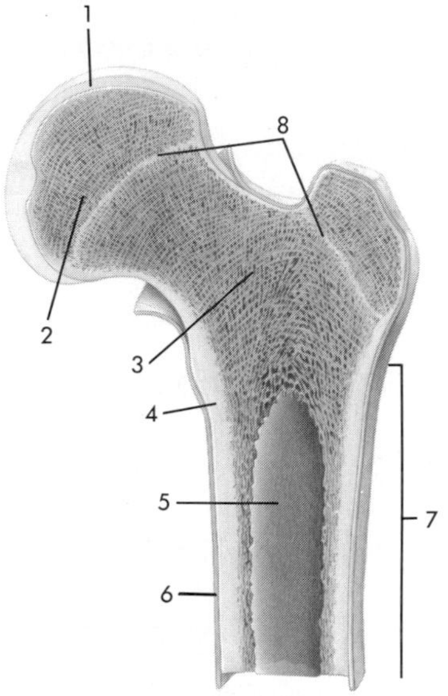

Figure 6-1

1. _____

2. _____

3. _____

4. _____

5. _____

6. _____

7. _____

8. _____

Compact Bone and Cancellous Bone

❝Compact bone is mostly solid matrix and cells with few spaces, whereas cancellous bone**❞** consists of a lacy network of bony plates and beams.

A. Using the terms provided, complete the following statements:

Circumferential lamellae	Haversian system (osteon)
Concentric lamellae	Trabeculae
Haversian (central) canals	Volkmann's (perforating) canals

Cancellous bone consists of interconnecting rods or plates of bone called _(1)_ . Compact bone is more dense, with blood vessels that run parallel to the long axis of the bone through _(2)_ , surrounded by _(3)_ . A _(4)_ consists of a single haversian canal, its contents,

1. _____

2. _____

3. _____

4. _____

and associated concentric lamellae and osteocytes. The blood vessels of the haversian canals are interconnected by a network of vessels contained within _(5)_, running perpendicular to the long axis of the bone. The outer surfaces of compact bone are covered by flat plates of bone called _(6)_.

5. _____

6. _____

B. Match these terms with the correct parts of compact bone and cancellous bone labeled in Figure 6-2:

Circumferential lamellae
Concentric lamellae
Haversian (central) canal
Haversian system (osteon)
Interstitial lamellae
Volkmann's (perforating) canal

1. _____

2. _____

3. _____

4. _____

5. _____

6. _____

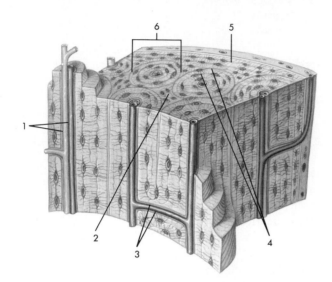

Figure 6-2

Bone Ossification

❝*Ossification is the formation of bone by osteoblasts.*❞

Match these terms with the correct statement or definition:

Endochondral
Intramembranous

_____ 1. Bone formation occurs within connective tissue membranes.

_____ 2. Ossification process that produces many skull bones and the clavicle.

_____ 3. Ossification process that produces most of the skeletal system.

 Both cancellous and compact bone are formed by endochondral and intramembranous ossification.

Intramembranous Ossification

Bone formation that occurs within connective tissue membranes is intramembranous.

Using the terms provided, complete the following statements:

Centers of ossification   Fontanels
Collagen   Osteoprogenitor cells

In the embryo, fibroblasts produce a (1) membrane in areas
where many skull bones are to develop. (2) cells differentiate to
become osteoblasts that lay down bone along the fibers of the
membrane. These areas of bone formation are called (3) . At
birth, some of the membrane is not ossified; these regions
are the (4) .

1. _____

2. _____

3. _____

4. _____

Endochondral Ossification

Bone formation in association with cartilage is endochondral.

Using the terms provided, complete the following statements:

Calcified cartilage   Osteoblasts
Chondrocytes   Periosteum
Diaphysis   Primary ossification center
Epiphysis   Secondary ossification center
Hyaline cartilage

In the embryo a (1) template of most of the bones of the skeleton
is formed. Osteoprogenitor cells become (2) that invade the
perichondrium and begin to lay down bone around the outside of
the template. The perichondrium is then called the (3) . In the
center of the template, chondrocytes hypertrophy and the matrix
between the chondrocytes becomes mineralized to form (4) , and
the (5) die. Blood vessels invade the area, and osteoblasts also
enter and begin to lay down bone, forming the (6) . In a long
bone this area of bone deposition develops in the (7) . Later, in
the epiphyses another area of bone formation, the (9) ,
completes the development of the bone.

1. _____

2. _____

3. _____

4. _____

5. _____

6. _____

7. _____

8. _____

9. _____

Bone Growth and Remodeling

*Bone growth differs from the growth of tendons, ligaments and cartilage
in that bones cannot grow by interstitial growth.*

Using the terms provided, complete the following statements:

Appositional growth   Osteoblasts
Endochondral growth   Osteoclasts
Epiphyseal line   Proliferating zone
Interstitial lamellae   Zone of resting cartilage

Bone growth can occur by either (1) , the formation of new bone
on the surface of bone, or by (2) , the growth of cartilage in the
epiphyseal plate and its eventual replacement by bone. The
chondrocytes nearest the epiphysis are arranged randomly within
the hyaline cartilage in the (3) and do not divide readily. The

1. _____

2. _____

3. _____

chondrocytes in the major portion of the epiphyseal plate, a region called the (4) , divide rapidly. The fusion of the epiphysis and diaphysis results in ossification of the epiphyseal plate, which becomes the (5) . (6) are constantly removing osteons and circumferential lamellae, and new osteons are being formed by osteoblasts. However, this process leaves portions of older osteons and circumferential lamellae, called (7) , between the newly developed osteons.

4. _____

5. _____

6. _____

7. _____

☞ Bone is the major storage site for calcium in the body.

Factors Affecting Bone Growth; Bone Disorders

"*The potential shape and size of a bone as well as final adult height are genetically determined,*" *but factors such as mechanical strain, nutrition, vitamins and hormones may greatly modify the expression of those genetic factors.*

A. Match these terms with the correct statement or definition:

Growth hormone
Osteomalacia
Osteomyelitis
Osteoporosis

Rickets
Vitamin C
Vitamin D

_____ 1. Necessary for normal absorption of calcium from the intestines.

_____ 2. Can occur when children have insufficient Vitamin D.

_____ 3. Can occur in adults if their bodies are unable to absorb fat, or if they have an unusual need for calcium.

_____ 4. Necessary for normal collagen synthesis and matrix mineralization by osteoblasts.

_____ 5. Can result from increased parathyroid hormone secretion, or post menopausal declines in estrogen production.

_____ 6. Excess quantities of this substance causes acromegaly or giantism.

☞ Mechanical stress applied to bone increases osteoblast activity.

B. Match these terms with the correct statement or definition:

Calcitonin
Growth hormone
Parathyroid hormone

Sex hormones
Thyroid hormone

_____ 1. Increase general tissue growth (two).

_____

_____ 2. Stimulates bone growth, but also stimulates ossification of epiphyseal plates.

_____ 3. Stimulates osteoclast activity and increases blood calcium level.

_____ 4. Inhibits osteoclast activity, and decreases blood calcium level.

Bone Repair

"Bone is a living tissue capable of repair."

Using the terms provided, complete the following statements:

Callus   Osteoblasts
Chondroblasts   Osteoclasts
Fibroblasts   Periosteum

When bone is damaged, blood vessels in the _(1)_ and in the bone bleed, and a clot is formed. Uncommitted cells from surrounding tissue invade the clot and become _(2)_ , which produce a fibrous network, and others become _(3)_ , which produce islets of fibrocartilage. The area of fiber and cartilage formation is called a _(4)_ . From adjacent bone, _(5)_ invade and form bone through intramembranous and endochondral ossification. Finally, the repaired bone is remodeled by _(6)_ .

1. _____
2. _____
3. _____
4. _____
5. _____
6. _____

QUICK RECALL

1. List three important components of the extracellular matrix of cartilage.

2. List the major organic and inorganic compounds found in the extracellular matrix of bone.

3. List three types of bone cells, depending on their function.

4. List three important structures visible in bone matrix.

5. Name, and distinguish between, two types of bone marrow.

6. List two types of bone depending upon its internal structure.

7. List three important components of a haversian system (osteon).

8. Name two types of bone ossification.

9. Name two types of bone growth.

10. List four factors affecting bone growth.

11. List two vitamins and the abnormal bone conditions that can occur with a dietary deficiency of each.

12. List five hormones that influence bone growth.

MASTERY LEARNING ACTIVITY

Place the letter corresponding to the correct answer in the space provided.

_____ 1. Which of the following is a function of bone?
   a. internal support and protection
   b. provide attachment for the muscles
   c. calcium and phosphate storage
   d. red blood cell formation
   e. all of the above

_____ 2. Tendons
   a. are surrounded by connective tissue called peritendineum.
   b. are composed of collagen fibers oriented in the same direction.
   c. are produced by lacuna cells.
   d. all of the above

_____ 3. Chondrocytes are mature cartilage cells found within the _____, and they are derived from _____.
   a. perichondrium, fibroblasts
   b. perichondrium, chondroblasts
   c. lacunae, fibroblast
   d. lacunae, chondroblasts

_____ 4. Cartilage
   a. often occurs in thin plates or sheets.
   b. receives nutrients and oxygen by diffusion.
   c. grows as a result of chondroblast formation by the perichondrium.
   d. all of the above

_____ 5. Which of the following substances make up the major portion of bone?
   a. collagen
   b. hydroxyapatite
   c. proteoglycan aggregates
   d. osteocytes
   e. osteoblasts

_____ 6. The flexibility of bone is due to
   a. osteoclasts.
   b. ligaments.
   c. calcium salts.
   d. collagen fibers.

7. Which of the following connective tissue membranes cover the surface of mature bones?
   a. periosteum
   b. perichondrium
   c. hyaline cartilage
   d. b and c

8. A fracture in the shaft of a bone would be a break in the
   a. epiphysis.
   b. perichondrium.
   c. diaphysis.
   d. articular cartilage.

9. The portion of a long bone that stores yellow marrow in adults is the
   a. medullary cavity.
   b. spongy bone.
   c. compact bone.
   d. epiphysis.

10. Haversian canals
    a. connect Volkmann's canals to canaliculi.
    b. connect spongy bone to compact bone.
    c. is where red blood cells are produced.
    d. are found only in spongy bone.

11. The type of lamellae found in osteons?
    a. circumferential lamellae
    b. concentric lamellae
    c. interstitial lamellae
    d. none of the above

12. Given the following events:
    1. osteoprogenitor cells become osteoblasts
    2. fibroblasts produce collagen fibers
    3. osteoblasts deposit bone

    Which sequence best describes intramembranous bone formation?
    a. 1, 2, 3
    b. 1, 3, 2
    c. 2, 1, 3
    d. 2, 3, 1
    e. 3, 2, 1

13. Given the following processes:
    1. chondrocytes die
    2. calcification of cartilage matrix
    3. chondrocyte hypertrophy
    4. osteoblasts deposit bone
    5. invasion by blood vessels

Which of the following sequences best represent the order in which they occur during endochondral bone formation?
   a. 3, 2, 1, 4, 5
   b. 3, 2, 1, 5, 4
   c. 5, 2, 3, 4, 1
   d. 3, 2, 5, 1, 4
   e. 3, 5, 2, 4, 1

14. Intramembraneous bone formation
    a. occurs at the epiphyseal plate.
    b. gives rise to the flat bones of the skull.
    c. is responsible for bone growth in diameter.
    d. occurs within a hyaline carrilage model.
    e. all of the above

15. The ossification regions formed during early embryonic development are
    a. secondary ossification centers.
    b. cartilage at the ends of bones.
    c. primary ossification centers.
    d. medullary cavities.

16. Articular cartilage
    a. is found on the ends of bone.
    b. is hyaline cartilage.
    c. is embryonic cartilage that does not ossify.
    d. all of the above

17. Growth in the length of bone occurs
    a. at the primary ossification center.
    b. beneath the periosteum.
    c. at the diaphysis.
    d. at the epiphyseal plate.

18. The prime function of osteoblasts is to
    a. prevent osteocytes from forming.
    b. resorb bone along the epiphyseal plate.
    c. inhibit the growth of bone.
    d. be involved in the production of calcium salts and collagen fibers.

19. The prime function of osteoclasts is to
    a. prevent osteoblasts from forming.
    b. breakdown bone.
    c. secrete calcium salts and collagen fibers.
    d. form compact bone.

20. In a bone that was no longer growing, where would one be most likely to find an osteoclast?
    a. under the periosteum
    b. epiphyseal plate
    c. articular cartilage
    d. Volkmann's canal

21. Destruction and remolding of bone may occur
    a. as bones change shape during growth.
    b. as bones are subjected to varying patterns of stress.
    c. during and following the healing of fractures.
    d. all of the above

22. Chronic vitamin D deficiency would result in which of the following?
    a. Bones would become brittle.
    b. Bones would become soft and pliable.
    c. The percentage of the bone composed of collagen would increase.
    d. The percentage of the bone composed of hydroxyapatite would increase.
    e. b and c

23. A 16-year-old female with delayed sexual maturation was found to have an adrenal tumor that produced large amounts of sex steroids (estrogen and testosterone). If that condition was allowed to go untreated, which of the following symptoms could be expected?
    a. She would experience a short period of rapid growth.
    b. As an adult she would be taller than a normal adult.
    c. As an adult she would be shorter than a normal adult.
    d. She would grow normally.
    e. a and b

24. A patient exhibiting the following symptoms:
    1. normal amounts of parathyroid hormone
    2. adequate dietary intake of calcium and vitamin D
    3. soft bones and rapid loss of teeth

    Could be suffering from
    a. vitamin C deficiency.
    b. vitamin B deficiency.
    c. hypersecretion of calcitonin.
    d. dwarfism.
    e. giantism.

25. In the healing of bone fractures
    a. a blood clot forms around the break.
    b. a callus is formed.
    c. both intramembranous and endochondral ossification are involved.
    d. the callus may eventually disappear.
    e. all of the above

Use a separate sheet of paper to complete this section.

1. A doctor tells an elderly friend of yours that the cartilage in his joints is degenerating and has lost its resiliency. Your friend asks you to explain what that means and what the consequences of it might be. What do you say?

2. When an x-ray film was taken to check for a broken leg in a patient, the physician observed that there was a transverse region of very dense bone below the epiphyseal line. How might this dense region have developed, what effect might it have on the height of the patient, and would you expect to see this region in the other leg?

3. Is the formation of bone that occurs when a bone increases in length most like intramembranous or endochondral ossification? Explain.

4. Is the repair of bone tissue following a fracture more like intramembranous or endochondral ossification? Explain.

5. A dental assistant overheard part of a conversation between a dentist and her patient. She was explaining to the patient why a tooth implant had come loose and stated that the metallic components of the the tooth implant were so strong that the underlying bone was "shielded." Can you complete the dentist's explanation and explain why this would cause the implant to come loose?

6. An adult patient has a digestive system disorder that inhibits the absorption of fats. What is most likely to develop, osteomalacia or osteoporosis? Explain.

ANSWERS TO CHAPTER 6

WORD PARTS

1. peritendineum; periosteum; perichondrium
2. lacunae
3. lamellae
4. cancellous
5. perichondrium; chondrocyte; chondroblast; endochondral
6. osteoblast; osteoclast; osteocyte; osteomalacia; osteoporosis; osteoprogenitor
7. osteoprogenitor
8. osteoprogenitor
9. hypertrophy
10. osteomalacia

CONTENT LEARNING ACTIVITY

Introduction

1. Bone
2. Cartilage
3. Tendon

General Histology

1. Collagen
2. Proteoglycan monomers
3. Proteoglycan aggregates

Tendons and Ligaments

1. Endotendineum
2. Fascicles
3. Peritendineum
4. Epitendineum
5. Fibroblasts
6. Fibrocyte
7. Appositional
8. Interstitial

Hyaline Cartilage

1. Chondroblasts
2. Chondrocytes
3. Perichondrium

Bone Matrix

1. Collagen
2. Osteoblast
3. Osteoclast
4. Lamellae
5. Canaliculi

Bone Anatomy

A. 1. Diaphysis
    2. Epiphyseal plate
    3. Epiphyseal line
    4. Medullary cavity
    5. Marrow
    6. Perforating (Sharpey's) fibers
    7. Endosteum

B. 1. Articular cartilage
    2. Epiphysis
    3. Cancellous bone
    4. Compact bone
    5. Medullary cavity
    6. Periosteum
    7. Diaphysis
    8. Epiphyseal line

Compact Bone and Cancellous Bone

A. 1. Trabeculae
    2. Haversian (central) canals
    3. Concentric lamellae
    4. Haversian system (osteon)
    5. Volkmann's (perforating) canals
    6. Circumferential lamellae

B. 1. Haversian (central) canal
    2. Interstitial lamella
    3. Volkmann's (perforating) canal
    4. Concentric lamella
    5. Circumferential lamella
    6. Haversian system (osteon)

Bone Ossification

1. Intramembranous
2. Intramembranous

3. Endochondral

Intramembranous Ossification

1. Collagen
2. Osteoprogenitor cells

3. Centers of ossification
4. Fontanels

Endochondral Ossification

1. Hyaline cartilage
2. Osteoblasts
3. Periosteum
4. Calcified cartilage

5. Chondrocytes
6. Primary ossification center
7. Diaphysis
8. Secondary ossification center

Bone Growth and Remodeling

1. Appositional growth
2. Endochondral growth
3. Zone of resting cartilage
4. Proliferating zone

5. Epiphyseal line
6. Osteoclasts
7. Interstitial lamellae

Factors Affecting Bone Growth; Bone Disorders

A. 1. Vitamin D
   2. Rickets
   3. Osteomalacia
   4. Vitamin C
   5. Osteoporosis
   6. Growth hormone

B. 1. Growth hormone; thyroid hormone
   2. Sex hormones
   3. Parathyroid hormone
   4. Calcitonin

Bone Repair

1. Periosteum
2. Fibroblasts
3. Chondroblasts

4. Callus
5. Osteoblasts
6. Osteoclasts

QUICK RECALL

1. Collagen, proteoglycans, and water
2. Organic: collagen; inorganic: hydroxyapatite
3. Osteoblasts, osteocytes, and osteoclasts
4. Lamellae, lacunae, and canaliculi
5. Yellow marrow - mostly fat; red marrow - site of blood cell formation
6. Compact bone and cancellous bone
7. Haversian (central) canal, concentric lamellae, and osteocytes

8. Intramembranous and endochondral
9. Appositional and endochondral
10. Mechanical strain, nutrition, vitamins, and hormones
11. Vitamin D: rickets in children, osteomalacia in adults; Vitamin C: osteoporosis
12. Growth hormone, thyroid hormone, sex hormones, parathyroid hormone, and calcitonin

MASTERY LEARNING ACTIVITY

1. E. Bone performs all of the functions listed plus acts as a lever system and a site of fat storage (yellow marrow).

2. B. Tendons are dense, regular connective tissue. Tendons are surrounded by the epitendineum. Fibroblasts produce tendons by appositional growth, and fibrocytes produce tendons by interstitial growth.

3. D. Chondroblasts become chondrocytes when they are surrounded by cartilage matrix and are found within a space called the lacunae.

4. D. Since most cartilage is avascular, nutrients can reach chondrocytes only by diffusing into the cartilage from blood vessels in the perichondrium. Diffusion is effective only over short distances; thus the chondrocytes must be close to the perichondrium (surface). This is accomplished by arranging the cartilage into thin plates or sheets.

   Chondroblasts, derived from the perichondrium, lay down new matrix along the surface of the cartilage, resulting in appositional cartilage growth. Chrondrocytes within the cartilage are responsible for interstitial cartilage growth.

5. B. For the average compact bone, salts (hydroxyapatite) make up about 70% of the bone, whereas the remaining 30% is a tough organic matrix. The organic matrix is composed mainly of collagen (95%). Also present are ground substance, osteoblasts, and osteoclasts. Proteoglycan aggregates are found only in cartilage.

6. D. Collagen fibers are responsible for bone flexibility. Calcium salts allow bones to withstand compression.

7. A. Whether mature bones are formed by intramembranous ossification or endochondral ossification, they are covered with a periosteum. With intramembraneous ossification, the periosteum is formed directly. In endochondral ossification the periosteum is derived from the perichondrium. Hyaline cartilage covers the articular surface of bones.

8. C. The shaft of a long bone is the diaphysis. It is covered by the periosteum. The perichondrium covers cartilage. The ends of long bones are the epiphyses; they are covered by the articular cartilage.

9. A. Yellow marrow or fat is stored within the medullary cavity of the diaphysis (shaft). Spongy bone in the epiphysis contains red bone marrow, which produces red blood cells.

10. A. Blood vessels, lymph nodes, and nerves from the periosteum or the medullary cavity enter bone by Volkmann's canals. Volkmann's canals join with haversian canals, which are connected by canaliculi to osteocytes. This system functions to provide osteocytes with nutrients and to remove waste products.

11. B. Concentric lamellae around haversian canals form the "rings" within osteons. Circumferential lamellae form the outer surface of bone, and interstitial lamellae are the remnants of partially destroyed concentric or circumferential lamellae.

12. C. Fibroblasts produce collagen fibers, forming a collagen membrane. Osteoprogenitor cells become osteoblasts that lay bone down along collagen fiber, forming centers of ossification.

13. B. Chondrocytes within the cartilage model hypertrophy, the cartilage matrix is calcified, and the chondrocytes die. Blood vessels invade the cartilage matrix, bringing osteoblasts that deposit bone on the calcified cartilage.

14. B. Intramembranous bone formation gives rise to the flat bones of the cranial roof, some facial bones, and the clavicle. The other terms describe endochondral bone formation. Endochondral ossification occurs within a hyaline cartilage model.

15. C. Primary ossification centers are formed during embryonic development. In many bones this occurs by the third month of fetal life. Most secondary ossification centers begin to differentiate sometime between birth and the first three years of life.

16. D. In endochondral bone formation the embryonic hyaline cartilage model is ossified, except for the articular cartilage (on the ends of bones within joints) and the epiphyseal plate.

17. D. The primary ossification center results in conversion of the diaphysis from hyaline cartilage into bone. Increase in length occurs at the epiphyseal plate. Increase in width occurs beneath the periosteum.

18. D. Osteoblasts are responsible for secreting calcium salts and collagen fibers, resulting in bone growth. Osteoblasts become osteocytes when surrounded by bone matrix.

19. B. Osteoclasts break down bone, and osteoblasts secrete calcium salts and collagen fibers (build bone). The interaction between these two types of cells is responsible for bone growth and remodeling.

20. A. Once the epiphyseal plate has united, the formation of new bone does not cease. There is a continual resorption (by osteoclasts) and deposition (by osteoblasts) of bone. This occurs primarily on the outer surface of the bone beneath the periosteum and within haversian canals.

21. D. Bone is continually broken down and rebuilt in response to growth or changes in stress or due to repair of fractures.

22. E. Lack of vitamin D leads to demineralization and subsequent softening of the bones. Vitamin D promotes the uptake of calcium and phosphorus from the intestine, making these elements available for bone formation. As the bone becomes demineralized, the percentage of bone composed of collagen increases.

23. E. Estrogen and testosterone both increase bone growth. However, they also both cause early uniting of the epiphyses of long bones, especially estrogen (which explains why growth in women stops several years earlier than in men). With the adrenal tumor, it is expected that the girl have a spurt of growth, followed by uniting of the epiphyses. Since sexual maturation was delayed, the girl is taller than normal because, during the period of delayed sexual maturation, she would have been slowly growing.

24. A. Vitamin C deficiency leads to osteoporosis and depression of bone deposition by osteoblasts. This could account for the soft bones and loss of teeth.

   b is incorrect. Vitamin B deficiency doesn't affect the bones.

   c is incorrect. Hypersecretion of calcitonin depresses osteoclast activity. Bone is not resorbed; thus the bones usually remain strong.

   d and e are incorrect. Dwarfism and giantism result from undersecretion and oversecretion, respectively, of growth hormone. The effect is on the closure of the epiphyseal plates.

25. E. Following a fracture, bleeding from blood vessels in the haversian system and the periosteum produces a blood clot. The clot becomes a callus when fibroblasts and chondroblasts invade the clot, the fibroblasts lay down fibers, and the chondroblasts form cartilage. Later the callus is ossified by intramembranous (fibers) and endochondral (cartilage) ossification.

 FINAL CHALLENGES

1. The major components of the hyaline (articular) cartilage found in joints are collagen and proteoglycan aggregates. The collagen provides structural strength, whereas the proteoglycan aggregates trap water and are responsible for the resilient property of cartilage. Resilience is the ability of the cartilage to resume its shape after being compressed. Aging results in a decrease in proteoglycan aggregates and a loss of resiliency. As a consequence, the cartilage is compressed, the collagen and other cartilage components are subjected to more wear and tear, and the cartilage degenerates.

2. The dense area of bone below the epiphyseal line indicates that there was a period of time during which the bone did not increase in length (or increased very slowly) and the bone tissue became more ossified than normal. This often happens when a person suffers from an extended period of malnutrition. One expects the other leg to show a similar area and the height of the individual to be shorter than it would have been otherwise. Alternatively, a break in the bone resulting in the formation of a callus could produce an area of dense bone. In this case the other leg does not show a similar area, and height is probably not affected.

3. It is like endochondral ossification. Cartilage is formed in the proliferating zone of the epiphyseal plate, the site of bone increase in length. The cartilage corresponds to the cartilage template formed during embryonic development. The chondrocytes (cartilage cells) hypertrophy, the matrix is calcified, the chondrocytes die, the area is invaded by blood vessels and osteoblasts, and bone is laid down and remodeled. These are the same events that occur in endochondral ossification.

4. Bone repair involves events that are similar to intramembranous ossification and endochondral ossification. Fibroblasts produce a fibrous net work in the clot that is to be ossified, similar to the ossification of the collagen membrane in intramembranous ossification. In the callus, chondrocytes produce cartilage islets that are later ossified. This is similar to the cartilage template of endochondral ossification.

5. Shielding the underlying bone reduces the stress on the bone. Without stress osteoblast activity decreases, but osteoclast activity continues. The result is a reduction in bone, weakening of the bone, and loss of the tooth implant.

6. Without absorption of fats there is inadequate absorption of vitamin D in the small intestine. Without vitamin D there is inadequate absorption of calcium in the small intestine. Therefore the bone is softer than normal, and osteomalacia would develop. Osteoporosis occurs when bone is broken down faster than it is formed and the bone is porous and brittle.

Skeletal System: Gross Anatomy

FOCUS: The skeletal system consists of the axial skeleton (skull, hyoid bone, vertebral column, and rib cage) and the appendicular skeleton (limbs and their girdles). The skull surrounds and protects the brain and the organs responsible for the primary senses (seeing, hearing, smelling, and tasting); the vertebral column supports the head and trunk and protects the spinal cord; and the rib cage protects the heart and lungs. The pectoral girdle attaches the upper limbs to the trunk and allows a wide range of movement of the upper limbs. The pelvic girdle attaches the lower limbs to the trunk and is specialized to support the weight of the body. The pelvis in women is broader and more open than the pelvis in men to facilitate delivery of a baby. The upper limbs are capable of detailed movements such as grasping and manipulating objects, whereas the lower limbs are best suited for locomotion.

WORD PARTS

Give an example of a new vocabulary word that contains each word part.

WORD PART	MEANING	EXAMPLE
foss-	ditch	1. _____
sphen-	wedge	2. _____
zyg-	paired	3. _____
lacr-	tears, weeping	4. _____
ethm-	a sieve	5. _____

WORD PART	MEANING	EXAMPLE
cervi-	the neck	6. _____
lumb-	the loin	7. _____
carp-	wrist	8. _____
meta-	beyond, change	9. _____
malle-	a hammer	10. _____

CONTENT LEARNING ACTIVITY

Introduction

66*The gross anatomy of the skeletal system includes those features of bones, cartilages,* 99
tendons, and ligaments that can be seen without the microscope.

Match these skeletal subdivisions
with the correct bones:

Appendicular skeleton
Axial skeleton

_____ 1. Vertebra

_____ 2. Frontal bone

_____ 3. Coxa

_____ 4. Femur

_____ 5. Rib

_____ 6. Carpal bone

Bone Shape

66_Individual bones can be classified according to their shape._**99**

Match these bone shapes
with the correct bones:

Flat bone	Long bone
Irregular bone	Short bone

_____ 1. Vertebra

_____ 2. Frontal bone

_____ 3. Coxa

_____ 4. Femur

_____ 5. Rib

_____ 6. Carpal bone

Bone Features

66Most of these features are based on the relationship between bone and associated soft tissues.**99**

Match these bone features
with the correct definition:

Canal (meatus)	Fossa
Condyle	Notch
Crest (crista)	Trochanter
Facet	Tubercle
Foramen	

_____ 1. Smooth, rounded articular surface

_____ 2. Small, flattened articular surface

_____ 3. Prominent ridge

_____ 4. Small, rounded process

_____ 5. A hole in a bone for passage of blood vessels or nerves

_____ 6. A tunnel running within a bone

_____ 7. General term for a depression

The Skull

"_The skull is composed of 28 separate bones._**"**

A. Match these terms with the
   correct parts of the skull
   labeled in Figure 7-1:

Coronal suture
External auditory meatus
Frontal bone
Lacrimal bone
Lambdoidal suture
Mandible
Mastoid process
Maxilla
Occipital bone
Parietal bone
Sphenoid bone
Squamosal suture
Styloid process
Temporal bone
Zygomatic arch
Zygomatic bone

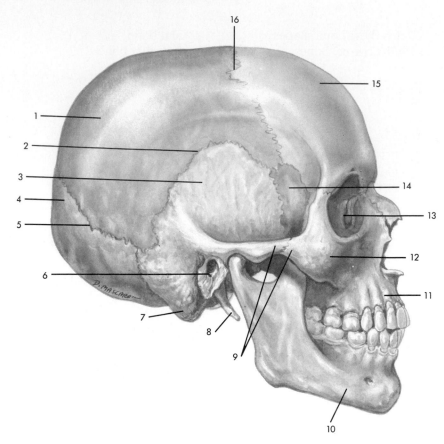

Figure 7-1

1. _____

2. _____

3. _____

4. _____

5. _____

6. _____

7. _____

8. _____

9. _____

10. _____

11. _____

12. _____

13. _____

14. _____

15. _____

16. _____

B. Match these structures with the
   correct description or definition:

Auditory ossicles   Nasal conchae
Cranial vault   Nasal septum
Crista galli   Orbit
Facial bones   Paranasal sinuses
Hard palate   Sella turcica
Hyoid bone

_____ 1. Six bones that function in hearing, found within the petrous part of the temporal bones.

_____ 2. Subdivision of the skull that protects the brain.

_____ 3. Bone that "floats" in the neck and is the attachment site for throat and tongue muscles.

_____ 4. Structure in the skull that encloses and protects the eye.

_____ 5. Three bony shelves of the nasal cavity that help to warm and moisten the air.

_____ 6. Air-filled cavities attached to the nasal cavity.

_____ 7. Connective tissue membrane (meninge) that holds the brain in place attaches to this prominent ridge.

_____ 8. Structure resembling a saddle that is occupied by the pituitary gland.

C. Match these bone parts
   with structures to which
   they contribute:

Horizontal plate of palatine bone   Temporal process of
Palatine process of maxilla   zygomatic bone
Perpendicular plate of ethmoid bone   Vomer
                                       Zygomatic process of
                                          temporal bone

_____ 1. Two parts that form the hard palate.

_____

_____ 2. Two parts that form the nasal septum.

_____

_____ 3. Two parts that form the zygomatic arch.

_____

D. Match these terms with the
correct parts of the skull
labeled in Figure 7-2A and 7-2B:

Hard palate
Horizontal plate of palatine bone
Palatine process of maxillary bone
Perpendicular plate of ethmoid bone
Septal cartilage

Temporal process of
 zygomatic bone
Vomer
Zygomatic arch
Zygomatic process of
 temporal bone

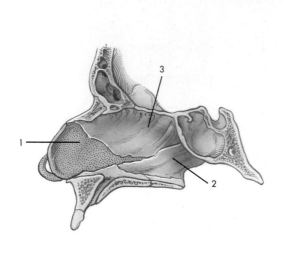

Figure 7-2A

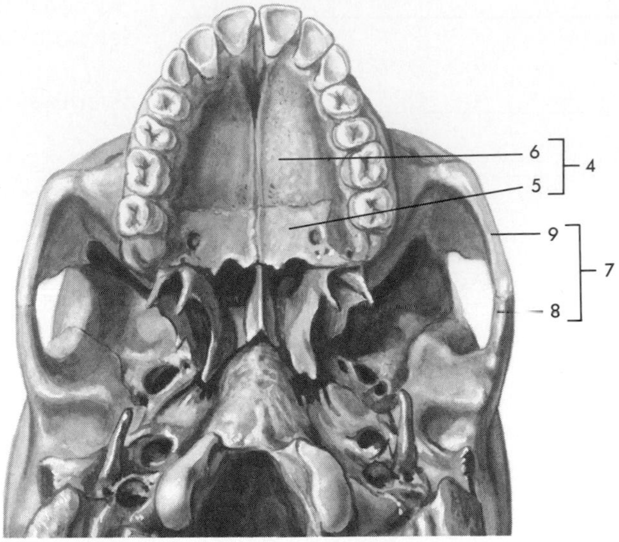

Figure 7-2B

1. _____

2. _____

3. _____

4. _____

5. _____

6. _____

7. _____

8. _____

9. _____

E. Match these bones with the
correct structures that
make up part of that bone:

Ethmoid
Hyoid
Mandible

Occipital
Sphenoid
Temporal

_____

_____

_____

_____

_____

_____

_____

_____

_____

_____

1. Coronoid process

2. Cribriform plate

3. Foramen magnum

4. Greater cornu

5. Lateral pterygoid process

6. Mandibular fossa

7. Mastoid process

8. Medial nasal concha

9. Ramus

10. Sella turcica

Muscles of the Skull

"_Many muscles attach to the skull._**"**

Match these muscle groups
with the correct description
of their attachments:

Eye movement muscles   Muscles of mastication
Muscles of facial expression   Neck muscles

_____ 1. Attach to the temporal bones, zygomatic arches, lateral pterygoid
processes of the sphenoid, and mandible.

_____ 2. Attach to the nuchal lines, external occipital protuberance, and
mastoid process.

_____ 3. Attach to the bones of the orbits.

_____ 4. Attach to the facial bones.

Openings of the Skull

"_There are many openings through the bones of the skull._**"**

Match the skull opening with
its function or with the
structures which it contains:

Carotid canal   Jugular foramen
External auditory meatus   Lacrimal canal
Foramen magnum   Olfactory foramina
Internal auditory meatus   Optic foramen

_____ 1. Openings in the cribriform plate that contain the nerves for the sense
of smell.

_____ 2. Contains the duct that carries tears from the eye to the nasal cavity.

_____ 3. Contains the nerve for the sense of hearing.

_____ 4. Most blood leaves the skull through this opening.

_____ 5. The two openings through which most blood reaches the brain.

_____

Joints of the Skull

"_The skull has several important joints._**"**

Match these skull joints with
their correct description:

Coronal suture   Occipital condyle
Lambdoidal suture   Sagittal suture
Mandibular fossa   Squamosal suture

_____ 1. Found between the parietal bones.

_____ 2. Found between the parietal bones and the occipital bone.

_____ 3. Found between the parietal bone and the temporal bone.

_____ 4. Articulation point between the skull and mandible.

_____ 5. Articulation point between the skull and the vertebral column.

Vertebral Column

The vertebral column provides flexible support and protects the spinal cord.

A. Match these terms with the correct statement or definition:

Cervical   Sacral and coccygeal
Kyphosis   Scoliosis
Lordosis   Thoracic
Lumbar

_____ 1. Two sections of the vertebral column that become convex anteriorly after birth.

_____ 2. An exaggerated convex curve, especially in the thorax.

_____ 3. An abnormal bending of the spine to one side.

B. Match these terms with the correct statement or definition:

Anulus fibrosus   Nucleus pulposus
Intervertebral disk

_____ 1. Structures located between the bodies of adjacent vertebrae.

_____ 2. External portion of an intervertebral disk.

_____ 3. Inner portion of an intervertebral disk.

_____ 4. Ruptured (herniated) disk is due to damage to this structure.

Vertebra Structure

A typical vertebra consists of a body, an arch and various processes.

A. Match these parts of a vertebra with their correct function or description:

Articular process   Pedicle
Body   Spinous process
Intervertebral foramina   Transverse process
Lamina   Vertebral foramen

_____ 1. Main weight-bearing portion of the vertebra.

_____ 2. Two parts that form the vertebral arch, which functions to protect the spinal cord.

_____ 3. Contains the spinal cord; all of them together form the vertebral canal.

_____ 4. Where the spinal nerves exit the vertebrae.

_____ 5. Connects one vertebra to another, increasing the rigidity of the vertebral column while allowing movement.

_____ 6. Midline point of muscle attachment.

B. Match these terms with the
correct parts of the vertebra
labeled in Figure 7-3:

Articular facet for tubercle of rib
Body
Lamina
Pedicle
Spinous process
Superior articular facet
Transverse process
Vertebral arch
Vertebral foramen

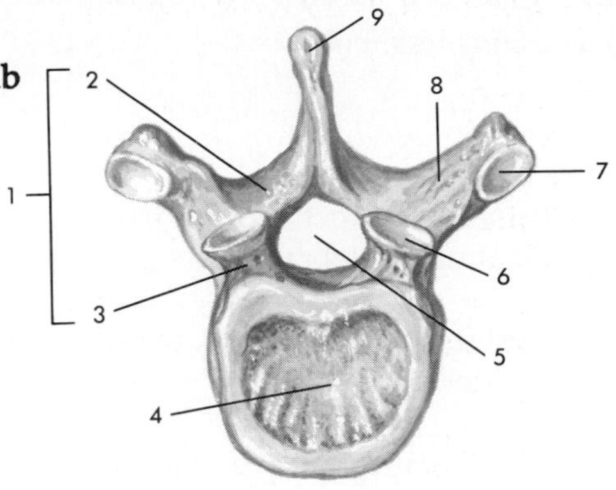

1. _____

2. _____

3. _____

4. _____

5. _____

6. _____

7. _____

8. _____

9. _____

Figure 7-3

Kinds of Vertebrae

❝_Several different kinds of vertebrae can be recognized._**❞**

Match the type of vertebra with
the characteristic unique to it:

Cervical vertebra
Coccygeal vertebra
Lumbar vertebra

Sacrum
Thoracic vertebra

_____

_____

_____

_____

_____

1. Transverse foramen through which the vertebral arteries going to the brain pass; spinous process is bifid (split).

2. Articular facet on the transverse process and on the body for the attachment of the rib.

3. Thick bodies and heavy, rectangular transverse and spinous processes; superior articular processes face medially, limiting movement of the vertebrae.

4. Transverse processes are fused to form the alae, which attaches to the pelvic bones.

5. No vertebral foramen.

 The first cervical vertebra, the atlas, is specialized to articulate with the occipital condyle of the skull. The second cervical vertebra, the axis, has a modified process (the dens) around which the atlas rotates.

Thoracic Cage

"*The thoracic cage or rib cage protects the vital organs within the thorax and prevents*** "*** the collapse of the thorax during respiration.*

A. Match these terms with the correct description or definition:

Body · Manubrium
Head · True (vertebrosternal) ribs
False (vertebrochondral) ribs · Tubercle
Floating (vertebral) ribs · Xiphoid process

_____ 1. First seven pairs of ribs, which articulate with both the vertebrae and the sternum.

_____ 2. Ribs that are attached to a common cartilage that, in turn, is attached to the sternum.

_____ 3. Eleventh and twelfth ribs, which have no attachment to the ribs.

_____ 4. Rounded process on a rib that articulates with the transverse process of one vertebra.

_____ 5. Middle part of the sternum.

_____ 6. Most inferior portion of the sternum.

B. Match these terms with the parts of the thoracic cage labeled in Figure 7-4:

Body · Manubrium
Costal cartilage · Sternum
False (vertebrochondral) ribs · True (vertebrosternal) ribs
Floating (vertebral) ribs · Xiphoid process

1. _____

2. _____

3. _____

4. _____

5. _____

6. _____

7. _____

8. _____

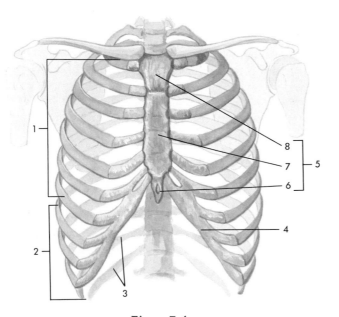

Figure 7-4

Pectoral Girdle

66*The pectoral, or shoulder, girdle consists of the scapulae and clavicles which attach the upper limb to the body.***99**

A. Match these terms with the correct description:

Acromion process
Clavicle
Coracoid process

Glenoid fossa
Scapular spine

_____ 1. Projection of scapula over the shoulder joint; articulates with the clavicle.

_____ 2. Large ridge extending from the acromion process across the scapula.

_____ 3. Projection on the anterior side of the scapula that provides a point of attachment for shoulder and arm muscles.

_____ 4. Shallow depression in the scapula, where it articulates with the humerus.

B. Match these terms with the correct parts of the scapula labeled in Figure 7-5:

Acromion process
Coracoid process
Glenoid fossa
Infraspinous fossa
Lateral border
Medial border
Scapular notch
Scapular spine
Supraspinous fossa

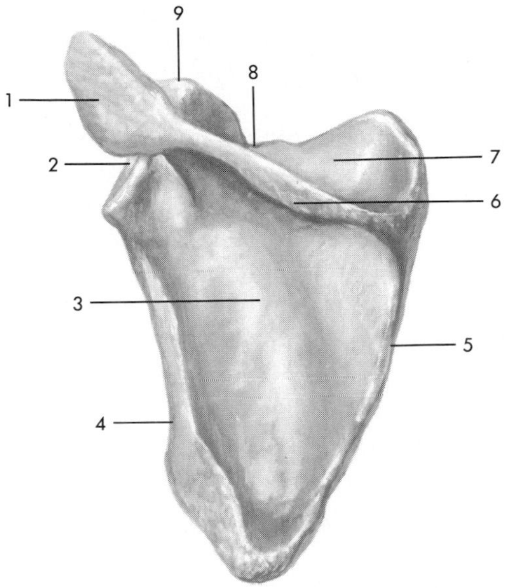

Figure 7-5

1. _____

2. _____

3. _____

4. _____

5. _____

6. _____

7. _____

8. _____

9. _____

Upper Limb

66*The arm has only one bone, the humerus, while the forearm has two bones, the radius and ulna.***99**

A. Match the bone parts that articulate with each other.

Head of humerus
Head of radius

Head of ulna
Trochlea

_____ 1. Glenoid fossa

_____ 2. Capitulum

_____ 3. Trochlear (semilunar) notch

_____ 4. Radial notch

_____ 5. Carpals

B. Match the bony part with
   its function:

Bicipital groove	Lesser tubercle
Deltoid tuberosity	Olecranon process
Epicondyles	Radial tuberosity
Greater tubercle	Styloid processes

_____ 1. Three places that shoulder muscles attach to the humerus.

_____

_____

_____ 2. Groove between the greater and lesser tubercle where one tendon of the biceps muscle passes.

_____ 3. Location where forearm muscles attach to the humerus.

_____ 4. Two places that arm muscles attach to the forearm.

_____

_____ 5. Ligaments of the wrist attach to the forearm bones here.

C. Match these groups of bones
   with the correct descriptions:

| Carpals | Phalanges |
| Metacarpals | |

_____ 1. There are five of these bones; they form the hand.

_____ 2. There are eight of these bones in two rows, forming the wrist.

_____ 3. Each finger has three of these bones, and the thumb has two.

Pelvic Girdle

“_The pelvic girdle is formed by the coxae and the sacrum. It serves as the point of attachment_**”** _of the vertebral column and the lower limbs._

A. Using the terms provided, complete the following statements:

Coxa	Iliac spines
False pelvis	Ischial tuberosity
Iliac crest	Obturator foramen
Iliac fossa	True pelvis

Each (1) is formed by the fusion of the ilium, ischium, and pubis during development. A large hole in the inferior half of the coxa through which nerves and blood vessels pass is the (2) . The superior portion of the ilium is called the (3) , whereas the large depression on the medial side of the ilium is the (4) . The anterior and posterior ends of the iliac crest are referred to as (5) , which serve as points of muscle attachment. A heavy (6) on the ischium provides a place for muscle attachment and a place to sit. The (7) is superior to the pelvic brim and is partially surrounded by bone, whereas the (8) is inferior to the pelvic brim and is completely surrounded by bone.

1. _____

2. _____

3. _____

4. _____

5. _____

6. _____

7 _____

8. _____

B. Match these terms with the correct descriptions:

Female pelvis
Male pelvis

_____ 1. Pelvic inlet tends to be heart-shaped.

_____ 2. Subpubic angle is more than 90 degrees.

_____ 3. Ischial spines closer together; ischial tuberosities turned medially.

_____ 4. Pelvic outlet is broader.

C. Match these terms with the correct parts of the coxa labeled in Figure 7-6:

Acetabulum
Anterior inferior iliac spine
Anterior superior iliac spine
Greater sciatic notch
Iliac crest
Ilium
Ischial spine
Ischial tuberosity
Ischium
Obturator foramen
Posterior inferior iliac spine
Posterior superior iliac spine
Pubis

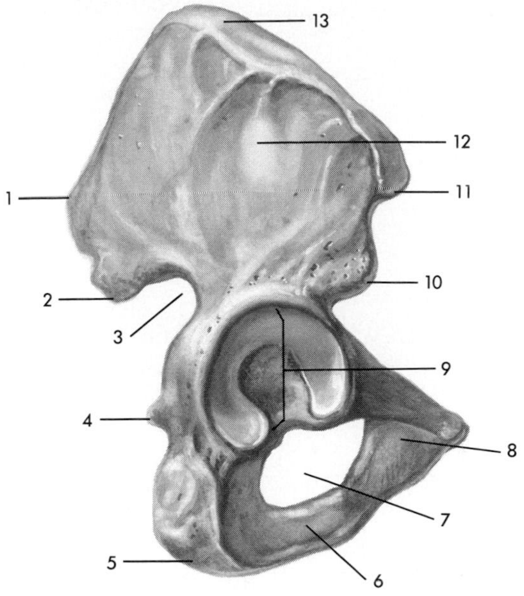

Figure 7-6

1. _____   6. _____   10. _____

2. _____   7. _____   11. _____

3. _____   8. _____   12. _____

4. _____   9. _____   13. _____

5. _____

Lower Limb

"The thigh has only one bone, the femur, whereas the leg has two bones, the tibia and the fibula."

A. Match the bone parts that articulate with each other:

Acetabulum   Patellar groove
Auricular surface   Pubic symphysis
Condyles   Talus
Head of fibula

_____ 1. Point of articulation between the two coxa.

_____ 2. Where the sacrum (ala) joins the coxa (sacroiliac joint).

_____ 3. Head of the femur articulates with the coxa.

_____ 4. Femur and tibia.

_____ 5. Tibia and proximal fibula.

_____ 6. Tibia and fibula with the ankle bone.

B. Match the bony part with its function:

Calcaneus   Patella
Condyles   Tibial tuberosity
Epicondyles   Trochanter (lesser and greater)

_____ 1. Large tuberosity lateral to the neck of the femur that serves as a site of hip muscle attachment.

_____ 2. These structures are medial and lateral to the condyles on the femur and serve as important sites of muscle attachment.

_____ 3. Large sesamoid bone located within the major anterior tendon of the thigh muscles; allows the tendon to turn the corner over the bone.

_____ 4. Point of attachment on the tibia for thigh muscles.

_____ 5. Serves as an attachment site for the calf muscles; heel bone.

C. Match these terms with the correct description:

Lateral malleolus   Phalanges
Medial malleolus   Tarsals
Metatarsals

_____ 1. Found on the distal end of the tibia; forms the ankle.

_____ 2. Seven of these bones constitute the ankle.

_____ 3. Five of these bones are located in the foot.

_____ 4. Three of these bones are in each toe, except the big toe.

1. List the two major subdivisions of the skull.

2. List the four major sutures found in the cranial vault.

3. List the four major curvatures of the vertebral column of the adult, and give the direction in which they curve.

4. Name the five types of vertebrae, and give the number of each found in the vertebral column.

5. Name the three types of ribs according to their attachment, and give the number of each type.

6. List the three parts of the sternum.

7. Name the bones of the pectoral and pelvic girdles.

8. Give the number of carpals, metacarpals, and phalanges in the upper limb, and give the number of tarsals, metatarsals, and phalanges in the lower limb.

Name the bone or part of the bone responsible for the described bony landmark:

_____ 9. Bump posterior and inferior to the ear

_____ 10. Part of the lower jaw used to pull the jaw anteriorly and open the air passages

_____ 11. Bump that can be felt anterior to the
                       ear when the jaw is moved
                       side to side

_____ 12. Bridge of the nose (two bones)

_____ 13. Prominent bump seen on the
                       midline of the neck when the neck
                       is bent anteriorly

_____ 14. Tip of the shoulder

_____ 15. Elbow projection located on the
                       medial side of the limb

_____ 16. Elbow projection located along the
                       midline of the limb

_____ 17. Bump at the distal end of the
                       forearm on the medial, posterior
                       surface

_____ 18. Knuckles

_____ 19. Ridge of bone felt when the hands
                       are placed on the hips

_____ 20. The bony prominence that one
                       sits upon

_____ 21. Kneecap

_____ 22. Shinbone

_____ 23. Bump on the anterior leg just distal
                       to the kneecap

_____ 24. Large protuberance on the lateral
                       surface of the ankle

MASTERY LEARNING ACTIVITY

Place the letter corresponding to the correct answer in the space provided.

_____ 1. Which of the following is part of the appendicular skeleton?
   a. cranium
   b. ribs
   c. clavicle
   d. sternum
   e. vertebra

_____ 2. A knob-like process on a bone?
   a. spine
   b. facet
   c. tuberosity
   d. sulcus
   e. ramus

_____ 3. The superior and medial nasal conchae are formed by projections of this bone?
   a. sphenoid bone
   b. vomer bone
   c. palatine process of maxillae
   d. palatine bone
   e. ethmoid bone

_____ 4. The crista galli
   a. separates the nasal cavity into two parts.
   b. attaches the hyoid bone to the skull.
   c. holds the pituitary gland.
   d. is an attachment site for the membrane (meninges) that surround the brain.
   e. contains the auditory ossicles.

_____ 5. The perpendicular plate of the ethmoid and the _____ form the nasal septum.
   a. palatine process of the maxilla
   b. horizontal plate of the palatine
   c. vomer
   d. a and b

_____ 6. Which of the following bones does NOT contain a paranasal sinus?
   a. ethmoid
   b. sphenoid
   c. temporal
   d. frontal
   e. maxillae

_____ 7. The mandible articulates with the skull at the
   a. styloid process.
   b. occipital condyle.
   c. mandibular fossa.
   d. zygomatic arch.
   e. medial pterygoid.

_____ 8. The nerves for the sense of smell pass through the
   a. lacrimal canal.
   b. internal auditory meatus.
   c. cribiform plate.
   d. optic foramen.
   e. orbital fissure.

_____ 9. The major blood supply to the brain enters through the
   a. foramen magnum.
   b. carotid canal.
   c. jugular foramen.
   d. a and b
   e. all of the above

_____ 10. Site of the sella turcica?
   a. sphenoid bone
   b. maxillae
   c. frontal bone
   d. ethmoid bone
   e. none of the above

_____ 11. Which of the following bones is NOT in contact with the sphenoid bone?
   a. maxilla
   b. inferior nasal concha
   c. ethmoid
   d. parietal
   e. vomer

_____ 12. The weight-bearing portion of a vertebrae is the
   a. vertebral arch.
   b. articular process.
   c. body.
   d. transverse process.
   e. spinous process.

_____ 13. Transverse foramina are found only in
a. cervical vertebrae.
b. thoracic vertebrae.
c. lumbar vertebrae.
d. the sacrum.
e. the coccyx.

_____ 14. Articular facets on the bodies and transverse processes are found only in
a. cervical vertebrae.
b. thoracic vertebrae.
c. lumbar vertebrae.
d. the sacrum.
e. the coccyx.

_____ 15. Medially facing, superior articulating processes and laterally facing, inferior articulating processes are found in
a. cervical vertebrae.
b. thoracic vertebrae.
c. lumbar vertebrae.
d. the sacrum.
e. the coccyx.

_____ 16. A ruptured (herniated) disk
a. occurs when the anulus fibrosus ruptures.
b. produces pain when the nucleus pulposus presses against spinal nerves.
c. occurs when the disk slips out of place.
d. a and b

_____ 17. Which of the following statements about vertebral column curvature is NOT true?
a. The cervical curvature develops before birth.
b. The thoracic curvature becomes exaggerated in kyphosis.
c. The lumbar curvature becomes exaggerated in lordosis.
d. The sacral curvature develops before birth.

_____ 18. Concerning ribs:
a. the true ribs attach directly to the sternum.
b. there are two pair of floating ribs.
c. the head of the rib attaches to the vertebra body.
d. a and b
e. all of the above

_____ 19. The point where the scapula and clavicle connect is the
a. coracoid process.
b. styloid process.
c. glenoid fossa.
d. acromion process.
e. capitulum.

_____ 20. Distal medial process of the humerus to which the ulna joins?
a. epicondyle
b. deltoid tuberosity
c. malleolus
d. capitulum
e. trochlea

_____ 21. Depression on the anterior surface of the humerus that receives part of the ulna when the forearm is flexed (bent)?
a. epicondyle
b. capitulum
c. coronoid fossa
d. olecranon fossa
e. radial fossa

_____ 22. Which of the following is NOT a point of muscle attachment on the pectoral girdle or upper limb?
a. epicondyles
b. styloid process
c. radial tuberosity
d. spine of scapula
e. greater tubercle

_____ 23. Which of the following parts of the upper limb is NOT correctly matched with the number of bones in that part?
a. arm: 1
b. forearm: 2
c. wrist: 10
d. palm of hand: 5
e. fingers: 14

_____ 24. The ankle bone that the tibia rests upon?
a. talus
b. calcaneous
c. metatarsals
d. navicular
e. phalanges

_____ 25. Place(s) where nerves or blood vessels pass from the trunk to the lower limb?
   a. acetabulum
   b. greater sciatic notch
   c. ischial tuberosity
   d. a and b
   e. all of the above

_____ 26. When comparing the shoulder girdle to the pelvic girdle, which of the following statements is true?

	Pectoral girdle	Pelvic girdle
a. relative mass	more	less
b. security of axial attachment	less	more
c. security of limb attachment	more	less
d. relative flexibility	less	more

_____ 27. When comparing a male pelvis to a female pelvis, which of the following statements is true?

	Female	Male
a. pelvic inlet	small, heart-shaped	large, circular
b. pubic arch	sharp, acute	rounded, obtuse
c. ischial spine	farther apart	closer together
d. sacrum	narrow, more curved	broad, less curved

_____ 28. Site of muscle attachment on the proximal end of the femur?
   a. greater trochanter
   b. lesser trochanter
   c. epicondyle
   d. a and b
   e. all of the above

_____ 29. Process forming the outer ankle?
   a. lateral malleolus
   b. lateral condyle
   c. lateral epicondyle
   d. lateral tuberosity
   e. none of the above

_____ 30. Projection on the pelvic girdle used as a landmark for finding an injection site?
   a. anterior superior iliac spine
   b. posterior superior iliac spine
   c. anterior inferior iliac spine
   d. posterior inferior iliac spine

Use a separate sheet of paper to complete this section.

1. The length of the lower limbs of a 5-year-old were measured, and it was determined that one limb was 2 cm shorter than the other limb. Would you expect the little girl to exhibit kyphosis, scoliosis, or lordosis? Explain.

2. Suppose the vertebral column were forcefully bent backward. List the vertebrae (e.g., cervical, thoracic, and lumbar) in the order that they are most likely to be damaged. Explain.

3. In what region of the vertebral column is a ruptured disk most likely to develop? Explain.

4. Can you suggest a possible advantage in having the coccyx attached to the sacrum by a flexible cartilage joint?

5. Suppose you needed to compare the length of one upper limb to another in the same individual. Using bony landmarks, suggest an easy way to accomplish the measurements.

6. During a fist fight a man wildly swings at his opponent, striking him on the jaw. The force of the blow resulted in a broken hand. What bone of the hand was most likely damaged? Explain.

7. Which bone is broken most often, the tibia or the fibula? Where would the break most likely occur?

ANSWERS TO CHAPTER 7

WORD PARTS

1. fossa
2. sphenoid
3. zygomatic
4. lacrimal
5. ethmoid

6. cervical
7. lumbar
8. carpal; metacarpal
9. metacarpal; metatarsal
10. malleolus

CONTENT LEARNING ACTIVITY

Introduction

1. Axial skeleton
2. Axial skeleton
3. Appendicular skeleton

4. Appendicular skeleton
5. Axial skeleton
6. Appendicular skeleton

Bone Shape

1. Irregular bone
2. Flat bone
3. Irregular bone

4. Long bone
5. Flat bone
6. Short bone

Bone Features

1. Condyle
2. Facet
3. Crest (crista)
4. Tubercle

5. Foramen
6. Canal (meatus)
7. Fossa

The Skull

A. 1. Parietal bone
2. Squamosal suture
3. Temporal bone
4. Occipital bone
5. Lambdoidal suture
6. External auditory meatus
7. Mastoid process
8. Styloid process
9. Zygomatic arch
10. Mandible
11. Maxilla
12. Zygomatic bone
13. Lacrimal bone

14. Sphenoid bone
15. Frontal bone
16. Coronal suture

B. 1. Auditory ossicles
2. Cranial vault
3. Hyoid bone
4. Orbit
5. Nasal conchae
6. Paranasal sinuses
7. Crista galli
8. Sella turcica

C. 1. Horizontal plate of the palatine bone; palatine process of the maxilla.
2. Perpendicular plate of the ethmoid bone; vomer.
3. Temporal process of the zygomatic bone; zygomatic process of the temporal bone.

D. 1. Septal cartilage
2. Vomer
3. Perpendicular plate of ethmoid bone
4. Hard palate
5. Horizontal plate of palatine bone
6. Palatine process of maxillary bone
7. Zygomatic arch

8. Zygomatic process of temporal bone
9. Temporal process of zygomatic bone

E. 1. Mandible
2. Ethmoid
3. Occipital
4. Hyoid
5. Sphenoid
6. Temporal
7. Temporal
8. Ethmoid
9. Mandible
10. Sphenoid

Muscles of the Skull

1. Muscles of mastication
2. Neck muscles

3. Eye movement muscles
4. Muscles of facial expression

Openings of the Skull

1. Olfactory foramina
2. Lacrimal canal
3. Internal auditory meatus

4. Jugular foramen
5. Foramen magnum; carotid canal

Joints of the Skull

1. Sagittal suture
2. Lambdoidal suture
3. Squamosal suture

4. Mandibular fossa
5. Occipital condyle

Vertebral Column

A. 1. Cervical; lumbar
2. Kyphosis
3. Scoliosis

B. 1. Intervertebral disk
2. Anulus fibrosus
3. Nucleus pulposus
4. Anulus fibrosus

Vertebra Structure

A.
1. Body
2. Lamina; pedicle
3. Vertebral foramen
4. Intervertebral foramina
5. Articular process
6. Spinous process

B.
1. Vertebral arch
2. Lamina
3. Pedicle
4. Body
5. Vertebral foramen
6. Superior articular facet
7. Articular facet for tubercle of rib
8. Transverse process
9. Spinous process

Kinds of Vertebrae

1. Cervical vertebra
2. Thoracic vertebra
3. Lumbar vertebra

4. Sacrum
5. Coccygeal vertebra

Thoracic Cage

A.
1. True (vertebrosternal) rib
2. False (vertebrochondral) rib
3. Floating (vertebral) ribs
4. Tubercle
5. Body
6. Xiphoid process

B.
1. True (vertebrosternal) ribs
2. False (vertebrochondral) ribs
3. Floating (vertebral) ribs
4. Costal cartilage
5. Sternum
6. Xiphoid process
7. Body
8. Manubrium

Pectoral Girdle

A.
1. Acromion process
2. Scapular spine
3. Coracoid process
4. Glenoid fossa

B.
1. Acromion process
2. Glenoid fossa
3. Infraspinous fossa

4. Lateral border
5. Medial border
6. Scapular spine
7. Supraspinous fossa
8. Scapular notch
9. Coracoid process

Upper Limb

A.
1. Head of humerus
2. Head of radius
3. Trochlea
4. Head of radius
5. Head of ulna

B.
1. Greater tubercle; lesser tubercle; deltoid tuberosity

2. Bicipital groove
3. Epicondyles
4. Olecranon process; radial tuberosity
5. Styloid processes

C.
1. Metacarpals
2. Carpals
3. Phalanges

Pelvic Girdle

A. 1. Coxa
   2. Obturator foramen
   3. Iliac crest
   4. Iliac fossa
   5. Iliac spines
   6. Ischial tuberosity
   7. False pelvis
   8. True pelvis

B. 1. Male pelvis
   2. Female pelvis
   3. Male pelvis
   4. Female pelvis

C. 1. Posterior superior iliac spine
   2. Posterior inferior iliac spine
   3. Greater sciatic notch
   4. Ischial spine
   5. Ischial tuberosity
   6. Ischium
   7. Obturator foramen
   8. Pubis
   9. Acetabulum
   10. Anterior inferior iliac spine
   11. Anterior superior iliac spine
   12. Ilium
   13. Iliac crest

Lower Limb

A. 1. Pubic symphysis
   2. Auricular surface
   3. Acetabulum
   4. Condyles
   5. Head of fibula
   6. Talus

B. 1. Trochanter
   2. Epicondyles
   3. Patella
   4. Tibial tuberosity
   5. Calcaneus

C. 1. Medial malleolus
   2. Tarsals
   3. Metatarsals
   4. Phalanges

QUICK RECALL

1. Cranial vault and facial bones
2. Coronal, squamosal, lambdoidal, and sagittal sutures
3. Cervical: convex anteriorly; thoracic: concave anteriorly; lumbar: convex anteriorly; sacrum and coccyx: concave anteriorly
4. Cervical: 7; thoracic: 12; lumbar: 5; sacrum: 1; coccyx: 1
5. Vertebrosternal (true) ribs: 7; vertebrochondral (false)ribs: 3; vertebral (floating) ribs: 2
6. Manubrium, body, and xiphoid process
7. Pectoral girdle: scapula and clavicle; pelvic girdle: coxae and sacrum
8. Upper limb: carpals 8; metacarpals 5; phalanges 14; lower limb: tarsals 7; metatarsals 5; phalanges 14
9. Mastoid process
10. Mandibular angle
11. Mandibular condyle
12. Nasal bones and maxillae
13. Vertebral prominens (spinous process of seventh cervical vertebra)
14. Acromion process of scapula
15. Medial epicondyle of humerus
16. Olecranon process
17. Head of ulna
18. Distal end of metacarpals
19. Iliac crest
20. Ischial tuberosity
21. Patella
22. Anterior crest of tibia
23. Tibial tuberosity
24. Lateral malleolus of fibula

MASTERY LEARNING ACTIVITY

1. C. The appendicular skeleton consists of the pectoral (clavicle, scapula) and pelvic (coxa) girdles plus the upper and lower limbs. The axial skeleton consists of the skull (cranium and auditory ossicles), hyoid, vertebrae, and ribcage (ribs and sternum).

2. C. A tuberosity is a knoblike process. Other processes include tubercules (usually smaller than a tuberosity) and trochanter (found only on the femur). A spine is a very high ridge, a facet is a small articular surface, a sulcus is a narrow depression, and a ramus is a branch off the main portion (body) of a bone.

3. E. The superior and medial nasal conchae are projections of the ethmoid bone. The inferior nasal conchae are separate bones.

4. D. The crista galli serve as a point of attachment for the meninges. The nasal septum divides the nasal cavity, the hyoid bone does not have a bony attachment to the skull, the sella turcica holds the pituitary gland, and the auditory ossicles are found in the petrous ridge of the temporal bone.

5. C. The vomer and perpendicular plate of the ethmoid form the bony part of the nasal septum. The anterior portion of the nasal septum is hyaline cartilage. The palatine process of the maxilla and the horizontal plate of the palatine form the hard palate.

6. C. The temporal bone does not have a paranasal sinus (cavity connected to the nasal cavity). The mastoid process of the temporal bone is filled with cavities called the mastoid air cells (sinuses).

7. C. The mandibular condyle articulates with the mandibular fossa. The occipital condyle articulates with the vertebral column. The styloid process, zygomatic arch, and medial pterygoids are points of muscle attachment.

8. C. The nerves for the sense of smell pass through the olfactory foramina of the cribriform plate. The nasolacrimal duct passes from the eye to the nasal cavity through the lacrimal canal, the nerve for the sense of hearing and balance passes through the internal auditory meatus, the nerve for the sense of vision passes through the optic foramen, and nerves of the orbit and face pass through the orbital fissure.

9. D. The vertebral arteries enter through the foramen magnum, and the carotid arteries enter by means of the carotid canal. Blood leaves the brain through the jugular foramen.

10. A. The sella turcica is part of the sphenoid bone.

11. B. The sphenoid is in contact with all the bones listed except the inferior nasal concha.

12. C. The body is the weight-bearing portion. The vertebral arch protects the spinal cord; the articular processes connect adjacent vertebrae, allowing movement but providing stability for the vertebral column; and the transverse and spinous processes are points of muscle attachment.

13. A. Only cervical vertebrae have transverse foramina.

14. B. The thoracic vertebrae have articular processes for the attachment of the ribs.

15. C. The arrangement of the articular processes in lumbar vertebrae provide greater stability to the lumbar region, which is subjected to greater weight and pressure than the thoracic and cervical vertebrae.

16. D. A herniated disk does not really involve a movement of the disk out of place. Instead, there is a rupture of the outer part of the disk, the anulus fibrosus. When the disk is subjected to pressure, its softer, inner portion, the nucleus pulposus, squeezes out of it. If the nucleus pulposes presses against the spinal cord or a spinal nerve, acute pain may result. Permanent damage to the nervous system is possible.

17. A. The cervical curvature develops after birth as the infant raises its head.

18. E. All of the statements are true.

19. D. The scapula articulates with the clavicle at the acromion process and with the humerus at the glenoid fossa. The coracoid process is a point of muscle attachment.

20. E. The humerus articulates with the ulna at the trochlea and with the radius at the capitulum.

21. C. When the forearm is flexed (bent), the coronoid process of the ulna fits into the coronoid fossa of the humerus. The head of the radius also fits into the radial fossa of the humerus. When the forearm is extended (straight), the olecranon process of the ulna fits into the olecranon fossa of the humerus.

22. B. The styloid processes of the ulna and radius are points of attachment for wrist ligaments.

23. C. There are eight carpal bones in the wrist.

24. A. The tibia and the fibula both articulate with the talus.

25. B. The sciatic nerve passes to the lower limb through the greater sciatic notch. The acetabulum is the point of articulation between the coxa and the femur, and the ischial tuberosity is a site of posterior thigh muscle attachment.

26. B. The pectoral girdle is attached to the axial skeleton only at the medial end of the clavicle. The pelvic girdle (coxa) is firmly attached to the sacrum at the sacroiliac joint.

27. C. The pelvic outlet in females is larger than in males, in part because the ischial spines are further apart. This makes delivery possible.

28. D. Hip muscles attach to the greater and lesser trochanter. The epicondyles are distal sites for the attachment of lower limb muscles.

29. A. The lateral malleolus (fibula) forms the outer ankle and the medial malleolus (tibia) forms the inner ankle.

30. A. The anterior superior iliac spine is used as a landmark for finding an injection site.

 FINAL CHALLENGES

1. Scoliosis would be expected, because with one lower limb shorter than the other limb, the hips would be abnormally tilted sideways. The vertebral column would curve laterally in an attempt to compensate.

2. The most likely order of damage would be the cervical, lumbar, and thoracic vertebrae. The cervical vertebrae are the most movable vertebrae; but, once they have exceeded their normal range of movement, they are more likely to be damaged than other vertebrae. The lumbar vertebrae are more massive, and the thoracic vertebrae are supported by the ribs. In addition, because the ribs limit movement of the thoracic vertebrae, the cervical and lumbar vertebrae are more likely to be the region of the vertebral column that is bent and damaged.

3. A ruptured disk is most likely to develop in the lumbar region because it supports more weight than the cervical or thoracic regions. The sacral vertebrae are fused.

4. The coccyx can bend out of the way (posteriorly) to increase the size of the pelvic outlet during delivery.

5. Measure from the acromion process to the head of the ulna.

6. Inexperienced fighters do not punch in a straight line. As a result they often strike their opponent with the medial knuckles of the hand. The fifth metacarpal is the smallest and least supported (by muscles and other bones) metacarpal, and the distal end is fractured. A skilled boxer punches so that the second and third metacarpals absorb the force of the blow. These larger and better supported metacarpals are less likely to fracture.

7. Despite its larger size, the tibia is more likely to fracture than the fibula. The tibia is the main weight-bearing bone of the leg and is subjected to the most stress. In contrast to the fibula, which is surrounded by muscles and supported by the tibia, the anteromedial portion of the tibia is unprotected by muscles or other bones. The distal one third of the tibia is relatively slender and is the likely location for a fracture to occur.

Articulations And Biomechanics Of Body Movement

FOCUS: An articulation or joint is a place where two bones come together. Depending upon the type of connective tissue that binds the bones together, joints can be classified as fibrous, cartilaginous, and synovial. Fibrous joints exhibit little movement, cartilaginous joints are slightly movable, and synovial joints are freely movable. The range or type of movement in a synovial joint is determined by the structure of the joint. Plane and hinge joints allow movements in one plane, saddle and elipsoid joints permit movement in two planes, ball-and-socket joints allow movement in all planes, and pivot joints restrict movement to rotation around a single axis.

WORD PARTS

Give an example of a new vocabulary word that contains each word part.

WORD PART	MEANING	EXAMPLE
art-	joint	1. _____
syn- (sym-)	with, together	2. _____
phy-	grow, produce	3. _____
odon-	a tooth	4. _____
burs-	bag or sac	5. _____

WORD PART	MEANING	EXAMPLE
ab-	off, away	6. _____
ad-	to, toward	7. _____
circum-	around	8. _____
duc-	lead	9. _____
plant-	sole of foot	10. _____

CONTENT LEARNING ACTIVITY

Classes of Joints

" *An articulation, or joint, is a place where two bones come together.* **"**

Match the class of joint
with its definition:

Cartilaginous   Synovial
Fibrous

_____

1. Consists of two bones that are united by fibrous tissue, has no joint cavity, and exhibits little or no movement.

_____

2. Unites two bones by means of either hyaline cartilage or fibrocartilage.

_____

3. Contains synovial fluid and allows considerable movement between articulating bones.

Fibrous and Cartilaginous Joints

" *Fibrous and cartilaginous joints are classified according to the major tissue type that binds the bones together.* **"**

A. Match the type of joint
with its definition:

Gomphosis   Synchondrosis
Suture   Syndesmosis
Symphysis   Synostosis

_____

1. Interdigitating bone joined by short, dense regular fibers.

_____

2. Joint with bones joined by ligaments; further apart than sutures.

_____

3. Joint with bones joined by periodontal ligaments; consists of pegs that fit into sockets.

_____

4. Joint with two bones joined by hyaline cartilage.

_____

5. Joint with bones joined by fibrocartilage.

B. Match the type of joint   Gomphosis   Synchondrosis
   with the correct example: Suture   Syndesmosis
                                  Symphysis   Synostosis

_____ 1. Temporal and parietal.

_____ 2. Epiphyseal line.

_____ 3. Styloid process of the temporal and the hyoid.

_____ 4. Ulna and radius.

_____ 5. Mandible and teeth.

_____ 6. Between the ribs and sternum.

_____ 7. Epiphyseal plate.

_____ 8. Between bodies of adjacent vertebrae.

Synovial Joints

66 *Synovial joints, those containing synovial fluid, allow considerable movement between articulating bones.* **99**

A. Match these terms with the   Articular cartilage   Synovial fluid
   correct statement or definition: Articular disk   Synovial membrane
                                    Bursa   Tendon sheath
                                    Fibrous capsule

_____ 1. Thin layer of hyaline cartilage that covers the articular surface of bones in synovial joints.

_____ 2. These structures containing fibrocartilage provide extra strength for joints such as the knee and temporomandibular joint.

_____ 3. One component of the joint capsule; an inner membrane that secretes fluid.

_____ 4. Fluid that consists of serum filtrate and secretions from the synovial cells.

_____ 5. Pocket or sac of the synovial membrane extending away from the rest of the joint cavity.

_____ 6. Pocket or sac containing synovial fluid extending along a tendon for some distance.

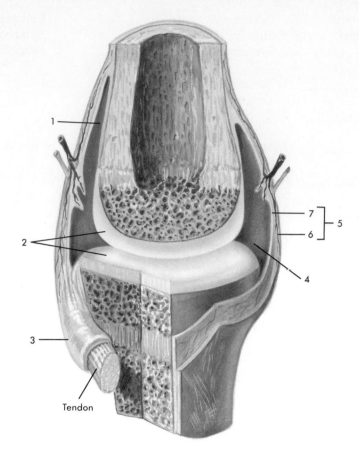

Figure 8-1

B. Match these terms with the correct parts of a synovial joint labeled in Figure 8-1:

Articular cartilage
Bursa
Fibrous capsule
Joint capsule

Joint cavity
Synovial membrane
Tendon sheath

1. _____

2. _____

3. _____

4. _____

5. _____

6. _____

7. _____

C. Match these synovial joints with the correct definition or example:

Ball-and-socket joint
Ellipsoid joint
Hinge joint

Pivot joint
Plane joint
Saddle joint

_____ 1. Two flat surfaces approximately equal in size; movement gliding or slight twisting.

_____ 2. Convex cylinder applied to a concave bone; movement in one plane only.

_____ 3. Cylindrical process that rotates within a ring; movement is rotation around one axis.

_____ 4. Round head fitting into a round depression; wide range of movement possible.

_____ 5. Modified ball-and-socket joint; range of motion nearly limited to one plane.

_____ 6. Articular processes of adjacent vertebrae.

_____ 7. Carpal and metacarpal of thumb.

_____ 8. Between phalanges.

_____ 9. Between the atlas and axis.

_____ 10. Coxageal (hip).

_____ 11. Atlas and occipital.

_____ 12. Metacarpals and phalanges.

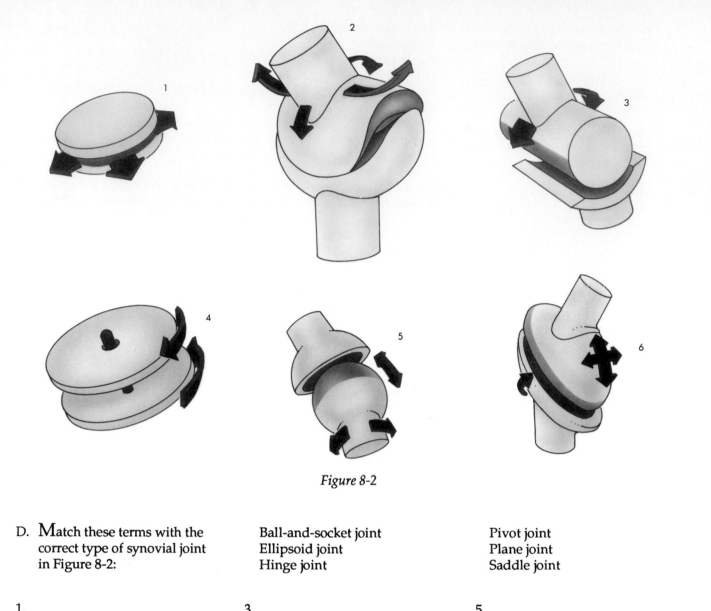

Figure 8-2

D. Match these terms with the correct type of synovial joint in Figure 8-2:

Ball-and-socket joint
Ellipsoid joint
Hinge joint

Pivot joint
Plane joint
Saddle joint

1. _____

2. _____

3. _____

4. _____

5. _____

6. _____

Movement in a Coronal Plane

66*The coronal plane divides the body into anterior and posterior portions.*99

Match these terms with the
correct statement or definition:

Dorsiflexion Plantar flexion
Extension Protraction
Flexion Retraction

_____ 1. Moving a structure in the anterior direction relative to the coronal plane.

_____ 2. Movement of the foot toward the plantar surface, such as standing on one's toes.

_____ 3. Movement of a structure toward the anterior surface in a straight horizontal line.

Movement in Relation to the Sagittal Plane

66*A sagittal plane divides the body into right and left halves.*99

Match these terms with
the correct definition:

Abduction Inversion
Adduction Lateral excursion
Eversion Medial excursion

_____ 1. Movement toward the midline.

_____ 2. Turning the ankle so that the plantar surface faces laterally.

_____ 3. Moving the mandible to the right or left of the midline.

Other Types of Movements

66*Other types of movements occur that are not specifically related to one plane.*99

Match these terms with
the correct definition:

Circumduction Pronation
Depression Rotation
Opposition Supination

_____ 1. Turning of a structure around its long axis.

_____ 2. Combination of flexion, extension, abduction and adduction.

_____ 3. Rotation of the palm so that it faces anteriorly.

_____ 4. Movement of a structure superiorly.

_____ 5. Movement returning the thumb and little finger to the anatomical position.

☞ Most movements that occur in the course of normal activities are combinations of movements.

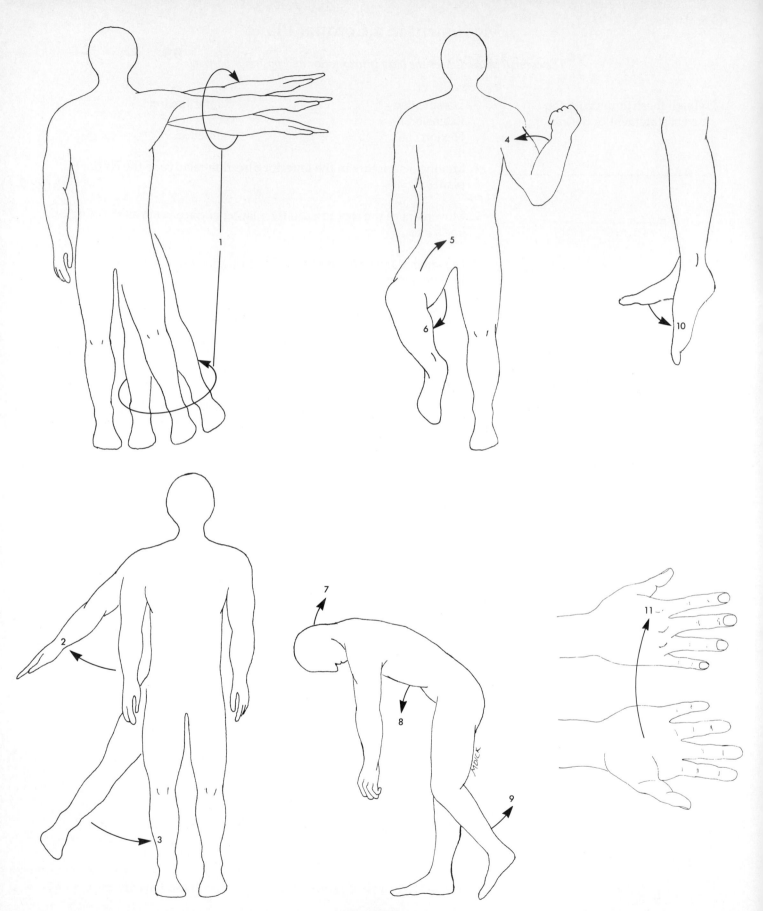

Figure 8-3

Movement Diagrams

Match these terms with
the types of movement
in Figure 8-3:

Abduct
Adduct
Circumduct
Dorsiflex
Extend
Flex
Plantar flex
Pronate
Supinate

1. _____

2. _____

3. _____

4. _____

5. _____

6. _____

7. _____

8. _____

9. _____

10. _____

11. _____

Temporomandibular Joint

"_The mandible articulates with the temporal bone to form the temporomandibular joint._**"**

Using the terms provided, complete the following statements:

Articular disk   Mandibular condyle
Combination plane and hinge   Mandibular fossa

The temporomandibular joint is the location where the _(1)_ from
the mandible articulates with the _(2)_ of the temporal bone. An
_(3)_ is interposed between the mandible and temporal bone, and
the joint is surrounded by a fibrous capsule. The
temporomandibular joint is a _(4)_ joint.

1. _____

2. _____

3. _____

4. _____

Shoulder Joint

"_The shoulder joint is a ball-and-socket joint in which stability is sacrificed somewhat for the sake of mobility._**"**

Match these terms with the
correct description or definition:

Biceps   Rotator cuff
Glenoid fossa   Subacromial and subscapular
Glenoid labrum   bursae

_____

_____

_____

_____

1. Fibrocartilage ring that builds up the rim of the glenoid fossa.

2. Sacs filled with synovial fluid that open into the shoulder joint cavity.

3. Four muscles, collectively, that pull the humeral head superiorly and
   medially.

4. Tendon of this muscle passes through the articular capsule and helps to
   hold the humerus against the scapula.

153

Hip Joint

"*The femoral head articulates with the relatively deep, concave acetabulum of the coxa,***"** *to form the coxageal, or hip joint.*

Match these terms with the correct description or definition:

Acetabular labrum
Articular capsule
Iliofemoral ligament

Ligamentum teres
Transverse acetabular ligament

_____ 1. Lip of fibrocartilage that strengthens and deepens the acetabulum.

_____ 2. Ligament that passes inferior to, and strengthens, the acetabulum.

_____ 3. Strong connective tissue structure that extends from the rim of the acetabulum to the neck of the femur, completely surrounding the femur head.

_____ 4. Ligament of the head of the femur; located inside the hip joint.

Knee Joint

"*The knee joint, or genu, has traditionally been classified as a hinge joint, although it is actually a***"** *complex ellipsoid joint that allows flexion, extension, and a small amount of rotation of the leg.*

Match these terms with the correct description or definition:

Bursae
Collateral ligaments
Cruciate ligaments

Intercondylar eminence
Menisci

_____ 1. Thick articular disks at the margins of the tibia.

_____ 2. Ligaments that extend between the fossa of the femur and the intercondylar eminence of the tibia; limit anterior and posterior movementof the joint.

_____ 3. Lateral and medial ligaments that limit movement of the joint from side to side.

_____ 4. Several sacs filled with synovial fluid that surround the knee.

Ankle Joint and Arches of the Foot

"*The ankle joint and arches of the foot provide support for the entire body.***"**

A. Using the terms provided, complete the following statements:

Arches
Dorsiflexion
Calcaneus

Hinge joint
Talocrural
Talus

The ankle, or _(1)_ joint, is a highly modified _(2)_ formed by the articulation of the tibia and fibula with the _(3)_ . Plantar flexion and _(4)_ , as well as limited inversion and eversion, can occur at this joint. There are three major _(5)_ in the foot that distribute the weight of the body between the _(6)_ and the ball of the foot.

1. _____

2. _____

3. _____

4. _____

5. _____

6. _____

B. Match the arches of the foot with the correct bones found in that arch:

Lateral longitudinal arch   Transverse arch
Medial longitudinal arch

_____ 1. Consists of the calcaneus, talus, navicular, cuneiforms, and three metatarsals.

_____ 2. Consists of the calcaneus, cuboid, and two metatarsals.

_____ 3. Consists of the cuboid and cuneiforms.

QUICK RECALL

1. List the three major classes of joints.

2. Name the three types of fibrous joints; give an example of each.

3. Name the two types of cartilaginous joints; give an example of each.

4. Name the six types of synovial joints; give an example of each.

5. Name six types of movement relative to the coronal plane.

6. List six types of movement relative to the sagittal plane.

7. Name the four types of circular movement.

MASTERY LEARNING ACTIVITY

Place the letter corresponding to the correct answer in the space provided.

_____ 1. Which of the following is used for classifying joints within the body?
   a. the material that connects the joints
   b. the way the joint moves
   c. the number of individual bones that articulate with one another
   d. a and b
   e. all of the above

_____ 2. Which of the following types of joints contain fibrous connective tissue?
   a. syndesmosis
   b. suture
   c. gomphosis
   d. a and b
   e. all of the above

_____ 3. Which of the following types of joints are held together by cartilage?
   a. symphysis
   b. synchondrosis
   c. synovial
   d. a and b
   e. all of the above

_____ 4. Which of the following joints is NOT correctly matched with the type of joint?
   a. parietal bone to occipital bone; suture
   b. between the coxae; symphysis
   c. humerus and scapula; synovial
   d. sternum and ribs; syndesmosis

_____ 5. The epiphyseal plate can be described as a type of joint. Choose the term that describes the joint before growth in the length of the bone has terminated.
   a. synchondrosis
   b. synostosis
   c. syndesmosis
   d. symphysis
   e. synovial

_____ 6. Which of the following types of joints are often temporary?
   a. syndesmoses
   b. synovial
   c. symphysis
   d. synchondroses

_____ 7. Which of the following joints are most movable?
   a. sutures
   b. syndesmoses
   c. symphysis
   d. synovial

_____ 8. The inability to produce the fluid which keeps most joints moist would likely be due to a disorder of the
   a. cruciate ligaments.
   b. synovial membrane.
   c. articular cartilage.
   d. bursae.
   e. none of the above

_____ 9. Which of the following is a characteristic of a synovial joint?
   a. articular surfaces covered with hyaline cartilage
   b. articular capsule
   c. synovial membrane
   d. synovial fluid
   e. all of the above

_____ 10. Synovial membranes form structures which, although not actually part of the synovial joint, are often associated with them. Which of the following is an example of those structures?
   a. bursae
   b. tendon sheaths
   c. articular disks
   d. a and b
   e. all of the above

_____ 11. Assume that a sharp object penetrated a synovial joint. From the following list of structures:
1. tendon or muscle
2. ligament
3. articular cartilage
4. articular capsule
5. skin
6. synovial membrane

Choose the order in which they would most likely be penetrated.
a. 5, 1, 2, 6, 4, 3
b. 5, 2, 1, 4, 3, 6
c. 5, 1, 2, 6, 3, 4
d. 5, 1, 2, 4, 3, 6
e. 5, 1, 2, 4, 6, 3

_____ 12. Which of the following do hinge joints and saddle joints have in common?
a. synovial joints
b. concave surface articulates with a convex surface
c. movement in only one plane
d. a and b
e. all of the above

_____ 13. Which of the following joints is correctly matched with the type of joint?
a. atlas to occipital; pivot
b. tarsals to metatarsals; saddle
c. femur to coxa; ellipsoid
d. tibia to talus; hinge

_____ 14. Once a door knob is grasped, what movement of the forearm is necessary to unlatch the door (turn the knob in a clockwise direction)?
a. pronation
b. rotation
c. supination
d. flexion
e. extension

_____ 15. After the door is unlatched, what movement of the forearm is necessary to open it?
a. protraction
b. retraction
c. flexion
d. extension

_____ 16. After the door is unlatched, what movement of the arm is necessary to open it?
a. flexion
b. extension
c. abduction
d. adduction

_____ 17. When grasping a door knob, the thumb and little finger will undergo
a. opposition.
b. reposition.
c. lateral excursion.
d. medial excursion.

_____ 18. Tilting the head to the side is
a. rotation.
b. depression.
c. abduction.
d. lateral excursion.
e. flexion.

_____ 19. A runner notices that the lateral side of her right shoe is wearing much more than the lateral side of her left shoe. This could mean that her right foot undergoes more _____ than her left foot.
a. inversion
b. eversion
c. plantar flexion
d. dorsiflexion

_____ 20. An articular disk is found in the
a. temporomandibular joint.
b. knee joint.
c. ankle joint.
d. a and b
e. all of the above

_____ 21. A lip (labrum) of fibrocartilage deepens the joint cavity of the
a. hip joint.
b. shoulder joint.
c. ankle joint.
d. a and b
e. all of the above

_____ 22. Which of the following joints have a tendon inside the joint cavity?
a. knee joint
b. shoulder joint
c. elbow joint
d. ankle joint

157

_____ 23. Which of the following joints have a ligament inside the articular capsule?
a. knee joint
b. hip joint
c. shoulder joint
d. a and b
e. all of the above

_____ 24. Bursitis of the subacromial bursa could result from
a. overuse of the shoulder joint.
b. leaning on the elbow.
c. kneeling.
d. running a long distance.

_____ 25. Which bone is found in the transverse arch of the foot?
a. cuneiforms
b. calcaneus
c. talus
d. a and b
e. a and c

☆——— FINAL CHALLENGES ———☆

Use a separate sheet of paper to complete this section.

1. Using an articulated skeleton, examine the joints listed below. Describe the type of joint and the movement(s) possible.
   a. the joint between the nasal bones and the frontal bone
   b. the head of a rib and the body of a vertebra
   c. between the ribs and the sternum
   d. between the body of the sternum and the manubrium

2. For each of the following joints, state what type of synovial joint it is and give the movements possible at the joint.
   a. between phalanges
   b. between phalanx and metatarsal
   c. between the occipital condyles and the atlas
   d. between the atlas and axis (dens)
   e. between the tibia and the talus

3. For each muscle described below, describe the motion(s) produced when the muscle contracts. It may be helpful to use an articulated skeleton.
   a. The triceps muscle attaches to the infraglenoid tubercle of the scapula and the olecranon process of the ulna. How does contraction of this muscle affect the forearm? The arm?

b. One of the muscles controlling movement of the index finger attaches to the distal third of the ulna. Its tendon runs over the dorsum (back) of the hand to the phalanx of the index finger. How does contraction of this muscle affect movement of the index finger? Of the wrist?
   c. The rectus abdominis attaches to the xiphoid process and the pubic crest. What movement of the vertebral column results when the rectus abdominis contracts?
   d. The semimembranosus muscle attaches to the ischial tuberosity and the medial condyle of the tibia. What movement of the thigh results when this muscle contracts? Of the leg?

4. What modifications to the skull and vertebrae would have to be made to make the joint between the skull and vertebra into a ball-and-socket joint?

5. During a karate match a man was standing with his left leg forward, flexed, and supporting 60% of his weight. He was kicked just below the knee on the tibia. What knee ligament was most likely damaged?

ANSWERS TO CHAPTER 8

WORD PARTS

1. articulate; articulation
2. synovial; synostosis; syndesmosis; synchondrosis; symphysis
3. symphysis; epiphysis; diaphysis
4. periodontal
5. bursa, bursitis
6. abduct; abduction
7. adduct; adduction
8. circumduction
9. abduction; adduction; circumduction
10. plantar

CONTENT LEARNING ACTIVITY

Classes of Joints

1. Fibrous
2. Cartilaginous
3. Synovial

Fibrous and Cartilaginous Joints

A. 1. Suture
2. Syndesmosis
3. Gomphosis
4. Synchondrosis
5. Symphysis

3. Syndesmosis
4. Syndesmosis
5. Gomphosis
6. Synchondrosis
7. Synchondrosis
8. Symphysis

B. 1. Suture
2. Synostosis

Synovial Joints

A. 1. Articular cartilage
2. Articular disks
3. Synovial membrane
4. Synovial fluid
5. Bursa
6. Tendon sheath

B. 1. Bursa
2. Articular cartilage
3. Tendon sheath
4. Joint cavity
5. Joint capsule
6. Fibrous capsule
7. Synovial membrane

C. 1. Plane joint
2. Hinge joint

3. Pivot joint
4. Ball and socket joint
5. Ellipsoid joint
6. Plane joint
7. Saddle joint
8. Hinge joint
9. Pivot joint
10. Ball-and-socket joint
11. Ellipsoid joint
12. Ellipsoid joint

D. 1. Plane joint
2. Saddle joint
3. Hinge joint
4. Pivot joint
5. Ball and socket joint
6. Ellipsoid joint

Movement in a Coronal Plane

1. Flexion
2. Plantar flexion

3. Protraction

Movement in Relation to the Sagittal Plane

1. Adduction
2. Eversion

3. Lateral excursion

Other Types of Movements

1. Rotation
2. Circumduction
3. Supination

4. Elevation
5. Reposition

Movement Diagrams

1. Circumduct
2. Abduct
3. Adduct
4. Flex
5. Flex
6. Extend

7. Extend
8. Flex
9. Flex
10. Plantar flex
11. Pronate

Temporomandibular Joint

1. Mandibular condyle
2. Mandibular fossa

3. Articular disk
4. Combination plane and hinge

Shoulder Joint

1. Glenoid labrum
2. Subacromial and subscapular bursae

3. Rotator cuff
4. Biceps

Hip Joint

1. Acetabular labrum
2. Transverse acetabular ligament

3. Articular capsule
4. Ligamentum teres

Knee Joint

1. Menisci
2. Cruciate ligaments

3. Collateral ligaments
4. Bursae

Ankle Joint and Arches of the Foot

A. 1. Talocrural
   2. Hinge joint
   3. Talus
   4. Dorsiflexion
   5. Arches
   6. Calcaneus

B. 1. Medial longitudinal arch
   2. Lateral longitudinal arch
   3. Transverse arch

QUICK RECALL

1. Fibrous, cartilaginous, and synovial joints
2. Sutures: any one of four examples; see Table 8-1 in textbook
   Syndesmosis: any one of six examples; see Table 8-1 in textbook
   Gomphosis: between the teeth and alveolar process in mandible
3. Synchondrosis: any one of four examples; see Table 8-1 in textbook
   Symphysis: any one of three examples; see Table 8-1 in textbook
4. Plane joint: any one of nine examples; see Table 8-1 in textbook
   Saddle joint: between carpal and metacarpal of thumb

Hinge joint: any one of four examples; see Table 8-1 in textbook
Pivot joint: between atlas and axis, or between radius and ulna
Ball-and-socket joint: between coxa and femur, or between the scapula and humerus
Ellipsoid joint: any one of four examples; see Table 8-1 in textbook
5. Flexion, extension, plantar flexion, dorsiflexion, protraction, and, retraction
6. Abduction, adduction, inversion, eversion, lateral excursion, and medial excursion
7. Rotation, circumduction, pronation, and supination

MASTERY LEARNING ACTIVITY

1. D. One way of classifying joints is by the type of material that connects them (fibrous, cartilaginous, or synovial). Synovial joints are often grouped according to the movement allowed by the joint (plane, hinge, pivot, ellipsoid, saddle, and ball-and-socket).

2. E. Sutures (between skull bones) and syndesmosis joints (interosseous membranes between the tibia and fibula) both contain fibers. Gomphoses are found between the teeth and jaws (periodontal membrane).

3. D. Symphyses are held together by fibrocartilage and synchondroses by hyaline cartilage. Synovial joints have articular cartilage, but it does not hold the bones together. Instead, it allows the bones to freely slide over each other.

4. D. The sternum and ribs are joined by a synchondrosis joint.

5. A. A joint that is held together by hyaline cartilage is classified as synchondrosis. The epiphyseal plate is hyaline cartilage.

6. D. Synchondrosis joints are held together by hyaline cartilage. The costal cartilages between the sternum and the ribs are examples of permanent synchondroses. Many synchondrosis joints are temporary, however, eventually being replaced by bone. For example, the epiphyseal plate is a synchondrosis joint where bone growth in length occurs. When growth stops, the epiphyseal plate is completely ossified and becomes a synostosis. It is then called the epiphyseal line.

7. D. Synovial joints are freely movable. Sutures and syndesmosis joints are held together by fibrous connective tissue. Symphysis joints are connected by fibrocartilage.

8. B. The synovial membrane, which lines the articular capsule of joints, produces the synovial fluid. This fluid acts as a lubricant between joint surfaces and provides nourishment to the articular cartilage.

9. E. The articular capsule surrounds and supports the joint. The articular cartilage covers the ends of bones within the joint. The synovial

membrane lines the inside of the joint and produces synovial fluid, which fills the joint cavity.

10. E. Bursae and tendon sheaths are synovial sacs that are filled with synovial fluid. Bursae act as a cushion between the structures they separate. Tendon sheaths surround tendons, forming a double-walled cushion for the tendon to slide through. Articular disks are an extension of the articular capsule; they provide extra support and can increase the depth of the joint cavity.

11. E. You must know the structure of a synovial joint, as well as the surrounding structures. List those structures from outside to inside in the correct order.

12. A. Hinge joints and saddle joints are both synovial joints. In hinge joints a concave surface articulates with a convex surface (cylindrical surface), whereas in saddle joints both articulatory surfaces are concave. Hinge joints are capable of movement in one plane, and saddle joints are capable of movement in two planes.

13. D. The tibia joins the talus as a hinge joint. The atlantooccipital joint is ellipsoid, the tarsometatarsal joint is plane, and the coxageal joint is ball-and-socket.

14. C. Supination of the forearm would cause the knob to turn clockwise, unlatching the door.

15. C. Flexion of the forearm would pull the door closer to the person, causing it to open.

16. B. Extension of the arm would pull the door closer to the person, causing it to open.

17. A. Opposition brings the thumb and little finger toward each other across the palm of the hand.

18. C. Abduction is movement away from the midline.

19. A. Inversion would turn the ankle so that more of the lateral surface of the foot would come into contact with the ground.

20. D. Both the temporomandibular joint and knee joint have articular disks.

21. D. The acetabular labrum is found in the hip joint, and the glenoid labrum is part of the shoulder joint.

22. B. The tendon of the biceps brachii muscle passes through the shoulder joint cavity.

23. D. The cruciate ligaments are found within the knee joint, and the ligament of the head of the femur is found within the hip joint.

24. A. The subacromial bursa is located in the shoulder.

25. A. The transverse arch consists of the cuboid and three cuneiforms. The calcaneous is part of the lateral and medial longitudinal arches, and the talus is part of the medial longitudinal arch.

 FINAL CHALLENGES

1. a. suture
   b. plane synovial
   c. synchondrosis
   d. symphysis

2. a. hinge; flexion and extension
   b. ellipsoid; flexion, extension, abduction, adduction, and circumduction
   c. ellipsoid; flexion, extension, abduction, adduction, and circumduction
   d. pivot; rotation
   e. hinge; plantar flexion, dorsiflexion, inversion, and eversion

3. a. extension of the forearm and arm
   b. extension of the finger and wrist
   c. flexion of the vertebral column
   d. extension of the thigh and flexion of the leg

4. The occipital condyles could be converted into the "ball" by enlarging and extending anteriorly and posteriorly. This would form a ball structure with a hole (the foramen magnum) in its center. The superior articular facet of the axis could be converted into the "socket" by deepening and extending anteriorly and posteriorly. The dens of the axis would have to be eliminated, and the atlas would no longer rotate around the axis.

5. The posterior cruciate prevents movement of the tibia in a posterior direction. The blow described would cause posterior movement of the tibia and likely damage the posterior cruciate ligament.

Membrane Potentials

FOCUS: The electric properties of cells result from the ionic concentration differences across the cell membrane and from the permeability characteristics of the cell membrane. There is a higher concentration of K^+ ions inside the cell than outside, and a higher concentration of Na^+ ions outside the cell than inside. For the most part, slight leakage of K^+ ions to the outside of the cell membrane is responsible for establishing the resting membrane potential (RMP), a charge difference between the inside and the outside of the unstimulated cell membrane. The outside of the cell membrane is positively charged compared to the inside. In response to a stimulus, an action potential is produced when the charge difference across the cell membrane reverses (due to the movement of Na^+ ions into the cell) and then returns to the resting condition (due to the movement of K^+ ions out of the cell). Following the action potential, the sodium-potassium exchange pump restores resting ion concentrations (Na^+ ions are pumped out of the cell and K^+ ions are pumped into the cell). Action potentials are propagated along the cell membrane of a cell and transferred across synapses to other cells. The strength of a stimulus determines the frequency of action potential generation, and the length of time the stimulus is applied determines the duration of action potential production, unless accommodation occurs.

WORD PARTS

Give an example of a new vocabulary word that contains each word part.

WORD PART	MEANING	EXAMPLE
sub-	under, below	1. _____
supra-	over, above	2. _____
maxi-	greatest	3. _____
stimul-	goad	4. _____
apsi-	juncture	5. _____
syn-	with, together	6. _____

propaga-	generate	7. _____
pol-	an axis	8. _____
de-	from, without	9. _____
re-	again, back	10. _____

CONTENT LEARNING ACTIVITY

Concentration Differences Across the Cell Membrane

"_The cell membrane is selectively permeable, allowing some, but not all, substances to pass through the membrane._**"**

Match these terms with the correct statement or definition:

Cl^- ions
K^+ ions
Na^+ ions

Proteins
Sodium-potassium exchange pump

_____ 1. These anions are repelled by negative charges within the cell and pass out through membrane channels.

_____ 2. These cations pass through the cell membrane channels easily.

_____ 3. The concentration of these cations increases inside the cell because of the attraction of negatively charged molecules.

_____ 4. There is a greater concentration of these cations outside the cell than inside.

_____ 5. These negatively charged molecules are too large and insoluble to diffuse across the cell membrane.

_____ 6. Movement of K^+ ions inward and Na^+ ions outward across the cell membrane; requires active transport (ATPs).

Resting Membrane Potential

"_The RMP exists when cells are in an unstimulated, or resting, state._**"**

Using the terms provided, complete the following statements:

Hyperpolarization
Hypopolarization (depolarization)
K^+ ions

Na^+ ions
Resting membrane potential
Selectively permeable

In nerve and skeletal muscle cells, the inside of the unstimulated cell membrane is negative when compared to the outside of the cell membrane. This potential difference is called the (1) and is approximately -85 mV. Unequal concentrations of charged molecules and ions that are separated by a (2) cell membrane are essential to RMP development. Because the cell membrane is somewhat permeable to (3) , these ions tend to diffuse from the inside to the outside of the cell. Their movement causes the outside of the cell membrane to become positively charged. If the extracellular concentration of K^+ ions increases, fewer K^+ ions diffuse out of the cell, resulting in (4) and a smaller RMP. Alternately, decreased K^+ ion concentration outside the cell causes more K^+ ions to diffuse outward and results in (5) . This effect can also be produced by an increase in permeability of the cell membrane to (6) . Changes in the concentration of (7) on either side of the cell membrane do not markedly affect the RMP.

1. _____

2. _____

3. _____

4. _____

5. _____

6. _____

7. _____

Movement of Ions Through the Cell Membrane

"*There are two separate systems for the movement of ions through the cell membrane,*"
ion channels and the sodium-potassium exchange pump.

A. Match these terms with the correct statement or definition:

Ca^{2+} ions   Gating proteins
Depolarization   Ion channels

_____

1. Pores in the cell membrane that allow one type of ion to pass through it but not others.

_____

2. Proteins that open and close ion channels.

_____

3. Extracellular concentration of these cations influence the gating proteins for Na^+ ions.

_____

4. Results in the opening of Na^+ ion channels.

B. Using the terms provided, complete the following statements:

ATPs
Concentration gradients
Into

Out of
Resting membrane potential

1. _____

2. _____

3. _____

4. _____

5. _____

The sodium-potassium exchange pump transports approximately three Na^+ ions (1) the cell for every two K^+ ions transported (2) the cell. Very little of the (3) is due to the sodium-potassium exchange pump, whose major function is to maintain the normal (4) across the cell membrane. The movement of Na^+ and K^+ ions is accomplished by active transport and requires (5) .

☞ The sodium-potassium exchange pump moves Na^+ and K^+ ions through the cell membrane against their concentration gradients.

Disorders

" *Several conditions provide examples of the clinical consequences of abnormal membrane potential.* **"**

Match these terms with the
correct description:

Hypocalcemia
Hypokalemia

_____ 1. Causes hyperpolarization of RMP, resulting in muscular weakness and
sluggish reflexes.

_____ 2. Causes action potentials to occur spontaneously, causing nervousness,
muscular spasms, and tetany.

QUICK RECALL

1. **List three causes of differences in intracellular and extracellular ion concentration.**

2. **List two factors of primary importance to the RMP.**

3. **List two separate systems for movement of ions through the cell membrane.**

4. **List three characteristics of a local potential.**

5. **List three characteristics of an action potential.**

6. **Define the all-or-none law.**

7. **List two factors that influence speed of propagation of an action potential.**

8. **List and define four different strengths of stimuli.**

MASTERY LEARNING ACTIVITY

Place the letter corresponding to the correct answer in the space provided.

_____ 1. Concerning concentration difference across the cell membrane there is
a. more K^+ ions outside the cell than inside.
b. more Na^+ ions inside the cell than outside.
c. more negatively charged proteins inside the cell than outside.
d. a and b
e. all of the above

_____ 2. Compared to the inside of the resting cell membrane, the outside surface of the membrane is
a. positively charged.
b. electrically neutral.
c. negatively charged.
d. continuously reversing its electrical charge so that it is positive one second and negative the next.
e. negatively charged whenever the sodium pump is in operation.

_____ 3. The resting membrane potential results when _____ accumulate on the outside of the cell membrane.
a. Na^+ ions
b. K^+ ions
c. Cl^- ions
d. negatively charged proteins

_____ 4. If K^+ ion leakage increased, the resting membrane potential would _____. This is called _____.
a. increase, hyperpolarization
b. increase, hypopolarization
c. decrease, hyperpolarization
d. decrease, hypopolarization

_____ 5. Which of the following statements are correctly matched?
a. hypopolarization: membrane potential becomes more negative
b. hyperpolarization: membrane potential becomes more positive
c. none of the above

_____ 6. Graphed below is a change in the resting membrane potential.

The indicated change represents a (an) _____ in the resting membrane potential and is called _____.
a. increase, hyperpolarization
b. increase, depolarization (hypopolarization)
c. decrease, hyperpolarization
d. decrease, depolarization (hypopolarization)

_____ 7. Which of the following statements about ion movement through the cell membrane is true?
a. Movement of Na^+ ions out of the cell requires energy (ATPs).
b. When Ca^{2+} ions bind to gating proteins the movement of Na^+ ions into the cell is inhibited.
c. There is a specific ion channel for regulating the movement of Na^+ ions.
d. all of the above

_____ 8. The MAJOR function of the sodium-potassium exchange pump is to
a. pump Na^+ ions into and K^+ ions out of the cell.
b. generate the resting membrane potential.
c. maintain the concentration gradients of Na^+ ions and K^+ ions across the cell membrane.
d. oppose any tendency of the cell to undergo hyperpolarization.

_____ 9. A local potential is
   a. all- or- none.
   b. not propagated for long distances.
   c. variable in magnitude.
   d. a and b
   e. b and c

_____ 10. If one could prevent Na^+ ions from diffusing into the cell,
   a. the membrane would hyperpolarize.
   b. there would be no local potential.
   c. there would still be a normal action potential.
   d. a and b

_____ 11. An action potential has the same intensity, whether it is triggered by a weak, but threshold stimulus, or by a strong, well above threshold stimulus. This is called
   a. an all-or-none response.
   b. a relative refractory response.
   c. an absolute refractory response.
   d. a latent period response.

_____ 12. During the depolarization phase of the action potential, the permeability of the membrane
   a. to K^+ ions is increased.
   b. to Na^+ ions is increased.
   c. to Ca^{2+} is decreased.
   d. is unchanged.

_____ 13. During depolarization the inside of the cell membrane becomes more _____ than the outside of the cell membrane.
   a. positive
   b. negative

_____ 14. Ms. Apeman went to the grocery store in time to take advantage of a sale on bananas. All morning and afternoon she ate bananas, which are high in potassium. That evening she didn't feel well. Which of the following symptoms is consistent with her overindulgence?
   a. increased heart rate
   b. hyperpolarization of her nerve cells
   c. action potentials that have a reduced amplitude in neurons
   d. action potentials that have an exaggerated amplitude in neurons

_____ 15. The repolarization phase of the action potential is due to the
   a. increased permeability of the membrane to Na^+ ions.
   b. decreased permeability of the membrane to Na^+ ions.

   c. increased permeability of the membrane to K^+ ions.
   d. b and c
   e. a and c

_____ 16. During repolarization of a cell membrane
   a. Na^+ ions rapidly move to the inside of the cell.
   b. Na^+ ions rapidly move to the out side of the cell.
   c. K^+ ions rapidly move to the inside of the cell.
   d. K^+ ions rapidly move to the outside of the cell.

_____ 17. The absolute refractory period
   a. limits how many action potentials can be produced in a given period of time.
   b. is the period of time when a strong stimulus can initiate a second action potential.
   c. is the period in which the permeability of the cell membrane to K^+ ions is increasing.
   d. b and c
   e. none of the above

_____ 18. Increasing the extracellular concentration of K^+ ions would affect the resting membrane potential by causing
   a. hyperpolarization.
   b. depolarization.
   c. no change.

_____ 19. Graphed below is an action potential that resulted when a nerve cell was stimulated.

During the interval from A to B
   a. K^+ ions moved into the cell.
   b. K^+ ions moved out of the cell.
   c. Na^+ ions moved into the cell.
   d. Na^+ ions moved out of the cell.
   e. none of the above

20. Increasing the extracellular concentration of Ca^{2+} ions (hypercalcemia) would greatly _____ the ability of _____ to cross the cell membrane.
    a. decrease, Na^+ ions
    b. increase, Na^+ ions
    c. decrease, K^+ ions
    d. increase, K^+ ions

21. What effect would decreasing the extracellular concentration of Ca^{2+} (hypocalcemia) have on the responsiveness of neurons to stimulus?
    a. A greater than normal stimulus would be required to initiate an action potential.
    b. A smaller than normal stimulus would be required to initiate an action potential.
    c. The neuron may generate action potentials spontaneously.
    d. b and c
    e. none of the above

22. Mr. Twitchell was diagnosed as having hyperparathyroidism. The parathyroid gland normally secretes a hormone that causes bone resorption and the release of bone minerals, including calcium, into the circulatory system. The symptom(s) that Mr. Twitchell's doctor probably observed was (were)
    a. increased heart rate.
    b. nervousness and tremors.
    c. tetany (sustained contraction of skeletal muscles).
    d. none of the above

23. A patient suffering from hypocalcemia experienced the following:
    1. tetany
    2. increased permeability of cell membranes to Na^+ ions
    3. increased action potential frequency in neurons
    4. increased rate of movement of Na^+ ions across cell membranes
    5. increased release of chemicals at nerve-muscle synapses

    Which of the following best represent the order in which the events occur?
    a. 2, 3, 4, 5, 1
    b. 2, 4, 3, 1, 5
    c. 2, 4, 3, 5, 1
    d. 3, 2, 4, 5, 1
    e. 2, 4, 5, 3, 1

24. Given the data below and what you know about action potentials:

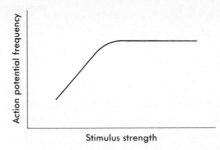

Choose the factor listed below that is primarily responsible for determining the frequency of action potentials at A for a neuron. Assume that the duration of the stimulus is constant and long enough that more than one action potential will be produced.
    a. the stimulus strength
    b. the absolute refractory period
    c. the relative refractory period
    d. the amplitude of the action potential
    e. none of the above

25. The strength-duration curve plots the combination of stimulus strength and length of time (duration) the stimulus is applied that will produce an action potential.

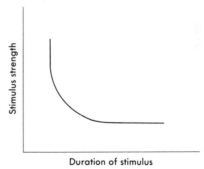

The strength-duration curve suggests that
    a. a stimulus must be applied for a minimum amount of time regardless of its strength.
    b. the greater the duration of a stimulus the smaller the amplitude (strength) of the stimulus required to initiate an action potential.
    c. the greater the amplitude (strength) of the stimulus the shorter the duration needed for it to initiate an action potential.
    d. a and b
    e. all of the above

Use a separate sheet of paper to complete this section.

1. Predict the consequence of low extracellular Na^+ ion concentration on the resting membrane potential.

2. A man exhibited numbness, muscular weakness, and occasional periods of paralysis. His physician prescribed potassium chloride tablets; their use resulted in dramatic improvement in the man's condition. Explain the symptoms and the reason for their improvement.

3. Predict the consequence of introducing an inhibitor of ATP synthesis into a nerve or muscle cell on its response to a prolonged threshold stimulus. Assume than accommodation does not occur.

4. Fluorescent lights flicker very rapidly, but they are perceived as producing a constant light. Explain how this can happen.

5. Cardiac muscle and smooth muscle are autorhythmic. Therefore they spontaneously generate action potentials. If you had the ability of simultaneously determining the membrane potential and varying the concentration of extracellular and intracellular ions, how could you experimentally determine the permeability characteristics of the muscle membranes that lead to autorhythmicity?

6. Cardiac muscle exhibits action potentials that have a prolonged depolarization phase, called the plateau phase. If you had the ability to measure the membrane potential and varying intracellular and extracellular potassium and sodium concentrations, design an experiment that would allow you to determine the permeability characteristics that give rise to the prolonged depolarization.

ANSWERS TO CHAPTER 9

WORD PARTS

1. subthreshold
2. supramaximal
3. maximal; supramaximal
4. stimulus; stimulation
5. synapse
6. synapse

7. propagation; propagated
8. polarization; depolarization; repolarization; hyperpolarization; hypopolarization
9. depolarized; depolarization
10. repolarization; refractory

CONTENT LEARNING ACTIVITY

Concentration Differences Across the Cell Membrane

1. Cl^- ions
2. K^+ ions
3. K^+ ions

4. Na^+ ions
5. Proteins
6. Sodium-potassium exchange pump

Membrane Resting Potential

1. Resting membrane potential
2. Selectively permeable
3. K^+ ions
4. Hypopolarization (depolarization)

5. Hyperpolarization
6. K^+ ions
7. Na^+ ions

Movement of Ions Through the Cell Membrane

A.
1. Ion channels
2. Gating proteins
3. Ca^{2+} ions
4. Depolarization

B.
1. Out of
2. Into
3. Resting Membrane Potential
4. Concentration gradient
5. ATPs

Electrically Excitable Cells

A.
1. Local potential
2. Local potential
3. Action potential
4. Action potential

B.
1. Depolarization
2. Repolarization
3. Repolarization
4. After potential
5. Relative refractory period

C.
1. Action potential
2. Depolarization
3. Repolarization
4. Absolute refractory period
5. Relative refractory period
6. After potential
7. Threshold

Propagation of Action Potentials

A. 1. Na$^+$ ion channels
    2. Myelin sheaths
    3. Chemical synapses
    4. Gap junctions

B. 1. Faster
    2. Slower
    3. Faster

Action Potential Frequency

1. Subthreshold stimulus
2. Threshold stimulus

3. Maximal stimulus
4. Accomodation

Disorders

1. Hypokalemia

2. Hypocalcemia

QUICK RECALL

1. Permeability characteristics of the cell membrane, presence of negatively charged proteins and other large anions within the cell, and the sodium-potassium exchange pump
2. Concentration difference of K$^+$ ions across the membrane and the permeability of the membrane to K$^+$ ions
3. Ion channels and the sodium-potassium exchange pump
4. Local potentials are graded, nonpropagated, and are hyperpolarizations.

5. Action potentials are all-or-none, propagated, and depolarizations.
6. Once the changes begin, they proceed without stopping ("all"); and, if the threshold is not reached, they do not occur at all (the "none" part).
7. The size of the neuron and the presence of a myelin sheath on the neuron
8. Subthreshold, threshold, maximal, and supramaximal stimuli

MASTERY LEARNING ACTIVITY

1. C. The inside of cells has more negatively charged proteins than the outside. The concentration of K$^+$ ions is greatest inside the cell, and the concentration of Na$^+$ ions is greatest outside the cell.

2. A. The outside of the resting cell membrane is positively charged compared to the inside.

3. B. The membrane is somewhat permeable to K$^+$ ions and they diffuse to the outside of the cell membrane, producing a potential difference across the membrane.

4. A. When membrane permeability to K$^+$ ions increases, K$^+$ ions diffuse from the inside of the

cell to the outside of the membrane, causing hyperpolarization.

5. C. During depolarization (hypopolarization) the membrane potential becomes more positive. During hyperpolarization the membrane potential becomes more negative.

6. D. The resting membrane potential decreases (gets closer to zero), and this is called hypopolarization.

7. D. Na$^+$ ions move out of the cell by active transport (sodium-potassium exchange pump) and diffuse into the cell through specific ion channels formed by proteins. Ca^{2+} ions binding to gating proteins in the ion channel inhibit the movement of Na$^+$ ions into the cell.

8. C. The major function of the sodium-potassium exchange pump is to maintain the sodium and potassium concentration gradients across the cell membrane. It does not play a significant role in directly establishing the resting membrane potential. However, the concentration gradients it generates are important in maintaining the resting membrane potential and supporting the action potential.

9. E. A subthreshold depolarization is a depolarization that is of insufficient magnitude to initiate an action potential. It is variable in magnitude (not all-or-none) in that its magnitude is proportional to the strength of the stimulus that is responsible for its generation. It is also not propagated in an all-or-none fashion. That is, as it moves further away from the point at which the subthreshold local potential is generated, its magnitude rapidly decreases, until it is not measurable within a few millimeters from its point of generation.

10. B. Depolarization occurs when Na^+ ions move into the cell. Preventing Na^+ ion movement would prevent depolarization.

11. A. Once stimulated to produce an action potential the cell responds maximally (all). If the stimulus is below threshold, there is no action potential (none).

12. B. During depolarization, membrane permeability to Na^+ ions increases, allowing an influx of positive charged Na^+ ions.

13. A. As positively charged ions move into the cell during depolarization the inside of the cell becomes more positive. Eventually enough Na^+ ions move into the cell to reverse the polarity across the cell membrane, and the inside of the cell becomes positively charged compared to the outside of the cell membrane.

14. A. To answer the question correctly one should recognize that Ms. Apeman's blood levels of K^+ ions were elevated due to the consumption of a large number of bananas. Therefore extracellular K^+ ion levels would be elevated, resulting in depolarization of excitable tissues. Depolarization of an excitable tissue to threshold results in the generation of action potentials. Since the heart is an excitable tissue an increased rate of action potential generation and an increased heart rate would be expected.

15. D. During the repolarization phase of the action potential, at least two significant permeability changes occur. The permeability of the

membrane to Na^+ ions decreases (it increased during the depolarization phase), and the permeability of the membrane to K^+ ions increases above normal for a short period of time.

16. D. Recall that the concentration of K^+ ions is higher inside the cell than outside the cell. Due to an increase in membrane permeability to K^+ ions, K^+ ions move from the inside to the outside of the cell during repolarization.

17. A. During the absolute refractory period, no stimulus, no matter how strong, will evoke another action potential. Thus a second action potential cannot be generated until AFTER the absolute refractory period. The relative refractory period is the time during which a strong stimulus can initiate a second action potential and is the period during which membrane permeability to K^+ ions is increasing.

18. B. Since the resting membrane potential is directly proportional to the potential for K^+ ions to diffuse out of the cell, an increase in the extra cellular concentration of K^+ ions reduces it (causes depolarization). Increasing the extracellular concentration of K^+ ions results in a smaller concentration gradient than normal for K^+ ions.

19. C. The part of the action potential from A to B is the depolarization phase. During this time the membrane permeability to Na^+ ions increases, and the Na^+ ions diffuse down their concentration gradient into the cell.

20. A. When the extracellular concentration of Ca^{2+} ions increases, the Ca^{2+} ions bind to gating proteins in Na^+ ion channels, reducing the ability of Na^+ ions to diffuse into the cell.

21. D. When calcium levels are low, less Ca^{2+} ions bind to gating protein in Na^+ ion channels, and the channels are more open. As a result, Na^+ ions can enter the cell with less difficulty, making it easier to initiate an action potential. If enough Na^+ ions leak into the cell, an action potential may be spontaneously generated.

22. D. Chronically increased parathyroid hormone secretion (hyperparathyroidism) may cause bone resorption at a rate great enough to elevate blood levels of calcium (calcium is a major constituent of bone). Therefore Mr. Twitchell suffers from symptoms associated with hypercalcemia. Elevated Ca^{2+} ion levels decrease the permeability of the plasma membranes to Na^+ ions. As a result, excitable tissues become less excitable (the probability that action potentials would be generated is decreased

because the depolarization that leads to an action potential is the result of an increase in the membrane's permeability to Na$^+$ ions). Therefore one would expect a decreased electrical activity in excitable tissues. One would not expect to see an increased heart rate, nervousness and tremors, or tetany.

23. C. Hypocalcemia causes an increased permeability of cell membranes to Na$^+$ ions, which results in an increased rate of sodium movement across the cell membrane. The result is an increased action potential frequency in neurons and an increased neurotransmitter released at the nerve-muscle synapse. Consequently, there would be an increased action potential frequency in skeletal muscle and tetany.

24. A. The frequency of action potentials depends on two factors, the frequency of the stimulus and the strength of the stimulus.

For a stimulus of constant length, the stronger the strength of the stimulus (up to a point), the greater the number of action potentials that will be generated. Recall that, during the relative refractory period, a stronger than normal stimulus will cause another action potential. By increasing stimulus strength, action potentials will be evoked sooner during the relative refractory period, resulting in an increased frequency of action potentials.

For a stimulus of constant strength, the faster the frequency of stimulation, the greater the number of action potentials that will be generated. The frequency of action potentials will eventually be limited by the absolute refractory period.

25. E. One must be able to interpret the graph.

 ## FINAL CHALLENGES

1. The resting membrane is relatively impermeable to Na$^+$, ions and therefore a change in Na$^+$ ion concentration has little effect on the resting membrane potential.

2. Hypopolarization of cell membranes could account for all the symptoms. Increasing extracellular K$^+$ ion concentration (potassium chloride tablets) would cause hypopolarization of cell membranes and increase the ability of cells to have action potentials because the resting membrane potential would be closer to threshold.

3. A prolonged threshold stimulus should produce a continuous series of action potentials. Inhibiting ATP synthesis would block the sodium-potassium exchange pump and prevent restoration of prestimulus ion concentrations across the cell membrane. However, so few ions move in and out of the cell during each action potential that it would take many action potentials (20,000 or more) before ion concentrations would be changed enough to affect the action potential.

4. When exposed to a flicker of light, the nerve cells in the eye must produce a local potential that causes a maximal rate of action potential generation. If the light flickers rapidly enough, the nerve cell will receive another flicker of light at the end of its refractory period and generate

another series of action potentials. The steady stream of action potentials is interpreted as a constant light.

5. A possible experimental approach is that one could measure the magnitude of the local potential at the site of action potential generation (pacemaker area). One could vary the extracellular concentration of ions that are likely to be involved in the production of the local potential (e.g. sodium, potassium, calcium). If increasing the extracellular Na$^+$ ion concentration leads to a larger local potential and increasing the intracellular concentration reduces the local potential, the data suggests that an increase in the permeability of the membrane to Na$^+$ ions is responsible for the spontaneous action potential. Other ions could be investigated in a similar fashion.

6. A possible approach is that one could monitor the magnitude of the plateau phase and the length of time that it lasts. One could then increase and decrease the intracellular and extracellular concentration of ions such as potassium, sodium, and calcium. If changing the concentrations of these ions affects the plateau, the data would suggest that the ion or ions that affect the plateau are involved in the plateau phase of the action potential and that the permeability of the membrane to the ions that affect the plateau are important in the plateau phase.

Muscular System: Histology And Physiology

FOCUS: Muscle tissue is specialized to contract with a force; it is responsible for the movement of body parts. The three types of muscle tissue are skeletal, smooth, and cardiac muscle. According to the sliding filament mechanism, the movement of actin myofilaments past myosin myofilaments results in the shortening of muscle fibers (cells) and therefore muscles. Muscle fibers respond to stimulation by contracting in an all-or-none fashion, but muscles show graded contractions due to multiple motor unit summation and multiple wave summation. Muscles use aerobic respiration to obtain the energy (ATPs) necessary for contraction but can also use anaerobic respiration for short periods of time. As a by-product of respiration muscles produce heat.

WORD PARTS

Give an example of a new vocabulary word that contains each word part.

WORD PART	MEANING	EXAMPLE
auto-	arising from within; self	1. _____
my-	muscle	2. _____
sarco-	flesh	3. _____
-plasm-	formed material	4. _____
lemma-	husk; covering	5. _____
reticul-	net-like	6. _____

WORD PART	MEANING	EXAMPLE
an-	without; not	7. _____
aero-	air	8. _____
tetan-	rigid; stretched	9. _____
troph-	nourish	10. _____

CONTENT LEARNING ACTIVITY

Characteristics of Muscle

66 *Muscle has four major functional characteristics: contractility, excitability, extensibility, and elasticity.* **99**

Match these terms with the correct statement or definition:

Elasticity   Extensibility
Excitability   Contractility

_____ 1. Ability to shorten forcefully.

_____ 2. Ability to be stretched to normal resting length or even beyond.

_____ 3. Ability to recoil to original resting length after being stretched.

Types of Muscle

66 *There are three types of muscle.* **99**

Match these terms with the correct statement or definition:

Cardiac   Smooth
Skeletal

_____ 1. Attaches to bone. Its cells are striated, long, and cylindrical with multiple, peripherally located nuclei. It is under voluntary control, is not capable of spontaneous contractions, and functions to move body parts such as the limbs.

_____ 2. Found only in the heart. Its cells are striated, cylindrical, and branched with a single centrally located nucleus. Cells are joined to each other by intercalated disks. It is under involuntary control, is capable of spontaneous contractions, and functions to pump blood throughout the body.

_____ 3. Forms the walls of hollow organs. Its cells are nonstriated, and spindle-shaped, with a single, centrally located nucleus. It is under involuntary control, is capable of spontaneous contractions, and functions to regulate the size of hollow organs.

Skeletal Muscle Structure

66*Skeletal muscles are composed of skeletal muscle fibers associated with smaller*99
amounts of connective tissue, blood vessels, and nerves.

A. **M**atch these terms with the   Endomysium   Fascia
   correct statement or definition: Epimysium   Perimysium

_____   1. Surrounds a bundle of muscle fibers (fasciculus).

_____   2. Surrounds groups of fasciculi (a whole muscle).

_____   3. Thick and fibrous; separates individual muscles and in some cases
                                      surrounds groups of muscles.

 A muscle fiber is a single muscle cell.

B. **M**atch these terms with the correct parts labeled in Figure 10-1:

Endomysium
Epimysium
Fascia
Muscle fasciculus (bundle)
Muscle fibers (cells)
Perimysium

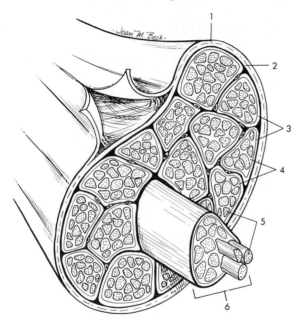

Figure 10-1

1. _____

2. _____

3. _____

4. _____

5. _____

6. _____

The Structure of Skeletal Muscle Fibers

66 *Skeletal muscle fibers are composed of myofibrils.* 99

Match these terms with the
correct statement or definition:

Myofibril   Sarcolemma
Myofilament   Sarcomere
Neuromuscular junction   Sarcoplasm

_____ 1. Area near the center of a muscle fiber where the nerve cell axon and
muscle fiber meet but do not touch.

_____ 2. Cytoplasmic material of a muscle cell (excluding the myofibrils).

_____ 3. Threadlike structure that extends from one end of a muscle fiber to the
other; consists of connected sarcomeres.

_____ 4. Highly organized unit of actin and myosin that extends from one Z-line
to another.

_____ 5. Two proteins, actin and myosin, that are organized into sarcomeres.

Elements of a Sarcomere

66 *The sarcomere is the functional unit of a skeletal muscle fiber.* 99

Match these terms with the
correct statement or definition:

A-band   M-line
Actin myofilaments   Myosin myofilaments
H-zone   Z-line
I-band

_____ 1. Also called thin myofilaments.

_____ 2. Zigzag line forming the boundary of a sarcomere, to which actin
myofilaments attach.

_____ 3. An area that extends from either side of a Z-line to the ends of the
myosin myofilaments.

_____ 4. Area of a sarcomere extending the entire length of myosin
myofilaments.

_____ 5. Area of a sarcomere composed of only myosin myofilaments; no
overlap of actin and myosin myofilaments.

_____ 6. Dark thin band in the middle of the H-zone composed of delicate
filaments that attach to myosin myofilaments and hold them in place.

Myofilaments

66 *Myofibrils are composed of myofilaments.* 99

A. Match these terms with the
correct statement or definition:

Cross bridge   Tropomyosin
G-actin   Troponin
Myosin head

_____ 1. Small globular units that combine to form a polymer of F-actin.

_____ 2. Covers active sites on the actin molecule.

_____ 3. Composed of three subunits, each of which binds with either actin, tropomyosin, or calcium.

_____ 4. Contains ATPase, an enzyme that splits ATP into ADP.

_____ 5. Structure that is formed when myosin binds with actin.

B. Match these terms with the correct parts labeled in Figure 10-2:

A-band
Actin
F-actin
H-zone
I-band
Myosin
Myosin head
Sarcomere
Tropomyosin
Troponin
Z-line

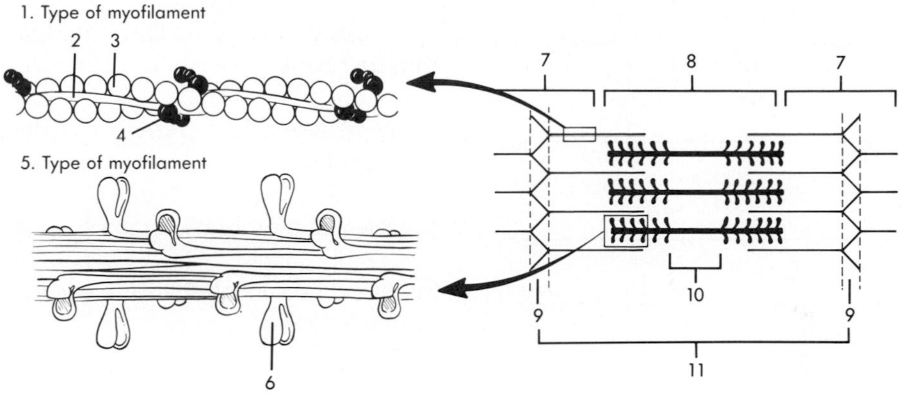

Figure 10-2

1. _____   5. _____   9. _____

2. _____   6. _____   10. _____

3. _____   7. _____   11. _____

4. _____   8. _____

Muscle Fiber Organelles

❝*The arrangement of myofibrils produces the striated appearance of muscle fibers.***❞**

Match these terms with the correct statement or definition:

Sarcoplasmic reticulum   Terminal cisterna
T tubules   Triad

_____ 1. Tubelike invaginations of the sarcolemma that project into the sarcoplasm and wrap around sarcomeres near the ends of the A bands.

_____ 2. Highly specialized, smooth endoplasmic reticulum, the membrane of which actively transports calcium ions from the sarcoplasm to the lumen of the sarcoplasmic reticulum.

_____ 3. Enlarged lumen of the sarcoplasmic reticulum next to a T tubule.

_____ 4. Grouping of a T tubule and its two adjacent terminal cisternae.

☞ Muscle fibers contain glycogen that can be broken down into glucose and used as a source of energy. Muscle fibers also contain mitochondria, which produce ATPs.

Neuromuscular Junction

66 *Skeletal muscle contracts in response to electrochemical stimuli.* 99

A. Match these terms to the correct statement or definition:

Acetylcholinesterase
Motor neuron
Neurotransmitter
Presynaptic terminal

Postsynaptic terminal
　(motor end plate)
Synaptic cleft
Synaptic vesicle

_____ 1. Specialized nerve cells that propagate action potentials to skeletal muscle fibers.

_____ 2. Space between the nerve fiber of the presynaptic terminal and the muscle fiber.

_____ 3. Muscle cell membrane in the area of the neuromuscular junction.

_____ 4. Spherical sacs located within each presynaptic terminal; contain acetylcholine.

_____ 5. Substance (such as acetylcholine) that is released from a presynaptic terminal, diffuses across the synaptic cleft, and stimulates (or inhibits) an action potential in the postsynaptic terminal.

_____ 6. Enzyme that breaks acetylcholine into acetic acid and choline; prevents accumulation of acetylcholine in the synaptic cleft.

B. Match these terms with the correct parts labeled in Figure 10-3:

Motor neuron
Muscle fiber

Synaptic cleft
Synaptic vesicles

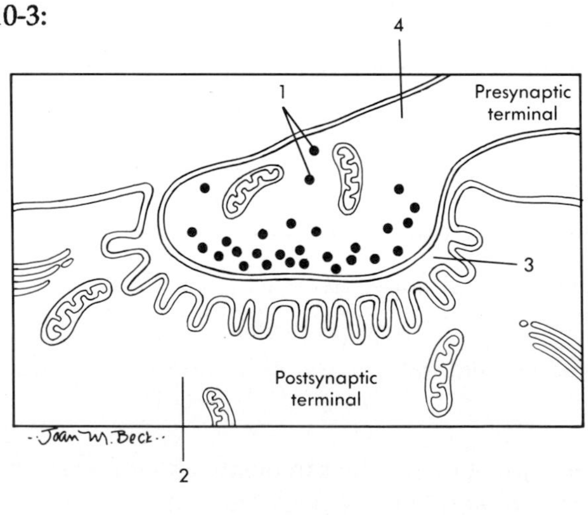

1. _____

2. _____

3. _____

4. _____

Figure 10-3

Excitation-Contraction Coupling

"_Production of an action potential in a skeletal muscle fiber leads to contraction of the muscle fiber._**"**

Use these terms to complete the following statements:

ATP
Calcium
Cross bridge
Myosin

Sarcoplasmic reticulum
T tubule
Tropomyosin
Troponin

Action potentials move into the _(1)_ system, causing the release of calcium from the _(2)_ . Calcium diffuses to the myofilaments and binds to _(3)_ , causing tropomyosin to move and expose actin to _(4)_ .Contraction occurs when actin and myosin bind, forming a _(5)_ ; myosin changes shape, and actin is pulled past myosin. Relaxation occurs when _(6)_ is taken up by the sarcoplasmic reticulum, _(7)_ binds to myosin, and the troponin- _(8)_ complex covers the actin active sites.

1. _____

2. _____

3. _____

4. _____

5. _____

6. _____

7. _____

8. _____

☞ Movement of the myosin head after the cross bridge is formed is the power stroke, and the return of the myosin head to its original position after cross bridge release is the recovery stroke.

Muscle Twitch

"_A muscle twitch is a contraction of a whole muscle in response to a stimulus that causes_**"** _an action potential in one or more muscle fibers._

Match these terms with the correct statement or definition:

Contraction phase
Lag phase

Relaxation phase

_____

_____

_____

_____

_____

_____

1. Time between the application of a stimulus and the actual onset of contraction.

2. An action potential causes the presynaptic terminal to release acetylcholine. The acetycholine crosses the synaptic cleft, binds to receptor molecules in the postsynaptic terminal, causing an action potential.

3. An action potential propagates down the T tubules, causing the release of calcium ions from the sarcoplasmic reticulum.

4. Calcium ions bind with troponin, the troponin-tropomyosin complex changes position, and active sites on actin molecules are exposed to the heads of the myosin molecules.

5. Cross bridges between actin and myosin molecules form, move, release, and reform, causing sarcomeres to shorten.

6. Calcium ions are actively transported into the sarcoplasmic reticulum, troponin-tropomyosin complexes inhibit cross bridge formation, and muscle fibers lengthen passively.

Stimulus Strength and Muscle Contraction

“An isolated muscle fiber or motor unit follows the all-or-none law of muscle contraction.**”**
However, whole muscles exhibit characteristics that are more complex.

A. Match these terms with the
correct statement or definition:

    Maximal stimulus   Subthreshold stimulus
    Multiple motor unit summation   Supramaximal stimulus
    Submaximal stimulus   Threshold stimulus

_____ 1. Stimulus just strong enough to produce an action potential in a single motor unit.

_____ 2. Stimulus strength between threshold and maximal values.

_____ 3. Stimulus that is stronger than necessary to activate all the motor units in a muscle.

_____ 4. Increasing stimulus strength, between threshold and maximum values, produces a graded increase in force of contraction of a muscle.

B. Match these terms with the
correct parts of the graph
in Figure 10-4:

    Maximal stimulus   Supramaximal stimulus
    Submaximal stimulus   Threshold stimulus
    Subthreshold stimulus

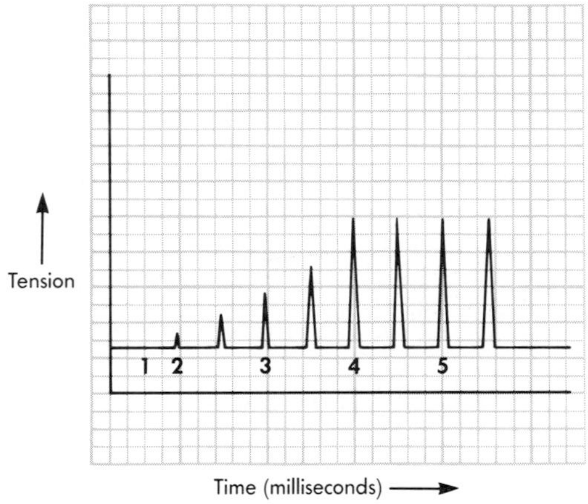

1. _____

2. _____

3. _____

4. _____

5. _____

Figure 10-4

Stimulus Frequency and Muscle Contraction

“The contractile mechanism does not exhibit a refractory period; muscle fibers do not have to**”**
completely relax before contracting again.

A. Match these terms with the
correct statement or definition:

    Complete tetanus   Multiple wave summation
    Incomplete tetanus   Treppe

_____ 1. Increase in the force of a muscle contraction due to an increased frequency of stimulation.

_____ 2. When stimuli occur so frequently there is no muscle relaxation.

_____ 3. Graded response occurring in muscle that has rested for a prolonged period of time. If the muscle is stimulated with a maximal stimulus at a frequency that allows complete relaxation between stimuli, the second contraction is of a slightly greater magnitude than the first, and the third is greater than the second. After a few stimuli, all contractions are of equal magnitude.

B. Match these terms with the correct parts labeled in Figure 10-5:

Complete tetanus   Multiple wave summation
Incomplete tetanus   Treppe

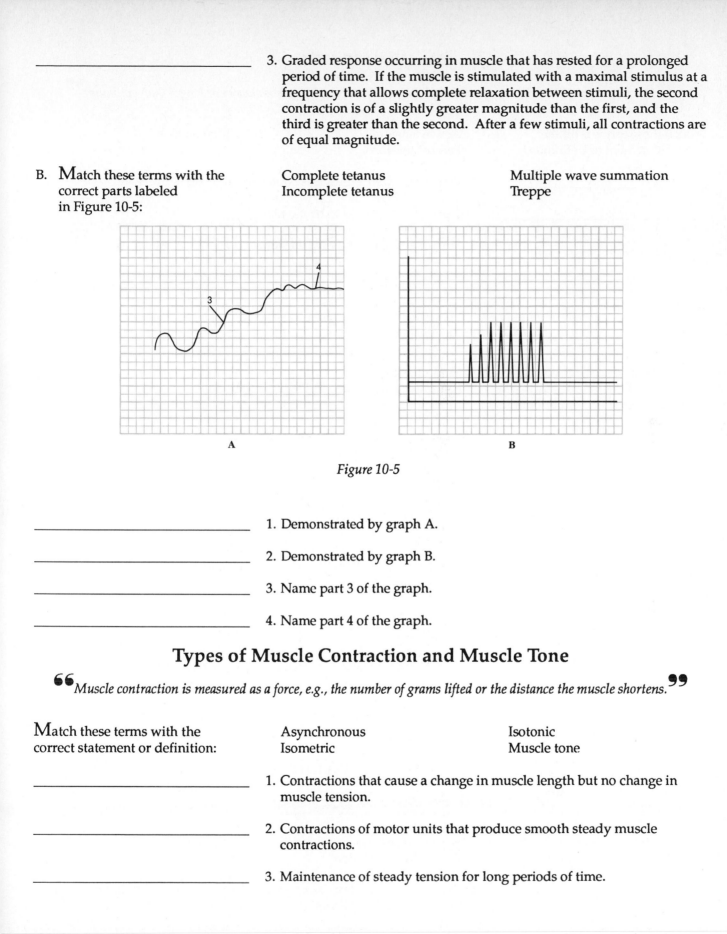

Figure 10-5

_____ 1. Demonstrated by graph A.

_____ 2. Demonstrated by graph B.

_____ 3. Name part 3 of the graph.

_____ 4. Name part 4 of the graph.

Types of Muscle Contraction and Muscle Tone

“_Muscle contraction is measured as a force, e.g., the number of grams lifted or the distance the muscle shortens._**”**

Match these terms with the correct statement or definition:

Asynchronous   Isotonic
Isometric   Muscle tone

_____ 1. Contractions that cause a change in muscle length but no change in muscle tension.

_____ 2. Contractions of motor units that produce smooth steady muscle contractions.

_____ 3. Maintenance of steady tension for long periods of time.

Length vs. Tension

"*Muscle contracts with less than maximum strength if its initial length is shorter or longer than optimum.***"**

Match these terms with the correct statement or definition:

Active tension	Total tension
Passive tension	

_____ 1. Produced when a muscle is stretched but is not stimulated.

_____ 2. Applied to an object to be moved.

_____ 3. Sum of active and passive tension.

Fatigue

"*Fatigue is the decreased capacity to do work and the reduced efficiency of performance***"** that normally follows a period of activity.

A. There are three common types of fatigue. Match these terms with the correct statement or definition:

Muscular fatigue	Synaptic fatigue
Psychological fatigue	

_____ 1. Involves the central nervous system. The muscles are capable of functioning, but the person "perceives" that additional work is not possible.

_____ 2. Result of ATP depletion. Without adequate ATP levels in muscle fibers, cross bridges cannot function normally.

_____ 3. Occurs in the neuromuscular junction when the rate of acetylcholine release is greater than the rate of acetylcholine synthesis. This form of fatigue is rare but it can occur after extreme exertion.

B. Match these terms with the correct statement or definition:

Physiological contracture
Rigor mortis

_____ 1. Extreme muscular fatigue in which a muscle can neither contract nor relax due to a lack of ATP.

_____ 2. Development of rigid muscles after death due to calcium leakage from the sarcoplasmic reticulum and too little ATP to allow relaxation.

Energy Sources

"*ATP is the immediate source of energy for muscular contraction.***"**

A. Match these terms with the correct statement or definition:

Aerobic respiration
Anaerobic respiration

_____ 1. Occurs in the absence of oxygen and results in the breakdown of glucose to yield ATP and lactic acid.

_____ 2. Requires oxygen and breaks down glucose to produce ATP, carbon dioxide, and water.

_____ 3. More efficient (produces the most ATPs for each molecule of glucose used) of the two types of respiration.

_____ 4. Faster (produces ATPs more rapidly) of the the two types of respiration.

_____ 5. More suited to short periods of intense exercise.

B. Use these terms to complete the following statements:

ATP   Glycogen
Creatine phosphate   Lactic acid
Fatty acids   Oxygen debt
Glucose

When beginning exercise, the most immediately usable form of energy is _(1)_. As it is used up, it is resynthesized by breaking the phosphate bond in an energy storage molecule, _(2)_. In addition, anaerobic respiration generates ATPs through the metabolism of glucose and _(3)_. An important non-carbohydrate source of energy for aerobic respiration is _(4)_. After exercise, the _(5)_ produced by anaerobic respiration is converted into _(6)_. This requires an energy molecule, _(7)_, produced through aerobic respiration. The oxygen required for this process is called the _(8)_.

1. _____

2. _____

3. _____

4. _____

5. _____

6. _____

7. _____

8. _____

Slow and Fast Fibers

“Not all skeletal muscles have identical functional capabilities.”

Match these terms with the correct statement or definition:

Fast-twitch muscle fibers
Slow-twitch muscle fibers

_____ 1. Have a more developed blood supply.

_____ 2. Have very little myoglobin and fewer and smaller mitochondria.

_____ 3. There is a greater concentration of this type of fiber in large postural muscles.

_____ 4. More fatigue-resistant.

_____ 5. Have large deposits of glycogen and are well adapted to perform anaerobic respiration.

The Effects of Exercise

66 *Neither fast-twitch nor slow-twitch muscle fibers can be converted to muscle fibers of another type. Nevertheless,* 99 *training can increase the capacity of both types of muscle fibers to perform more efficiently.*

Use these terms to complete the following statements:

Aerobic   Fatigue-resistant
Anaerobic   Hypertrophy

An increase in muscle strength is due to an increase in the size of
muscle fibers and is called _(1)_ . Intense exercise with _(2)_
respiration has the greatest effect on fast-twitch muscle fibers,
causing them to hypertrophy. The blood supply to both
fast-twitch and slow-twitch muscle fibers is increased by
endurance exercise requiring _(3)_ respiration, making both types
of muscle fibers more _(4)_ . Endurance exercises also cause
hypertrophy, mainly of slow-twitch fibers.

1. _____

2. _____

3. _____

4. _____

☞ Exercise increases heat production and body temperature; shivering is a response to
a decline in body temperature and helps raise body temperature by rapid
skeletal muscle contractions.

Structure of Smooth Muscle Fibers

66 *Although smooth muscle has characteristics common to skeletal muscle, it also has some unique properties.* 99

Use these terms to complete the following statements:

Calcium ions   Single
Calmodulin   Spindle-shaped
Caveolae   Striated
Gap junctions

Skeletal muscle fibers are cylindrical cells, whereas
smooth-muscle fibers are _(1)_ . Skeletal muscle fibers have many
nuclei per cell, whereas smooth muscle fibers have a _(2)_ nucleus
per cell. Skeletal muscle fibers appear to be _(3)_ but smooth-
muscle fibers are not. In visceral smooth muscle the cells are
joined by _(4)_ and function as a single unit. Although smooth
muscle does not have a T tubule system, it does have _(5)_ .
Smooth muscle lacks an extensive sarcoplasmic reticulum, so
_(6)_ must enter the cell from the extracellular fluid to initiate
contraction. Instead of binding to troponin, contraction may be
initiated when calcium binds to _(7)_ .

1. _____

2. _____

3. _____

4. _____

5. _____

6. _____

7. _____

Types of Smooth Muscles

66 *There are two types of smooth muscle: multiunit and visceral.* 99

Match the type of smooth muscle   Multiunit smooth muscle
with the correct statement   Visceral smooth muscle
or definition:

_____ 1. Has few gap junctions and normally contracts only when stimulated by nerves or hormones; each cell acts as an independent unit.

_____ 2. Occurs in sheets (blood vessels), small bundles (arector pili), or single cells (spleen capsule).

_____ 3. Usually found in sheets (digestive, reproductive, and urinary tracts).

_____ 4. Often (but not always) autorhythmic, has numerous gap junctions, and acts as a single unit.

Properties of Smooth Muscle

❝_Smooth muscle has several properties not seen in skeletal muscle._**❞**

Use these terms to complete the following statements:

Autorhythmic	Involuntary
Gradual	Sudden

Spontaneous generation of action potentials in smooth muscle occurs because of the _(1)_ leakage of sodium and calcium ions into the cell. The action potential produces a contraction; therefore smooth muscle is _(2)_ . Due to a _(3)_ increase in length, smooth muscle contracts. Despite a _(4)_ increase or decrease in length, smooth muscle maintains a constant tension and amplitude of contraction. Smooth muscle is innervated by the autonomic nervous system and is therefore under _(5)_ control.

1. _____

2. _____

3. _____

4. _____

5. _____

Cardiac Muscle

❝_Cardiac muscle is found only in the heart._**❞**

Match these terms with the correct statement or definition:

Autorhythmic	One
Intercalated disk	Striated
Involuntary	

_____ 1. Specialized cell-to-cell attachments between cardiac muscle cells.

_____ 2. Spontaneous, repetitive contraction of cardiac muscle cells.

_____ 3. Type of nervous system control of cardiac muscle.

_____ 4. Usual number of nuclei in a cardiac muscle cell.

Muscle Disorders

❝_Muscle disorders are caused by several factors._**❞**

A. Match these terms with the correct description:

Flaccid paralysis
Spastic paralysis

_____ 1. Condition produced by the action of organophosphates at the neuromuscular junction.

_____ 2. Condition produced by the action of curare at the neuromuscular junction.

189

B. Match these terms with the correct description:

Atrophy   Muscular dystrophy
Fibrositis

_____ 1. Condition that may be caused by inactivity or denervation.

_____ 2. Inherited myopathy that destroys skeletal muscle tissue and leads to eventual replacement by fatty tissue.

_____ 3. Inflammation of fibrous connective tissue caused by muscular strain and resulting in stiffness, pain, and soreness.

QUICK RECALL

1. List the four functional characteristics of muscle.

2. List the four connective tissue structures associated with skeletal muscle.

3. List the parts of a sarcomere found in the I-band, A-band, and H-zone.

4. List the three substances with which troponin can combine.

5. List the events that result in the transfer of an action potential from a neuron to a skeletal muscle.

6. List three roles of ATP in muscle contraction and relaxation.

7. List the three phases of a muscle twitch.

8. State the all-or-none law of skeletal muscle contraction.

9. List two factors that increase the force of contraction during multiple wave summation.

10. List two types of muscle contraction.

11. List three types of muscle fatigue.

12. List three molecules used as energy sources for muscle contraction. Start with the most immediately available molecule, and end with the molecule most commonly used at the beginning of anaerobic and aerobic respiration.

13. List two types of skeletal muscle.

14. List two types of smooth muscle.

MASTERY LEARNING ACTIVITY

Place the letter corresponding to the correct answer in the space provided.

_____ 1. Which of the following is true of skeletal muscle?
a. spindle-shaped cells
b. under involuntary control
c. many peripherally located nuclei per muscle cell
d. forms the walls of hollow internal organs
e. all of the above

_____ 2. The connective tissue sheath that surrounds a muscle fasciculus is the
a. perimysium.
b. endomysium.
c. epimysium.
d. hypomysium.

_____ 3. Given the following structures:
1. whole muscle
2. muscle cell (fiber)
3. myofilament
4. myofibril
5. muscle bundle

Choose the arrangement that lists the structures in the correct order from the outside to the inside of a skeletal muscle.
a. 1, 2, 5, 3, 4
b. 1, 2, 5, 4, 3
c. 1, 5, 2, 3, 4
d. 1, 5, 2, 4, 3
e. 1, 5, 4, 2, 3

191

_____ 4. Each myofibril
a. is made up of many muscle fibers.
b. contains sarcoplasmic reticulum.
c. is made up of many sarcomeres.
d. contains T tubules.
e. is equivalent to a muscle fiber.

_____ 5. Myosin
a. is attached to the Z line.
b. is found primarily in the I band.
c. is thinner than actin.
d. none of the above

_____ 6. Which of the following statements about the molecular structure of myofilaments is true?
a. Tropomyosin has a binding site for Ca^{2+} ions.
b. The head of the myosin molecule binds to an active site on G-actin.
c. ATPase is found on troponin.
d. Troponin binds to the rodlike portion of myosin.

_____ 7. The part of the sarcolemma that invaginates into the interior of a skeletal muscle cell is the
a. T tubule system.
b. sarcoplasmic reticulum.
c. myofibrils.
d. none of the above

_____ 8. Given the following events:
1. acetylcholine broken down into acetic acid and choline
2. acetycholine moves across the synaptic cleft
3. action potential reaches the terminal branch of a motor neuron
4. acetylcholine combines with a receptor molecule
5. action potential produced on muscle cell membrane

Choose the arrangement that lists the events in the order they occur at a neuromuscular junction.
a. 2, 3, 4, 1, 5
b. 3, 2, 4, 5, 1
c. 3, 4, 2, 1, 5
d. 4, 5, 2, 1, 3

_____ 9. Bob Canner improperly processed some home-grown vegetables. As a result, he contracted botulism poisoning upon eating the vegetables. Symptoms included difficulty in swallowing and breathing. Eventually he died of respiratory failure (his respiratory muscles relaxed and would not contract). Which of the following could the botulism toxin have caused?
a. increased calcium transport into muscles
b. decreased potassium leakage from muscle cell membranes
c. decreased acetylcholine release
d. decreased acetylcholinesterase release
e. none of the above

_____ 10. Given the following events:
1. sarcoplasmic reticulum releases calcium
2. sarcoplasmic reticulum takes up calcium
3. calcium diffuses into the myofibrils
4. action potential moves down the T tubule
5. sarcomere shortens
6. muscle relaxes

Choose the arrangement that lists the events in the order they occur following a single stimulation of a skeletal muscle cell.
a. 1, 3, 4, 5, 2, 6
b. 2, 3, 5, 4, 6, 2
c. 4, 1, 3, 5, 2, 6
d. 4, 2, 3, 5, 1, 6
e. 5, 1, 4, 3, 2, 6

_____ 11. Given the following events:
1. calcium combines with tropomyosin
2. calcium combines with troponin
3. tropomyosin pulls away from actin
4. troponin pulls away from actin
5. tropomyosin pulls away from myosin
6. troponin pulls away from myosin
7. myosin binds to actin

Choose the arrangement that lists the events in the order they occur during muscle contraction.
a. 1, 4, 7
b. 2, 3, 7
c. 1, 3, 7
d. 2, 4, 7
e. 2, 5, 6

_____ 12. Skeletal muscles
   a. require energy in order to relax.
   b. require energy in order to contract.
   c. shorten with a force.
   d. b and c
   e. all of the above

_____ 13. Which of the following events occur during the latent phase of muscle contraction?
   a. cross-bridge formation
   b. active transport of calcium ions into the sarcoplasmic reticulum
   c. calcium ions combine with troponin
   d. the sarcomere shortens

_____ 14. Which of the following bands shorten during skeletal muscle contraction?
   a. A
   b. I
   c. H
   d. A and I
   e. I and H

_____ 15. A motor unit is
   a. a nerve fiber and all of the muscle fibers it innervates.
   b. a nerve fiber and the muscle fiber it innervates.
   c. a muscle fiber and the nerve fiber that innervates it.
   d. a muscle fiber and the nerve fibers that innervate it.
   e. none of the above

_____ 16. Under normal circumstances, at the the level of a single skeletal muscle fiber, there are
   a. graded contractions, depending on stimulus input.
   b. several rapid contractions resulting in the main contraction.
   c. all-or-none responses.

_____ 17. Considering the force of contraction of a muscle cell, multiple wave summation occurs because of
   a. increased frequency of action potentials on the cell membrane.
   b. increased number of cross bridges formed.
   c. increase in calcium concentration around the myofibrils.
   d. all of the above

_____ 18. Graphed below is the response of an isolated skeletal muscle to a single stimulus of increasing strength.

Given the preceding information and what you know about skeletal muscle, choose the appropriate response(s) listed below.
   a. Skeletal muscles respond in an all-or-none fashion.
   b. Between stimulus strength A and B the number of sarcomeres that contract increases.
   c. Between stimulus strength B and C the number of sarcomeres that contract increases.
   d. a and b
   e. a and c

_____ 19. Each sarcomere in a muscle fiber responds in an all-or-none fashion. However, when maximal stimuli are applied to a skeletal muscle at an increasing frequency, the amplitude of contraction increases. Which of the following statements are consistent with the above observation?
   a. Tetanic stimuli cause a greater degree of contraction of the sarcomeres.
   b. Early in a tetanic contraction the connective tissue components of muscle are stretched, allowing it to apply a greater tension to the load.
   c. A greater concentration of Ca^{2+} ions in the sarcoplasm is achieved.
   d. a and c
   e. all of the above

_____ 20. A weight-lifter attempts to lift a weight from the floor, but the weight is so heavy he is unable to move it. The type of muscle contraction the weight-lifter used was mostly
   a. isometric
   b. isotonic
   c. isokinetic

21. Given the following graph:

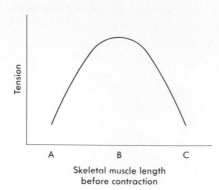

Tension

A   B   C

Skeletal muscle length
before contraction

Choose the most appropriate response.
a. The maximum sarcomere length occurs at B.
b. It is an advantage for one to greatly stretch one's muscles before attempting to lift a heavy object.
c. The overlap of actin and myosin is greater at B than at A.
d. It would be an advantage if the resting length of one's muscles was near B.
e. all of the above

22. Sam Macho was demonstrating to his friends the number of push-ups he could accomplish. After 30 push-ups he could do no more. After a short resting period he tried again. This time his friends vigorously yelled encouragement and he was able to do 35 push-ups. Which of the following statements is most appropriate?
a. Fatigue during the first attempt was primarily due to ATP depletion in muscle.
b. Fatigue during the first attempt was primarily due to "fatigue" within the central nervous system.
c. Following the depletion of creatine phosphate during the first attempt, the rate of anaerobic metabolism increased to provide a greater amount of ATP during the second attempt.
d. After a vigorous "warm up" period one can nearly always perform better.

23. Marvin Amskray broke his school record of 19 seconds in the 200-yard dash. During his record breaking run which of the following provided the greatest amount of energy for muscular contraction?

a   reactions that require oxygen
b. reactions that do not use oxygen
c. non-ATP compounds that have a phosphate group bound to them
d. a and c
e. b and c

24. In response to a homework assignment a group of students in an anatomy and physiology class had to run a quarter of a mile as fast as possible. Following the race, which of the following conditions would be expected to exist in their skeletal muscles immediately after the run?
a. elevated oxygen consumption
b. formation of creatine phosphate
c. increased production of lactic acid
d. a and b
e. all of the above

25. Which of the following conditions would one expect to find within the leg muscles of an avid long-distance bicycle rider?
a. myoglobin-poor
b. contract very quickly
c. creatine phosphate is a major supplier of energy for contraction
d. primarily anaerobic
e. none of the above

26. Aerobic exercise training can
a. increase the fatigue resistance of slow-twitch muscle fibers.
b. increase the number of myofibrils in slow-twitch fibers.
c. convert fast-twitch muscle fibers into slow-twitch muscle fibers.
d. a and b
e. all of the above

27. Body temperature following exercise is higher than body temperature before exercise because
a. the chemical reactions necessary for contraction release energy in the form of heat.
b. the chemical reactions necessary to pay back the oxygen debt release energy in the form of heat.
c. the rate of chemical reactions in contracting muscles increases.
d. all of the above

_____ 28. After contraction has occurred in smooth muscle, the calcium
   a. is destroyed by acetylcholinesterase.
   b. is chemically bound to the filaments.
   c. is pumped out of the cell by active transport.
   d. is secreted by the Golgi apparatus.
   e. combines with ATP.

_____ 29. Which of the following often have spontaneous contractions?
   a. multiunit smooth muscle
   b. visceral smooth muscle
   c. skeletal muscle
   d. a and b
   e. a and c

_____ 30. Compared to skeletal muscle, visceral smooth muscle
   a. has the same ability to be stretched.
   b. when stretched, loses the ability to contract forcefully.
   c. maintains about the same tension even when stretched.
   d. cannot maintain long, steady contractions.

✩——— **FINAL CHALLENGES** ———✩

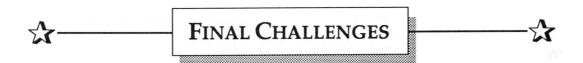

Use a separate sheet of paper to complete this section.

1. Given the characteristics listed below for a muscle tissue:
   1. multinucleated cells
   2. spindle-shaped cells
   3. striations
   4. under involuntary control
   5. cells branch

   List the minimum number of characteristics that would be sufficient to conclude that the type of muscle tissue was skeletal muscle.

2. A patient was poisoned with an unknown chemical. The symptoms and data listed below were exhibited:
   1. tetany of skeletal muscles
   2. elevated levels of acetylcholine in the circulatory system
   3. curare counteracted the tetany and caused flaccid paralysis
   4. the patient was then put on a respirator until he recovered

   The physician concluded that the toxic substance inhibited acetylcholinesterase and therefore caused tetany. Do the data support that conclusion? Explain.

3. A single muscle cell was placed between two small clamps so that it was under slight tension. A very short stimulus was applied, and the force of contraction was measured as an increase in tension. After allowing the muscle cell to recover, the process was repeated. Each time the stimulus strength was increased slightly. The results are graphed below:

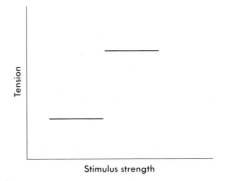

What conclusion about the response of a muscle cell to stimulation can you make on the basis of this experiment?

4. Define the all-or-none law in skeletal muscle, and provide an explanation for it.

5. A very short stimulus was applied to the nerve supplying a whole muscle. A muscle twitch was produced, and the force of contraction was measured. After allowing the muscle to recover, the procedure was repeated. Each time the stimulus strength was increased slightly. The results are graphed below:

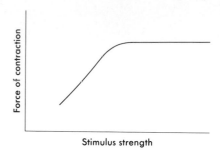

Given that the all-or-none law of skeletal muscle contraction is true, explain how the force of contraction gradually increases and then remains constant.

6. Given the following graph of a muscle twitch:

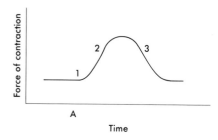

If the muscle were stimulated at time A, during which part of the graph (1, 2, or 3) is there increased calcium uptake into the sarcoplasmic reticulum? What is the name for this part of a muscle twitch?

7. Predict the response of a muscle to a supramaximal strength stimulus applied to the nerve of the muscle at a very high frequency.

8. Two different muscle preparations, A and B, were stimulated. The stimuli were applied as a single stimulus to the nerve that supplies the muscle. The force of contraction that each muscle preparation generated in response to a single stimulus of increasing strength was determined, and the results graphed below:

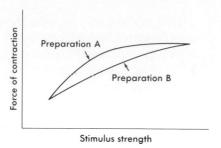

Explain why the force of contraction of muscle preparation A increased more rapidly than the force of contraction of muscle preparation B.

9. Sally Gorgeous, an avid jogger, is running down the beach when she meets Sunny Beachbum, an avid weight lifter. Sunny flirts with Sally, who decides he has more muscles than brains. She runs down the beach, but Sunny runs after her. After about a half mile Sunny tires and gives up. Explain why Sally was able to outrun Sunny (i.e., do more muscular work) despite the fact that she obviously is less muscular.

10. The following experiments were performed in an anatomy and physiology laboratory. The rate and depth of respiration for a resting student was determined. In experiment A the student ran in place for 30 seconds, immediately sat down, and relaxed and respiration rate and depth was again determined. Experiment B was just like experiment A, except that the student held her breath while running in place. What differences in respiration would you expect for the two different experiments? Explain the basis for your predictions.

ANSWERS TO CHAPTER 10

1. autorhythmic
2. myoblast; endomysium; epimysium; perimysium; myofilaments; myofibril
3. sarcolemma; sarcomere; sarcoplasm; sarcoplasmic
4. sarcoplasm; sarcoplasmic reticulum
5. sarcolemma
6. sarcoplasmic reticulum
7. anisotropic; anaerobic
8. aerobic; anaerobic
9. tetanus; tetany
10. hypertrophy; atrophy

CONTENT LEARNING ACTIVITY

Characteristics of Muscle

1. Contractility
2. Extensibility
3. Elasticity

Types of Muscle

1. Skeletal
2. Cardiac
3. Smooth

Skeletal Muscle Structures

A. 1. Perimysium
    2. Epimysium

B. 1. Fascia
    2. Epimysium
    3. Perimysium

4. Endomysium
5. Muscle fibers (cells)
6. Fasciculus

The Structure of Skeletal Muscle Fibers

1. Neuromuscular junction
2. Sarcoplasm
3. Myofibril
4. Sarcomere
5. Myofilament

Elements of a Sarcomere

1. Actin myofilaments
2. Z-line
3. I-band
4. A-band
5. H-zone
6. M-line

Myofilaments

A.
1. G-actin
2. Tropomyosin
3. Troponin
4. Myosin head
5. Cross bridge

B.
1. Actin
2. Tropomyosin
3. F-actin

4. Troponin
5. Myosin
6. Myosin head
7. I-band
8. A-band
9. Z-line
10. H-zone
11. Sarcomere

Muscle Fiber Organelles

1. T tubules
2. Sarcoplasmic reticulum

3. Terminal cisterna
4. Triad

Neuromuscular Junction

A.
1. Motor neuron
2. Synaptic cleft
3. Postsynaptic terminal (motor end plate)
4. Synaptic vesicles
5. Neurotransmitter
6. Acetylcholinesterase

B.
1. Synaptic vesicles
2. Muscle fiber
3. Synaptic cleft
4. Motor neuron

Excitation-contraction Coupling

1. T tubule
2. Sarcoplasmic reticulum
3. Troponin
4. Myosin

5. Cross bridge
6. Calcium
7. ATP
8. Tropomyosin

Muscle Twitch

1. Lag phase
2. Lag phase
3. Lag phase

4. Lag phase
5. Contraction phase
6. Relaxation phase

Stimulus Strength and Muscle Contraction

A.
1. Threshold stimulus
2. Submaximal stimulus
3. Supramaximal stimulus
4. Multiple motor unit summation

B.
1. Subthreshold stimulus
2. Threshold stimulus
3. Submaximal stimulus
4. Maximal stimulus
5. Supramaximal stimulus

Stimulus Frequency and Muscle Contraction

A. 1. Multiple wave summation
2. Complete tetanus
3. Treppe

B. 1. Multiple wave summation
2. Treppe
3. Incomplete tetanus
4. Complete tetanus

Types of Muscle Contraction and Muscle Tone

1. Isotonic
2. Asynchronous

3. Muscle tone

Length vs. Tension

1. Passive tension
2. Active tension

3. Total tension

Fatigue

A. 1. Psychological fatigue
2. Muscular fatigue
3. Synaptic fatigue

B. 1. Physiological contracture
2. Rigor mortis

Energy Sources

A. 1. Anaerobic respiration
2. Aerobic respiration
3. Aerobic respiration
4. Anaerobic respiration
5. Anaerobic respiration

B. 1. ATP
2. Creatine phosphate

3. Glycogen
4. Fatty acids
5. Lactic acid
6. Glucose
7. ATP
8. Oxygen debt

Slow and Fast Fibers

1. Slow-twitch muscle fibers
2. Fast-twitch muscle fibers
3. Slow-twitch muscle fibers

4. Slow-twitch muscle fibers
5. Fast-twitch muscle fibers

The Effects of Exercise

1. Hypertrophy
2. Anaerobic

3. Aerobic
4. Fatigue-resistant

The Structure of Smooth Muscle Fibers

1. Spindle-shaped
2. Single
3. Striated
4. Gap junctions

5. Caveolae
6. Calcium ions
7. Calmodulin

Types of Smooth Muscles

1. Multiunit smooth muscle
2. Multiunit smooth muscle

3. Visceral smooth muscle
4. Visceral smooth muscle

Properties of Smooth Muscle

1. Gradual
2. Autorhythmic
3. Sudden

4. Gradual
5. Involuntary

Cardiac Muscle

1. Intercalated disks
2. Autorhythmic

3. Involuntary
4. One

Muscle Disorders

A. 1. Spastic paralysis
   2. Flaccid paralysis

B. 1. Atrophy
   2. Muscular dystrophy
   3. Fibrositis

QUICK RECALL

1. Contractility, excitability, extensibility, and elasticity
2. Endomysium, perimysium, epimysium, and fascia
3. I-band:　a Z-line and the actin myofilaments that extend from either side of the Z-line to the ends of the myosin myofilaments

   A-band:　extends the length of the myosin myofilaments in the sarcomere

   H-zone:　only myosin myofilaments present; no overlapping actin filaments
4. Actin, tropomyosin, and calcium ions
5. a.　An action potential arrives at the presynaptic terminal of the axon of a motor neuron.

b. Ca^{2+} ion channels in the axon's cell membrane open, allowing calcium ions to diffuse into the cell.
c. Ca^{2+} ions cause the acetycholine in a few synaptic vesicles to be secreted into the synaptic cleft.
d. Acetylcholine molecules diffuse across the cleft and bind to receptor molecules of the postsynaptic terminal.
e. The membrane of the postsynaptic terminal becomes permeable to Na^+ ions.
f. A local depolarization occurs that leads to an action potential in the muscle cell.

6. a. ATP must bind to myosin before cross bridges can form.
   b. ATP is required for movement of the cross bridges.
   c. ATP is required for the cross bridges to be released.
   d. ATP is required for Ca^{2+} ion uptake into the sarcoplasmic reticulum.
7. Lag, contraction, and relaxation phase

8. For a given condition the skeletal muscle fiber contracts maximally or not at all.
9. More motor units are recruited, and motor units contain different numbers of fibers.
10. Isometric contractions and isotonic contractions
11. Psychological, muscular, and synaptic fatigue
12. ATP, fatty acids, and glucose
13. Fast-twitch and slow-twitch muscle fibers
14. Multiunit smooth and visceral smooth muscle

MASTERY LEARNING ACTIVITY

1. C. Skeletal muscle consists of long, cylindrical cells that are under voluntary control. They have many peripherally located nuclei. Smooth muscle or cardiac muscle make up the walls of hollow internal organs.

2. A. By definition the connective tissue sheaths that surround the entire muscle, muscle fasciculus, and muscle fiber are the epimysium, perimysium, and endomysium, respectively.

3. D. Whole muscles are made up of muscle bundles. In each muscle bundle are many muscle fibers (cells). Each muscle fiber is composed of many myofibrils. The myofibrils consist of the myofilaments actin and myosin.

4. C. A myofibril consists of many sarcomeres lined up end-to-end. There are many myofibrils within each muscle fiber. Each of the sarcomeres is made up of myofilaments (actin and myosin). The T tubule system is an extension of the sarcolemma (cell membrane of the muscle fiber) into the cell.

5. D. Myosin is thicker than actin and is not attached to the Z-line. Myosin is found in the A-band.

6. B. When the head of the myosin molecule binds to an active site on G-actin, a cross bridge is formed. Troponin has a binding site for Ca^{2+} ions, tropomyosin, and actin. ATPase is found on the myosin head.

7. A. The muscle cell membrane, the sarcolemma, dips into the cell to form the T tubule system. The sarcoplasmic reticulum is a specialized type of endoplasmic reticulum and is not part of the T tubule system.

8. B. At the neuromuscular junction the motor neuron produces acetylcholine that diffuses across the synaptic cleft. There is no direct contact between the neuron and the skeletal muscle cell. The acetylcholine combines with a receptor site in the skeletal muscle cell membrane. This leads to the production of an action potential in the skeletal muscle membrane. Acetylcholinesterase deactivates acetylcholine by splitting it into acetic acid and choline.

9. C. Decreased acetylcholine release would prevent action potentials in muscle cells. Thus muscles, especially respiratory muscles, would relax and not contract. All the other choices would increase the likelihood of action potentials in muscle cells.

10. C. First, an action potential is formed in the muscle cell membrane at the neuromuscular junction. The action potential moves along the membrane and down into the T tubule system. This stimulates the sarcoplasmic reticulum to release calcium. The calcium diffuses into the myofibril, initiates contraction, and the sarcomere shortens. Finally, the sarcoplasmic reticulum takes up calcium by active transport and the muscle relaxes.

11. B. Calcium enters myofibrils and combines with troponin. This causes a reaction that results in tropomyosin pulling away from (exposing) actin. Myosin then binds to actin and there is cross bridge movement.

12. E. Energy (ATP) is required for active transport of calcium ions into the sarcoplasmic reticulum so relaxation can occur. Energy (ATP) is required for the movement of actin over myosin during contraction. Due to contraction, muscle is capable of shortening with a force.

13. C. Calcium ions released from the sarcoplasmic reticulum combine with troponin during the latent phase, leading to contraction.

a is incorrect because crossbridge formation occurs during the contraction phase.

b is incorrect because active transport of calcium ions into the sarcoplasmic reticulum occurs during the relaxation phase.

d is incorrect because the sarcomere shortens during the contraction phase.

14. E. When sarcomeres shorten, the Z-line is brought closer to the end of the myosin myofilament. Thus the I-band must shorten. In addition, the ends of the actin myofilaments attached to a Z-line approach the ends of the actin myofilaments of the other Z-line in the same sarcomere. Thus the H-band must shorten.

15. A. By definition, a motor unit is a single nerve fiber and all of the muscle fibers it innervates.

16. C. A single skeletal muscle fiber responds in an all-or-none fashion. This means that, for a given condition, the fiber will respond maximally or not at all. Under favorable conditions a single skeletal muscle fiber may contract more forceably than under less favorable circumstances.

17. D. Increased frequency of action potentials results in multiple wave summation, which leads to increased release of calcium from the sarcoplasmic reticulum. As calcium concentrations increase, the number of cross bridges that form are increased. If frequency of stimulation is fast enough, tetany results.

18. B. As the stimulus strength increases, the number of motor units and therefore sarcomeres responding to the stimulus increases. The increase in tension between A and B is consistent with that relationship. After all of the sarcomeres are responding (at point B), a further increase in the stimulus strength causes no more sarcomeres to respond.

19. E. At least two hypotheses explain the phenomenon. First, the connective tissue and cytoplasm of muscles have elastic characteristics. During a single muscle twitch, some of the tension generated by the muscle is used to stretch the elastic components. During tetanic contraction the elastic component is stretched early, and then the tension generated by the muscle is applied to the load. The result is a greater measured tension. Second, the action potentials responsible for tetanic contraction may cause the release of enough calcium ion by the sarcoplasmic reticulum to result in a greater than normal concentration of calcium ion in the sarcoplasm. The greater concentration of calcium ions may make the contraction of the sarcomeres more complete.

20. A. A contraction in which the length of the muscle does not change is isometric. Since the weight did not move, the weight-lifter's limbs did not change position, and therefore neither did the length of his muscles.

21. D. If the resting length of one's muscles was near B, one's muscles would be "adjusted" to achieve the maximum amount of tension nearly all of the time.

a is incorrect. The maximum sarcomere length is at C.

b is incorrect. It would be an advantage to stretch one's muscles to B but not to C and there fore not an advantage to greatly stretch one's muscles.

c is incorrect. The greatest amount of overlap for actin and myosin occurs at A. The optimal amount of overlap for contraction occurs at B.

22. B. In an intact animal, fatigue is often the result of the central nervous system. That encouragement increased the number of push-ups that same person was able to perform is consistent with central nervous system fatigue.

23. E. For approximately a 20-second period, two nonaerobic sources provide adequate energy to maintain muscular contraction. Creatine phosphate breaks down to ATP and creatine, and anaerobic metabolism produces ATPs.

24. D. Following vigorous exercise the oxygen consumption remains elevated until the oxygen debt is "repaid." Events that occur during that process include the formation of creatine phosphate and the conversion of lactic acid into glucose.

25. E. An avid bicycle rider would certainly ride long distances. His leg muscles would probably be adapted for long-term exercise. Characteristics of such muscles are myoglobin-rich, contract relatively slowly, and depend primarily on aerobic metabolism for energy. Although creatine phosphate may be present, it does not, over long distances, supply a major proportion of the energy required for muscle contraction.

26. D. Aerobic exercise training increases the blood supply to muscle, making both slow-twitch and fast-twitch muscle fibers more resistant to fatigue. Aerobic training causes hypertrophy of slow-twitch fibers due to an increase in the number of myofibrils per cell. Exercise does not convert fast-twitch into slow-twitch fibers, or vice versa.

27. D. During exercise chemical reactions in contracting muscles increase. Since the reactions release heat, body temperature increases and will remain elevated until the excess heat is lost following exercise. In addition, the chemical reactions involved with the oxygen debt also release heat and increase body temperature.

28. C. Smooth muscle has a very poorly developed sarcoplasmic reticulum. Calcium, which initiates the contractile process, comes from the extracellular fluid. After contraction, calcium is pumped out of the cell by active transport.

29. B. Visceral smooth muscle and cardiac muscle are capable of spontaneous contraction. Skeletal muscle and multiunit smooth muscle are stimulated by the somatic and autonomic nervous systems, respectively.

30. C. Even when stretched smooth muscle can maintain the same tension. The force of contraction in skeletal muscle depends on muscle length. Compared to skeletal muscle, smooth muscle can be stretched or shorten considerably and still maintain the ability to contract forcefully. This is a useful characteristic for muscle found in hollow organs that change shape. Also, smooth muscle can maintain a steady tension for long periods of time.

 FINAL CHALLENGES

1. Given that the tissue was muscle, the presence of multinucleated cells would be sufficient to identify the tissue as skeletal muscle, since smooth muscle and cardiac muscle have one nucleus per cell. As for the other characteristics listed:

   A. Smooth muscle has spindle-shaped cells, whereas cardiac and skeletal muscle have cylindrical-shaped cells.

   B. Striations are found in skeletal and cardiac muscle but not in smooth muscle.

   C. Skeletal muscle is under voluntary control, whereas cardiac and smooth muscle are under involuntary control.

   D. Branching cells are typical only of cardiac muscle.

2. The data do support the conclusion that the toxic substance inhibited acetylcholinesterase. However, there are other conclusions that are consistent with the data. For example, if the frequency of action potentials of motor neurons were increased by the toxic substance, one would expect to see a similar response.

3. In response to a stimulus, a single muscle cell either does not contract, or it contracts with the same force. In other words, this experiment demonstrates the all-or-none law of skeletal muscle contraction.

4. The all-or-none law states that a skeletal muscle cell will either contract maximally or not at all.

   A. The all part of the all-or-none law can be explained as follows: once an action potential is generated in the muscle cell membrane, it travels down the T tubule system causing the release of calcium ions from the sarcoplasmic reticulum. The calcium initiates the contraction process. Since an action potential is of the same magnitude (for a given condition), no matter how strong the stimulus, the effect on the sarcoplasmic reticulum will be the same, and the force of contraction will be the same (maximal).

   B. The none part of the all-or-none law can also be explained on the basis of the action potential. A subthreshold stimulus produces no action potential and therefore no contraction. Once the stimulus is threshold or stronger, an action potential is generated, and contraction occurs.

5. As stimulus strength to the muscle nerve increases, the number of motor units stimulated increases (multiple motor unit summation), causing a gradual increase in the force of contraction. Once all the motor units are recruited, the force of contraction remains constant.

6. Part 1 of the graph is the lag phase, part 2 is the contraction phase, and part 3 is the relaxation phase of muscle contraction. During the relaxation phase there is increased calcium ion uptake by active transport into the sarcoplasmic reticulum. The removal of calcium ions from troponin allows tropomyosin to move between actin and myosin, and relaxation occurs.

7. A supramaximal stimulus (or a maximal stimulus) would stimulate all the motor units in a muscle, causing a muscle twitch. A high frequency of supramaximal stimuli would cause multiple wave summation and result in tetany.

8. Since the force of contraction is eventually the same for both muscle preparations it is reasonable to conclude that both preparations have the same number of muscle fibers. As stimulus strength increases, the number of motor units (and therefore the number of muscle fibers) recruited increases, resulting in a greater force of contraction. If preparation A has fewer motor units than preparation B, preparation A has more muscle fibers per motor unit than preparation B. For each motor unit recruited, the force of contraction will therefore be greater in preparation A than in preparation B, and the force of contraction will increase faster for preparation A.

9. Sally's aerobic exercise program of jogging has developed her slow-twitch muscle fibers and increased the fatigue resistance of her fast-twitch muscle fibers. Sunny's weight-lifting program consists of intense, but short, periods of exercise. This increases strength (hypertrophy of fast-twitch fibers); but, because it relies on anaerobic respiration, it does not develop aerobic, fatigue-resistant abilities. Sunny needs to do some aerobic exercises, which are included in most modern day weight-lifting programs. It is also possible that Sally was less affected by psychological fatigue.

10. In experiment A the student used anaerobic respiration as she started to run in place, but aerobic respiration also increased to meet most of her energy needs. When she stopped, respiration rate would be increased over resting levels because of repayment of the oxygen debt due to anaerobic respiration. In experiment B almost all of her energy would come from anaerobic respiration because she is holding her breath while running in place. Consequently, she would have a much larger oxygen debt. One would predict that following running in place in experiment B her respiration rate would be greater than in experiment A, or that her respiration rate would be elevated for a longer period of time in experiment A.

Muscular System: Gross Anatomy

FOCUS: Muscles cause movement by shortening. The movable end of the muscle is called the insertion, the fixed end is the origin. The force of contraction of a muscle is applied to a bone that acts as a lever across a movable joint (fulcrum). There are three classes of lever systems in the human body. A muscle that causes a particular movement such as flexing the forearm is called an agonist. Since muscles can only lengthen passively, the opposite movement (extending the forearm) is achieved by a different muscle, the anatgonist. The interaction between agonist and antagonist allows precise control of movements. Often, several muscles work together to produce a movement; they are termed synergists. The study of muscle actions can be approached by examining groups of muscles first and then individual muscles within a group.

WORD PARTS

Give an example of a new vocabulary word that contains each word part.

WORD PART	MEANING	EXAMPLE
sphin-	squeeze, strangle	1. _____
infra-	below, beneath	2. _____
hyoid	U-shaped	3. _____
pteryg-	wing	4. _____
lun-	the moon	5. _____
semi-	half	6. _____

WORD PART	MEANING	EXAMPLE
thenar	palm of hand	7. _____
peri-	around	8. _____
-ineum	discharge	9. _____
perone-	fibula	10. _____

CONTENT LEARNING ACTIVITY

General Principles

66*Most muscles extend from one bone to another and cross at least one joint.*99

Match these terms with the
correct statement or definition:

Antagonist   Origin (head)
Belly   Prime mover
Fixators   Synergists
Insertion

_____ 1. End of the muscle that moves the least.

_____ 2. Largest portion of a muscle, between its origin and insertion.

_____ 3. Muscles that work together to cause movement.

_____ 4. One muscle that plays the major role in accomplishing a desired movement.

_____ 5. Muscles that stabilize the origin of the prime mover.

Muscle Shapes

66*Muscles come in a wide variety of shapes.*99

Match the type of muscle with
the correct description:

Circular   Parallel
Convergent   Pennate

_____ 1. Fasciculi arranged around a tendon like barbs of a feather arranged around the feather shaft.

_____ 2. Fasciculi organized in direct line with the long axis of the muscle.

_____ 3. Best able to shorten, but shorten with less force than pennate muscles.

_____ 4. Acts as a sphincter to close an opening.

Nomenclature

"Muscles are named according to many different criteria."

Match the muscle nomenclature with the correct meaning:

Brachialis
Brevis
Deltoid
Latissimus
Masseter
Orbicularis

Pectoralis
Quadriceps
Rectus
Teres
Vastus

_____ 1. Breastbone

_____ 2. Arm

_____ 3. Huge

_____ 4. Short

_____ 5. Round

_____ 6. Small circle

_____ 7. Wide

_____ 8. Straight

_____ 9. Four heads

_____ 10. Chewer

Movements Accomplished by Muscles

"When muscles contract, force is applied to levers (bones), resulting in movements of the lever about a fulcrum (joint)."

Match these terms with the correct statement or definition:

Class I lever system
Class II lever system

Class III lever system

_____ 1. An example of this system is a child's seesaw; the fulcrum is between the force and the weight.

_____ 2. Weight is located between the fulcrum and the force; a wheelbarrow is an example.

_____ 3. Most common type of lever system in the body; the force is located between the fulcrum and the weight.

Head Movement

" *The muscles of the neck that are attached to the skull provide the force for a wide variety of movements.* **"**

Match the neck muscle group with
the correct description or example:

Anterior neck muscles   Posterior neck muscles
Lateral neck muscles

_____ 1. Flexion of the head is largely caused by gravity but is assisted by this
                                   muscle group.

_____ 2. Mainly responsible for the extension of the head.

_____ 3. Involved in rotation and abduction of the head.

_____

_____ 4. Sternocleidomastoid is an example.

Facial Expression

" *The skeletal muscles of the face are somewhat unique in that they are not necessarily attached to the skeleton.* **"**

Match these muscles with
the correct function:

Corrugator   Occipitofrontalis
Depressor anguli oris   Orbicularis oculi
Levator labii superioris   Orbicularis oris
Nasalis   Zygomaticus major and minor

_____ 1. Raises the eyebrows and furrows the skin of the forehead.

_____ 2. Closes the eyelids.

_____ 3. One of the kisssing muscles.

_____ 4. Smiling.

_____ 5. This muscle makes sneering possible.

_____ 6. Muscle for frowning or pouting.

Mastication

" *The muscles of mastication and the hyoid muscles move the mandible.* **"**

Match these muscles with the
correct mandibular movements:

Infrahyoid muscles   Medial pterygoid muscle
Lateral pterygoid muscle   Suprahyoid muscles
Masseter muscle   Temporalis muscle

_____ 1. Three muscles that elevate the mandible.

_____

_____

_____ 2. Muscle of mastication that depresses the mandible.

_____ 3. Group of muscles that elevate or protract the hyoid.

The tongue and buccinator muscles hold food in place between the teeth.

Tongue Movements

"_The tongue is very important in mastication and speech._**"**

Match the type of tongue muscle
with the correct description:

Extrinsic tongue muscles
Intrinsic tongue muscles

_____   1. Muscles that make up the tongue.

_____   2. Cause the shape of the tongue to change, as in rolling the tongue into
a tube.

_____   3. Muscles that attach the tongue to the mandible, hyoid and skull.

_____   4. Responsible for moving the tongue about, as in sticking out the tongue.

Swallowing and the Larynx

"_The hyoid muscles and muscles of the soft palate, pharynx, and larynx have several important functions._**"**

A. Match the type of hyoid muscle
with the correct description:

Infrahyoid muscles
Suprahyoid muscles

_____   1. Depress the mandible.

_____   2. Attach the hyoid bone to the scapula, sternum and larynx.

_____   3. Fix the hyoid bone when the mandible is depressed.

B. Match the muscle group
with its function.

Laryngeal muscles   Soft palate muscles
Pharyngeal muscles

_____   1. Close the posterior opening to the nasal cavity during swallowing.

_____   2. Open auditory tube to equalize pressure between middle ear and the
atmosphere.

_____   3. Constrict to force food into the esophagus.

_____   4. Prevent food from entering the larynx; shorten vocal cords to raise the
pitch of the voice.

Movements of the Eyeball

"The movements of the eyeball are made possible by six muscles.**"**

Match these terms with their correct location in the following statements:

Globe   Orbit
Oblique   Rectus

The eyeball muscles have their origin on the _(1)_ and insertion on the _(2)_. The four _(3)_ muscles cause the eyeball to move superiorly, inferiorly, medially, or laterally. The two _(4)_ muscles cause lateral-superior and lateral-inferior movements.

1. _____

2. _____

3. _____

4. _____

Muscles Moving the Vertebral Column

"The muscles along the spine can be divided into a deep group and a superficial group.**"**

Match the muscle group of the back with the correct description:

Deep back muscles
Superficial back muscles

_____   1. Also called the erector spinae group.

_____   2. Produce the mass of muscles seen lateral to the midline of the back; extend from the coxae and sacrum to the ribs, from the vertebrae to the ribs or from rib to rib.

_____   3. Extend from one vertebra to the next adjacent vertebra.

☞ The superficial and deep back muscles extend, abduct, and rotate the vertebral column.

Thoracic Muscles

"The muscles of the thorax are involved almost entirely in the process of breathing.**"**

Match the muscle groups with their correct functions or descriptions:

Diaphragm   Internal intercostals
External intercostals   Scalene

_____   1. Elevates the first two ribs during inspiration.

_____   2. Elevate the ribs and sternum, increasing the diameter of the thorax.

_____   3. Depress the ribs and sternum, forcing expiration.

_____   4. Responsible for the majority of volume change in the thoracic cavity during quiet breathing; dome-shaped when relaxed.

Abdominal Wall

66The muscles of the anterior abdominal wall flex and rotate the vertebral column.99

Match these terms with the
correct definition or description:

External abdominal oblique   Rectus abdominis
Internal abdominal oblique   Tendinous inscriptions
Linea alba   Transversus abdominis
Linea semilunaris

_____ 1. Tendinous area that produces a vertical line from the xiphoid process
through the navel.

_____ 2. Thick muscle, on either side of the midline, with fasciculi oriented
vertically in a straight line.

_____ 3. Connective tissue that transects the rectus abdominis, causing it to
appear segmented.

_____ 4. Most superficial lateral abdominal wall muscle.

_____ 5. Deepest lateral abdominal wall muscle.

Pelvic Floor and Perineum

*66The inferior opening of the pelvis is closed by a muscular floor through which the anus99
and urogenital openings penetrate.*

Match these terms with the
correct description or definition:

Pelvic diaphragm   Urogenital diaphragm
Perineum

_____ 1. Mostly formed by the coccygeus and levator ani muscles; forms the
pelvic floor.

_____ 2. The diamond-shaped area superficial and inferior to the pelvic floor.

_____ 3. Consists of the deep transverse perineus muscle and the sphincter
urethrae muscle.

Scapular Movements

66The major scapular muscles fix or rotate the scapula.99

Match these muscles with
the correct function:

Levator scapulae   Serratus anterior
Pectoralis minor   Trapezius
Rhomboideus major and minor

_____ 1. Elevates, depresses, rotates, and fixes the scapula.

_____ 2. Retracts and fixes the scapula.

_____ 3. Rotates and protracts the scapula.

_____ 4. Depresses the scapula.

 The major connection of the upper limb to the body is accomplished by muscles.

Arm Movements

"The arm is attached to the thorax and scapula by several muscles."

Match these muscles with
the correct description:

Coracobrachialis	Subscapularis
Deltoid	Supraspinatus
Infraspinatus	Teres major
Latissimus dorsi	Teres minor
Pectoralis major	

_____ 1. Two muscles that abduct the arm.

_____

_____ 2. Two muscles that can flex or extend the arm.

_____

_____ 3. Two muscles that attach to the trunk and adduct the arm.

_____

_____ 4. Three muscles that attach to the scapula and adduct the arm.

_____

_____

Forearm Movements

"Numerous muscles accomplish the movements of the forearm."

Match these muscles with
the correct description:

Anconeus	Pronator quadratus
Biceps brachii	Pronator teres
Brachialis	Supinator
Brachioradialis	Triceps brachii

_____ 1. Two muscles that extend the forearm.

_____

_____ 2. Three muscles that flex the forearm.

_____

_____

_____ 3. Two muscles that supinate the forearm.

_____

_____ 4. The muscle that flexes the arm.

_____ 5. The muscle that extends the arm.

Wrist, Hand, and Finger Movements

66 *The forearm muscles can be divided into anterior and posterior groups.* **99**

A. Match the muscle group
with the correct statement:

Anterior forearm muscles
Posterior forearm muscles

_____ 1. Most of these muscles originate on the medial epicondyle of the humerus; responsible for flexion of the wrist and fingers.

_____ 2. Most of these muscles originate on the lateral epicondyle of the humerus; responsible for extension of the wrist and fingers.

B. Match these terms with the
correct description or definition:

Extrinsic hand muscles   Opponens muscles
Hypothenar eminence   Retinaculum
Intrinsic hand muscles   Thenar eminence

_____ 1. Muscles located in the forearm, but with tendons that extend into the hand.

_____ 2. Strong band of fibrous connective tissue that covers the flexor and extensor tendons.

_____ 3. General term for muscles located entirely within the hand.

_____ 4. These muscles allow the tip of the little finger to touch the tip of the thumb.

_____ 5. Fleshy prominence at the base of the thumb.

Thigh Movements

66 *Muscles located in the hip and thigh are responsible for thigh movements.* **99**

A. Match the hip muscles
with the correct function:

Deep thigh rotators   Iliacus
Gluteus maximus   Psoas major
Gluteus medius   Tensor fasciae latae
Gluteus minimus

_____ 1. Two anterior hip muscles that flex the thigh.

_____ 2. Posterolateral hip muscle that extends, adducts, and laterally rotates the thigh.

_____ 3. Two posterolateral hip muscles that abduct and medially rotate the thigh.

_____ 4. Muscle that inserts on the tibia through the iliotibial tract; stabilizes the knee and abducts the thigh.

_____ 5. Group of hip muscles that laterally rotate the thigh.

B. Match the thigh muscles   Anterior thigh muscles   Posterior thigh muscles
    with their function: Medial thigh muscles

_____ 1. Thigh muscles that flex the thigh.

_____ 2. Thigh muscles that extend the thigh.

_____ 3. Thigh muscles that adduct the thigh.

Leg Movements

"_The muscles of the thigh that move the leg are contained in anterior, posterior, and medial groups._**"**

A. Match the muscle group   Anterior thigh muscles   Posterior thigh muscles
    with the correct statement: Medial thigh muscles

_____ 1. Muscle group responsible for extending the leg (except for the sartorius muscle).

_____ 2. Muscle group mostly responsible for flexing the leg.

_____ 3. Muscle group that mainly functions to adduct and flex the thigh.

_____ 4. Rectus femoris belongs to this group.

_____ 5. Biceps femoris belongs to this group.

B. Match the thigh muscle   Adductor brevis   Sartorius
    with its function: Adductor longus   Semimembranosus
                                    Adductor magnus   Semitendinosis
                                    Biceps femoris   Vastus intermedius
                                    Gracilis   Vastus lateralis
                                    Pectineus   Vastus medialis
                                    Rectus femoris

_____ 1. Four muscles that extend the leg; comprise most of the anterior thigh muscle.

_____

_____

_____ 2. Five muscles that flex the leg.

_____

_____

_____

_____

_____ 3. Five muscles that adduct the thigh; comprise the medial compartment
_____   of the thigh.

_____

_____

_____

_____ 4. Two muscles of the anterior thigh compartment that flex the thigh.

_____

_____ 5. Three muscles of the posterior thigh compartment that extend the thigh.

_____

_____

☞ The quadriceps femoris consists of the rectus femoris, vastus lateralis, vastus intermedius, and vastus medialis.

Ankle, Foot, and Toe Movements

"_Extrinsic foot muscles move the ankle, foot, and toes, whereas intrinsic foot muscles move the toes._**"**

A. Match the muscle group with the correct statement:

Anterior leg muscles   Lateral leg muscles
Intrinsic foot muscles   Posterior leg muscles

_____ 1. Acts to dorsiflex the foot and extend the toes.

_____ 2. Acts to plantar flex and invert the foot and flex the toes.

_____ 3. Acts to plantar flex and evert the foot.

_____ 4. Muscles found within the foot that flex, extend, abduct, and adduct the toes.

_____ 5. The peroneus longus belongs to this group.

B. Match the leg muscle with the correct description:

Gastrocnemius, soleus, and plantaris   Tibialis anterior and posterior
Peroneus brevis, longus and tertius

_____ 1. Muscles that invert the foot.

_____ 2. Muscles that evert the foot.

_____ 3. Muscles that join to form the calcaneal (Achilles) tendon; act to plantar flex the foot.

215

Location of Superficial Muscles

A. Match these muscles with the correct parts labeled in Figure 11-1:

Adductors of thigh
Biceps brachii
Deltoideus
External abdominal oblique
Flexors of wrist and fingers
Gastrocnemius
Orbicularis oculi
Pectoralis major
Rectus abdominis
Rectus femoris
Sartorius
Serratus anterior
Sternocleidomastoideus
Vastus lateralis
Vastus medialis

1. _____

2. _____

3. _____

4. _____

5. _____

6. _____

7. _____

8. _____

9. _____

10. _____

11. _____

12. _____

13. _____

14. _____

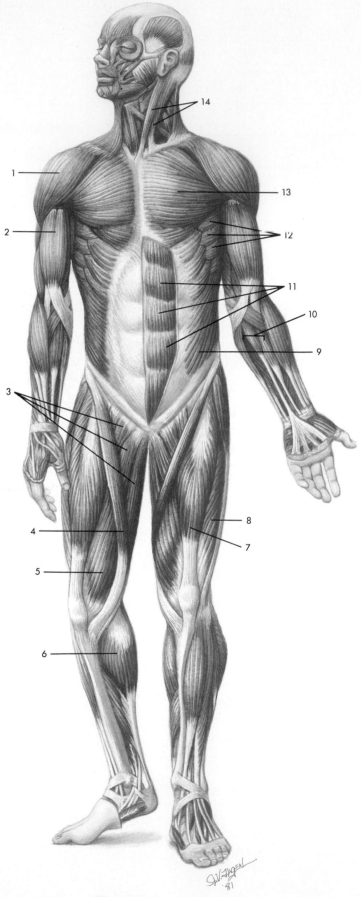

Figure 11-1

B. **M**atch these muscles with
the correct parts labeled
in Figure 11-2:

Biceps femoris
Deltoideus
Extensors of wrist and fingers
Gastrocnemius
Gluteus maximus
Infraspinatus
Latissimus dorsi
Semitendinosus
Trapezius
Triceps brachii

1. _____

2. _____

3. _____

4. _____

5. _____

6. _____

7. _____

8. _____

9. _____

10. _____

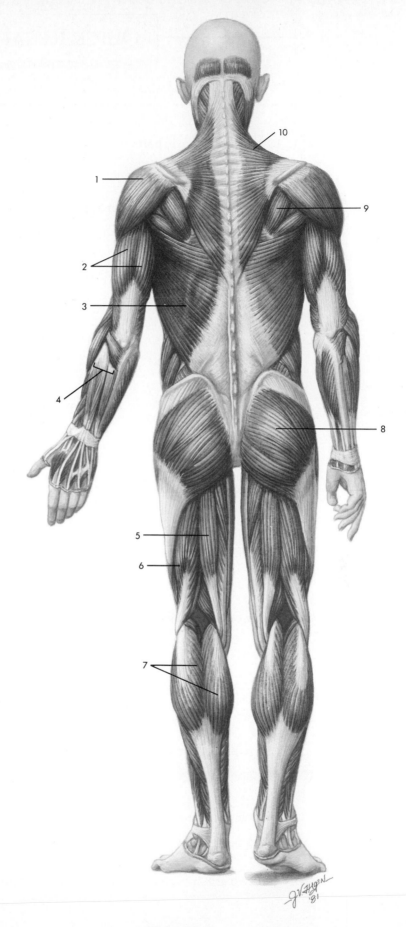

Figure 11-2

1. List the three basic parts of a muscle.

2. List four classes of muscle shape.

3. Name, and give and example of, the three classes of levers.

4. List the three groups of muscles that move the head.

5. Name the two muscles of facial expression that are circular muscles, and give their function.

6. List the muscles of mastication.

7. List the two major types of muscle groups that act on the vertebral column.

8. Name three thoracic muscles that are involved with inspiration.

9. List four muscles in the abdominal wall.

10. List two functions for the muscles that attach the scapula to the thorax.

11. Define the rotator cuff.

12. List two arm muscles that flex and two arm muscles that extend theforearm.

13. List the two major muscle groups of the forearm, and give their function.

14. Define extrinsic and intrinsic hand muscles.

15. List the three groups of hip muscles and the three groups of thigh muscles that cause movement of the thigh.

16. List the three groups of thigh muscles that cause movement of the leg.

17. List the three muscle groups of the leg that act on the foot and toes.

18. Define extrinsic and intrinsic foot muscles.

MASTERY LEARNING ACTIVITY

Place the letter corresponding to the correct answer in the space provided.

_____ 1. Muscles that oppose one another are
   a. synergists.
   b. hateful.
   c. levers.
   d. antagonists.

_____ 2. The most movable attachment of a muscle is its
   a. origin.
   b. insertion.
   c. fascia.
   d. fulcrum.

_____ 3. Which of the following muscles is correctly matched with its type of fascicle orientation? (Hint: see diagrams of the muscles).
   a. pectoralis major - pennate
   b. transversus abdominis - circular

   c. temporalis - convergent
   d. biceps femoris - parallel

_____ 4. The muscle whose name means the muscle is large and round?
   a. gluteus maximus
   b. vastus lateralis
   c. teres major
   d. latissimus dorsi
   e. adductor magnus

_____ 5. In a class three lever system the
   a. fulcrum is located between the force and the weight.
   b. weight is located between the fulcrum and the force.
   c. the force is located between the fulcrum and the weight.

6. A prominent lateral muscle of the neck that can cause flexion of the head or rotate the head toward the shoulder is the
   a. digastric.
   b. mylohyoid.
   c. sternocleidomastoid.
   d. buccinator.
   e. platysma.

7. Harry Wolf has just picked up his date for the evening. She is very attractive. Harry shows his appreciation by moving his eyebrows up and down, winking, smiling, and finally kissing her. Given the muscles listed below:
   1. zygomaticus
   2. levator labii superioris
   3. occipitofrontalis
   4. orbicularis oris
   5. orbicularis oculi

   In which order did he use the muscles?
   a. 2, 3, 4, 1
   b. 2, 5, 3, 1
   c. 2, 5, 4, 3
   d. 3, 5, 1, 4
   e. 3, 5, 2, 4

8. An aerial circus performer who supports herself only by her teeth while spinning around and around should have strong
   a. temporalis muscles.
   b. masseter muscles.
   c. buccinator muscles.
   d. a and b
   e. all of the above

9. The tongue curls and folds PRIMARILY because of the action of the
   a. intrinsic tongue muscles.
   b. extrinsic tongue muscles.

10. The infrahyoid muscles
    a. elevate the mandible.
    b. move the mandible from side to side.
    c. fix (prevent movement of) the hyoid.
    d. a and b
    e. all of the above

11. The soft palate muscles
    a. prevent food from entering the nasal cavity.
    b. force food into the esophagus.
    c. prevent food from entering the larynx.
    d. open the auditory tube to equalize pressure between the middle ear and atmosphere.

12. The erector spinae muscles can cause
    a. extension of the vertebral column.
    b. abduction of the vertebral column.
    c. rotation of the vertebral column.
    d. all of the above

13. Which of the following muscles are responsible for flexion of the vertebral column (below the neck)?
    a. deep back muscles
    b. superficial back muscles (erector spinae)
    c. rectus abdominis
    d. a and b
    e. all of the above

14. In addition to the diaphragm, which of the following muscles is involved with inspiration of air?
    a. external intercostals
    b. scalenes
    c. transversus thoracis
    d. a and b
    e. all of the above

15. Given the following muscles:
    1. external abdominal oblique
    2. internal abdominal oblique
    3. transversus abdominis

    Choose the arrangement that lists the muscles from most superficial to most deep.
    a. 1, 2, 3
    b. 1, 3, 2
    c. 2, 1, 3
    d. 2, 3, 1
    e. 3, 1, 2

16 The tendinous inscriptions
    a. attach the rectus abdominis muscles to the xiphoid process.
    b. divide the rectus abdominis muscles into subsections.
    c. separate the abdominal wall from the leg.
    d. are the site of exit of blood vessels and nerves from the abdomen into the leg.

17. Which of the following muscles is a synergist of the trapezius?
    a. rhomboids
    b. levator scapulae
    c. serratus anterior
    d. a and b
    e. all of the above

18. Which of the following muscles is a fixator (prevents abduction) of the scapula?
    a. trapezius
    b. levator scapulae
    c. pectoralis minor
    d. serratus anterior

19. Which of the following muscles adduct the arm (humerus)?
    a. latissimus dorsi
    b. teres major
    c. pectoralis major
    d. a and b
    e. all of the above

20. Which of the following muscles abduct the arm (humerus)?
    a. supraspinatus
    b. infraspinatus
    c. teres minor
    d. subscapularis
    e. all of the above

21. Which of the following muscles would one expect to be especially well developed in a boxer?
    a. biceps brachii
    b. brachialis
    c. deltoid
    d. triceps brachii
    e. supinator

22. Antagonist of the triceps brachii?
    a. biceps brachii
    b. brachialis
    c. anconeus
    d. a and b
    e. all of the above

23. The posterior group of forearm muscles is responsible for
    a. flexion of the wrist.
    b. flexion of the fingers.
    c. extension of the fingers.
    d. a and b

24. The intrinsic hand muscles that move the thumb?
    a. thenar muscles
    b. hypothenar muscles
    c. flexor pollicis longus
    d. extensor pollicis longus

25. Which of the following muscles can extend the femur?
    a. gluteus maximus
    b. gluteus minimus
    c. gluteus medius
    d. a and b
    e. all of the above

26. Given the following muscles:
    1. iliacus
    2. psoas major
    3. rectus femoris
    4. sartorius

    Which of the muscles act to flex the thigh?
    a. 1, 2
    b. 1, 3
    c. 3, 4
    d. 2, 3, 4
    e. 1, 2, 3, 4

27. Which of the following muscles is found in the medial compartment of the thigh?
    a. rectus femoris
    b. sartorius
    c. gracilis
    d. vastus medialis
    e. semitendinosus

28. Which of the following muscles would be well developed in a football player whose speciality is kicking field goals?
    a. sartorius
    b. semitendinosus
    c. gastrocnemius
    d. quadriceps femoris

29. Which of the following muscles can flex the leg?
    a. biceps femoris
    b. sartorius
    c. gastrocnemius
    d. a and b
    e. all of the above

30. Which of the following muscles would one expect to be especially well developed in a ballerina?
    a. gastrocnemius
    b. gracilis
    c. gluteus maximus
    d. quadratus lumborum
    e. triangularis

_____ 31. Which of the following muscles are responsible for everting the foot?
a. peroneus brevis
b. peroneus longus
c. peroneus tertius
d. a and b
e. all of the above

_____ 32. Which of the following muscles cause plantar flexion of the foot?
a. tibialis anterior
b. extensor digitorum longus
c. peroneus tertius
d. soleus

_____ 33. Which of the following muscles are attached to the coracoid process of the scapula?
a. pectoralis minor
b. coracobrachialis

c. biceps brachii
d. a and b
e. all of the above

_____ 34. The anterior group of forearm muscles MOSTLY have their origin on the
a. medial epicondyle of the humerus.
b. lateral epicondyle of the humerus.
c. distal portion of the radius.
d. distal portion of the ulna.

_____ 35. Which of the following muscles spans two joints?
a. gastrocnemius
b. biceps femoris
c. sartorius
d. rectus femoris
e. all of the above

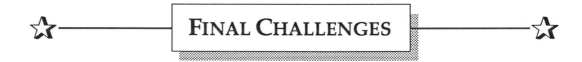

FINAL CHALLENGES

Use a separate sheet of paper to complete this section.

1. Describe the movement each of these muscles produces: quadratus lumborum, levator scapulae, latissimus dorsi, brachioradialis, gluteus minimus, semimembranosus, adductor magnus, and tibilias anterior. Name the muscle that acts as synergist and antagonist for each movement of the muscle.

2. What kind of lever system is in operation when the rectus femoris muscle moves the leg? When the semimembranosus muscle moves the leg?

3. In general, the origin of limb muscles is proximal, and the insertion is distal. What would happen if the opposite arrangement were true?

4. In Bell's palsy there is damage to the facial nerve. One of the symptoms is that the eye on the affected side remains open at all times. Explain why this happens.

5. Observe the actions of the abdominal muscles when you take a normal breath. Do they contract or relax? Why do they do this? What

do the abdominal muscles do when you expel all possible air out of your lungs? Why do they do this?

6. While trying to catch a baseball, the most distal digit of Bobby Bumblefingers index finger was forcefully flexed, tearing a muscle tendon. What muscle was damaged, and what symptoms would Bumblefingers exhibit?

7. According to an advertisement, a special type of sandal makes the calf look better because the calf muscles are exercised when the toes curl and grip the sandal while walking. Explain why you believe or disbelieve this claim.

8. A patient suffered damage to a nerve that supplies the leg muscles. Consequently her left foot was plantar flexed and slightly inverted, creating a tendency for the toes to drag when she walked. To compensate she would raise her left leg higher than normal then slap the foot down. What muscles of the leg were no longer functional?

ANSWERS TO CHAPTER 11

1. sphincter
2. infraspinatus
3. hyoid; suprahyoid; infrahyoid
4. pterygoid
5. semilunaris

6. semilunaris
7. thenar; hypothenar
8. perineum
9. perineum
10. peroneus

CONTENT LEARNING ACTIVITY

General Principles

1. Origin (head)
2. Belly
3. Synergists

4. Prime mover
5. Fixators

Muscle Shapes

1. Pennate
2. Parallel

3. Parallel
4. Circular

Nomenclature

1. Pectoralis
2. Brachialis
3. Vastus
4. Brevis
5. Teres

6. Orbicularis
7. Latissimus
8. Rectus
9. Quadriceps
10. Masseter

Movements Accomplished by Muscles

1. Class I lever system
2. Class II lever system

3. Class III lever system

Head Movement

1. Anterior neck muscles
2. Posterior neck muscles

3. Posterior neck muscles; lateral neck muscles
4. Lateral neck muscles

Facial Expression

1. Occipitofrontalis
2. Orbicularis oculi
3. Orbicularis oris

4. Zygomaticus major and minor
5. Levator labii superioris
6. Depressor anguli oris

Mastication

1. Temporalis muscle; masseter muscle; medial pterygoid muscle
2. Lateral pterygoid muscle

3. Suprahyoid muscles

Tongue Movements

1. Intrinsic tongue muscles
2. Intrinsic tongue muscles

3. Extrinsic tongue muscles
4. Extrinsic tongue muscles

Swallowing and the Larynx

A. 1. Suprahyoid muscles
    2. Infrahyoid muscles
    3. Infrahyoid muscles

B. 1. Soft palate muscles
    2. Pharyngeal muscles
    3. Pharyngeal muscles
    4. Laryngeal muscles

Movements of the Eyeball

1. Orbit
2. Globe

3. Rectus
4. Oblique

Muscles Moving the Vertebral Column

1. Superficial back muscles
2. Superficial back muscles

3. Deep back muscles

Thoracic Muscles

1. Scalene
2. External intercostals

3. Internal intercostals
4. Diaphragm

Abdominal Wall

1. Linea alba
2. Rectus abdominis
3. Tendinous inscriptions

4. External abdominal oblique
5. Transversus abdominis

Pelvic Floor and Perineum

1. Pelvic diaphragm
2. Perineum

3. Urogenital diaphragm

Scapular Movements

1. Trapezius
2. Rhomboideus major and minor

3. Serratus anterior
4. Pectoralis minor

Arm Movements

1. Deltoid; supraspinatus
2. Deltoid; pectoralis major

3. Latissimus dorsi; pectoralis major
4. Coracobrachialis; teres major; teres minor

Forearm Movements

1. Triceps brachii; anconeus
2. Brachialis; biceps brachii; brachioradialis
3. Supinator; biceps brachii

4. Biceps brachii

5. Triceps brachii

Wrist, Hand and Finger Movements

A. 1. Anterior forearm muscles
   2. Posterior forearm muscles

B. 1. Extrinsic hand muscles
   2. Retinaculum

3. Intrinsic hand muscles
4. Opponens muscles
5. Thenar eminence

Thigh Movements

A. 1. Iliacus; psoas major
   2. Gluteus maximus
   3. Gluteus medius; gluteus minimus
   4. Tensor fasciae latae
   5. Deep thigh rotators

B. 1. Anterior thigh muscles
   2. Posterior thigh muscles
   3. Medial thigh muscles

Leg Movements

A. 1. Anterior thigh muscles
   2. Posterior thigh muscles
   3. Medial thigh muscles
   4. Anterior thigh muscles
   5. Posterior thigh muscles

B. 1. Rectus femoris; vastus intermedius; vastus
      lateralis; vastus medialis

2. Biceps femoris; semimembranosus;
   semitendinosus; sartorius; gracilis
3. Adductor brevis; adductor longus; adductor
   magnus; gracilis, pectineus
4. Rectus femoris; sartorius
5. Biceps femoris; semimembranosus;
   semitendinosus

Ankle, Foot, and Toe Movements

A. 1. Anterior leg muscles
   2. Posterior leg muscles
   3. Lateral leg muscles
   4. Intrinsic foot muscles
   5. Lateral leg muscles

B. 1. Tibialis anterior and posterior
   2. Peroneus brevis, longus and tertius
   3. Gastrocnemius, soleus, and plantaris

Location of Superficial Muscles

A. 1. Deltoideus
   2. Biceps brachii
   3. Adductors of thigh
   4. Sartorius
   5. Vastus medialis
   6. Gastrocnemius
   7. Rectus femoris
   8. Vastus lateralis
   9. External abdominal oblique
   10. Flexors of wrist and fingers
   11. Rectus abdominis
   12. Serratus anterior
   13. Pectoralis major
   14. Sternocleidomastoideus

B. 1. Deltoideus
   2. Triceps brachii
   3. Latissimus dorsi
   4. Extensors of wrist and fingers
   5. Semitendinosus
   6. Biceps femoris
   7. Gastrocnemius
   8. Gluteus maximus
   9. Infraspinatus
   10. Trapezius

QUICK RECALL

1. Origin (head), insertion, and belly
2. Pennate, parallel, convergent, and circular
3. Class I lever: a child's seesaw
   Class II lever: a wheelbarrow
   Class III lever: a shovel
4. Anterior, lateral, and posterior neck muscles
5. Orbicularis oculi: closing the eyelids; orbicularis oris: puckering the mouth
6. Temporalis, masseter, and medial and lateral pterygoid
7. Superficial and deep back muscles
8. External intercostals, internal intercostals, scalene, serratus posterior, transversus thoracis, and diaphragm
9. Rectus abdominis, external oblique, internal oblique, and transversus abdominis
10. Elevation, depression, retraction, protraction or fixation of the scapula
11. Four muscles that bind the humerus to the scapula and form a cuff or cap over the proximal humerus

12. Flex forearm: biceps brachii, brachialis, and brachioradialis
    Extend forearm: triceps brachii and anconeus
13. Anterior forearm: Flex fingers, thumb, and wrist
    Posterior forearm: Extend fingers, thumb, and wrist; abduct of adduct wrist
14. Extrinsic hand muscles are in the forearm, but have tendons that extend into the hand.
    Intrinsic hand muscles are entirely within the hand.
15. Hip muscles: Anterior; posterior and lateral; and deep thigh rotators
    Thigh muscles: Anterior, posterior, and medial
16. Anterior, posterior, and medial
17. Anterior, posterior, and lateral compartment
18. Extrinsic foot muscles are in the leg, but have tendons that extend into the foot.
    Intrinsic foot muscles are located entirely in the foot.

1. D. By definition two or more muscles that oppose one another are antagonists. Muscles that aid each other are synergists.

2. B. By definition the most movable attachment of a muscle is the insertion. The origin is the "fixed" end of the muscle.

3. C. The temporalis has a broad origin with fascicles that converge to insert on the mandible. The pectoralis major is convergent, the transversus abdominis is parallel, and the biceps femoris is pennate.

4. C. Teres means round, and major means larger.

5. C. The most common type of lever system in the body is a class III lever system in which the force is located between the fulcrum and the weight.

6. C. The muscles and their functions are:

Muscle	Action
Sternocleidomastoid	Flex and rotate head
Digastric	Depress mandible, elevate hyoid
Mylohyoid	Depress mandible, elevate hyoid
Buccinator	Flatten cheek
Platysma	Depress lower lip, wrinkle skin of neck

7. D. The muscles and their functions are:

Muscle	Function
Occipitofrontalis	Move eyebrows
Orbicularis oculi	Winking
Zygomaticus	Smiling
Orbicularis oris	Kissing
Levator labii superioris	Sneering

8. D. The temporalis and masseter muscles close the jaw. The buccinator flattens the cheeks, acting to keep food between the teeth.

9. A. The intrinsic tongue muscles curl and fold the tongue, whereas the extrinsic tongue muscles move the tongue about as a unit.

10. C. The infrahyoid muscles fix the hyoid during mandibular depression. The suprahyoid muscles depress the mandible. The medial and lateral pterygoid muscles move the mandible from side to side.

11. A. The soft palate muscles prevent food from entering the nasal cavity. The pharyngeal muscles open the auditory tube and force food into the esophagus. The laryngeal muscles prevent food from entering the larynx.

12. D. The erector spinae and the deep back muscles can cause extension, abduction, and rotation of the vertebral column.

13. C. The back muscles do not flex the vertebral column; that is accomplished by the abdominal muscles (rectus abdominis, external abdominal oblique, and internal abdominal oblique).

14. D. The external intercostals elevate the ribs, and the scalenes elevate the first and second ribs. This increases thoracic volume and results in inspiration. The transverse abdominis depresses the ribs.

15. A. From superficial to deep, the muscles of the lateral abdominal wall are the external abdominal oblique, internal abdominal oblique, and transversus abdominis.

16. B. The tendinous inscriptions subdivide the rectus abdominis, giving the abdomen a segmented appearance.

17. E. The trapezius and rhomboids abduct the scapula, the trapezius and levator scapulae elevate the scapula, and the trapezius and serratus anterior rotate (upward) the scapula.

18. A. The trapezius and rhomboids fix the scapula.

19. E. The latissimus dorsi and pectoralis major attach to the trunk and cause adduction of the arm. The teres major attaches to the scapula and causes adduction of the arm.

20. A. The supraspinatus and deltoid abduct the arm.

21. D. Extension of the forearm during a punch is critical to a boxer: the triceps brachii performs that function.

22. D. The triceps brachii extends the forearm. Its antagonists (biceps brachii and brachialis) flex the forearm. The anconeus is a synergist of the triceps brachii and helps to extend the forearm.

23. C. The posterior group of forearm muscles is responsible for extension of the wrist and fingers, whereas the anterior group of forearm muscles is responsible for flexion of the wrist and fingers.

24. A. The thenar muscles are intrinsic (located with the hand) muscles that move the thumb. The hypothenar muscles are intrinsic hand muscles that move the little finger. The flexor pollicis longus and extensor pollicis longus both move the thumb, but they are extrinsic (located in the forearm) hand muscles.

25. A. The gluteus maximus extends, adducts, and laterally rotates the thigh. The gluteus minimus and medius abduct and medially rotate the thigh.

26. E. The iliacus and psoas major (the iliopsoas) are hip muscles that flex the thigh. The rectus femoris and sartorius are thigh muscles that flex the thigh.

27. C. The gracilis is found in the medial compartment; the semitendinosus in the posterior compartment; and the rectus femoris, vastus medialis, and sartorius in the anterior compartment.

28. D. The quadriceps femoris is composed of four muscles: the rectus femoris, vastus intermedius, vastus medialis and vastus lateralis. These muscles are responsible for extending the leg. The rectus femoris also flexes the thigh.

29. E. The biceps femoris (posterior thigh), sartorius (anterior thigh), and gastrocnemius (posterior leg) all flex the leg.

30. A. Since a ballerina often dances on her toes, plantar flexion of the foot is necessary. The gastrocnemius muscles perform that function.

31. E. All of the peroneal muscles can cause eversion of the foot.

32. D. The soleus and gastrocnemius are the two muscles primarily responsible for plantar flexion of the foot. The other muscles listed cause dorsiflexion of the foot.

33. E. The pectoralis minor inserts on the coracoid process of the scapula, and the coracobrachialis and biceps brachii (short head) originate on the scapular coracoid process.

34. A. The anterior forearm muscles mostly originate on the medial humeral epicondyle, whereas the posterior forearm muscles mostly originate on the lateral humeral epicondyle.

35. E. The biceps femoris, sartorius, and rectus femoris extend from the coxa to the tibia (across the hip and knee joint). The gastrocnemius extends from the femur to calcaneus (across the knee and ankle joint).

1.

Muscle	Action	Synergist	Antagonist
Quadratus lumborum	Abducts vertebral column	Erector spinae same side	Erector spinae same side; quadratus lumborum opposite side
Levator scapulae	Elevates scapula	Trapezius	Pectoralis major
Latissimus dorsi	Adducts, extends, and medially rotates arm	Teres major; pectoralis major	Pectoralis major (flexes arm)
Brachioradialis	Flexes forearm	Brachialis; biceps brachii	Anconeus; triceps brachii
Gluteus minimus	Abducts and medially rotates thigh	Gluteus medius	Gluteus maximus
Semimembranousus	Flexes leg and extends thigh	Semitendinousus	Rectus femoris
Adductor magnus	Adducts, extends, and laterally rotates thigh	Gluteus maximus	Gluteus minimus
Tibialis anterior	Dorsiflexes foot	Peroneus tertius	Tibialis posterior; gastrocnemius; soleus

2. The foot is the weight part of the lever system, and the knee joint is the fulcrum. The rectus femoris (the force) inserts on the tibial tuberosity between the weight and the fulcrum. Therefore it is a class III lever system.

The semimembranosus inserts on the lateral condyle of the tibia; thus the force it generates is also between the weight and fulcrum, forming a class III lever system.

3. Limb movement would still be possible; but, when the muscles contracted, they would cause movement proximal to the muscle and the joint. This means the muscle would have to move (lift) itself, as well as part of the limb. In addition, we would all have limbs like Popeye the sailor!

4. Without stimulation by means of the facial nerve, the orbicularis oculi muscle is completely relaxed (flaccid paralysis), and the eye remains open.

5. During inspiration the abdominal muscles relax, allowing the abdominal organs to make room for the diaphragm as it descends inferiorly. This increases thoracic volume and the amount of air inspired. During forced expiration, the abdominal muscles push the abdominal organs and diaphragm superiorly, forcing more air out of the lungs.

6. Since the digit was forcefully flexed, the tendon of the muscle that opposes flexion would be stretched and torn (i.e.; the extensor digitorum). Damage to this muscle would make him unable to extend the last phalanx of the index finger. In fact, unopposed contraction of the flexor digitorum profundus would keep the last digit of the index finger in a flexed position.

7. Disbelieve this advertisment. The muscles of the leg that are responsible for flexing the toes include the flexor hallucis longus and flexor digitorum longus. These muscles are small compared to the gastrocnemius and soleus muscles, which make up the bulk of the leg muscles. In addition, the toe flexor muscles are deep to the gastrocnemius and soleus, and enlargement of the toe flexor muscles due to exercise would not have a noticeable affect on the appearance of the leg.

8. The muscles of the posterior compartment of the leg that cause plantar flexion and inversion of the foot are still functioning; but their antagonist, the muscles of the anterior and lateral compartments of the leg that cause dorsiflexion and eversion of the foot, are not functioning. Consequently the foot is plantar flexed and inverted.

Functional Organization Of Nervous Tissue

FOCUS: The nervous system can be divided into the central nervous system (brain and spinal cord) and the peripheral nervous system (nerves and ganglia). Cells of the nervous system are neurons (multipolar, bipolar, or unipolar) and glial cells (astrocytes, ependymal cells, microglia, oligodendrocytes, and Schwann cells). The neurons produce and conduct action potentials, whereas the glial cells support and protect the neurons. Neurons connect with each other by means of synapses, and action potentials can be transferred in only one direction across the synapse. Neurotransmitters are chemicals released by presynaptic neurons that cross the synapse and either excite or inhibit the postsynaptic neuron. The amount of neurotransmitter released can be increased through spatial or temporal summation. Neurons are arranged to form reflexes, convergent circuits, divergent circuits, and oscillating circuits.

WORD PARTS

Give an example of a new vocabulary word that contains each word part.

WORD PART	MEANING	EXAMPLE
af- (ad-)	to	1. _____
ef-	out	2. _____
fer- (pher-)	to bear	3. _____
gli-	glue	4. _____
dendr-	a tree	5. _____
telo-	an end	6. _____

WORD PART	MEANING	EXAMPLE
gemm-	a bud	7. _____
astr-	a star	8. _____
endym-	a garment	9. _____
salt-	to leap or dance	10. _____

CONTENT LEARNING ACTIVITY

Divisions of the Nervous System

❝*Each subdivision has structural and functional features that separate it from the other subdivisions.*❞

Match these terms with the
correct statement or definition:

Afferent division
Autonomic nervous system
Central nervous system
Efferent division

Parasympathetic nervous system
Peripheral nervous system
Somatic nervous system

_____ 1. Consists of the brain and spinal cord.

_____ 2. Consists of all the nerves and ganglia; located outside the CNS.

_____ 3. Subdivision of the PNS that transmits action potentials from sensory organs to the CNS.

_____ 4. Subdivision of the efferent division that transmits impulses from the CNS to skeletal muscle.

_____ 5. Subdivision of the efferent division that transmits impulses from the CNS to smooth muscle, cardiac muscle, and glands.

_____ 6. Autonomic subdivision that regulates vegetative functions.

☞ The nerves of the PNS can be subdivided into 12 pairs of cranial nerves, which originate in the brain, and 31 pairs of spinal nerves, which originate in the spinal cord.

Neurons

❝*Neurons, or nerve cells, receive stimuli and transmit action potentials to other neurons or effector organs.*❞

A. Match these terms with the
correct statement or definition:

Axon
Dendrite

Neuron cell body (soma)

_____ 1. Portion of a neuron that contains the nucleus and other organelles such as rough endoplasmic reticulum.

_____ 2. Short, often highly branched cytoplasmic extension that is tapered from the neuron cell body to its tip.

_____ 3. Structures that are specialized to conduct action potentials toward the presynaptic terminals.

B. Match these terms with the correct statement or definition:

Axolemma   Nissl bodies
Axon hillock   Telodendria
Axoplasm   Terminal boutons
Dendritic spines (gemmules)   (presynaptic terminals)

_____ 1. Areas of rough endoplasmic reticulum found in the cytoplasm of the neuron cell body and base of dendrites; location of most protein synthesis in a neuron.

_____ 2. Small extensions from dendrites; the location of synapses with axons of other neurons.

_____ 3. Enlarged area of each neuron cell body from which the axon arises.

_____ 4. Cytoplasm of the axon.

_____ 5. Branching ends of axons.

_____ 6. Enlarged ends of telodendria; contain vesicles with neurotransmitters.

C. Match these terms with the correct parts of the diagram labeled in Figure 12-1:

Axon
Axon hillock
Dendrite
Neuron cell body (soma)
Node of Ranvier
Schwann cell
Telodendria

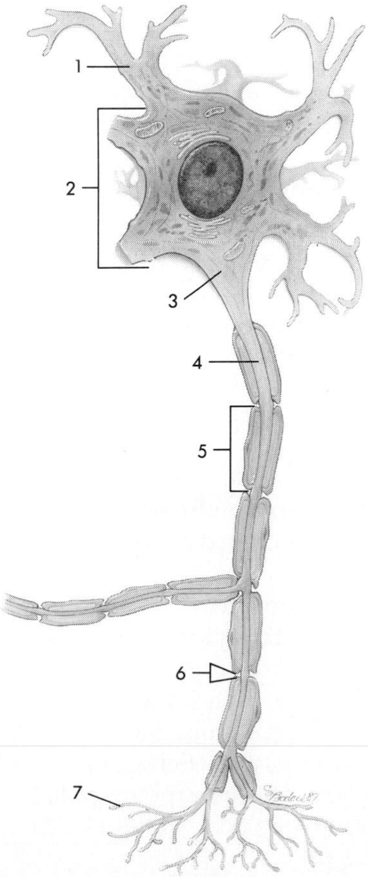

1. _____

2. _____

3. _____

4. _____

5. _____

6. _____

7. _____

Figure 12-1

233

Neuron Types

" *Three categories of neurons exist based on their shape.* **"**

Match these terms with the
correct statement or definition:

Bipolar neuron   Unipolar neuron
Multipolar neuron

_____ 1. Neuron with several dendritic processes and a single axon.

_____ 2. Neuron with a single dendrite and a single axon.

_____ 3. Most neurons in the CNS, including motor neurons, are of this type.

_____ 4. Most sensory neurons are of this type.

Neuroglia

" *There are five types of neuroglia, each with unique structural and functional characteristics.* **"**

A. Match these types of neuroglial
cells with the correct statement
or definition:

Astrocytes   Oligodendrocytes
Ependymal cells   Satellite cells
Microglia   Schwann cells

_____ 1. Neuroglia that are star-shaped; their cell processes and the endothelium of the blood vessels form the blood-brain barrier.

_____ 2. Neuroglia that line the ventricles of the brain; specialized within choroid plexuses to secrete cerebrospinal fluid.

_____ 3. Small cells that become mobile and phagocytic in response to inflammation.

_____ 4. Neuroglia with cytoplasmic extensions that form myelin sheaths around axons in the CNS.

_____ 5. Neuroglial cells that form myelin sheaths around axons in the PNS.

Axon Sheaths

" *Axons, surrounded by sheaths, are structurally similar in both the PNS and CNS.* **"**

Match these terms with their correct location in the following statements:

1. _____

Internodes   Nodes of Ranvier
Myelin sheath   Saltatory conduction
Myelinated axons   Unmyelinated axons

2. _____

3. _____

 (1) rest in an invagination of the oligodendrocytes or the
Schwann cells, but the axons remain outside the cytoplasm of the
cells that ensheathe them. Action potentials are propagated along
the entire axon membrane. In (2) of the PNS, each Schwann cell
ensheathes only one axon and repeatedly wraps around a
segment of an axon to form the (3) . Gaps between Schwann
cells are called (4) , whereas the areas between the gaps are called
 (5) . Conduction of action potentials from one node of Ranvier to
another in myelinated neurons is called (6) .

4. _____

5. _____

6. _____

Organization of Nervous Tissue

66 *Nervous tissue is organized so that axons form bundles, and nerve cell bodies and their relatively* 99
short dendrites are organized into groups.

A. Match these terms with the correct statement or definition:

Endoneurium	Nerve fascicle
Epineurium	Nerve tract
Ganglia	Nuclei
Gray matter	Perineurium
Nerve	White matter

_____ 1. Bundles of parallel axons with their associated sheaths that appear whitish in color.

_____ 2. Conduction pathway formed by white matter of the CNS.

_____ 3. Bundle of axons and their sheaths in the PNS.

_____ 4. Collection of nerve cell bodies in either the CNS or PNS that appears gray in color.

_____ 5. Collections of gray matter within the brain.

_____ 6. Heavier connective tissue layer surrounding groups of peripheral nerve axons and sheaths.

_____ 7. Group of axons surrounded by the perineurium.

_____ 8. Layer of connective tissue binding nerve fascicles together to form a nerve.

B. Match these terms with the correct parts of the diagram labeled in Figure 12-2:

Axon
Endoneurium
Epineurium
Myelin sheath
Nerve
Nerve fasicle
Perineurium

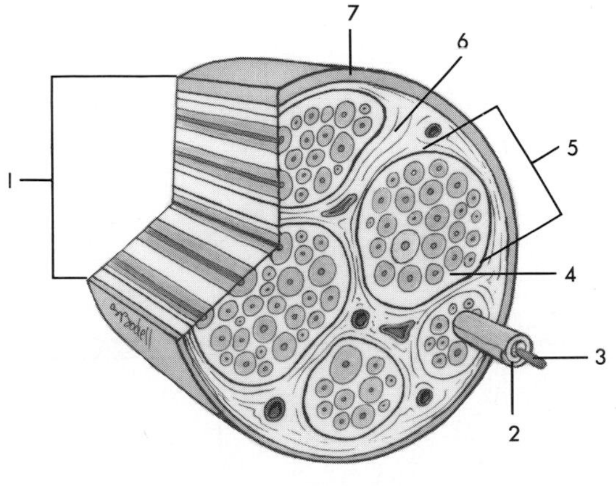

1. _____

2. _____

3. _____

4. _____

5. _____

6. _____

7. _____

Figure 12-2

The Synapse

"The synapse is the membrane-to-membrane contact between one nerve cell and: another nerve cell,**"** an effector such as a muscle or gland, or a sensory receptor cell.

Match these terms with the
correct statement or definition:

Acetylcholinesterase	Presynaptic terminal
Neuromodulator	Synaptic cleft
Neurotransmitter	Synaptic vesicle
Postsynaptic terminal	

_____ 1. End of a nerve cell that contains synaptic vesicles.

_____ 2. Space separating two neurons at a synapse.

_____ 3. Membrane at this location contains receptor sites for neurotransmitters.

_____ 4. Membrane-bound organelle that contains neurotransmitters.

_____ 5. Chemical that influences the sensitivity of neurons to neurotransmitters, but neither stimulates nor inhibits the postsynaptic terminal.

EPSP's and IPSP's

"The combination of neurotransmitters with their specific receptor causes either an EPSP or an IPSP.**"**

A. Match these terms with the
correct statement or definition:

Excitatory neuron	Inhibitory neuron
Excitatory postsynaptic potential (EPSP)	Inhibitory postsynaptic potential (IPSP)

_____ 1. Result of depolarization of the postsynaptic terminal.

_____ 2. Caused by an increase in permeability of the cell membrane, primarily to Na^+ ions.

_____ 3. Result of hyperpolarization of the postsynaptic terminal.

_____ 4. Caused by an increase in permeability of the cell membrane to only K^+ and Cl^- ions.

_____ 5. Neuron that causes EPSP's.

B. Match these terms with the
correct parts of the diagram
labeled in Figure 12-3:

EPSP
IPSP

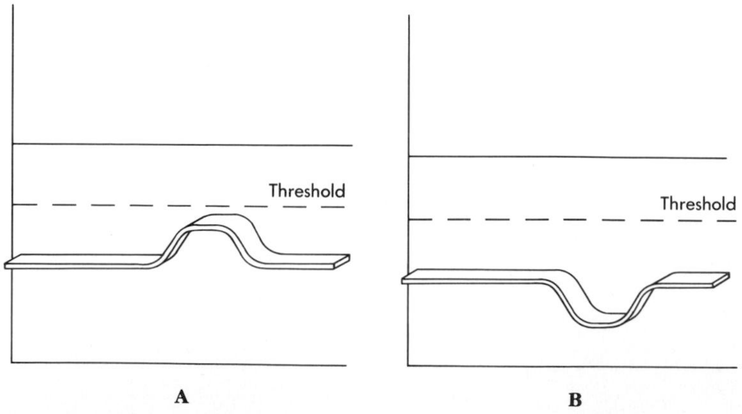

Figure 12-3

_____ 1. The change in membrane potential seen in graph A.

_____ 2. The change in membrane potential seen in graph B.

The Synapse and Integration

"The synapse is an essential structure for the process of integration carried out by the CNS."

Match these terms with their correct location in the following statements:

IPSPs   Temporal
Local potentials   Threshold
Spatial

A series of presynaptic action potentials causes a series of _(1)_ in the postsynaptic neuron. These combine in a process called summation at the axon hillock of the postsynaptic neuron. If this combination of local potentials exceeds _(2)_ at the axon hillock, an action potential is produced. _(3)_ summation results when two action potentials arrive simultaneously at two different presynaptic terminals that synapse with the same postsynaptic neuron. _(4)_ summation occurs when two action potentials arrive in very close succession at a single presynaptic terminal. Excitatory and inhibitory neurons may synapse with a single postsynaptic neuron. Summation occurs in the postsynaptic neuron, and the _(5)_ tend to cancel the EPSP's.

1. _____

2. _____

3. _____

4. _____

5. _____

Reflexes

"The reflex arc, or reflex, is the basic functional unit of the nervous system."

A. Match these terms with their correct location in the following statements:

Afferent neurons   Efferent neurons
Association neurons   Sensory receptors
Effector organ

In a reflex arc, stimuli are detected by _(1)_ , causing the production of action potentials that are carried to the central nervous system by _(2)_ . Within the central nervous system, afferent neurons usually synapse with _(3)_ . These neurons synapse with _(4)_ , which carry action potentials to the _(5)_ .

1. _____

2. _____

3. _____

4. _____

5. _____

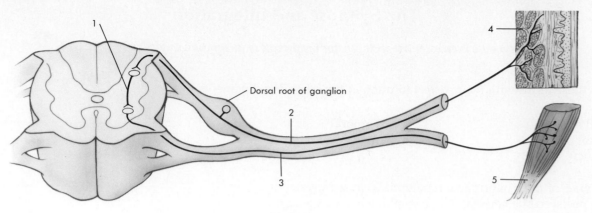

Dorsal root of ganglion

Figure 12-4

B. Match these terms with the
correct parts of the diagram
labeled in Figure 12-4:

Afferent neuron
Association neuron
Effector organ
Efferent neuron
Sensory receptor

1. _____
2. _____
3. _____
4. _____
5. _____

C. Match these terms with their correct location in the
following statements:

Do
Do not

Reflexes _(1)_ require conscious thought, and they _(2)_ result in
consistent and predictable responses. Reflexes _(3)_ function to
maintain homeostasis and so they _(4)_ operate in isolation within
the nervous system.

1. _____
2. _____
3. _____
4. _____

Neuronal Circuits

"Neurons are organized within the CNS to form circuits."

Match these terms with the correct statement or definition:

After discharge
Convergent circuit

Divergent circuit
Oscillating circuit

_____ 1. Circuit type with many neurons that come together and synapse with a smaller number of neurons.

_____ 2. Innervation of motor neurons in the spinal cord is an example.

_____ 3. Afferent neurons carrying action potentials from pain receptors are an example.

_____ 4. Circuits with neurons arranged in a circular fashion.

_____ 5. More than one action potential is produced by a neuron in response to a circular arrangement of neurons.

_____ 6. Type of circuit responsible for an activity that occurs periodically such as respiration.

QUICK RECALL

1. **Name two divisions of the efferent division of the nervous system. List two important characteristics of each.**

2. **List three types of neurons based on their shape. Give an example of each type.**

3. **List five types of neuroglia. Give a function for each type.**

4. **Contrast structural and functional characteristics of myelinated and unmyelinated axons.**

5. List two differences between white matter and gray matter. Give an example of each in the CNS and the PNS.

6. Name three types of connective tissue coverings associated with nerves.

7. List three components of the synapse.

8. Distinguish between an EPSP and an IPSP.

9. List two types of summation and distinguish between them.

10. List the five components of a reflex arc.

11. Describe three types of neuronal circuits.

MASTERY LEARNING ACTIVITY

Place the letter corresponding to the correct answer in the space provided.

_____ 1. The peripheral nervous system includes the
  a. somatic nervous system.
  b. brain.
  c. spinal cord.
  d. nuclei.

_____ 2. The branch of the nervous system that controls smooth muscle, cardiac muscle and glands is the
  a. somatic nervous system.
  b. autonomic nervous system.
  c. skeletal division.
  d. sensory nervous system.

_____ 3. Neurons have cytoplasmic extensions that connect one neuron to another neuron. Given the following structures:
  1. axon
  2. dendrite
  3. gemmule
  4. terminal bouton

Choose the arrangement that lists the structures in the order they would be found at the synapse between two neurons.
a. 1, 4, 2, 3
b. 1, 4, 3, 2
c. 4, 1, 2, 3
d. 4, 1, 3, 2

_____ 4. A neuron with many short dendrites and a single long axon is a
a. multipolar neuron.
b. unipolar neuron.
c. bipolar neuron.
d. none of the above

_____ 5. Most sensory neurons are _____ neurons.
a. unipolar
b. bipolar
c. multipolar
d. efferent
e. a and b

_____ 6. Substances that pass from the blood vessels of the central nervous system to neurons must pass through a cellular membrane derived from which of the following cell types?
a. astrocytes
b. oligodendrocytes
c. microglia
d. all of the above
e. a and b

_____ 7. Neuroglia that act as phagocytes (are phagocytic) within the central nervous system are
a. oligodendrocytes.
b. microglia.
c. ependymal cells.
d. astrocytes.
e. somas.

_____ 8. Axons within nerves may have which of the following associated with them?
a. Schwann cells
b. nodes of Ranvier
c. oligodendrocytes
d. a and b
e. all of the above

_____ 9. Action potentials are conducted more rapidly
a. in small-diameter axons than in large-diameter axons.
b. in unmyelinated axons than in myelinated axons.
c. along axons that have nodes of Ranvier.
d. all of the above

_____ 10. Clusters of nerve cell bodies within the peripheral nervous system are called
a. ganglia.
b. fasicles.
c. nuclei.
d. laminae.

_____ 11. Gray matter contains primarily
a. myelinated fibers.
b. neuron cell bodies.
c. Schwann cells.
d. all of the above

_____ 12. Given the following connective tissue structures:
1. endoneurium
2. epineurium
3. perineurium

Choose the arrangement that lists the structures in order from the outside of a nerve to the inside.
a. 1, 2, 3
b. 1, 3, 2
c. 2, 1, 3
d. 2, 3, 1
e. 3, 2, 1

_____ 13. Neurotransmitter substances are stored in vesicles that are located primarily in specialized portions of the
a. soma.
b. axon.
c. dendrite.
d. none of the above

_____ 14. Within a synapse,
a. synaptic vesicles are found in both the presynaptic and the postsynaptic terminals.
b. action potentials are conducted only from presynaptic to postsynaptic membranes.
c. action potentials are conducted only from postsynaptic to presynaptic membranes.
d. a and b

15. Given the following list of information concerning synaptic transmission:
    1. presynaptic action potentials
    2. local depolarization
    3. release of neurotransmitter
    4. either metabolism or reabsorption of neurotransmitter
    5. combination of neurotransmitter with receptors
    6. postsynaptic action potentials

    Choose the sequence listed below that best represents the order in which the events occur:
    a. 1, 3, 2, 4, 5, 6
    b. 1, 3, 5, 2, 6, 4
    c. 4, 3, 5, 2, 1, 6
    d. 1, 3, 5, 6, 2, 4
    e. 1, 3, 2, 5, 6, 4

16. An inhibitory neuron affects the neuron it synapses with by
    a. producing an IPSP in the neuron.
    b. hyperpolarizing the cell membrane of the neuron.
    c. causing K^+ ions to diffuse out of the neuron.
    d. causing Cl^- ions to diffuse into the neuron.
    e. all of the above

17. Summation
    a. is due to a combining of several local potentials.
    b. can occur when two action potentials arrive simultaneously at two different presynaptic terminals.
    c. can occur when two action potentials arrive in very close succession at a single presynaptic terminal.
    d. all of the above

18. Which of the following structures is a component of the reflex arc?
    a. afferent neuron
    b. sensory receptor
    c. efferent neuron
    d. effector organ
    e. all of the above

19. A convergent circuit
    a. is a positive feedback system that produces many action potentials.
    b. makes possible a reflex response and awareness of the reflex stimulus.
    c. occurs when many neuron synapse with a smaller number of neurons.
    d. all of the above

20. The output of a convergent circuit could be
    a. an IPSP.
    b. an EPSP.
    c. an action potential.
    d. no action potential.
    e. all of the above

Use a separate sheet of paper to complete this section.

1. Describe the consequences of the following situation in a neuronal circuit. Several neurons converge on a single neuron. One of the presynaptic neurons exhibits a constant frequency of action potentials and secretes a stimulatory neuromodulator. The other presynaptic neurons secrete stimulatory neuro-transmitters. What effect does a drug that blocks the receptor molecules for the neuromodulator have on the response of the postsynaptic neuron to stimulation by the presynaptic neurons?

2. Given two series of neurons, explain why action potentials could be propagated along one series more rapidly than the other series.

3. In the diagram below, A and B represent different neuronal circuits. For each 10 times neuron 1 is stimulated, neuron 2 has 10 action potentials and neuron 3 has one action potential. Propose an explanation for the different number of action potentials produced in neurons 2 and 3.

4. The Hering-Breur reflex limits overexpansion of the lungs during inspiration. Given that an oscillating circuit is responsible for the continual stimulation of respiratory muscles during inspiration and that stretch receptors in the lungs produce action potentials when the lungs are stretched, draw a simple neuronal circuit that would account for the Hering-Breur reflex.

5. Nerve fibers that conduct the sensation of sharp, pricking, localized pain are myelinated fibers with a diameter of 2 to 6 micrometers. Nerve fibers that conduct the sensation of burning, aching, not well-localized pain are unmyelinated with a diameter of 0.5 to 2 micrometers. Describe the sensations produced when a painful stimuli is applied. Explain why there are two different systems for transmitting pain.

ANSWERS TO CHAPTER 12

WORD PARTS

1. afferent
2. efferent
3. afferent; efferent
4. glial; neuroglia; microglia
5. oligodendrocytes; telodendria

6. telodendria
7. gemmules
8. astrocytes
9. ependymal
10. saltatory

CONTENT LEARNING ACTIVITY

Divisions of the Nervous System

1. Central nervous system
2. Peripheral nervous system
3. Afferent division

4. Somatic nervous system
5. Autonomic nervous system
6. Parasympathetic nervous system

Neurons

A. 1. Neuron cell body (soma)
2. Dendrite
3. Axon

B. 1. Nissl bodies
2. Dendritic spines (gemmules)
3. Axon hillock
4. Axoplasm
5. Telodendria
6. Terminal boutons (presynaptic terminals)

C. 1. Dendrite
2. Neuron cell body (soma)
3. Axon hillock
4. Axon
5. Schwann cell
6. Node of Ranvier
7. Telodendria

Neuron Types

1. Multipolar neuron
2. Bipolar neuron

3. Multipolar neuron
4. Unipolar neuron

Neuroglia

1. Astrocytes
2. Ependymal cells
3. Microglia

4. Oligodendrocytes
5. Schwann cells

Axon Sheaths

1. Unmyelinated axons
2. Myelinated axons
3. Myelin sheath

4. Nodes of Ranvier
5. Internodes
6. Saltatory conduction

Organization of Nervous Tissue

A. 1. White matter
2. Nerve tract
3. Nerve
4. Gray matter
5. Nuclei
6. Perineurium
7. Nerve fascicles
8. Epineurium

B. 1. Nerve
2. Myelin sheath
3. Axon
4. Endoneurium
5. Nerve fascicle
6. Perineurium
7. Epineurium

The Synapse

1. Presynaptic terminal
2. Synaptic cleft
3. Postsynaptic terminal

4. Synpatic vesicle
5. Neuromodulator

EPSPs and IPSPs

A. 1. Excitatory postsynaptic potential (EPSP)
2. Excitatory postsynaptic potential (EPSP)
3. Inhibitory postsynaptic potential (IPSP)
4. Inhibitory postsynaptic potential (IPSP)
5. Excitatory neuron

B. 1. EPSP
2. IPSP

The Synapse and Integration

1. Local potentials
2. Threshold
3. Spatial

4. Temporal
5. IPSPs

Reflexes

A. 1. Sensory receptor
2. Afferent neuron
3. Association neuron
4. Efferent neuron
5. Effector organ

3. Efferent neuron
4. Sensory receptor
5. Effector organ

C. 1. Do not
2. Do
3. Do
4. Do not

B. 1. Association neuron
2. Afferent neuron

Neuronal Circuits

1. Convergent circuit
2. Convergent circuit
3. Divergent circuit

4. Oscillating circuit
5. After discharge
6. Oscillating circuit

QUICK RECALL

1. Somatic nervous system: innervates skeletal muscle, under voluntary control; Autonomic nervous system: innervates cardiac muscle, smooth muscle, and glands, and is involuntary.
2. Multipolar neuron - motor neurons
   Bipolar neuron - rod and cone cells
   Unipolar neuron - sensory neurons
3. Astrocytes: form the blood-brain barrier
   Microglia: macrophages that phagocytize foreign or necrotic tissue
   Ependymal cells: produce cerebrospinal fluid
   Schwann cells: form myelin sheaths around axons in PNS
   Satellite cells: support and nourish neuron cell bodies in ganglia
4. Myelinated neuron: wrapped by several layers of oligodendrocytes or Schwann cells, saltatory conduction (rapid)
   Unmyelinated neuron: rests in invagination of oligodendrocytes or Schwann cells, conducts action potentials slowly
5. White matter: myelinated axons, propagates action potentials, forms nerve tracts and nerves

Gray matter: neuron cell bodies, site of integration (synapses), forms nuclei and ganglia
6. Endoncurium, perineurium, and epineurium
7. Presynaptic terminal, synaptic cleft, and post synaptic terminal
8. EPSP: depolarization of postsynaptic terminal due to increase in membrane permeability to sodium and potassium ions
   IPSP: hyperpolarization of the postsynaptic terminal due to an increase in membrane permeability to chlorine or potassium ions
9. Spatial summation occurs when two or more presynaptic terminals simultaneously stimulate a postsynaptic terminal.
   Temporal summation occurs when two or more action potentials arrive in succession at a single presynaptic terminal
10. Sensory receptor, afferent neuron, association neuron, efferent neuron, effector organ
11. Convergent, divergent, and oscillating circuits

MASTERY LEARNING ACTIVITY

1. A. The nervous system is divided into the central nervous system (brain and spinal cord) and the peripheral nervous system, which includes the afferent (sensory) and efferent (somatic and autonomic) divisions.

2. B. The autonomic nervous system regulates the activities of smooth muscle, cardiac muscle, and glands. The somatic nervous system controls skeletal muscle.

3. B. Axons extend toward other neurons, and the ends of the axons are the terminal boutons (presynaptic terminals). The terminal boutons are separated by the synaptic cleft from gemmules (postsynaptic terminals) on the dendrites of the neuron.

4. A. Multipolar neurons have many dendrites and a single axon. Bipolar neurons have a single dendrite and a single axon. Unipolar neurons have a single axon and no dendrites.

5. E. Efferent (away from the central nervous system) neurons are multipolar motor neurons. Afferent (toward the central nervous system) neurons are sensory neurons. Most sensory neurons are unipolar, although a few are bipolar. Association neurons are multipolar and are found only within the central nervous system.

6. A. Astrocytes have cytoplasmic extensions that form a barrier between blood vessels and neurons. The barrier is thought to be the "blood-brain" barrier. Therefore substances that

pass from capillaries of the central nervous system to neurons must pass through the cellular membrane derived from the astrocytes.

7. B. Microglia are neuroglia that are phagocytic within the central nervous system.

8. D. Nerves are part of the peripheral nervous system. Nerves contain axons that are surrounded by Schwann cells. In myelinated axons the Schwann cells give rise to the myelin sheath. The spaces between adjacent Schwann cells are called nodes of Ranvier. Oligodendrocytes are neuroglia cells found only within the central nervous system, where they form myelin sheaths around axons.

9. C. Saltatory conduction from one node of Ranvier to the next node of Ranvier is the most rapid method of conducting action potentials. Myelinated axons are faster than unmyelinated axons, and large-diameter axons are faster than small-diameter axons.

10. A. Clusters of nerve cell bodies within the peripheral nervous system are called ganglia. Clusters of nerve cell bodies within the central nervous system are called nuclei.

11. B. Gray matter consists of unmyelinated neuron cell bodies, dendrites, and neuroglia cells. Myelinated neurons make up white matter. In the central nervous system myelin is produced by oligodendrocytes, and in the peripheral nervous system by Schwann cells.

12. D. Nerves are surrounded by the epineurium. Nerve fascicles within the nerve are surrounded by the perineurium. Axons within the nerve fascicles are surrounded by the endoneurium.

13. B. The presynaptic terminal that contains the neurotransmitter within synaptic vesicles is a specialized part of the axon. Release of neurotransmitter from the axon only ensures one way transmission of nerve impulses.

14. C. Since only presynaptic terminals have synaptic vesicles with neurotransmitter, the action potential is always transferred from the presynaptic membrane to the postsynaptic membrane.

15. B. Synaptic transmission must involve (1) an action potential in the presynaptic neuron that (3) initiates the release of a neurotransmitter, which then diffuses across the synaptic cleft and (5) combines with receptors of the postsynaptic membrane. The combination of the receptor with neurotransmitter initiates (2) a local depolarization, which (6) initiates postsynaptic action potentials if it is sufficiently large. Although metabolism or reabsorption of the neurotransmitter (4) probably begins as soon as the transmitter is released, it does not become a factor until after the postsynaptic response has occurred.

16. E. The inhibitory neuron releases a neurotransmitter that changes the permeability of the neuron with which it synapses to K^+ and Cl^- ions. Movement of K^+ ions out of the neuron and Cl^- ions into the neuron cause hyperpolarization of the neuron cell membrane. Since this increases the difference between the resting membrane potential and threshold, an IPSP (inhibitory postsynaptic potential) is produced.

17. D. Summation occurs when local potentials combine to reach threshold and produce an action potential. Spatial summation occurs when simultaneous stimulation from two or more presynaptic terminals cause the local potentials to combine. Temporal summation is due to two or more successive stimulations from a single presynaptic terminal.

18. E. The following components make up the reflex arc: sensory receptor, afferent neuron, association neuron (absent in some reflex arcs), efferent neuron, and effector organ.

19. C. A convergent circuit is the synapsing of many neurons with a smaller number of neurons. A divergent circuit occurs when a few neurons synapse with a larger number of neurons. Input to a reflex and to the brain is an example. Oscillating circuits are positive feedback systems that produce many action potentials.

20. E. The simplest example of a convergent circuit is two neurons synapsing with a third neuron. If one of the two were an inhibitory neuron, an IPSP would be produced. If one of the two were an excitatory neuron, an EPSP would be produced. If both neurons were simultaneously stimulated there could be two possible outcomes. If the IPSP effects dominate, the third neuron's membrane would hyperpolarize and there would be no action potential. If the EPSP effects dominate, the third neuron's membrane would depolarize to threshold level and an action potential would result.

1. The continual release of the stimulatory neuro-modulator makes the postsynaptic neuron more sensitive to the neurotransmitter. Blocking the receptors for the neuromodulator would result in less sensitivity to the neurotransmitter, making it more difficult to produce an action potential.

2. If one series of neurons had more neurons, it would have more synapses, which would slow down the rate of action potential propagation. Or, if one series were unmyelinated and the other myelinated, or if one series had smaller-diameter axons, the rate would also be slower.

3. Suppose that the neurons in circuit B require temporal summation. Thus, for an action potential to be passed from neuron to neuron in circuit B, several action potentials would be necessary to cause the release of adequate amounts of neurotransmitter. The requirement for temporal summation could be due to small amounts of neurotransmitter release, reduced numbers of neurotransmitter receptors, or neuromodulators in circuit B.

4. The oscillating circuit diagrammed below provides continual stimulation of the respiratory muscles until it is inhibited by an inhibitory neuron activated when the lungs overstretch.

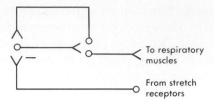

5. The large-diameter, myelinated axons transmit action potentials more rapidly (6 to 30 m/second) than the small-diameter, unmyelinated fibers (0.5 to 2 m/second). Therefore a painful stimulus produces two kinds of sensation: a sharp, localized, pricking pain followed by a burning, aching pain that is not as well localized. The fast pain conducting fibers allow us to be quickly aware of pain, where the pain is coming from, and quickly respond to new pain stimuli. For example, the hand can be rapidly removed from a hot object. However, the fast conducting pain fibers take up space because of their large diameter and associated myelin sheath. The slow conducting fibers take up less space and allow us to be aware of chronic pain for which a quick response is not necessary.

A. Match the structure with th
   part of the brainstem in
   which it is located:

_____

_____

_____

_____

_____

_____

 The reticular f

B. Match these parts of the br
   with the correct function:

_____

_____

_____

_____

_____

Central Nervous System

FOCUS: The central nervous system consists of the brain and spinal cord. The brain can be subdivided into the the brainstem (medulla oblongata, pons, and midbrain), diencephalon (thalamus, hypothalamus, and others), cerebrum, and cerebellum. The brainstem contains important centers for regulating circulatory and respiratory functions, contains the nuclei for most of the cranial nerves, and functions as a tract system to connect the spinal cord to different areas of the brain. The diencephalon functions as a relay center between the brainstem and cerebrum, controls the activities of the pituitary gland, and regulates autonomic functions. The cerebrum is the site of conscious sensation, thoughts, and control of body movements, whereas the cerebellum is involved with balance and smooth, coordinated movements. The brain contains hollow spaces, the ventricles, which are filled with cerebrospinal fluid. The cerebrospinal fluid circulates and enters the meninges, membranes which surround and protect the brain. The spinal cord consists of neuron cell bodies (gray matter organized into horns) and ascending and descending nerve tracts formed by neuron axons (white matter). Each tract carries specific information such as fine touch (medial lemniscal system), pain and temperature (lateral spinothalamic tract), and voluntary motor activity (lateral corticospinal tract). Reflexes are stereotyped responses to stimuli that are integrated within the spinal cord and brain. Important reflexes include the stretch reflex (stretching a muscles results in muscle contraction), the Golgi tendon reflex (excessive tension in a muscle causes the muscle to relax), and the withdrawal reflex (a limb is removed from a painful stimulus).

WORD PARTS

Give an example of a new vocabulary word that contains each word part.

WORD PART	MEANING	EXAMPLE
medull-	under; beneath	1. _____
pon-	a bridge	2. _____
cort-	bark; shell	3. _____
collicul-	a little hill	4. _____

249

tegmen-

pedunc-

gangli-

corp-

menin-

arachn-

_____

A. Match these terms w
correct statement or

_____

_____

_____

_____

B. Match these terms
correct statement or

_____

_____

_____

C. **M**atch these terms with the correct parts of the diagram labeled in Figure 13-1:

Inferior colliculus
Medulla oblongata
Midbrain
Pineal body
Pons
Superior colliculus
Thalamus

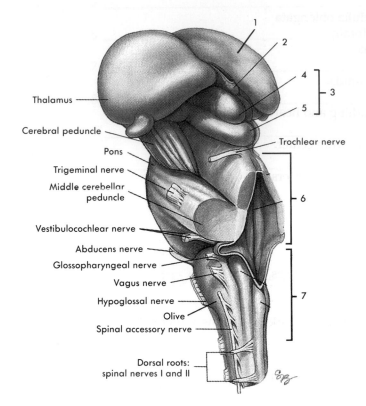

Thalamus

Cerebral peduncle

Pons

Trigeminal nerve

Middle cerebellar peduncle

Vestibulocochlear nerve

Abducens nerve

Glossopharyngeal nerve

Vagus nerve

Hypoglossal nerve

Olive

Spinal accessory nerve

Dorsal roots: spinal nerves I and II

Trochlear nerve

1. _____

2. _____

3. _____

4. _____

5. _____

6. _____

7. _____

Figure 13-1

Diencephalon

❝_The diencephalon is the part of the brain between the brainstem and the cerebrum._**❞**

A. **M**atch these parts of the diencephalon with the correct descriptions or functions:

Epithalamus
Hypothalamus

Subthalamus
Thalamus

_____ 1. Most sensory input projects to this point.

_____ 2. Involved in controlling motor functions.

_____ 3. Contains habenular nuclei, which are responsible for emotional and visceral responses to odors.

_____ 4. Contains the pineal body, which plays a role in the onset of puberty.

_____ 5. Contains mamillary bodies, which are involved in olfactory reflexes.

_____ 6. Regulates the secretion of hormones by the pituitary gland.

_____ 7. Involved in control of many autonomic functions such as movement of food through the digestive tract.

_____ 8. Contains hunger and thirst control centers.

B. Match these terms with the correct parts of the diagram labeled in Figure 13-2:

Corpus callosum
Hypothalamus
Intermediate mass
Mamillary body
Midbrain
Optic chiasma

Pineal body (epithalamus)
Pituitary gland
Pons
Subthalamus
Thalamus

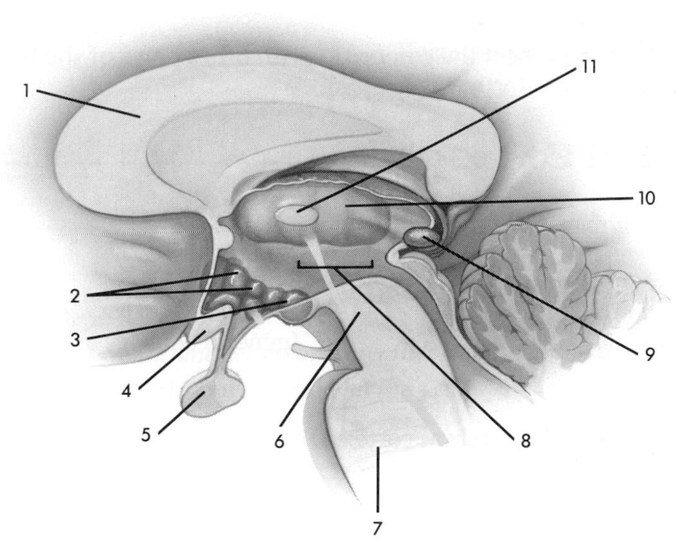

1. _____
2. _____
3. _____
4. _____
5. _____
6. _____
7. _____
8. _____
9. _____
10. _____
11. _____

Figure 13-2

Cerebrum

❝*The cerebrum is the largest portion of the brain.*❞

A. Match these terms with the correct statement or definition:

Central sulcus
Gyri
Lateral fissure

Longitudinal fissure
Sulci

_____ 1. Deep groove that separates the right and left cerebral hemispheres.

_____ 2. Raised folds or ridges on the surface of the cerebrum.

_____ 3. Sulcus that separates the frontal and parietal lobes.

_____ 4. Deep groove that separates the temporal lobe from the rest of the cerebrum.

B. Match these lobes of the cerebrum with the correct primary function:

Frontal lobe
Occipital lobe

Parietal lobe
Temporal lobe

_____ 1. Voluntary motor function, motivation, aggression, and mood.

_____ 2. Evaluation of most sensory input (excluding smell, hearing and vision).

_____ 3. Reception and integration of visual input.

_____ 4. Reception and integration of olfactory and auditory input.

_____ 5. Abstract thought and judgment ("psychic cortex").

Memory

❝Memory can be classified in several ways.**❞**

Match these terms with the correct location in the following statements:

Calpain   Memory engrams
Cerebellum   Procedural memory
Declarative memory   Sensory memory
Hippocampus and amygdala   Short-term memory
Long-term memory

1. _____

2. _____

3. _____

4. _____

5. _____

6. _____

7. _____

8. _____

9. _____

 (1) is the very short-term retention of sensory input while something is scanned, evaluated, and acted upon. However, data may be considered valuable enough to move into (2) , where information is retained for a few seconds to a few minutes. Certain pieces of information are transferred into (3) , which may involve a physical change in neuron shape. In the latter process a calcium influx into the cell activates an enzyme called (4) , which aids in the change of dendrite shape. A whole series of neurons, called (5) , are probably involved in the long-term retention of a given piece of information. There are two types of long-term memory. (6) involves the retention of names, dates, and places, whereas (7) involves the learning of a skill such as riding a bicycle. Declarative memory is stored in the (8) , whereas procedural memory is stored in the (9) .

Right and Left Hemispheres

❝Language and perhaps some other functions are not shared equally between the two hemispheres.**❞**

Match these terms with the correct statement or definition:

Corpus callosum   Right cerebral hemisphere
Left cerebral hemisphere

_____ 1. Controls muscular activity in and receives sensory input from the right half of the body.

_____ 2. In most people the analytical hemisphere involved in mathematics and speech.

_____ 3. In most people the aesthetic hemisphere involved in spatial perception and musical ability.

_____ 4. The largest commissure between the right and left hemisphere.

Basal Ganglia

❝The basal ganglia are a group of functionally related nuclei in the inferior cerebrum, diencephalon, and midbrain.**❞**

Match these terms with the correct statement or definition:

Caudate nucleus   Substantia nigra
Corpus striatum   Subthalamic nucleus
Lentiform nucleus

_____ 1. Located in the diencephalon.

_____ 2. Located in the midbrain.

_____ 3. Collectively, the nuclei in the cerebrum.

_____ 4. Two nuclei that are part of the corpus striatum.

_____

☞ The basal ganglia are involved in coordinating motor movements and posture.

Limbic System

❝_Portions of the cerebrum and diencephalon are grouped together under the title limbic system._**❞**

Match these terms with the correct location in the following statements:

Basal ganglia   Hypothalamus
Fornix   Olfactory cortex
Habenular nuclei   Olfactory nerves
Hippocampus

Structurally, the limbic system consists of many parts. These parts consist of certain cerebral cortical areas, including the cingulate gyrus and (1) ; various nuclei, including (2) in the epithalamus; parts of the (3) ; the (4) , especially the mamillary bodies; the (5) , associated with the sense of smell, and tracts such as the (6) connecting these areas. One of the major sources of sensory input into the limbic system is the (7) .

1. _____

2. _____

3. _____

4. _____

5. _____

6. _____

7. _____

☞ Emotions, the visceral responses to these emotions, motivation, mood, and sensations of pain and pleasure are all influenced by the limbic system.

Cerebellum

❝_The major function of the cerebellum is motor coordination._**❞**

A. Match these terms with the   Cerebellar peduncles   Lateral hemispheres
   correct statement or definition: Flocculonodular lobe   Vermis

_____ 1. Nerve tracts that communicate with other areas of the CNS.

_____ 2. Small, inferior portion of the cerebellum involved in balance.

_____ 3. Narrow central portion of the cerebellum.

_____ 4. Paired structures involved in fine motor coordination.

☞ The cerebellum is organized like the cerebrum, with gray matter on the inside as nuclei and outside as cortex.

B. Match these terms with the correct location in the following statements:

1. _____

Cerebellum   Motor cortex
Comparator   Proprioceptors

2. _____

3. _____

Impulses that initiate voluntary movements are sent from the motor cortex to skeletal muscles and to the _(1)_ . At the same time, _(2)_ send information about body position to the cerebellum. If a difference is detected between the intended movement and the actual movement, impulses are sent from the cerebellum to the _(3)_ to correct the difference. Thus a major function of the cerebellum is that of a _(4)_ , resulting in smooth, coordinated movements.

4. _____

Spinal Cord

❝*The spinal cord is extremely important to the overall function of the nervous system.***❞**

A. Match these terms with the correct statement or definition:

Cauda equina   Filum terminale
Cervical enlargement Medullary cone
Lumbar enlargement

_____ 1. Region where nerves that supply the arms enter and exit the spinal cord.

_____ 2. Inferior end of the spinal cord (extends to L2).

_____ 3. Connective tissue filament that anchors the inferior end of the spinal cord to the coccyx.

_____ 4. Medullary cone and numerous nerves extending inferiorly from it.

B. Match these terms with the correct statement or definition:

Anterior (ventral) horns   Gray and white commissures
Dorsal root   Lateral horn
Dorsal root ganglion   Posterior (dorsal) horn
Fasciculi   Ventral root

_____ 1. Columns of white matter in the spinal cord; consist of funiculi, or tracts.

_____ 2. Axons of many sensory neurons synapse with neurons in this structure.

_____ 3. Contains the cell bodies of autonomic neurons.

_____ 4. Allows communication between the right and left halves of the spinal cord.

_____ 5. Structure that conveys efferent nerve processes away from the cord.

_____ 6. Contains the cell bodies of sensory neurons.

C. Match these terms with the correct parts labeled in Figure 13-3:

Anterior (ventral) horn   Lateral horn
Dorsal root   Posterior (dorsal) horn
Dorsal root ganglion   Ventral root
Funiculi

1. _____

2. _____

3. _____

4. _____

5. _____

6. _____

7. _____

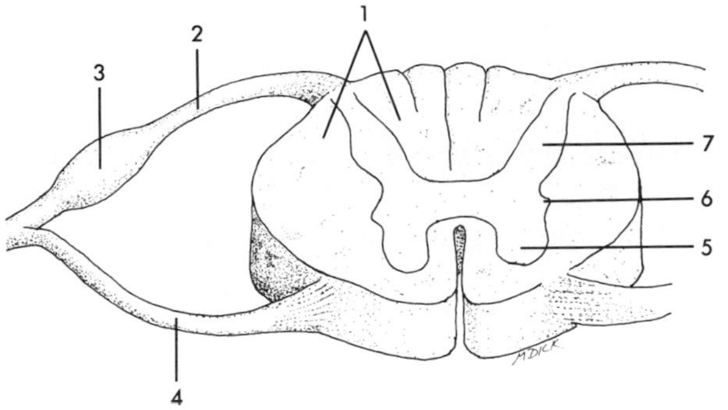

Figure 13-3

Spinal Reflexes

66 *Automatic reactions to stimuli that occur without conscious thought are reflexes.* **99**

A. Match these terms with the correct description or function:

Crossed extensor reflex   Stretch reflex
Golgi tendon reflex   Withdrawal reflex
Reciprocal innervation

_____

_____

_____

_____

1. When too much tension is produced in a muscle, this reflex causes the muscle to relax.

2. Removes a limb from a painful stimulus.

3. Causes the antagonist of a muscle to relax.

4. Prevents falling when the withdrawal reflex occurs in one leg.

B. Match these terms with the correct statement or definition:

Alpha motor neuron   Golgi tendon organ
Gamma motor neuron   Muscle spindle

_____

_____

_____

_____

1. Sensory receptor for the stretch reflex.

2. These innervate the muscle in which a muscle spindle is embedded.

3. Impulses carried by these cause muscle spindles to contract.

4. Stimulation of these causes inhibition of alpha motor neurons.

Ascending Spinal Pathways

"*There are several major ascending tracts in the spinal cord.***"**

A. Match these spinal pathways
   with the correct function:

    Anterior spinothalamic   Spino-olivary
    Lateral spinothalamic   Spinoreticular
    Medial lemniscal   Spinotectal
    Spinocerebellar

_____ 1. Pain and temperature from most of the body.

_____ 2. Light touch, pressure, tickle, and itch.

_____ 3. Fine touch, proprioception, pressure and vibration.

_____ 4. Proprioception to the cerebellum.

_____ 5. Contributes information to coordination of movement associated with balance.

☞ The medial lemniscal system can be divided into the fasiculus gracilis (inferior to midthorax) and the fasiculus cuneatus (superior to midthorax). Pain and temperature from the face and teeth are conducted by the trigeminothalamic tract.

B. Match the nerve tract with the
   location where it crosses over
   (decussates) from one side of
   the body to the other:

    Pons
    Medulla oblongata
    Spinal cord
    Uncrossed

_____ 1. Lateral spinothalamic tract.

_____ 2. Anterior spinothalamic tract.

_____ 3. Medial lemniscal system.

_____ 4. Posterior spinocerebellar tract.

_____ 5. Anterior spinocerebellar tract (crosses twice).

_____

C. Match the type of neuron with
   the correct location of the
   neuron cell body:

    Primary neuron
    Secondary neuron
    Tertiary neuron

_____ 1. Dorsal root ganglion.

_____ 2. Spinal cord.

_____ 3. Medulla oblongata.

_____ 4. Thalamus.

Descending Spinal Pathways

66 *Most of the descending pathways are involved in the control of motor functions.* **99**

A. Match these terms with the
correct statement or definition:

Extrapyramidal system
Pyramidal system

_____
1. Involved in muscle tone and fine movements such as dexterity.

_____
2. Involved in less precise control of motor functions, especially ones
associated with overall body coordination and cerebellar function.

_____
3. Includes the corticospinal and corticobulbar tracts.

_____
4. Includes the rubrospinal, vestibulospinal, and reticulospinal tracts.

B. Match the nerve tract with
the correct function:

Corticobulbar tract   Rubrospinal tract
Corticospinal tract   Vestibulospinal tract
Reticulospinal tract

_____
1. Primarily, direct control of upper limb movement.

_____
2. Direct control of head and neck movement.

_____
3. Cerebellar-like function in control of distal hand and arm movements.

_____
4. Primarily involved in maintenance of upright posture.

☞ The descending pathways reduce the transmission of pain impulses by secreting
endorphins (natural analgesics).

C. Match the nerve tract with the
location where it crosses over
(decussates) from one side
of the body to the other:

Medulla oblongata
Spinal cord

_____
1. Lateral corticospinal tract.

_____
2. Anterior corticospinal tract.

D. Match the type of neuron with
the correct location of the
neuron cell body:

Lower motor neuron
Upper motor neuron

_____
1. Cerebral cortex, cerebellum, and brainstem.

_____
2. Spinal cord.

_____
3. Cranial nerve nuclei.

Menges

❝_Three connective tissue layers surround and protect the spinal cord._**❞**

A. **M**atch these terms with the correct statement or definition:

Arachnoid layer	Epidural space
Denticulate ligaments	Pia mater
Dura mater	Subarachnoid space

_____ 1. Most superficial and thickest meningeal layer; continuous with the periosteum of the cranial vault.

_____ 2. Divides to form the dural sinuses.

_____ 3. Space that separates the dura mater from the periosteum of the vertebral canal.

_____ 4. Middle, wispy meningeal layer.

_____ 5. Space between the arachnoid layer and the pia mater.

_____ 6. Connective tissue strands that hold the spinal cord in place in the vertebral canal.

B. **M**atch these terms with the correct parts of the diagram labeled in Figure 13-4:

Arachnoid granulations   Pia mater
Arachnoid layer   Subarachnoid space
Dura mater   Superior sagittal sinus

1. _____

2. _____

3. _____

4. _____

5. _____

6. _____

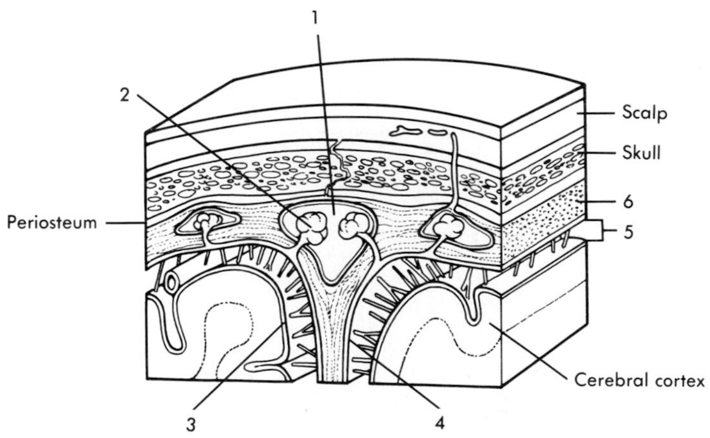

Figure 13-4

Ventricles

❝_The CNS is a hollow tube that may be quite reduced in some areas and expanded in others._**❞**

A. **M**atch these terms with the correct statement or definition:

Central canal	Lateral ventricle
Cerebral aqueduct	Septa pellucida
Fourth ventricle	Third ventricle
Interventricular foramen	

_____ 1. Separates the lateral ventricles from each other.

_____ 2. Midline cavity located between the lobes of the thalamus.

_____ 3. Connects the lateral ventricles to the third ventricle.

_____ 4. Cavity located in the inferior pontine and superior medullary regions.

_____ 5. Connects the third and fourth ventricles.

_____ 6. Continuation of the fourth ventricle into the spinal cord.

B. Match these terms with the correct parts of the diagram labeled in Figure 13-5:

Arachnoid granulation
Central canal
Cerebral aqueduct
Choroid plexus
Fourth ventricle

Interventricular foramen
Lateral ventricles
Subarachnoid space
Superior sagittal sinus
Third ventricle

1. _____

2. _____

3. _____

4. _____

5. _____

6. _____

7. _____

8. _____

9. _____

10. _____

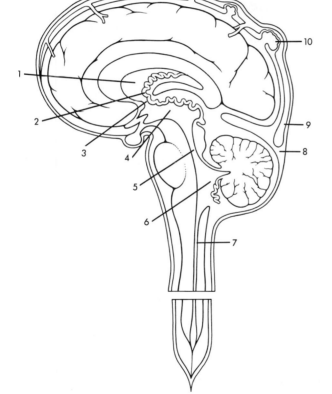

Figure 13-5

Cerebrospinal Fluid

66 *Cerebrospinal fluid provides a protective cushion around the CNS and provides some nourishment to CNS tissue.* 99

Match these terms with the
correct statement or definition:

Arachnoid granulations   Lateral and median foramina
Cerebrospinal fluid   Subarachnoid space
Choroid plexuses   Venous sinus

_____ 1. Fills the ventricles and subarachnoid space of the brain and spinal cord.

_____ 2. Site of production of cerebrospinal fluid, which consists of ependymal cells, their support tissue, and associated blood vessels.

_____ 3. Cerebrospinal fluid passes from the fourth ventricle and enters this.

_____ 4. Allow cerebrospinal fluid to pass from the fourth ventricle into the sub arachnoid space.

_____ 5. Space filled with blood; point where cerebrospinal fluid reenters the bloodstream.

_____ 6. Cerebrospinal fluid passes into venous sinuses through these.

Clinical Applications

66 *There are many different causes of central nervous system disorders.* 99

Match these terms with the
correct statement or definition:

Dyskinesias   Reye's syndrome
Encephalitis   Rabies
Meningitis   Stroke
Multiple sclerosis

_____ 1. Includes cerebral palsy, Huntington's chorea, and Parkinson's disease.

_____ 2. Results from demyelination of the brain and spinal cord.

_____ 3. May develop in children following an influenza or chickenpox infection.

_____ 4. Inflammation of the brain caused by a virus, or sometimes a bacteria.

_____ 5. Hemorrhage, thrombosis, embolism, or vasospasm of cerebral blood vessels that result in an infarct.

1. Complete the following table:

STRUCTURE	FUNCTION
Cerebral cortex	
Cerebral medulla	
Basal ganglia	
Thalamus	
Hypothalamus	
Limbic system	
Midbrain	
Medulla oblongata	
Pons	
Cerebellum	

2. Name the three parts of the brainstem.

3. List four major parts of the diencephalon.

4. List the four largest lobes of the cerebrum, and list their major functions.

5. List the three types of nerve tracts found in the cerebral medulla.

6. List three major reflexes.

7. Complete the following table for Ascending Spinal Pathways:

SPINAL PATHWAY	FUNCTION
Spinothalamic system	
Medial lemniscal system	
Spinocerebellar system	

8. Distinguish between the pyramidal and extrapyramidal systems.

9. List the three meninges and two spaces associated with them.

10. List the four ventricles of the brain and the openings that connect them to each other.

MASTERY LEARNING ACTIVITY

Place the letter corresponding to the correct answer in the space provided.

_____ 1. If a section was made that separated the brainstem from the rest of the brain, the cut would be made between the
a. medulla and pons.
b. pons and midbrain.
c. midbrain and diencephalon.
d. thalamus and cerebrum.

_____ 2. Important centers for vasoconstriction and cardiac control are located in the
a. cerebrum.
b. medulla.
c. cerebellum.
d. basal ganglia.

_____ 3. An important respiratory center is located in the
a. cerebrum.
b. cerebellum.
c. pons.
d. sacral plexus.

_____ 4. Our conscious state is maintained by activity generated in the
a. cerebellum.
b. reticular formation.
c. limbic system.
d. medulla.

_____ 5. The major relay station for impulses going to and from the cerebral cortex is the
a. hypothalamus.
b. thalamus.
c. pons.
d. cerebellum.
e. cranial nerves.

_____ 6. The furrows on the surface of the cerebrum are called the
a. gyri.
b. sulci.
c. commissures.
d. tracts.
e. none of the above

_____ 7. A cutaneous nerve to the hand is severed at the elbow. The proximal end of the nerve at the elbow is then stimulated. The subject reports
a. no sensation because the receptors are gone.
b. a sensation limited to the region of the elbow.
c. a sensation localized in the region of the hand.
d. a vague sensation throughout the side of the body containing the cut nerve.

_____ 8. General sensory inputs (pain, touch, temperature) to the cerebrum terminate in the
a. precentral gyrus.
b. postcentral gyrus.
c. central sulcus.
d. arachnoid.

_____ 9. Concerning the visual centers of the brain,
a. they are located in the parietal lobes of the cerebrum.
b. the primary visual centers relate past to present visual experiences and evaluate what is being seen.
c. the visual association areas interpret shape and color.
d. none of the above

_____ 10. Given the following areas of the cerebrum:
1. superior part of the primary motor area
2. inferior part of the primary motor area
3. prefrontal area
4. premotor area

Suppose Harry Wolf wanted to flirt with a girl by winking and then smiling. In what order would the areas be used?
a. 4, 1
b. 3, 4, 1
c. 3, 4, 2
d. 4, 1, 2
e. 4, 1, 2, 3

_____ 11. Given the following areas of the cerebral cortex:
1. Broca's area
2. premotor area
3. primary motor area
4. Wernicke's area

If a person hears and understands a word and then says the word out loud, in what order would the areas be used?
a. 1, 4, 2, 3
b. 1, 4, 3, 2
c. 4, 1, 2, 3
d. 4, 1, 3, 2

_____ 12. During meditation an adult would generate
a. alpha waves.
b. gamma waves.
c. delta waves.
d. theta waves.

_____ 13. Long-term memory appears to involve
a. a change in the shape of neurons.
b. movement of calcium into the neuron.
c. activation of the enzyme calpain.
d. all of the above

_____ 14. The main connection between the left and right hemisphere of the cerebrum is the
a. intermediate mass.
b. corpus callosum.
c. vermis.
d. unmyelinated nuclei.

_____ 15. Concerning the basal ganglia,
   a. the corpus striatum is an example.
   b. they are gray matter located in the cerebrum, diencephalon, and mid brain.
   c. they can act to inhibit muscle movement.
   d. all of the above

_____ 16. A football player exhibited the following symptoms after a third-quarter play in which he was injured:
   1. uncontrolled rhythmic contraction of skeletal muscles
   2. abnormally great tension in muscles

   The injury probably involved what area of the brain?
   a. medulla oblongata
   b. basal ganglia
   c. cerebral cortex
   d. cerebellum
   e. hypothalamus

_____ 17. The limbic system is involved in the control of
   a. sleep and wakefulness.
   b. maintaining posture.
   c. higher intellectual processes.
   d. emotion.

_____ 18. A nurse is caring for a patient that exhibits the following symptoms:
   1. inability to stand and walk normally
   2. normal intelligence
   3. capable of performing voluntary movements although the movements are not smooth and precise
   4. normal tension in skeletal muscles
   5. no obvious palsy

   The patient is probably suffering from a condition that affected the
   a. cerebral ganglia.
   b. cerebellum.
   c. cerebral cortex.
   d. medulla oblongata.
   e. pons.

_____ 19. The spinal cord extends from the
   a. medulla to coccyx.
   b. level of the third cervical vertebra to coccyx.
   c. level of the axis to the lowest lumbar vertebra.
   d. medulla to level of the second lumbar vertebra.
   e. axis to sacral vertebra 5.

_____ 20. All sensory neurons entering the spinal cord
   a. enter through the dorsal horn.
   b. have their cell bodies in the dorsal root ganglia.
   c. are part of a spinal nerve.
   d. all of the above

_____ 21. Which of the following events occur when a person steps on a tack with their right foot?
   a. The right foot is pulled away from the tack due to the Golgi tendon reflex.
   b. The left leg is extended to support the body due to the stretch reflex.
   c. The flexor muscles of the thigh contract, and the extensor muscles of the thigh relax due to reciprocal innervation.
   d. all of the above

_____ 22. Several of the events that occurred between the time that a physician struck a patient's patellar ligament with a rubber hammer and the time his quadriceps femoris contracted (knee jerk reflex) are listed below:
   1. increased frequency of action potentials in the afferent neurons
   2. stretch of the muscle spindles
   3. increased frequency of action potentials in the alpha motor neurons
   4. stretch of the quadriceps femoris
   5. contraction of the quadriceps femoris

   Which of the following lists most closely describes the sequence of events as they normally occur?
   a. 4, 1, 2, 3, 5
   b. 4, 1, 3, 2, 5
   c. 1, 4, 3, 2, 5
   d. 4, 2, 1, 3, 5
   e. 4, 2, 3, 1, 5

_____ 23. The extrapyramidal system is mainly concerned with
   a. skilled, learned movements.
   b. water balance.
   c. gross movements and posture.
   d. transmission of motor signals from the medulla to the cerebrum.

_____ 24. Most fibers of the pyramidal system
   a. decussate in the medulla.
   b. synapse in the pons.
   c. descend in the rubrospinal tract.
   d. do not decussate in the brain.
   e. a and c

_____ 25. If one severed the lateral spinothalamic
        tract on the right side of the spinal cord,
   a. pain sensations below the damaged
      area on the right side would be
      eliminated.
   b. pain sensations below the damaged
      area on the left side would be
      eliminated.
   c. temperature sensation would be
      unaffected.
   d. none of the above

_____ 26. A person with a spinal cord injury is
        suffering from paresis (partial paralysis)
        in her right lower limb. Which of the
        following spinal pathways is probably
        involved?
   a. left lateral corticospinal tract
   b. right lateral corticospinal tract
   c. left medial lemniscus system
   d. right medial lemniscus system

_____ 27. Given the following symptoms:
   1. loss of sensations from the left upper
      limb
   2. loss of sensations from the right
      upper limb
   3. ataxia (failure of motor coordination)
      on the left upper limb
   4. ataxia (failure of motor coordination)
      on the right upper limb

   Which of the symptoms is (are)
   consistent with damage to the thalamus
   on the right side?
   a. 1
   b. 4
   c. 1, 3
   d. 1, 4
   e. 2, 4

_____ 28. The most superficial of the meninges is
        a thick, tough membrane called the
   a. pia mater.
   b. arachnoid.
   c. dura mater.
   d. epidural mater.

_____ 29. The ventricles of the brain are
        interconnected. Which of the following
        ventricles are NOT correctly matched
        with the structures that connect them?
   a. lateral ventricle to the third ventricle
      through the interventricular foramina
   b. left lateral ventricle to right lateral
      ventricle through the central canal
   c. third ventricle to fourth ventricle
      through the cerebral aqueduct
   d. fourth ventricle to subarachnoid
      space through the median and lateral
      foramina

_____ 30. Cerebrospinal fluid is produced by the
        _____, circulates through the ventri-
        cles, and is reabsorbed by the _____.
   a. choroid plexus, arachnoid villi
   b. arachnoid villi, choroid plexus
   c. dural sinus, dura mater
   d. dura mater, dural sinus

FINAL CHALLENGES

Use a separate sheet of paper to complete this section.

1. How could one distinguish between damage to the hypothalamus and damage to the medulla?

2. Although alcohol has effects on other areas of the brain, it has a considerable effect on cerebellar function. What kinds of motor tests would reveal the drunken condition?

3. Would a patient with Parkinson's disease be expected to have reduced (hyporeflexive) or exaggerated (hyperreflexive) reflexes. Explain.

4. In pneumoencephalography the ventricles are partially filled with air and X-rays are taken. Abnormalities in the shape of the ventricles can be used to locate tumors or areas of atrophy. One way this procedure is accomplished is by lumbar puncture. If the patient is sitting up with the back and neck flexed when the air is injected, trace the route taken by an air bubble into the ventricles. If the patient lies supine after the air is injected, what part of the ventricles would fill with air?

5. A patient suffered a small lesion in the center of the spinal cord at the level of the sixth thoracic vertebrae. What symptoms would it be possible to observe?

6. A patient suffered a loss of fine touch and pressure on the right side of the body. Voluntary movement of muscles was not affected, and pain and thermal sensations were normal. Is it possible to conclude that the right side of the spinal cord was damaged?

ANSWERS TO CHAPTER 13

WORD PARTS

1. medulla
2. pons
3. cortex
4. colliculus, colliculi
5. tegmentum

6. peduncles
7. ganglion, ganglia
8. corpus, corpora
9. meninges; meningitis
10. arachnoid

CONTENT LEARNING ACTIVITY

Development

A. 1. Neural plate
    2. Neural tube
    3. Neural crest
    4. Central canal

B. 1. Telencephalon
    2. Diencephalon
    3. Metencephalon
    5. Myelencephalon

Brainstem

A. 1. Medulla oblongata
    2. Medulla oblongata
    3. Medulla oblongata
    4. Pons; medulla oblongata
    5. Midbrain
    6. Midbrain
    7. Midbrain

B. 1. Pyramids
    2. Olives
    3. Colliculi

    4. Cerebral peduncles
    5. Reticular formation

C. 1. Thalamus
    2. Pineal body
    3. Midbrain
    4. Superior colliculus
    5. Inferior colliculus
    6. Pons
    7. Medulla oblongata

Diencephalon

A. 1. Thalamus
    2. Subthalamus
    3. Epithalamus
    4. Epithalamus
    5. Hypothalamus
    6. Hypothalamus
    7. Hypothalamus
    8. Hypothalamus

B. 1. Corpus callosum
    2. Hypothalamus
    3. Mamillary body
    4. Optic chiasma
    5. Pituitary gland
    6. Midbrain
    7. Pons
    8. Subthalamus
    9. Pineal body (epithalamus)
    10. Thalamus
    11. Intermediate mass

Cerebrum

A. 1. Longitudinal fissure
    2. Gyri
    3. Central sulcus
    4. Lateral fissure

B. 1. Frontal lobe
    2. Parietal lobe
    3. Occipital lobe
    4. Temporal lobe
    5. Temporal lobe

C. 1. Gray matter
    2. White matter
    3. Association fibers
    4. Projection fibers

Cerebral Cortex

A. 1. Primary sensory areas
    2. Somesthetic cortex (general sensory area)
    3. Projection
    4. Association areas
    5. Association areas

B. 1. Primary motor area
    2. Primary motor area
    3. Premotor area
    4. Prefrontal area

Speech

1. Wernicke's area
2. Broca's area

3. Premotor area
4. Primary motor area

Brain Waves

1. Alpha wave
2. Beta wave

3. Delta wave
4. Theta wave

Memory

1. Sensory memory
2. Short-term memory
3. Long-term memory
4. Calpain
5. Memory engrams

6. Declarative memory
7. Procedural memory
8. Hippocampus and amygdala
9. Cerebellum

Right and Left Hemispheres

1. Left cerebral hemisphere
2. Left cerebral hemisphere

3. Right cerebral hemisphere
4. Corpus callosum

Basal Ganglia

1. Subthalamic nucleus
2. Substantia nigra

3. Corpus striatum
4. Caudate nucleus; lentiform nucleus

Limbic System

1. Hippocampus
2. Habenular nuclei
3. Basal ganglia
4. Hypothalamus

5. Olfactory cortex
6. Fornix
7. Olfactory nerves

Cerebellum

A. 1. Cerebellar peduncles
2. Flocculonodular lobe
3. Vermis
4. Lateral hemispheres

B. 1. Cerebellum
2. Proprioceptors
3. Motor cortex
4. Comparator

Spinal Cord

A. 1. Cervical enlargement
2. Medullary cone
3. Filum terminale
4. Cauda equina

B. 1. Fasciculi
2. Posterior (dorsal) horn
3. Lateral horn
4. Gray and white commissures
5. Ventral root
6. Dorsal root ganglion

C. 1. Funiculi
2. Dorsal root
3. Dorsal root ganglion
4. Ventral root
5. Anterior (ventral) horn
6. Lateral horn
7. Posterior (dorsal) horn

Spinal Reflexes

A. 1. Golgi tendon reflex
2. Withdrawal reflex
3. Reciprocal innervation
4. Crossed extensor reflex

B. 1. Muscle spindle
2. Alpha motor neuron
3. Gamma motor neuron
4. Golgi tendon organ

Ascending Spinal Pathways

A. 1. Lateral spinothalamic
2. Anterior spinothalamic
3. Medial lemniscal
4. Spinocerebellar
5. Spino-olivary

B. 1. Spinal cord
2. Spinal cord
3. Medulla oblongata
4. Uncrossed
5. Pons; spinal cord

C. 1. Primary neuron
2. Secondary neuron
3. Secondary neuron
4. Tertiary neuron

Descending Spinal Pathways

A. 1. Pyramidal
   2. Extrapyramidal
   3. Pyramidal
   4. Extrapyramidal

B. 1. Corticospinal tract
   2. Corticobulbar tract
   3. Rubrospinal tract
   4. Vestibulospinal tract
   5. Reticulospinal tract

C. 1. Medulla oblongata
   2. Spinal cord

D. 1. Upper motor neuron
   2. Lower motor neuron
   3. Lower motor neuron

Meninges

A. 1. Dura mater
   2. Dura mater
   3. Epidural space
   4. Arachnoid layer
   5. Subarachnoid space
   6. Denticulate ligaments

B. 1. Superior sagittal sinus
   2. Arachnoid granulation
   3. Pia mater
   4. Arachnoid layer
   5. Subarachnoid space
   6. Dura mater

Ventricles

A. 1. Septa pellucida
   2. Third ventricle
   3. Interventricular foramen
   4. Fourth ventricle
   5. Cerebral aqueduct
   6. Central canal

B. 1. Lateral ventricles
   2. Choroid plexus
   3. Interventricular foramen
   4. Third ventricle
   5. Cerebral aqueduct
   6. Fourth ventricle
   7. Central canal
   8. Subarachnoid space
   9. Superior sagittal sinus
   10. Arachnoid granulation

Cerebrospinal fluid

1. Cerebrospinal fluid
2. Choroid plexuses
3. Subarachnoid space

4. Lateral and median foramina
5. Venous sinus
6. Arachnoid granulations

Clinical Applications

1. Dyskinesias
2. Multiple sclerosis
3. Reye's syndrome

4. Encephalitis
5. Stroke

1. Cerebral cortex: interprets sensations, initiates motor activities, thoughts, reasoning, speech, and other higher brain functions

   Cerebral medulla: tract system that connects parts of a cerebral hemisphere with other parts of the same hemisphere, with the opposite cerebral hemisphere, or other parts of the brain and spinal cord

   Basal ganglia: coordinates motor movement and posture, inhibits unwanted muscular activity

   Thalamus: relay center for sensory and motor functions

   Hypothalamus: endocrine control, autonomic control, temperature regulation, hunger, thirst, and emotions

   Limbic system: emotions

   Midbrain: visual and auditory reflexes (colliculi) motor pathways (cerebral peduncles), connects forebrain with hindbrain

   Medulla oblongata: motor pathways (pyramids), balance (olives), autonomic reflexes (e.g., heart rate, breathing, swallowing), ascending and descending tracts, consciousness (reticular activating system)

   Pons: connects cerebellum to brain (cerebellar peduncles), regulates respiration, and contains ascending and descending tracts

   Cerebellum: muscle coordination and balance
2. Medulla oblongata, pons, and mid-brain
3. Thalamus, subthalamus, hypothalamus, and epithalamus
4. Frontal lobe: voluntary motor function, motivation, aggression and mood

   Parietal lobe: major center for reception and evaluation of most sensory information

   Occipital lobe: reception and integration of visual input.

   Temporal lobe: reception and evaluation of olfactory and auditory input; memory
5. Association, commissural, and projection traits
6. Stretch reflex, Golgi tendon reflex, withdrawal reflex, reciprocal innervation, and crossed extensor reflex
7. Spinothalamic system: ascending pathway carrying pain, temperature, light touch, pressure, tickle and itch sensations

   Medial lemniscal system: ascending pathways carrying senses of proprioception, two-point discrimination, and vibration

   Spinocerebellar system: ascending pathways carrying unconscious proprioception, coordination, eye reflexes, and arousing consciousness
8. Pyramidal system: conscious control of skeletal muscle, especially concerned with fine movements and speed; increases muscle tone

   Extrapyramidal system: conscious and unconscious control of skeletal muscle; involved with error correcting function of cerebellum; maintenance of posture and balance; control of large movements (e.g., in the trunk and limbs); inhibits muscle activity
9. Dura mater, arachnoid layer, and pia mater. The subdural space separates the dura mater and arachnoid layer; the subarachnoid space separates the arachnoid layer and the pia mater.
10. Lateral ventricles are connected to the third ventricle by the interventricular foramen; the third ventricle is connected to the fourth ventricle by the cerebral aqueduct.

MASTERY LEARNING ACTIVITY

1. C. The brainstem consists of the medulla, pons, and midbrain. The midbrain connects to the diecephalon (thalamus, hypothalamus, subthalamus, and epithalamus).

2. B. The medulla contains many nuclei that are responsible for many basic life functions. Severe damage to the medulla usually results in death. Other medullary functions include swallowing, vomiting, sneezing, and coughing.

3. C. The pons functions as a connection between the cerebellum and the brainstem and between lower and upper levels of the central nervous system. It also contains two respiratory centers.

4. B. The reticular formation is a loose network of nerve fibers and nuclei scattered throughout the medulla and extending up through the brainstem to the thalamus. When the reticular formation is active, it stimulates other areas of

the brain, resulting in a conscious or awake state. Barbiturates and general anesthetics depress the activity of the reticular formation.

5. B. The thalamus is the major relay center for impulses going to and from the cerebral cortex. All sensory tracts, except for olfaction, synapse in the thalamus.

6. B. There are many rounded ridges or convolutions on the surface of the cerebrum called gyri. Separating the ridges are furrows called sulci. Fissures divide the brain into major parts. The longitudinal fissure divides the cerebrum into left and right halves. The transverse fissure separates the cerebrum from the cerebellum. The white matter of the brain makes up nerve tracts. There are three basic kinds of tracts in the cerebrum:
   1. Association tracts: limited to one cerebral hemisphere.
   2. Commissural tracts: cross from one cerebral hemisphere to another (corpus callosum)
   3. Projection tracts: from the cerebral hemispheres to other parts of the nervous system.

7. C. Normally, when a specific body part such as the hand is stimulated, action potentials are generated that travel to the brain. There, through projection, awareness that the hand was stimulated occurs. When the severed nerve was stimulated, action potentials were generated in the axon supplying the hand. When these action potentials reached the brain, they were projected to the hand, and it felt as though the hand was stimulated.

8. B. General sensory input terminates in the postcentral gyrus (primary somesthetic cortex). The precentral gyrus (primary motor area) is concerned with the control of skeletal muscles. The central sulcus is the furrow that separates the precentral gyrus from the postcentral gyrus. The arachnoid is one of the meninges.

9. D. The primary and association visual areas are located in the occipital lobe. The primary visual centers interpret shape and color whereas the association visual areas relate past to present visual experiences.

10. C. The motivation and foresight to initiate movement of the facial muscle occur in the pre-frontal area. Then the premotor area determines which muscles contract and the order in which they contract. Finally the primary motor area

sends action potentials to each specific muscle. The inferior part of the primary motor area controls the face.

11. C. Wernicke's area is necessary for understanding and formulating speech. Broca's area initiates the series of movements necessary for speech. Then the premotor area programs the movements and finally muscle movement is initiated in the primary motor area.

12. A. Alpha waves are produced in an awake, but quietly resting person. Theta waves usually occur in children, and delta waves occur in deep sleep. There are no gamma waves.

13. D. Calcium movement into the the neuron activates calpain, which partially degrades the neuron cytoskeleton, causing a change in shape.

14. B. The left and right cerebral hemispheres are connected by the corpus callosum. The intermediate mass connects the left and right thalamus. The vermis connects the left and right cerebellar hemispheres.

15. D. The basal ganglia are nuclei (gray matter) located in the cerebrum, diencephalon, and mid brain. The corpus striatum in the cerebrum is an example. The basal ganglia function to plan and coordinate motor movements and posture. One major effect of the basal ganglia is to inhibit unwanted muscular activity.

16. B. One function of the basal ganglia is to inhibit unwanted muscular activity. Loss of this inhibitory effect could result in uncontrolled rhythmic contractions of skeletal muscle and the greater than normal tension in skeletal muscle.

17. D. The limbic system is comprised of those structures that affect emotional responses. This includes some hypothalamic and thalamic nuclei, parts of the basal ganglia, the olfactory bulbs, parts of the cerebral cortex, the fornix, and the mamillary bodies.

18. B. The symptoms listed suggest that the patient has the ability to perform voluntary movements and has normal tension in the skeletal muscles. Therefore the cerebrum does not appear to be involved. Since the patient's ability to perform coordinated movements is affected, it suggests the comparator function of the cerebellum is involved. None of the data suggest that the medulla or pons are involved.

19. D. During fetal development the vertebral column grows faster than the spinal cord. As a result, the spinal cord reaches only to the level of the first or second lumbar vertebrae. Below that region is a mass of descending nerves called the cauda equina. Lumbar punctures and spinal anesthesia are usually administered between the third and fourth lumbar vertebrae.

20. D. Sensory neuron cell bodies are located in the dorsal root ganglia, which are in the dorsal roots of spinal nerves. The sensory neuron axons enter the spinal cord through the dorsal horn. In contrast, motor neuron cell bodies are located in the ventral horn of the spinal cord. They exit through the ventral root of the spinal nerve.

21. C. The right foot is removed from the tack as the flexor muscles contract (withdrawal reflex) and the extensor muscles relax (reciprocal innervation). The left leg is extended due to the crossed extensor reflex. The Golgi tendon reflex causes muscles under excessive strain to relax.

22. D. The rubber hammer striking the patella ligament causes the quadriceps femoris to stretch (4), which also results in the stretch of the muscle spindles (2). The stretch of the muscle spindles increases the frequency of afferent neuron impulses (1), stimulating alpha motor neurons (3), that stimulate the quadriceps femoris to contract (5).

23. C. The extrapyramidal system is involved with gross movement and posture. The extrapyramidal system is complex, including parts of the cerebrum, cerebellum, basal ganglia, thalamus, and brainstem. The pyramidal system consists of motor areas in the cerebral cortex and the nerve tracts leading to skeletal muscles. It is concerned with voluntary, skilled, learned movements.

24. A. Most (75%-85%) pyramidal fibers decussate in the medulla and descend in the lateral corticospinal tracts. Some (15%-25%) of pyramidal fibers do not decussate in the brain and descend in the anterior corticospinal tract. The upper motor neurons of the pyramidal system synapse with lower motor neurons in the spinal cord.

25. B. The spinothalamic tract carries pain and thermal sensations, and it decussates (crosses) within the spinal cord. Interruption of the spinothalamic tract on the right side will result in the loss of pain and thermal sensations on the left side below the damaged area.

26. B. Disruption of the lateral corticospinal pathway results in paresis. Since it decussates within the medulla, disruption of the right lateral corticospinal pathway affects the right side of the body. The patient is probably suffering from damage to her right lateral corticospinal tract.

27. C. Since all sensory information passes through the thalamus on the side opposite to the side where the sensory receptors are located, 1 and 3 are correct. Ataxia on the left side would result from the loss of proprioception sensations from that side, since the medial lemniscal system carries most of the proprioceptive information from the upper limb to the cerebrum.

28. C. There are three layers of connective tissue called meninges, which cover the entire central nervous system. From outside to inside, they are the dura mater, arachnoid, and pia mater.

29. B. The lateral ventricles are separated by the septum pellucidum and are not interconnected. The central canal is a continuation of the fourth ventricle into the spinal cord.

30. A. The pia mater and the arachnoid are modified to form the choroid plexus, which produces cerebrospinal fluid. The arachnoid villi reabsorb the cerebrospinal fluid from the sub-arachnoid space. It then enters the dural sinus.

★ FINAL CHALLENGES ★

1. One strategy is to look for easily observed abnormalities associated with the hypothalamaus but not the medulla: e.g., abnormal emotional behavior (fear, rage), hunger, thirst, sweating, shivering, or urine production. Observation of functions that could be affected by either the hypothalamus or medulla such as heart rate, sleep (consciousness), or blood vessel diameter, would not be useful.

2. Since the cerebellum acts to match intended movements with actual movements, reduced cerebellar functions results in an inability to point precisely to an object (such as one's nose). It also results in poor balance.

3. Parkinson's disease results in decreased activity of inhibitory neurons. Therefore the neurons to muscles become overstimulated, resulting in tremors. The same hyperexcited state also results in exaggerated reflexes.

4. In a lumbar puncture air is injected into the sub-arachnoid space around the spinal cord. The air would travel superiorly and enter the fourth ventricle through the lateral foramina. From the fourth ventricle air can pass through the cerebral aqueduct to the third ventricle and from there into the lateral ventricles through the interventricular foramina. In the supine position, the anterior portions of the lateral ventricles are filled with air.

5. Since the lesion is in the center of the spinal cord, only those tracts that cross over in the spinal cord are affected. Therefore one expects to see loss of pain and temperature sensation (lateral spinothalamic tract). The loss occurs bilaterally at the level of the injury, but not below the injury. Since the medial lemniscal system (fine touch) and lateral corticospinal tract (voluntary motor activity) cross over in the medulla, they would be unaffected by a lesion in the center of the spinal cord.

6. It is possible that the medial lemniscal system within the right side of the spinal cord is damaged. However, it is also possible that this tract system could be damaged within the medulla, where neurons synapse and cross over to the left side of the brain, or within the tracts on the left side that ascend from the medulla to the thalamus. Another possibility is damage to the cerebral cortex on the left side. Additional information is needed to decide exactly where the injury is located.

Peripheral Nervous System

FOCUS: The peripheral nervous system consists of 31 pairs of spinal nerves and 12 pairs of cranial nerves. The spinal nerves branch to form dorsal rami, which supply the dorsal trunk, and ventral rami. The ventral rami give rise to the intercostal nerves or join with each other to form plexuses. The major plexuses are the cervical, brachial, lumbar, sacral, and coccygeal plexuses. Within a plexus, the fibers from different spinal nerves join together to form nerves that leave the plexus. The cranial nerves all originate within nuclei of the brain and supply the structures of the head and neck, with the exception of the vagus nerve, which supplies the visceral organs of the thorax and abdomen. The cranial nerves function as sensory nerve (e.g., optic nerve for vision), somatomotor and proprioception (e.g., oculomotor nerve, which controls movements of the eyeball), and parasympathetic (e.g. oculomotor nerve which controls the size of the pupil). Both spinal and cranial nerves have specialized nerve endings associated with afferent fibers. These can be grouped as cutaneous receptors (tactile information from the skin), visceroreceptors (information from internal organs), and proprioceptors (information about body position and movement).

WORD PARTS

Give an example of a new vocabulary word that contains each word part.

WORD PART	MEANING	EXAMPLE
plex-	a network	1. _____
ram-	a branch	2. _____
phren-	the diaphragm	3. _____
cost-	the rib	4. _____
brachi-	the arm	5. _____
axilla	the armpit	6. _____

WORD PART	MEANING	EXAMPLE
sur-	calf of the leg	7. _____
pudend-	female external genitalia	8. _____
proprio-	one's own	9. _____
recept-	a receiver	10. _____

┤ **CONTENT LEARNING ACTIVITY** ├

Spinal Nerves

❝The spinal nerves arise through numerous rootlets along the dorsal and ventral surfaces of the spinal cord.❞

A. Match these terms with the correct statement or definition:

Dermatome   Plexuses
Dorsal rami   Ventral rami
Dorsal root   Ventral root
Dorsal root ganglion

_____

_____ 1. These two join just lateral to the spinal cord to form the spinal nerve.

_____ 2. Collection of nerve cell bodies located in the dorsal root of the spinal nerve.

_____ 3. Area of skin supplied by a given pair of spinal nerves.

_____ 4. Innvervate most of the deep muscles of the dorsal trunk.

_____ 5. Branches of spinal nerves that are distributed as intercostal nerves or plexuses.

_____ 6. Networks of nerves that divide and reconnect.

B. Match these terms with the correct parts of the diagram labeled in Figure 14-1:

Dorsal ramus
Dorsal root
Dorsal root ganglion
Spinal nerve
Ventral ramus
Ventral root

1. _____

2. _____

3. _____

4. _____

5. _____

6. _____

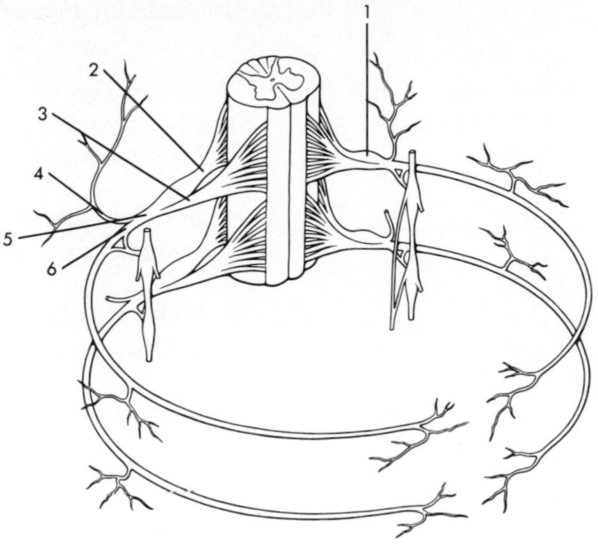

Figure 14-1

C. Match the type of spinal nerve with the number of pairs of that spinal nerve:

| Cervical | Lumbar |
| Coccygeal | Thoracic |

_____ 1. One pair.

_____ 2. Five pairs.

_____ 3. Eight pairs.

_____ 4. 12 pairs.

Cervical Plexus

" *The cervical plexus is a relatively small plexus originating from spinal nerves C1 to C4.* **"**

Match these nerves with their correct function or innervation:

| Ansa cervicalis | Superior cervical nerve |
| Phrenic nerve | |

_____ 1. Supply the hyoid muscles.

_____

_____ 2. Innervates the diaphragm for breathing.

☞ The cutaneous innervation of the cervical plexus is to the neck and posterior portion of the head.

Brachial Plexus

The brachial plexus originates from spinal nerves C5 to T1.

A. Match these nerves with their correct motor function or innervation:

Axillary nerve   Radial nerve
Median nerve   Ulnar nerve
Musculocutaneous nerve

_____ 1. Innervates the deltoid and teres minor muscles; abduction and lateral rotation of the arm.

_____ 2. Innervates all the extensor muscles of the upper limb, supinator muscle, and two muscles that flex the forearm.

_____ 3. Innervates the anterior muscles of the arm.

_____ 4. Innervates two forearm muscles and most of the intrinsic hand muscles.

☞ Additional brachial plexus nerves innervate the shoulder muscles.

B. Match these nerves with their correct sensory innervation:

Axillary nerve   Radial nerve
Median nerve   Ulnar nerve
Musculocutaneous nerve

_____ 1. Sensory innervation to the skin over the shoulder.

_____ 2. Sensory innervation to the posterior arm, forearm, and lateral two thirds of the dorsum of the hand.

_____ 3. Sensory innervation to the lateral surface of the forearm.

_____ 4. Sensory innervation to the lateral two thirds of the palm and thumb, and the surface of the index, middle, and lateral half of the ring finger.

☞ Additional brachial plexus nerves supply the skin of the arm and forearm.

Lumbar and Sacral Plexuses

The lumbar and sacral plexuses originate from L1 to S4, and are often considered as a single lumbosacral plexus.

A. Match these nerves with the correct statement:

Common peroneal nerve   Tibial nerve
Sciatic nerve

_____ 1. Tibial and common peroneal nerves bound together in the same connective tissue sheath.

_____ 2. Branches to form the medial plantar, lateral plantar, and sural nerves.

_____ 3. Branches to form the deep and superficial peroneal nerves.

B. Match these nerves with their   Common peroneal nerve   Obturator nerve
    correct motor function   Femoral nerve   Tibial nerve
    or innervation: Medial and lateral plantar nerves

_____   1. Innervates the muscles that adduct the thigh.

_____   2. Innervates the iliopsoas, sartorius, and quadriceps femoris muscles.

_____   3. Innervates most of the posterior thigh and leg muscles.

_____   4. Innervates the anterior and lateral muscles of the leg and foot.

☞ Several other nerves from the lumbosacral plexus innervate the lower abdominal,
   gluteal, upper thigh, and perineal muscles.

C. Match these nerves with their   Common peroneal nerve   Obturator nerve
    correct sensory innervation: Femoral nerve   Tibial nerve

_____   1. Sensory innervation to the superior medial side of the thigh.

_____   2. Sensory innervation to the anterior and lateral thigh, medial leg,
                                      and foot.

_____   3. Sensory innervation to the lateral and posterior one third of the leg and
                                      the sole of the foot.

☞ Several other nerves from the lumbosacral plexus provide sensory innervation for the
   lower abdomen, upper thigh, buttock, and perineum regions. The coccygeal plexus is a
   very small plexus (arising from spinal nerves S4, S5, and the coccygeal nerve) that supplies
   motor innervation to the pelvic floor and cutaneous innervation to the skin over the coccyx.

Cranial Nerves

“*The 12 cranial nerves have several different combinations of function.***”**

A. Match these terms with the   Parasympathetic   Sensory
    correct statement or definition: Proprioception   Somatomotor

_____   1. Includes vision, touch, and pain.

_____   2. Involves control of skeletal muscles through motor neurons.

_____   3. Information is gathered concerning the position of the body and
                                      its parts.

_____   4. Involves the regulation of glands, smooth muscle, and cardiac muscle.

B. Match the name of the cranial nerve with its number:

Abducens   Olfactory
Accessory   Optic
Facial   Trigeminal
Glossopharyngeal   Trochlear
Hypoglossal   Vagus
Oculomotor   Vestibulocochlear

_____ I.

_____ II.

_____ III.

_____ IV.

_____ V.

_____ VI.

_____ VII.

_____ VIII.

_____ IX.

_____ X.

_____ XI.

_____ XII.

👉 Some helpful mnemonics: On Old Olympus Towering Top, A Finn And* German Viewed A Hop

*Please note that VIII was formerly called the Auditory rather than the Vestibulocochlear.

C. Match these cranial nerves with their correct sensory function:

Facial (VII)   Trigeminal (V)
Glossopharyngeal (IX)   Vagus (X)
Olfactory (I)   Vestibulocochlear (VIII)
Optic (II)

_____ 1. Sensory from face, teeth, upper and lower jaw, and oral cavity.

_____ 2. Sense of taste from anterior two thirds of tongue; sensory from palate and external ear.

_____ 3. Sense of hearing and balance.

_____ 4. Sense of taste from posterior one third of tongue; sensory from pharynx and middle ear.

_____ 5. Sensory from the thoracic and abdominal organs; sense of taste from posterior tongue.

D. Match these cranial nerves with their correct motor function:

Abducens (VI)
Accessory (XI)
Facial (VII)
Glossopharyngeal (IX)
Hypoglossal (XII)

Oculomotor (III)
Trigeminal (V)
Trochlear (IV)
Vagus (X)

_____ 1. Motor to four extrinsic eye muscles.

_____ 2. Motor to superior oblique eye muscle.

_____ 3. Motor to the muscles of mastication, soft palate, and throat.

_____ 4. Motor to the muscles of facial expression and the throat.

_____ 5. Motor to the soft palate, pharynx and laryngeal muscles (voice production).

_____ 6. Motor to the soft palate, pharynx, sternocleidomastoid, and trapezius muscles.

_____ 7. Motor to the tongue and throat muscles.

E. Match these cranial nerves with their correct parasympathetic function:

Facial (VII)
Glossopharyngeal (IX)

Oculomotor (III)
Vagus (X)

_____ 1. Parasympathetic to the pupil of the eye and ciliary muscle of the lens.

_____ 2. Parasympathetic to the submandibular and sublingual salivary glands and lacrimal glands.

_____ 3. Parasympathetic to the parotid salivary gland.

_____ 4. Parasympathetic to the thoracic and abdominal viscera.

Types of Afferent Nerve Endings

❝_There are eight major types of sensory nerve endings._**❞**

A. Match these types of receptors with the correct definition:

Cutaneous receptors
Proprioceptors

Visceroreceptors

_____ 1. Associated with the skin; provides tactile information about the external environment.

_____ 2. Associated with the internal organs; provides information about the internal environment.

_____ 3. Associated with joints, tendons, and connective tissue; provides information about body position and movement.

285

B. Match these types of nerve ending with their correct function:

Free nerve endings   Merkel's disks
Golgi tendon apparatus   Muscle spindle
Hair follicle receptors   Pacinian corpuscle
Meissner's corpuscles   Ruffini's end organs

_____ 1. Tactile disks associated with light touch and superficial pressure.

_____ 2. Complex nerve endings resembling an onion; respond to deep cutaneous pressure and vibration.

_____ 3. Distributed through the dermal papillae and involved in two-point discrimination.

_____ 4. Located in the dermis; respond to skin displacement in continuous touch or pressure.

_____ 5. Proprioception associated with tendon movement.

_____ 6. Helps control muscle tone.

Clinical Applications

❝_There are many PNS disorders._**❞**

A. Match these terms with the correct statement or definition:

Anesthesia   Neuritis
Hyperesthesia   Paresthesia
Neuralgia

_____ 1. Increased sensitivity to stimuli.

_____ 2. General term for inflammation of a nerve.

_____ 3. Severe spasms of throbbing pain along a nerve pathway; an example is sciatica.

_____ 4. Abnormal spontaneous sensations such as tingling, prickling, or burning.

B. Match these terms with the correct statement or definition:

Herpes simplex I   Myasthenia gravis
Herpes simplex II   Poliomyelitis
Herpes zoster

_____ 1. Viral disease that causes fever blisters or cold sores.

_____ 2. Viral disease that causes chicken pox and shingles.

_____ 3. An autoimmune disorder resulting in a decrease of acetylcholine receptors at neuromuscular junctions.

_____ 4. Viral disease that affects motor neurons in the anterior horn of gray matter of the spinal cord.

QUICK RECALL

1. List three types of rami that branch from spinal nerves, and state what type of nerve fiber (sensory, motor, or autonomic) is found in each.

2. Name five major plexuses formed by the spinal nerves, and list the level of the spinal cord each from which plexus arises.

3. List the five major nerves originating from the brachial plexus.

4. List the four major nerves originating from the lumbosacral plexus.

5. List the 12 cranial nerves.

6. List the three basic functions of the cranial nerves.

7. List the eight types of sensory nerve endings and their functions.

Place the letter corresponding to the correct answer in the space provided.

_____ 1. Which of the following would be a correct count of the spinal nerves?
   a. nine cervical, 12 thoracic, five lumbar, five sacral, one coccygeal
   b. eight cervical, 12 thoracic, five lumbar, five sacral, one coccygeal
   c. seven cervical, 12 thoracic, five lumbar, five sacral, one coccygeal
   d. eight cervical, 11 thoracic, four lumbar, six sacral, one coccygeal
   e. none of the above

_____ 2. Given the following structures:
   1. dorsal ramus
   2. dorsal root
   3. plexus
   4. ventral ramus
   5. ventral root

   Choose the arrangement that lists the structures in the order that an action potential would pass through them, given that the action potential originated in the spinal cord and propagated to a peripheral nerve.
   a. 2, 1, 3
   b. 2, 3, 1
   c. 3, 4, 5
   d. 5, 3, 4
   e. 5, 4, 3

_____ 3. Damage to the dorsal ramus of a spinal nerve would result in
   a. loss of sensation.
   b. loss of motor control.
   c. a and b

_____ 4. A collection of spinal nerves that join together after leaving the spinal cord is called a
   a. ganglion.
   b. nucleus.
   c. projection nerve.
   d. plexus.

_____ 5. A dermatome
   a. is the area of skin supplied by a pair of cranial nerves.
   b. may be supplied by more than one nerve from a plexus.
   c. can be used to locate the site of spinal cord injury.
   d. all of the above

_____ 6. Anesthetic injected into the cervical plexus would
   a. prevent pain in the leg.
   b. prevent pain in the thigh and leg.
   c. prevent pain in the arm.
   d. interfere with the ability of the patient to breathe.

_____ 7. The skin of index and middle fingers are supplied by the
   a. median nerve.
   b. radial nerve.
   c. ulnar nerve.
   d. a and b
   e. all of the above

_____ 8. The extensor muscles of the upper limb are supplied by the
   a. musculocutaneous nerve.
   b. radial nerve.
   c. median nerve.
   d. ulnar nerve.
   e. all of the above

_____ 9. The flexor muscles of the upper limb are supplied by the
   a. musculocutaneous nerve.
   b. ulnar nerve.
   c. median nerve.
   d. radial nerve.
   e. all of the above

_____ 10. The intrinsic hand muscles, other than those that move the thumb, are supplied by the
   a. musculocutaneous nerve.
   b. axillary nerve.
   c. popliteal nerve.
   d. intrinsic nerve.
   e. ulnar nerve.

11. The sciatic nerve is actually two nerves combined within the same sheath. The two nerves are the
    a. femoral and obturator.
    b. femoral and gluteal.
    c. common peroneal and tibial.
    d. common peroneal and obturator.
    e. tibial and gluteal.

12. The muscles of the anterior thigh compartment are supplied by the
    a. obturator nerve.
    b. gluteal nerve.
    c. sciatic nerve.
    d. femoral nerve.
    e. ilioinguinal nerve.

13. The muscles of the posterior compartment of the leg and the intrinsic foot muscles are supplied by the
    a. femoral nerve.
    b. obturator nerve.
    c. common peroneal nerve.
    d. tibial nerve.
    e. saphenous nerve.

14. The cranial nerve involved in chewing food?
    a. trochlear (IV)
    b. trigeminal (V)
    c. abducens (VI)
    d. facial (VII)
    e. vestibulocochlear (VIII)

15. The cranial nerve responsible for focusing the eye (innervates the ciliary muscle of the eye)?
    a. optic (III)
    b. oculomotor (III)
    c. trochlear (IV)
    d. abducens (V)
    e. facial (VII)

16. The cranial nerve involved in feeling a toothache?
    a. trochlear (IV)
    b. trigeminal (V)
    c. abducens (VI)
    d. facial (VII)
    e. vestibulocochlear (VIII)

17. From the following list of cranial nerves:
    1. olfactory (I)
    2. optic (II)
    3. oculmotor (III)
    4. abducens (VI)
    5. vestibulocochlear (VIII)

    Select the nerves that are sensory only.
    a. 1, 2, 3
    b. 1, 2, 5
    c. 2, 3, 4
    d. 2, 3, 5
    e. 3, 4, 5

18. From the following list of cranial nerves:
    1. optic (II)
    2. oculomotor (III)
    3. trochlear (IV)
    4. trigeminal (V)
    5. abducens (VI)

    Select the nerves that are involved in "rolling" the eyes.
    a. 1, 2, 3
    b. 1, 2, 4
    c. 2, 3, 4
    d. 2, 4, 5
    e. 2, 3, 5

19. From the following list of cranial nerves:
    1. trigeminal (V)
    2. facial (VII)
    3. glossopharyngeal (IX)
    4. vagus (X)
    5. hypoglossal (XII)

    Select the nerves that are involved in the sense of taste.
    a. 1, 2, 3
    b. 1, 4, 5
    c. 2, 3, 4
    d. 2, 3, 5
    e. 3, 4, 5

20. From the following list of cranial nerves:
    1. trigeminal (V)
    2. facial (VII)
    3. glossopharyngeal (IX)
    4. vagus (X)
    5. hypoglossal (XII)

    Select the nerves that innervate the salivary glands.
    a. 1, 2
    b. 2, 3
    c. 3, 4
    d. 4, 5
    e. 3, 5

21. From the following list of cranial nerves:
    1. occulomotor (III)
    2. trigeminal (V)
    3. facial (VII)
    4. vestibulocochlear (VIII)
    5. glossopharyngeal (IX)
    6. vagus (X)

    Select the nerves that are part of the parasympathetic nervous system.
    a. 1, 2, 4, 5
    b. 1, 3, 5, 6
    c. 1, 4, 5, 6
    d. 2, 3, 4, 5
    e. 2, 3, 5, 6

22. After examining a patient, the patient's physician concluded the following symptoms were indicative of a lesion in the trigeminal nerve on the right side:
    1. Opened jaw deviated to the right.
    2. The temporalis and masseter muscles were atrophied.
    3. There were no sensations on the right side of the scalp.
    4. There was no corneal sensation on the right side.
    5. There was no sensation to pin pricks in the right side of the oral cavity.

    Which of the above symptoms suggest that the maxillary and mandibular branches of the trigeminal nerve are involved?
    a. 1, 2, 4
    b. 1, 2, 5
    c. 1, 2, 3, 4
    d. 1, 2, 3, 5
    e. 2, 3, 4, 5

23. Receptors that are associated with joints and tendons and provide information about body position and movement are called
    a. tensoceptors.
    b. cutaneous receptors.
    c. visceroreceptors.
    d. proprioceptors.

24. The type of nerve ending associated with pain and itch sensations?
    a. Merkel's disks
    b. Pacinian corpuscles
    c. Ruffini's end-organs
    d. free nerve endings

25. The type of receptor involved with fine discriminative touch (two-point discrimination)?
    a. Merkel's disks
    b. Pacinian corpuscles
    c. Ruffini's end organs
    d. Meissner's corpuscles

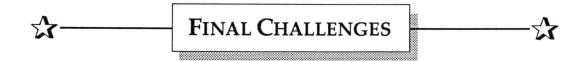

FINAL CHALLENGES

Use a separate sheet of paper to complete this section.

1. A patient has a severe case of the hiccoughs (spasmodic contractions of the diaphragm). The physician injects an anesthetic solution into the neck about an inch above the clavicle. What nerve was injected?

2. A woman slips on a wet kitchen floor. As she falls, she forcefully strikes the edge of the kitchen table with her elbow. Name the nerve she is most likely to damage. What symptoms would you expect her to develop?

3. Jock Player was tackled and slammed into the ground while playing football, resulting in a dislocated shoulder. After the humerus was reset into the glenoid cavity of the scapula, it was discovered that Jock could not abduct his arm. Explain. Where would you expect there to be a loss of sensation in the skin?

4. A stab wound in the proximal, medial part of the arm produced impairment of flexion and supination of the forearm. There was a loss of sensation from the lateral surface of the forearm. What nerve was damaged? Explain why there was only impairment and not complete loss of flexion and supination.

5. A skier breaks his ankle. As part of his treatment, the ankle and leg are placed in a plaster cast. Unfortunately, the cast is too tight about the proximal portion of the leg and presses in against the neck of the fibula. Where would you predict the patient would experience tingling or numbness in the leg? Explain.

6. One day you and a friend observe a man walking down the street in a peculiar fashion. He seems to raise his left foot much higher than normal to keep his toes from dragging on the ground. Apparently he is unable to dorsiflex his left foot. When he places the foot on the ground, it makes a clopping or flapping sound. You recognize this condition as foot drop, and, turning to your friend, you comment that the man must have tibial nerve damage. Your friend disagrees. Who is right?

7. A patient was diagnosed as having tic douloureaux. In this condition, a light touch to the upper lip, cheek, or under the eye, or chewing, produces a sudden sensation of severe pain in and around the area touched. Although the cause of tic douloureaux is unknown, what nerve must be involved? Can you determine which branch of the nerve is most likely affected in this patient?

8. Skip Puck slid into some other players while playing hockey and was severely injured by a skate blade, producing a deep cut in the neck on the left side just lateral to the larynx. After the wound was sutured, the following observations were made: the tongue, when protruded, deviated to the left; the left shoulder was drooping; and Skip could not turn his head to the right side. Name the nerves that were damaged, and explain why this damage produced the observed symptoms.

ANSWERS TO CHAPTER 14

1. plexus
2. ramus, rami
3. phrenic
4. costal
5. brachial

6. axillary
7. sural
8. pudendal
9. proprioception; proprioceptor
10. receptor; proprioceptor; visceroreceptor

CONTENT LEARNING ACTIVITY

Spinal Nerves

A. 1. Dorsal root; ventral root
2. Dorsal root ganglion
3. Dermatome
4. Dorsal rami
5. Ventral rami
6. Plexuses

4. Dorsal ramus
5. Spinal nerve
6. Ventral ramus

C. 1. Coccygeal
2. Lumbar
3. Cervical
4. Thoracic

B. 1. Dorsal root ganglion
2. Dorsal root
3. Ventral root

Cervical Plexus

1. Ansa cervicalis; superior cervical nerve

2. Phrenic nerve

Brachial Plexus

A. 1. Axillary nerve
2. Radial nerve
3. Musculocutaneous nerve
4. Ulnar nerve

B. 1. Axillary nerve
2. Radial nerve
3. Musculocutaneous nerve
4. Median nerve

Lumbar and Sacral Plexuses

A. 1. Sciatic nerve
2. Tibial nerve
3. Common peroneal nerve

C. 1. Obturator nerve
2. Femoral nerve
3. Tibial nerve

B. 1. Obturator nerve
2. Femoral nerve
3. Tibial nerve
4. Common peroneal nerve

Cranial Nerves

A. 1. Sensory
    2. Somatomotor
    3. Proprioception
    4. Parasympathetic

B. I. Olfactory
    II. Optic
    III. Oculomotor
    IV. Trochlear
    V. Trigeminal
    VI. Abducens
    VII. Facial
    VIII. Vestibulocochlear
    IX. Glossopharyngeal
    X. Vagus
    XI. Accessory
    XII. Hypoglossal

C. 1. Trigeminal (V)
    2. Facial (VII)
    3. Vestibulocochlear (VIII)
    4. Glossopharyngeal (IX)
    5. Vagus (X)

D. 1. Oculomotor (III)
    2. Trochlear (IV)
    3. Trigeminal (V)
    4. Facial (VII)
    5. Vagus (X)
    6. Accessory (XI)
    7. Hypoglossal (XII)

E. 1. Oculomotor (III)
    2. Facial (VII)
    3. Glossopharyngeal (IX)
    4. Vagus (X)

Types of Afferent Nerve Endings

A. 1. Cutaneous receptors
    2. Visceroreceptors
    3. Proprioceptors

B. 1. Merkel's disks
    2. Pacinian corpuscle
    3. Meissner's corpuscles
    4. Ruffini's end organs
    5. Golgi tendon apparatus
    6. Muscle spindle

Clinical Applications

A. 1. Hyperesthesia
    2. Neuritis
    3. Neuralgia
    4. Paresthesia

B. 1. Herpes simplex I
    2. Herpes zoster
    3. Myasthenia gravis
    4. Poliomyelitis

1. Dorsal rami: both sensory and motor nerves
   Ventral rami: both sensory and motor nerves
   Autonomic rami: autonomic nerves
2. Cervical plexus: C1 to C4
   Brachial plexus: C5 to T1
   Lumbar plexus: L1 to L4
   Sacral plexus: L4 to S4
   Coccygeal plexus: S4, S5 and coccygeal nerve
3. Axillary, radial, musculocutaneous, ulnar, and median nerves
4. Obturator, femoral, tibial, and common peroneal
5. Olfactory (I), Optic (II), Oculomotor (III), Trochlear (IV), Trigeminal (V), Abducens (VI), Facial (VII), Vestibulocochlear (VIII), Glossopharyngeal (IX), Vagus (X), Accessory (XI), and Hypoglossal (XII)
6. Sensory, somatomotor and proprioception, and parasympathetic
7. Free nerve endings: pain, tickle, itch, temperature, joint movement and proprioception
   Merkel's disks: light touch and superficial pressure
   Hair follicle: light touch; bending of hair
   Pacinian corpuscle: deep cutaneous pressure, vibration and proprioception
   Meissner's corpuscle: two-point discrimination
   Ruffini's end organs: continuous touch or pressure or stretch of skin
   Golgi tendon apparatus: proprioception associated with tendon movement
   Muscle spindle: detects stretch, controls muscle tone

MASTERY LEARNING ACTIVITY

1. B. One must know the correct count.

2. E. The efferent axon from the spinal cord passes through the ventral root into a spinal nerve. The ventral rami of spinal nerves join to form plexuses from which the peripheral nerves arise.

3. C. The dorsal ramus is a branch of a spinal nerve. Spinal nerves contain both afferent (sensory) and efferent (motor) fibers.

4. D. A plexus is a collection of spinal nerves. Ganglia are aggregates of cell bodies in the peripheral nervous system. Nuclei are aggregates of cell bodies in the central nervous system.

5. D. The distribution of a pair of spinal nerves in the skin is a dermatome. Spinal nerves pass through a plexus, and their fibers can exit the plexus through two or more peripheral nerves. Therefore a dermatome can be supplied by more than one nerve. Loss of sensation in a dermatome could indicate damage to the segment of the spinal cord (or more superior segments) associated with a spinal nerve.

6. D. The phrenic nerves, which supply the diaphragm, arise from the cervical plexuses.

7. D. The median nerve supplies the lateral two thirds of the hand, mostly on the anterior surface, whereas the radial nerve supplies the lateral two thirds of the hand, mostly on the posterior surface.

8. C. The radial nerve supplies the muscles that extend the forearm, wrist, and digits.

9. E. The muscles that control flexion of the forearm are supplied by the musculocutaneous and radial nerves. The muscles that flex the wrist and digits are innervated by the ulnar and median nerves.

10. E. The intrinsic hand muscles are only found within the hand, and they are mostly supplied by the ulnar nerve. The median nerve supplies most of the intrinsic hand muscles that move the thumb.

11. C. The common peroneal and tibial nerves form the sciatic nerve within the posterior thigh.

12. D. The femoral nerve supplies the anterior thigh compartment, the obturator nerve the medial thigh compartment, and the sciatic nerve (mostly the tibial nerve) supplies the posterior thigh compartment.

13. D. The tibial nerve and its branches supply the posterior leg and intrinsic foot muscles. The common peroneal nerve supplies the anterior and lateral compartments of the leg.

14. B. The trigeminal nerves supply the muscles of mastication. The facial nerves supply the muscles of facial expression.

15. B. Parasympathetic innervation through the oculomotor nerve causes the eye to focus and the pupil to constrict. Sympathetic stimulation causes the pupil to open.

16. B. The trigeminal nerve is the major sensory nerve of the head. The maxillary branch supplies the teeth of the upper jaw, and the mandibular branch the teeth of the lower jaw.

17. B. The olfactory (sense of smell), optic (sense of vision), and vestibulocochlear (sense of hearing and balance) are sensory only.

18. E. The oculomotor supplies the inferior oblique and the superior, medial, and inferior rectus muscles. The trochlear supplies the superior oblique, and the abducens supplies the lateral rectus.

19. C. The sense of taste from the anterior two thirds of the tongue is carried by the facial nerve and from the posterior one third of the tongue by the glossopharyngeal nerve. The vagus nerve also is involved with the sense of taste from the posterior tongue.

20. B. The facial nerves supply the submandibular and sublingual glands, and the glossopharyngeal nerves supply the parotid glands.

21. B. The oculomotor, facial, glossopharyngeal, and vagus nerves have parasympathetic fibers.

22. B. The temporalis and mandibular muscles are innervated by the maxillary and mandibular branches of the trigeminal nerve. A deviated jaw with atrophied masseter and temporalis muscles suggests the involvement of these branches. Additionally, the maxillary and mandibular branches carry pain sensations from the oral cavity to the central nervous system. Therefore 1, 2, and 5 are correct.

Sensations from the scalp and cornea are carried by the opthalmic branch of the trigeminal nerve. Therefore 3 and 4 are not correct.

23. D. Proprioceptors detect body position and movement. Cutaneous receptors provide tactile information from the skin, and visceroreceptors provide information about internal organs. There is no such thing as a tensoceptor.

24. D. Free nerve endings are responsible for pain, itch, tickle, temperature, and proprioception sensation.

25. D. Distributed throughout the dermal papillae, Meissner's corpuscles are involved with fine, discriminative touch.

 ## FINAL CHALLENGES

1. The phrenic nerve, which supplies the diaphragm, was injected.

2. The ulnar nerve is the nerve most likely to be damaged. One might expect a loss of sensation from the medial one third of the hand. If motor function is affected, one might expect an inability to abduct/adduct the medial four fingers, adduct the wrist, or flex the distal interphalangeal joints of the fourth and fifth digits. There could be impairment of the wrist flexion due to loss of function of the flexor carpiulnaris muscle.

3. When the shoulder was dislocated, the axillary nerve was stretched and damaged, resulting in loss of function of the deltoid muscle and therefore the ability to abduct the arm. Loss of sensation in the skin of the shoulder, over the deltoid, would be expected.

4. The musculocutaneous nerve was damaged, causing the loss of function of the biceps brachii and part of the brachialis muscle. There is only impairment because the radial nerve supplies part of the brachialis and the brachioradialis muscles, which cause flexion. Although the supinator function of the biceps brachii is lost, the supinator muscle, supplied by the radial nerve, can still function.

5. The plaster cast is pressing against the common peroneal nerve at the neck of the fibula. Tingling would be expected along the lateral and anterior leg and the dorsum of the foot.

6. Your friend is right. Injury to the common peroneal nerve results in the loss of dorsiflexion. Tibial nerve damage produces an inability to plantar flex the foot and toes.

7. The maxillary branch of the trigeminal nerve.

8. Damage to the hypoglossal nerve to the tongue produces the deviation of the tongue. Damage to the accessory nerves could cause the shoulder to droop (trapezius muscle) and result in an inability to turn the head (sternocleidomastoid muscle).

Autonomic Nervous System

FOCUS: The autonomic nervous system regulates the activities of smooth muscle, cardiac muscle, and glands and can be divided into sympathetic and parasympathetic divisions. The cell bodies of the preganglionic neurons of the sympathetic division are located in the lateral horns of the spinal cord. Their axons pass through spinal nerves T1 to L2 into white rami communicantes to the sympathetic chain ganglia to synapse with postganglionic neurons. The axons of the postganglionic neurons reenter the spinal nerves through gray rami communicantes or exit the sympathetic chain ganglia as small nerves. In some cases the preganglionic neurons pass through the sympathetic chain ganglia (forming splanchnic nerves) and synapse in collateral ganglia or in the adrenal medulla. The parasympathetic division arises from the brain as part of the cranial nerves (III, VII, IX, and X) or in the lateral horns of the spinal cord from S2 to S4 (giving rise to the pelvic nerves). The preganglionic neurons synapse with postganglionic neurons within terminal ganglia that are near or on their effector organ. The preganglionic neurons of both divisions secrete acetylcholine that binds to nicotinic receptors of the postganglionic neurons. All the postganglionic neurons of the parasympathetic division and some of the postganglionic neurons of the sympathetic division secrete acetylcholine that binds to muscarinic receptors of the effector organs. Most sympathetic postganglionic neurons secrete norepinephrine that binds to adrenergic receptors (alpha and beta). Generally the parasympathetic division maintains homeostasis on a day-to-day basis, whereas the sympathetic division prepares the body for activity. Both divisions are capable of producing excitatory or inhibitory effects, and, when they innervate the same organ, the two divisions usually produce opposite effects.

WORD PARTS

Give an example of a new vocabulary word that contains each word part.

WORD PART	MEANING	EXAMPLE
auto-	self	1. _____
-nomic	law	2. _____
soma-	body	3. _____

B. Match these terms with the parts labeled on the diagram in Figure 15-2:

Collateral ganglion
Gray rami communicantes
Splanchnic nerve
Sympathetic chain ganglion
White rami communicantes

1. _____

2. _____

3. _____

4. _____

5. _____

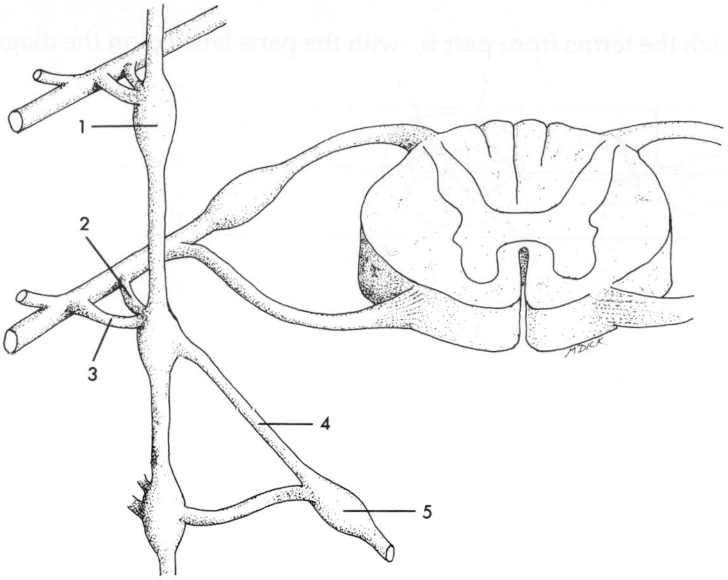

Figure 15-2

Parasympathetic Division

66 *Parasympathetic stimulation activates vegetative functions such as digestion, defecation, and urination.* **99**

Match these terms with the correct location in the following statements:

Cranial nerve nuclei   Postganglionic
Craniosacral   Preganglionic
Lateral horns   Terminal ganglia

1. _____

2. _____

3. _____

4. _____

5. _____

Preganglionic cell bodies of the parasympathetic division are either within _(1)_ or within the _(2)_ of the gray matter from S2 to S4 in the spinal cord. Axons of the _(3)_ neurons pass through cranial and pelvic nerves to _(4)_ either near or embedded in the wall of the organ innervated. The axons of the _(5)_ neurons extend the relatively short distance from the parasympathetic ganglia to the target organ.

☞ The cranial nerves that carry parasympathetic fibers include the oculomotor (III), facial (VII), glossopharyngeal (IX), and vagus (X) nerves.

Comparison of the Differences Between Sympathetic and Parasympathetic Divisions

66*There are a number of structural differences between the sympathetic and parasympathetic divisions of the ANS.*99

Match these autonomic divisions with the correct description:

Parasympathetic division
Sympathetic division

_____ 1. The ratio of postganglionic to preganglionic neurons is much less for this division.

_____ 2. Activity of this division tends to have a more generalized effect rather than a highly localized effect.

_____ 3. Cell bodies of the preganglionic neurons are located in the thoracic and lumbar regions of the spinal cord.

_____ 4. Cell bodies of the postganglionic neurons are located in chain ganglia and collateral ganglia.

_____ 5. Cranial nerves III, VII, IX, X and pelvic nerves are included in this division.

Neurotransmitter Substances and Receptors

66*The sympathetic and parasympathetic nerve endings secrete one of two transmitters.*99

Match these terms with the correct statement or definition:

Adrenergic fiber   Muscarinic receptors
Adrenergic receptor (alpha or beta)   Nicotinic receptors
Cholinergic fiber

_____ 1. Neuron that secretes acetylcholine.

_____ 2. All preganglionic neurons of the autonomic division are of this type.

_____ 3. All postganglionic neurons of the parasympathetic division are of this type.

_____ 4. Most postganglionic neurons of the sympathetic division are of this type.

_____ 5. Found in postganglionic neurons, these cholinergic receptors produce an excitatory response to acetylcholine.

_____ 6. Found in effector organs, these cholinergic receptors may produce either an excitatory or inhibitory response to acetylcholine.

_____ 7. Found in effector organs, norepinephrine binds to and activates these receptors.

Functional Generalizations About the Autonomic Nervous System

❝Most generalizations about the function of the ANS on effector organs have exceptions.❞

Match these divisions of the ANS with the correct statement or description:

Parasympathetic division
Sympathetic division

_____ 1. Sweat glands and blood vessels are innervated almost exclusively by this division.

_____ 2. This division stimulates digestive glands and contracts the gall bladder.

_____ 3. This division has the greatest activity under conditions of physical activity or stress.

_____ 4. Metabolism is increased, and glucose released from the liver when this division is stimulated.

_____ 5. Stimulation of this division causes an increase in heart rate and force of contraction and causes dilation of the pupils of the eyes.

 Autonomic reflexes allow an individual to maintain homeostasis by mediating responses to stressful conditions and changes in physical and mental activity.

1. Complete the following table:

CHARACTERISTIC	SOMATOMOTOR	AUTONOMIC
Number of neurons		
Effector organs		
Neurotransmitter		
Conscious vs. unconscious control		

2. Ganglia are collections of neuron cell bodies. Complete the following table by indicating what kind of cell body is found in each ganglion.

LOCATION	PARASYMPATHETIC OR SYMPATHETIC CELL BODY	PREGANGLIONIC OR POSTGANGLIONIC CELL BODY
Cranial nuclei		
Lateral horns T1 to L2		
Lateral horns S2 to S4		
Chain ganglia		
Collateral ganglia		
Terminal ganglia		

3. Name two types of neurotransmitters released by the ANS and indicate where each is released in both the sympathetic and parasympathetic neurons.

4. List four types of postganglionic receptors found in the ANS and give their location.

5. List three functional generalizations concerning the ANS.

MASTERY LEARNING ACTIVITY

Place the letter corresponding to the correct answer in the space provided.

_____ 1. Given the following statements:
1. neuron cell bodies in the motor nuclei of cranial nerves
2. neuron cell bodies in the lateral horn of the spinal cord
3. two synapses between the CNS and effector organs
4. effector organs include smooth muscle

Which of the statements are true for the autonomic division, but not the somato-motor division, of the peripheral nervous system?
a. 1, 3
b. 2, 4
c. 1, 2, 3
d. 2, 3, 4
e. 1, 2, 3, 4

_____ 2. Given the the following structures:
1. gray ramus communicantes
2. white ramus communicantes
3. spinal nerve
4. sympathetic chain ganglion

Choose the arrangement that lists the structures in the order action potentials pass through them on the way to an effector organ.
a. 1, 4, 2, 3
b. 2, 4, 1, 3
c. 3, 1, 4, 2
d. 3, 4, 1, 2

_____ 3. Given the following structures:
1. collateral ganglion
2. sympathetic chain ganglion
3. white ramus communicantes
4. splanchnic nerve

Choose the arrangement that lists the structures in the order action potentials pass through them on the way to an effector organ.
a. 1, 3, 2, 4
b. 1, 4, 2, 3
c. 3, 1, 4, 2
d. 3, 2, 4, 1

_____ 4. The white ramus contains
a. presynaptic sympathetic fibers.
b. postsynaptic sympathetic fibers.
c. presynaptic parasympathetic fibers.
d. postsynaptic parasympathetic fibers.

_____ 5. The cell bodies of the postganglionic neurons of the sympathetic nervous system are located in the
a. sympathetic chain ganglia.
b. collateral ganglia.
c. lateral horns of the thoracic and lumbar regions of the spinal cord.
d. dorsal root ganglia.
e. a and b

_____ 6. Splanchnic nerves
a. are part of the parasympathetic nervous system.
b. have preganglionic neurons that synapse in the collateral ganglia.
c. exit from the cervical region of the spinal cord.
d. all of the above

_____ 7. Which of the following nerves are part of the parasympathetic nervous system?
a. cranial nerves III, VII, IX, and X
b. pelvic nerves
c. thoracic nerves
d. lumbar spinal nerves one and two
e. a and b

_____ 8. Concerning the preganglionic neurons of the autonomic nervous system,
a. in the parasympathetic nervous system they secrete acetylcholine.
b. in the sympathetic nervous system they secrete acetylcholine.
c. in the sympathetic nervous system they secrete norepinephrine.
d. a and b
e. a and c

_____ 9. Generally speaking, the parasympathetic nervous system
   a. has more postganglionic neurons per preganglionic neuron than does the sympathetic nervous system.
   b. has fewer postganglionic neurons per preganglionic neuron that does the sympathetic nervous system.
   c. has longer preganglionic neurons than does the sympathetic nervous system.
   d. a and c
   e. b and c

_____ 10. A cholinergic neuron
   a. secretes acetylcholine.
   b. reacts with acetylcholine.
   c. secretes norepinephrine.
   d. reacts with norepinephrine.

_____ 11. Which of the neurotransmitters is correctly matched with its receptor?
   a. acetylcholine - nicotinic receptor
   b. acetylecholine - muscarinic receptor
   c. norepinephrine - alpha adrenergic receptor
   d. norepinephrine - beta adrenergic receptor
   e. all of the above

_____ 12. Nicotinic receptors are located in
   a. postganglionic neurons of the parasympathetic nervous system.
   b. postganglionic neurons of the sympathetic nervous system.
   c. effector organs of the parasympathetic nervous system.
   d. a and b
   e. all of the above

_____ 13. Alpha adrenergic receptors
   a. can be excitatory or inhibitory.
   b. are found on preganglionic neurons.
   c. are found on postganglionic neurons.
   d. are part of the parasympathetic nervous system.

_____ 14. The sympathetic nervous system
   a. is always stimulatory.
   b. is always inhibitory.
   c. is usually under conscious control.
   d. generally opposes the actions of the parasympathetic nervous system.
   e. a and c

_____ 15. Which of the following is expected if the sympathetic nervous system is stimulated?
   a. Blood flow to the visceral organs increases.
   b. Blood flow to skeletal muscles increases.
   c. Heart rate decreases.
   d. Glucose release from the liver decreases.
   e. a and b

Use a separate sheet of paper to complete this section.

1. When trying to remember whether or not the sympathetic or parasympathetic systems have an excitatory or inhibitory effect on a particular organ, what generalization is useful?

2. A patient with Horner's syndrome exhibits the following symptoms on the left side of the face: pupillary constriction, flushing of the skin, and absence of sweating. Injury (loss of function) of what part of autonomic nervous system could produce these symptoms?

3. Given the following parts of the autonomic nervous system:
   1. vagus nerves
   2. splanchnic nerves
   3. pelvic nerves
   4. cranial nerves
   5. outflow of gray rami

   Match the part of the autonomic nervous system that, if damaged, produces the following symptoms:
   A. inability to produce "goose flesh"
   B. flushed skin
   C. increased heart rate
   D. inability to defecate (move feces through the inferior end of the large intestine)
   E. dry eyes

4. A man is accidentally poisoned with mushrooms that contain muscarine. Which of the following symptoms do you expect to observe: a dry mouth, diarrhea or even involuntary defecation, and contracted pupils?

5. For which of the following conditions would epinephrine be effective: nasal congestion, asthma attack, high blood pressure, and tachycardia? Explain.

ANSWERS TO CHAPTER 15

WORD PARTS

1. Autonomic
2. Autonomic
3. Somatomotor
4. Somatomotor
5. Preganglionic

6. Postganglionic
7. Adrenal
8. Adrenal
9. Intramural
10. Intramural

CONTENT LEARNING ACTIVITY

Contrasting the Somatomotor and Autonomic Nervous Systems

A. 1. Autonomic division
   2. Somatomotor division
   3. Autonomic division
   4. Somatomotor division
   5. Autonomic division

B. 1. Preganglionic neuron
   2. Postganglionic neuron
   3. Ganglionic synapse

4. Autonomic ganglia
5. Neuroeffector synapse

C. 1. Preganglionic neuron
   2. Autonomic ganglion
   3. Postganglionic neuron
   4. Ganglionic synapse
   5. Neuroeffector synapse

Sympathetic Division

A. 1. Sympathetic chain ganglia
   2. White rami communicantes
   3. Postganglionic
   4. Gray rami communicantes
   5. Splanchnic nerves
   6. Collateral ganglia
   7. Adrenal medulla

B. 1. Sympathetic chain ganglion
   2. Gray rami communicantes
   3. White rami communicantes
   4. Splanchnic nerve
   5. Collateral ganglion

Parasympathetic Division

1. Cranial nerve nuclei
2. Lateral horns
3. Preganglionic

4. Ganglia
5. Postganglionic

Comparison of the Differences Between Sympathetic and Parasympathetic Divisions

1. Parasympathetic division
2. Sympathetic division
3. Sympathetic division

4. Sympathetic division
5. Parasympathetic division

Neurotransmitter Substances and Receptors

1. Cholinergic fiber
2. Cholinergic fiber
3. Cholinergic fiber
4. Adrenergic fiber

5. Nicotinic receptors
6. Muscarinic receptors
7. Adrenergic receptors (alpha or beta)

Functional Generalizations About the Autonomic Nervous System

1. Sympathetic division
2. Parasympathetic division
3. Sympathetic division

4. Sympathetic division
5. Sympathetic division

QUICK RECALL

1. Characteristic

	Somatomotor	Autonomic
Number of neurons	One	Two
Effector organs	Skeletal muscle	Smooth muscle, cardiac muscle, and glands
Neurotransmitter	Acetylcholine	Acetylcholine or norepinephrine
Conscious vs. unconscious control	Conscious	Unconscious

2. Locations

	Parasympathetic or sympathetic cell body	Preganglionic or postganglionic cell body
Cranial nuclei	Parasympathetic	Preganglionic
Lateral horns T1-L2	Sympathetic	Preganglionic
Lateral horns S2-S4	Parasympathetic	Preganglionic
Chain ganglia	Sympathetic	Postganglionic
Collateral ganglia	Sympathetic	Postganglionic
Terminal ganglia	Parasympathetic	Postganglionic

3. All preganglionic fibers: acetylcholine
   Postganglionic fibers of parasympathetic: acetylcholine
   Postganglionic fibers of sympathetic: mostly norepinephrine, some release acetylcholine
4. Alpha and beta adrenergic receptors: effector organs of sympathetic division
   Nicotinic cholinergic receptors: postganglionic receptors for both sympathetic and parasympathetic
   Muscarinic cholinergic receptors: effector organ receptors of parasympathetic and some sympathetic

5. In most cases the influence of the two autonomic divisions is opposite on structures that receive dual innervation; each division can produce inhibitory or excitatory effects; and the parasympathetic division is consistent with resting conditions, whereas the sympathetic division is consistent with physical activity or stress.

1. D. Motor neurons of the somatomotor division are located in cranial nerve nuclei and the ventral horns of the spinal cord. The preganglionic neurons of the autonomic division are located in cranial nerve nuclei and the lateral horns of the spinal cord. Since the preganglionic neuron synapses with the postganglionic neuron and the postganglionic neuron synapses with the effector organ, there are two synapses. The effector organs for the autonomic division include smooth muscle, cardiac muscle and glands, whereas the effector organ for the somatomotor division is skeletal muscle.

2. B. From the spinal nerve, preganglionic axons pass through the white ramus into a sympathetic chain ganglion. At the same or a different level the preganglionic axons synapse with a postganglionic neuron, the axons of which pass through a gray ramus communicantes into a spinal nerve.

3. D. Preganglionic fibers pass through the white ramus communicantes, the sympathetic chain ganglion, and splanchnic nerve to synapse with a postganglionic neuron in the collateral ganglion.

4. A. The white ramus contains presynaptic sympathetic fibers, which are generally myelinated (thus the white color). The gray ramus contains postsynaptic sympathetic fibers that are unmyelinated (hence the gray color).

5. E. Sympathetic postganglionic neuron cell bodies are found in sympathetic chain ganglia and in collateral ganglia. Sympathetic preganglionic neurons are found in the lateral horns of the spinal cord. Sensory neuron cell bodies are found in the dorsal root ganglia.

6. B. Splanchnic nerves have preganglionic fibers that synapse in collateral ganglia. They are part of the sympathetic nervous system and originate between T5 and T12 of the spinal cord.

7. E. The parasympathetic nerves include four cranial nerves (III, VII, IX, and X) and the pelvic nerves. The thoracic spinal nerves and the first two lumbar spinal nerves are part of the sympathetic nervous system.

8. D. All preganglionic neurons of the autonomic nervous system secrete acetylcholine.

9. E. The parasympathetic nervous system has fewer preganglionic neurons per postganglionic neuron than does the sympathetic nervous system. However, the preganglionic neurons of the parasympathetic system are longer because they extend to ganglia on the effector organ, whereas sympathetic preganglionic neurons extend to the sympathetic chain ganglia (or collateral ganglia).

10. A. By definition, cholinergic neurons secrete acetylcholine, and adrenergic neurons secrete norepinephrine.

11. E. Acetycholine binds to nicotinic or muscarinic receptors. Norepinephrine binds to both alpha and beta receptors, although it has a greater affinity for alpha receptors.

12. D. Postganglionic neurons of both divisions of the autonomic nervous system have nicotinic receptors. The effector organs of the parasympathetic system have muscarinic receptors.

13. A. Alpha and beta receptors can be either excitatory or inhibitory. They are found on effector organs as part of the sympathetic nervous system.

14. D. When innervating the same organ, the sympathetic system generally produces the opposite effect of the parasympathetic system. The sympathetic system or the parasympathetic system can produce stimulatory or inhibitory effects (i.e., they have a stimulatory effect in one organ and an inhibitory effect in another organ). The sympathetic and parasympathetic systems are under involuntary control.

15. B. The sympathetic nervous system prepares the body for activity. Blood flow to muscle increases, heart rate increases, and blood sugar levels increase. Meanwhile, activities not immediately necessary for activity are inhibited such as decreasing blood flow to visceral organs.

1. In general, the parasympathetic system has a greater effect under resting conditions, whereas the sympathetic system has a major role under conditions of physical activity or stress. The sympathetic system increases heart rate, causes vasodilation of skeletal muscle blood vessels and vasoconstriction of visceral blood vessels, promotes glucose release from the liver, and decreases intestinal tract activities.

2. Horner's syndrome results from interruption of sympathetic nerves. Consequently there is an inability to sweat in the area affected; and blood vessels vasodilate, producing flushing of the skin. Without sympathetic stimulation of the pupil, the parasympathetic system dominates and the pupil constricts.

3. A. gray rami
   B. gray rami
   C. vagus nerves
   D. pelvic nerves
   E. cranial nerves

4. Muscarine poisoning should result in the activation of muscarinic receptors on effector organs of the parasympathetic nervous system, causing increased salivation, increased motility of the intestine (diarrhea or involuntary defecation), and contracted pupils.

5. Epinephrine would produce similar effects to sympathetic stimulation. Epinephrine would cause vasoconstriction of blood vessels and, by reducing blood flow, could help relieve nasal congestion (reduce fluid loss into the nasal cavity). Epinephrine could also cause dilation of the airway system of the lungs and help relieve asthma symptoms. Epinephrine would increase blood pressure by causing vasoconstriction and would increase heart rate (tachycardia).

Special Senses

FOCUS: The special senses include olfaction, taste, vision, hearing, and balance. Odors are detected by olfactory neurons in the nasal cavity. These neurons send action potentials through the olfactory nerves to the olfactory bulbs. The olfactory tracts carry action potentials from the olfactory bulbs to the olfactory cortex. Taste bud cells on the tongue detect the four basic tastes: sour, salty, bitter, and sweet. The sense of taste is carried by the facial nerves (anterior two thirds of the tongue) and the glossopharyngeal nerves (posterior one third of the tongue). The eye is responsible for the sense of vision. Light entering the eye is refracted by the cornea and the lens so that an image is focused on the macula lutea. Changes in the shape of the lens (accommodation) allow proper focusing of near and distant objects. The retina has cones, which respond best to bright light and are responsible for visual acuity and color vision, and rods, which respond best to low levels of light. The optic nerve carries action potentials from the retina to the optic chiasma, where some of the optic nerve fibers cross over to the opposite side of the brain. Optic nerve fibers synapse in the thalamus, and the thalamic neurons extend to the visual cortex of the brain through the visual radiations. The ear is divided into three parts. The auricle and external auditory meatus of the external ear funnel sound waves to the tympanic membrane. In the middle ear the ear ossicles transmit vibrations of the tympanic membrane to the perilymph of the inner ear. Sound waves in the perilymph cause the vestibular membrane to vibrate, which produces waves in the endolymph of the cochlear duct. These waves cause the basilar membrane to vibrate, producing action potentials in the spiral organ of Corti. The action potentials are transmitted to the brain through the vestibulocochlear nerve. The saccule and the utricle of the inner ear are responsible for detecting the position of the head relative to the ground, whereas the semicircular canals detect the rate of movement of the head in different planes.

WORD PARTS

Give an example of a new vocabulary word that contains each word part.

WORD PART	MEANING	EXAMPLE
olfact-	smell	1. _____
gusta-	taste	2. _____

WORD PART	MEANING	EXAMPLE
scler-	hard	3. _____
vitr-	glassy	4. _____
retin-	a network	5. _____
opsi-	sight	6. _____
fore-	a pit	7. _____
lacr-	tears; weeping	8. _____
oto-	ear	9. _____
lith	a stone	10. _____

CONTENT LEARNING ACTIVITY

Olfaction

"_Olfaction occurs in response to odors that enter the extreme superior portion of the nasal cavity._**"**

A. Match these terms with the correct statement or definition:

Basal cells
Olfactory bulbs
Olfactory epithelium
Olfactory recess
Olfactory tracts
Olfactory vesicles

_____ 1. Specialized cells lining the olfactory recess.

_____ 2. Axons of olfactory neurons project through the cribriform plate to these structures.

_____ 3. These structures project from the olfactory bulbs to the cerebral cortex.

_____ 4. Bulbous enlargements of the dendrites of olfactory neurons.

_____ 5. Lost olfactory neurons in the olfactory epithelium are replaced by these cells.

B. Using the terms provided, complete the following statements:

Association neurons | Medial olfactory area
Intermediate olfactory area | Mitral cells
Lateral olfactory area | Olfactory nerves

Axons from olfactory neurons form the (1) , which enter the olfactory bulbs, where the axons synapse with (2) . The latter cells relay olfactory information to the brain through olfactory tracts and synapse with (3) in the olfactory bulb. Each olfactory tract terminates in the olfactory cortex of the brain, which has three distinct areas. The (4) is involved in the conscious perception of smell, whereas the (5) is responsible for visceral and emotional reactions to odors. The (6) receives considerable input from the other two areas of the olfactory cortex, as well as the olfactory bulb, causing sensory information to be modulated.

1. _____

2. _____

3. _____

4. _____

5. _____

6. _____

Taste

❝Taste buds are specialized sensory structures that detect gustatory, or taste stimuli.**❞**

A. Match these terms with the correct statement or definition:

Circumvallate papillae | Foliate papillae
Filiform papillae | Fungiform papillae

_____ 1. Largest, but least numerous papillae; surrounded by a groove or valley.

_____ 2. Mushroom-shaped papillae; appear as small red dots scattered irregularly over the tongue.

_____ 3. Leaf-shaped papillae; distributed over the sides of the tongue and containing the most sensitive taste buds.

_____ 4. Filament-shaped papillae; most numerous papillae, but with no taste buds.

B. Match these terms with the correct statement or definition:

Epithelial cells | Taste buds
Gustatory (taste) cells | Taste (gustatory) pores
Gustatory hairs

_____ 1. Oval structures, consisting of two types of cells, embedded in the epithelium of the tongue and mouth.

_____ 2. About 40 of these cells are found internally in each taste bud.

_____ 3. Microvilli found on a gustatory cell.

_____ 4. An opening in the epithelium through which gustatory hairs extend.

☞ The tastes detected by the taste buds can be divided into four basic types: sour, salty, bitter, and sweet.

C. Match these nerves with
   their function or description:

   Facial nerve   Vagus nerve
   Glossopharyngeal nerve

_____ 1. Branch of this nerve crosses the tympanic membrane.

_____ 2. Taste from the anterior two thirds of the tongue is carried in this nerve.

_____ 3. Taste from the posterior one third of the tongue is carried by this nerve.

Anatomy of the Eye

❝_The eye is composed of three coats or tunics._**❞**

A. Match these terms with the
   correct statement or definition:

   Cornea   Sclera
   Fibrous tunic   Vascular tunic
   Nervous tunic

_____ 1. Outer layer of the eye, consisting of the sclera and cornea.

_____ 2. Middle layer of the eye, consisting of the choroid, ciliary body, and iris.

_____ 3. Inner layer of the eye, consisting of the retina.

_____ 4. Firm, opaque, white outer layer of the posterior five sixths of the eye.

_____ 5. Avascular, transparent structure that allows light to enter the eye.

B. Match these parts of the vascular
   tunic with the correct statement
   or definition:

   Choroid   Dilator pupillae
   Ciliary body   Iris
   Ciliary processes Pupil
   Ciliary ring   Sphincter pupillae

_____ 1. Vascular tunic associated with the scleral portion of the eye.

_____ 2. Structure attached to the anterior margin of the choroid and the lateral margin of the iris.

_____ 3. Part of the ciliary body that contains smooth muscles (intrinsic eye muscles); attaches to the lens by the suspensory ligaments and changes the shape of the lens.

_____ 4. These structures are a complex of capillaries and cuboidal epithelium that produce aqueous humor and the suspensory ligaments.

_____ 5. Contractile structure consisting mainly of smooth muscle surrounding the pupil.

_____ 6. Radial group of muscles in the iris.

C. Match these terms with the
   correct statement or definition:

   Fovea centralis   Pigmented retina
   Macula lutea   Rods and cones
   Optic disc   Sensory retina

_____ 1. Outer portion of the retina that contains simple cuboidal epithelium cells.

_____ 2. Photoreceptor cells in the retina.

_____ 3. Small yellow spot near the center of the posterior retina.

_____ 4. Small pit that is the portion of the retina with the greatest visual acuity.

_____ 5. Blind spot of the eye; place where blood vessels and nerve processes pass through the outer two tunics of the eye.

D. Match these terms with the correct parts of the diagram labeled in Figure 16-1:

Choroid   Optic nerve
Ciliary body   Pupil
Cornea   Retina
Iris   Sclera
Lens   Suspensory ligaments

1. _____

2. _____

3. _____

4. _____

5. _____

6. _____

7. _____

8. _____

9. _____

10. _____

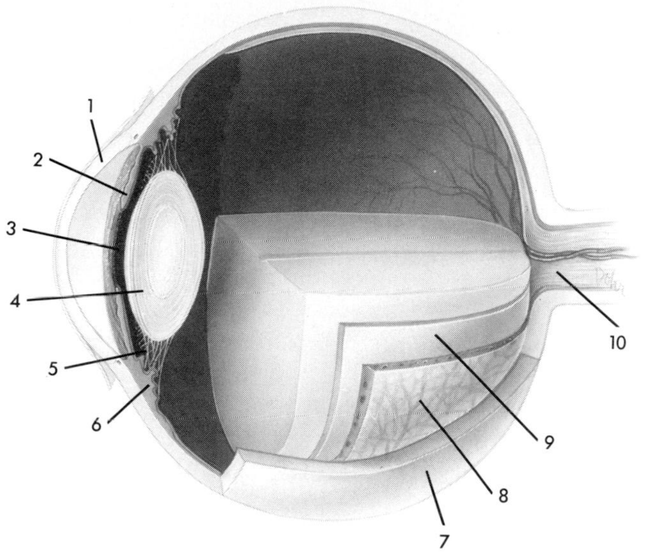

Figure 16-1

E. Match these terms with the correct statement or definition:

Anterior chamber   Posterior chamber
Aqueous humor   Vitreous humor
Canal of Schlemm

_____ 1. Part of the anterior eye cavity that lies between the iris and the lens.

_____ 2. Fluid that fills the anterior and posterior chambers of the eye.

_____ 3. Venous ring that returns aqueous humor to the circulatory system.

_____ 4. Transparent jellylike substance that fills the posterior cavity of the eye.

F. Using the terms provided, complete the following statements:

Capsule   Lens fibers
Crystallines   Suspensory ligaments
Cuboidal epithelial

1. _____

2. _____

3. _____

4. _____

5. _____

The lens consists of a layer of _(1)_ cells on its anterior surface and a posterior portion of very long columnar epithelial cells called _(2)_. Lens fibers lose their nuclei and other cellular organelles and accumulate a special set of proteins called _(3)_. The lens is covered by a highly elastic transparent _(4)_. The lens is suspended between the two eye compartments by the _(5)_, which are connected to the lens capsule and to the ciliary body.

Accessory Structures

❝_Accessory structures aid the function of the eye in many ways._**❞**

A. Match these terms with the correct statement or definition:

Bulbous conjunctiva   Meibomian glands
Canthi   Palpebrae
Caruncle   Palpebral conjunctiva
Chalazion   Sty
Ciliary glands   Tarsal plate

_____ 1. Eyelids.

_____ 2. Angles where the eyelids join at their lateral and medial margins.

_____ 3. Reddish-pink mound in the medial canthus.

_____ 4. Layer of dense connective tissue that helps maintain the shape of the eyelid.

_____ 5. Modified sweat glands that open into the eyelash follicles.

_____ 6. Inflamed ciliary gland.

_____ 7. Glands near the inner margins of the eyelids that produce sebum, an oily semifluid that lubricates the eyelids.

_____ 8. Infection or blockage of a meibomian gland; a meibomian cyst.

_____ 9. Mucous membrane covering the inner surface of the eyelids.

B. Match these terms with the correct statement or definition:

Lacrimal apparatus   Lacrimal sac
Lacrimal canaliculi   Nasolacrimal duct
Lacrimal gland   Punctum
Lacrimal papilla

_____ 1. Gland that produces tears.

_____ 2. Passageway in the medial corner of the eye into which excess tears flow.

_____ 3. Opening into the lacrimal canaliculi.

_____ 4. Small lump upon which the punctum is located.

_____ 5. Structure located between the lacrimal canaliculus and the nasolacrimal duct.

_____ 6. Passageway that opens into the inferior meatus of the nasal cavity.

☞ Tears moisten the surface of the eye, lubricate the eyelids, wash away foreign objects, and contain lysozyme, which kills certain bacteria.

C. Match these terms with the correct parts of the diagram labeled in Figure 16-2:

Lacrimal canaliculi
Lacrimal gland
Lacrimal sac
Nasolacrimal duct
Puncta

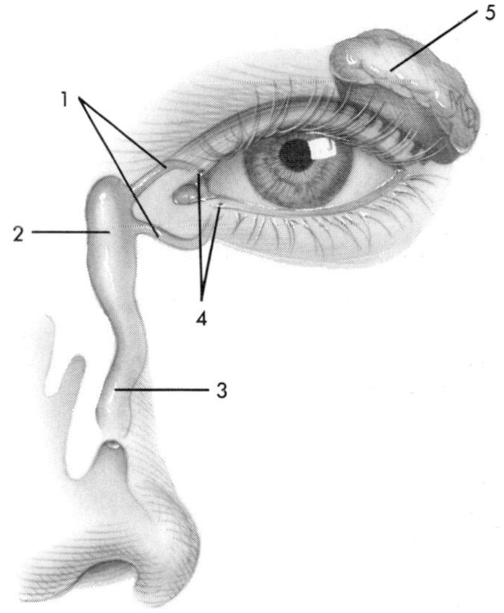

1. _____

2. _____

3. _____

4. _____

5. _____

Figure 16-2

D. Match these extrinsic eye muscles with the correct description:

Oblique muscles
Rectus muscles

_____ 1. Extrinsic eye muscles that run more or less straight anteroposteriorly.

_____ 2. Extrinsic eye muscles that are placed at an angle to the eye.

_____ 3. There are four of these muscles.

Function of the Complete Eye

❝ *The eye functions much like a camera.* **❞**

A. Match these terms with the correct statement or definition:

Concave lens surface   Reflection
Convex lens surface   Refraction
Focal point   Visible light
Focusing

_____ 1. Portion of the electromagnetic spectrum that can be detected by the eye.

_____ 2. Bending of light rays as they pass into a new medium, such as light passing from air into water.

_____ 3. Causes light rays to converge.

_____ 4. Point where convergent light rays cross.

_____ 5. Act of causing light rays to converge.

_____ 6. Light rays that bounce off a nontransparent object.

B. Match these terms with the correct statement or definition:

Accommodation   Far point of vision
Ciliary muscles   Near point of vision
Emmetropia   Suspensory ligaments

_____ 1. Structures that maintain elastic pressure on the lens.

_____ 2. Normal resting, flattened condition of the lens.

_____ 3. Structures that contract and reduce the tension on the lens.

_____ 4. Process of allowing the lens to assume a more spherical (convex) shape.

_____ 5. Point beyond which accommodation is not required; usually 20 feet or more from the eye.

☞ During accommodation the lens becomes more spherical, the pupil constricts, and the eyeballs turn medially.

Structure and Function of the Retina

❝ *The retina consists of several distinct layers.* **❞**

A. Match these parts of the retina with the correct description:

Pigmented retina
Sensory retina

_____ 1. This portion of the retina has three layers of neurons.

_____ 2. This portion of the retina is a layer of cells filled with melanin.

☞ Melanin pigment enhances visual acuity by isolating individual photoreceptors in a black matrix.

B. Match these terms with the correct statement or definition:

Bleaching
Cone cells
Fovea centralis
Iodopsin

Retinal
Rhodopsin
Rod cells

_____ 1. Bipolar photoreceptor cells that cannot detect color, most important in night vision.

_____ 2. Molecule containing opsin in loose chemical combination with retinal; found in rod cells.

_____ 3. Pigment molecule derived from vitamin A.

_____ 4. Process that occurs when rhodopsin is exposed to light, resulting in the separation of retinal and opsin.

_____ 5. Bipolar photoreceptor cells that are sensitive to blue, red, or green light, most important in visual acuity.

_____ 6. Visual pigment in cones.

_____ 7. Part of the retina that has only cone cells.

☞ Away from the fovea centralis, the concentration of cones decreases, and the concentration of rods increases.

C. Match the type of adaptation to changing light conditions with the correct statement:

Dark adaptation
Light adaptation

_____ 1. Amount of rhodopsin increases.

_____ 2. Pupil constricts.

_____ 3. Rod function increases, cone function decreases.

_____ 4. More rapid of the two processes.

D. Match these terms with the correct statement or definition:

Association neurons
Bipolar cells
Ganglion cells

Optic nerve
Photoreceptor cells

_____ 1. Neurons in the outermost layer of the retina.

_____ 2. Innermost layer of neurons in the sensory retina.

_____ 3. Collection of all the ganglion cells as they exit the eye.

_____ 4. Neurons that modify the signal from photoreceptor cells before they leave the retina.

E. Using the terms provided, complete the following statements:

Bipolar   Increases
Decreases   Spatial summation
Ganglion

In the retina numerous rods usually synapse with one _(1)_ cell. Further, many of these cells then synapse with one _(2)_ cell. This arrangement results in _(3)_ that _(4)_ light sensitivity and _(5)_ visual acuity. In comparison, cones exhibit little convergence. This decreases light sensitivity and increases visual acuity.

1. _____
2. _____
3. _____
4. _____
5. _____

Neuronal Pathways

66*There are several neuronal pathways from the eyes through the brain.***99**

A. Using the terms provided, complete the following statements:

Optic chiasma   Thalamus
Optic nerve   Visual cortex
Optic tract   Visual radiations
Superior colliculi

Ganglion cells from the sensory retina converge and exit the eye in the _(1)_, then pass into the cranial vault through the optic foramen. Just inside the vault the optic nerves are connected at the _(2)_, where some of the axons cross to the opposite side of the brain. Beyond the optic chiasma, the route of the ganglionic axons is called the _(3)_. Most of the optic tract axons end in the _(4)_, although some separate from the optic tract and terminate in the _(5)_, the center for reflexes initiated by visual stimuli. Neurons of the thalamus form the _(6)_, projecting to the _(7)_ in the occipital lobe.

1. _____
2. _____
3. _____
4. _____
5. _____
6. _____
7. _____

B. Match the visual fields with the location to which images from that visual field project:

Nasal visual field
Temporal visual field

_____ 1. Images from this visual field fall on the nasal portion of the retina.

_____ 2. Images from this field do not cross to the opposite side of the brain.

☞ Images from the right side of both visual fields (left nasal visual field and the right temporal visual field) project to the left side of the brain, and vice versa.

Clinical Applications

66*There are many common disorders of the eye.***99**

A. Match these terms with the   Astigmatism   Myopia
correct statement or definition: Hyperopia   Presbyopia

_____ 1. Occurs when the lens and cornea are optically too strong, or when the eyeball is too long.

_____ 2. Occurs when the eye becomes less able to accommodate as a result of aging.

_____ 3. Occurs when the cornea or lens is not uniformly curved.

_____ 4. Corrected by a convex lens.

_____ 5. Corrected by reading glasses or bifocals.

_____ 6. Corrected by lenses with curvature opposite to the gradation defect of the eye.

B. Match these terms with the correct statement or definition:

Cataract Glaucoma
Color blindness Strabismus
Diabetes Trachoma

_____ 1. Lack of parallelism of light paths through the eye.

_____ 2. Disease due to increased intraocular pressure because of the buildup of aqueous humor.

_____ 3. Absence or deficiency of one or more cone pigments, inherited as a recessive X-linked trait.

_____ 4. Clouding of the lens resulting from a buildup of proteins.

_____ 5. Disease that causes defective circulation to the eye and is a leading cause of blindness.

_____ 6. Bacterial infection of the corneal epithelial cells.

Auditory Structures and Their Functions

❝_The external, middle, and inner ear are all involved in hearing._**❞**

A. Match these parts of the external ear with the correct description or definition:

Cerumen Pinna
External auditory meatus Tympanic membrane

_____ 1. Fleshy part of the external ear on the outside of the head.

_____ 2. Passageway from outside to the eardrum.

_____ 3. Modified sebum commonly called earwax, that helps to prevent foreign objects from reaching the tympanic membrane.

_____ 4. Thin, semitransparent, nearly oval, three-layered membrane that vibrates in response to sound waves.

B. Match the parts of the middle ear with the correct description or definition:

Auditory tube
Incus
Malleus
Mastoid air cells

Oval window
Round window
Stapes

_____ 1. Membrane-covered opening with the foot plate of the stapes attached.

_____ 2. Membrane-covered opening on the medial side of the middle ear with nothing attached.

_____ 3. Spaces in the mastoid processes of the temporal bones.

_____ 4. Structure that allows air pressure to equalize between the middle ear and the outside air.

_____ 5. Auditory ossicle that is attached to the tympanic membrane.

_____ 6. Middle auditory ossicle.

C. Match these terms with the correct parts of the diagram labeled in Figure 16-3:

Auditory tube
Auricle
Cochlea
External auditory meatus
External ear
Incus
Inner ear
Malleus
Middle ear
Oval window
Round window
Semicircular canals
Stapes
Tympanic membrane
Vestibule

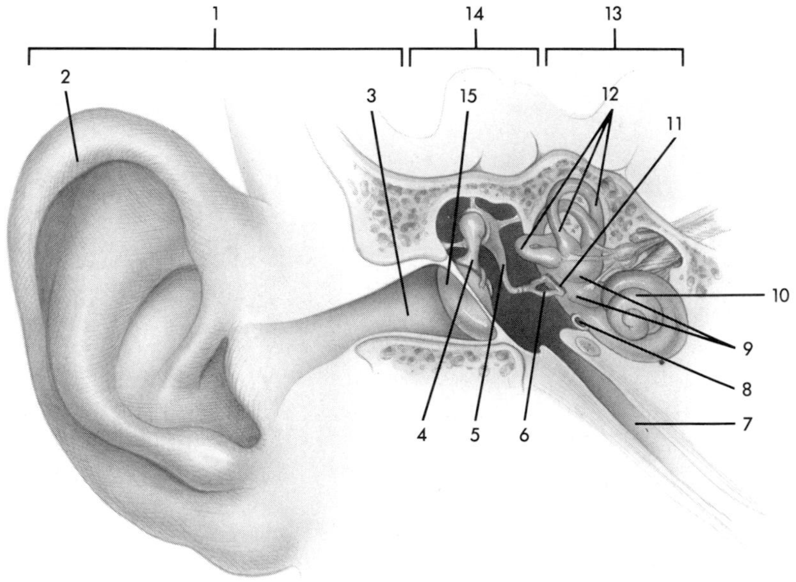

Figure 16-3

1. _____   6. _____   11. _____

2. _____   7. _____   12. _____

3. _____   8. _____   13. _____

4. _____   9. _____   14. _____

5. _____   10. _____   15. _____

D. Match these parts of the inner ear with the correct description:

Basilar membrane
Bony labyrinth
Cochlea
Cochlear duct
Helicotrema
Membranous labyrinth

Modiolus
Scala tympani
Scala vestibuli
Spiral lamina
Spiral ligament
Vestibular membrane

_____ 1. Interconnecting tunnels and chambers in the petrous portion of the temporal bone.

_____ 2. Part of the inner ear that can be divided into scala tympani, scala vestibuli, and cochlear duct.

_____ 3. Cochlear chamber that contains perilymph and extends from the oval window to the helicotrema.

_____ 4. Cochlear chamber that contains perilymph and extends from the helicotrema to the round window.

_____ 5. Wall of the membranous labyrinth bordering the scala vestibuli.

_____ 6. Wall of the membranous labyrinth bordering the scala tympani.

_____ 7. Interior of the membranous labyrinth; space between the vestibular and basilar membrane.

_____ 8. Bony core of the cochlea.

E. Using the terms provided, complete the following statements:

Cochlear ganglion
Cochlear nerve
Hair cells

Organ of Corti
Tectorial membrane
Vestibulocochlear nerve

1. _____

2. _____

3. _____

The cells inside the cochlear duct are highly modified to form a structure called the _(1)_, or spiral organ. This structure contains supporting epithelial cells and specialized sensory cells called _(2)_, which have specialized hairlike projections at their ends. The apical ends of the hairs are embedded within an acellular gelatinous shelf called the _(3)_ which is attached to the spiral lamina. Hair cells have no axons, but the basilar region of each hair cell is covered by synaptic terminals of sensory neurons, the cell bodies of which are grouped into a _(4)_ within the cochlear modiolus. The proximal, afferent fibers of these neurons join to form the _(5)_, which then joins the vestibular nerve to become the _(6)_.

4. _____

5. _____

6. _____

F. Using the terms provided, complete the following statements:

Amplitude
Attenuation reflex
Auditory ossicles
Basilar membrane
Endolymph

Frequency
Scala vestibuli
Tectorial membrane
Tympanic membrane

1. _____

2. _____

3. _____

4. _____

5. _____

6. _____

7. _____

8. _____

9. _____

Vibrations are propagated through the air as sound waves. Volume (loudness) of sound is a function of the _(1)_ (height) of the waves, and pitch is a function of _(2)_ (how far the waves are apart). Sound waves are collected by the auricle and are conducted through the external auditory meatus toward the _(3)_. Sound waves strike the tympanic membrane and cause it to vibrate, which in turn causes the _(4)_ to vibrate. Two small skeletal muscles attached to the ossicles reflexively dampen excessively loud sounds. The contraction of these muscles, called the _(5)_, protects the delicate ear structures from being damaged by loud noises. The vibration of the auditory ossicles is transferred to the oval window and causes vibration in the perilymph of the _(6)_. This produces waves in the perilymph, which causes the vestibular membrane to vibrate, and this vibration creates waves in the _(7)_ and vibration of the _(8)_. As the basilar membrane vibrates, the hairs embedded in the _(9)_ become bent, inducing action potentials in cochlear neurons.

G. Match these terms with the correct parts of the diagram labeled in Figure 16-4:

Basilar membrane
Cochlear duct
Helicotrema
Organ of corti
Oval window
Round window
Scala tympani
Scala vestibuli
Spiral organ
Tectorial membrane
Vestibular membrane

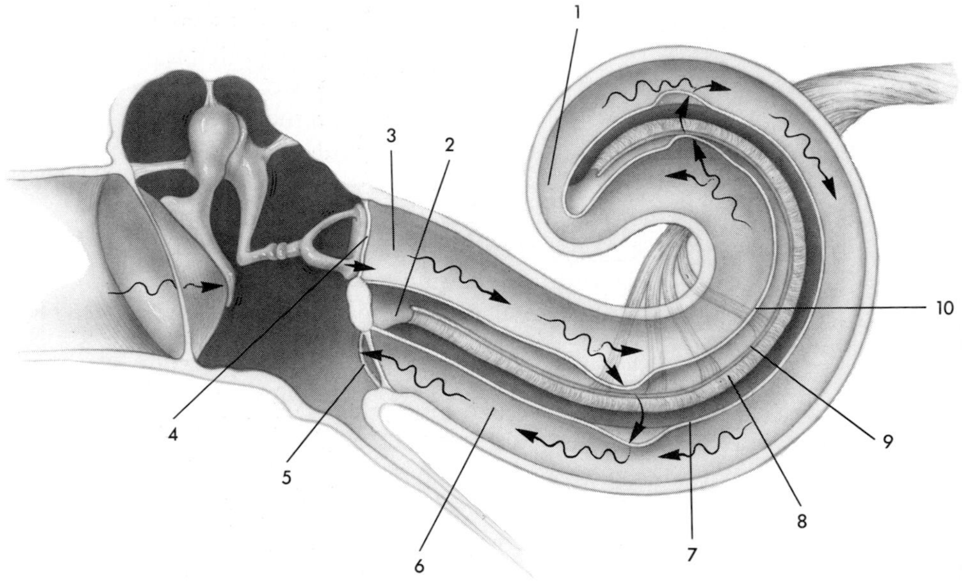

Figure 16-4

1. _____

2. _____

3. _____

4. _____

5. _____

6. _____

7. _____

8. _____

9. _____

10. _____

Neuronal Pathways for Hearing

66*The special senses of hearing and balance are both transmitted by the vestibulocochlear nerve.***99**

Using the terms provided, complete the following statements::

Auditory cortex   Superior colliculus
Inferior colliculi   Superior olivary nucleus
Medulla   Thalamus

The neurons from the cochlear ganglion synapse with the central nervous system neurons in the dorsal or ventral cochlear nucleus in the _(1)_ . These neurons in turn either synapse in or pass through the _(2)_ . Ascending neurons from this point travel in the lateral lemniscus. All ascending fibers synapse in the _(3)_ , and neurons from there project to the medial geniculate nucleus of the _(4)_ , where they synapse with neurons that terminate in the _(5)_ in the dorsal portion of the temporal lobe. Neurons from the inferior colliculus also project to the _(6)_ and cause reflexive turning of the head and eyes toward a loud sound.

1. _____

2. _____

3. _____

4. _____

5. _____

6. _____

Balance

66*The organs of balance are found within the inner ear.***99**

Match these terms with the correct statement or definition:

Ampulla   Macula
Crista ampullaris   Otoliths
Cupula   Semicircular canals
Kinetic labyrinth   Static labyrinth

_____ 1. Structure consisting of utricle and saccule; involved in evaluating the position of the head relative to gravity or linear acceleration.

_____ 2. Specialized patch of epithelium in the utricle and saccule; similar to the organ of Corti.

_____ 3. Structures composed of protein and calcium carbonate that add weight to the gelatinous mass that embeds hair cells.

_____ 4. Structures arranged in three planes, enabling a person to detect movement in all directions.

_____ 5. Specialized sensory epithelium found in each ampulla and similar to a macula.

_____ 6. Curved gelatinous mass suspended over the crest of the crista ampullaris.

Neuronal Pathways for Balance

66 *Balance is a complex process not simply confined to one kind of input.* **99**

Using the terms provided, complete the following statements:

Nystagmus   Vestibular nuclear complex
Proprioceptive nerves   Vestibular nucleus

Neurons synapsing on the hair cells of the maculae and cristae
ampullari converge into the _(1)_ , where their cell bodies are
located. Afferent fibers from these neurons terminate in the
_(2)_ within the medulla. In addition to vestibular sensory input,
the vestibular nuclear complex receives input from the _(3)_
throughout the body. Reflex pathways exist between the kinetic
portion of the vestibular system and the nuclei controlling the
extrinsic eye muscles. Spinning the head causes the eyes to track
slowly in the direction of motion and return with a rapid recovery
movement. This oscillation of the eyes is called _(4)_ .

1. _____

2. _____

3. _____

4. _____

Clinical Applications

66 *There are some disorders that affect the ear.* **99**

Match these terms with the   Otitis media   Tinnitus
correct statement or definition: Otosclerosis

_____   1. Disorder caused by spongy bone growth over the oval window,
                               immobilizing the stapes.

_____   2. Disorder characterized by noises in the ears.

_____   3. Infections of the middle ear.

QUICK RECALL
---

1. **Name the three functional areas of the olfactory cortex and the function of each.**

2. **List the four basic tastes detected by taste buds.**

3. **List the three coats (tunics) of the eye.**

4. Name the two types of photoreceptor cells in the retina, and list two important differences between them.

5. List the two major compartments of the eye and the substances that fill each.

6. List the three layers of neurons in the sensory retina.

7. List the visual pigments found in rod and cone cells.

8. List the three types of cone cells.

9. Name the three auditory ossicles found in the middle ear.

10. List the three subdivisions of the body labyrinth, and give their function.

11. List the three cochlear chambers and the fluid found in each.

12. List the two functional parts of the organs of balance.

13. Name the structures that relieve pressure in the middle ear and the inner ear.

Mastery Learning Activity

Place the letter corresponding to the correct answer in the space provided.

_____ 1. Olfactory neurons
  a. have many projections called cilia.
  b. have axons which combine to form the olfactory nerves.
  c. connect to the olfactory bulb.
  d. have receptors that react with molecules dissolved in fluid.
  e. all of the above

_____ 2. The olfactory cortex can be divided into three areas.
  a. The lateral olfactory area is involved with the emotional reactions to smell.
  b. The medial olfactory area is involved in the conscious perception of smell.
  c. The intermediate olfactory area modifies input from the olfactory bulb.
  d. all of the above

_____ 3. Taste cells
  a. are found only on the tongue.
  b. extend through tiny openings called taste buds.
  c. release neurotransmitter when stimulated.
  d. have axons that extend to the taste area of the cerebral cortex.
  e. all of the above

_____ 4. Which of the following is NOT one the basic tastes?
  a. spicy
  b. sweet
  c. sour
  d. salt
  e. bitter

_____ 5. The fibrous tunic of the eye includes the
  a. conjunctiva.
  b. sclera.
  c. choroid.
  d. iris.
  e. retina.

_____ 6. The ciliary body
  a. contains smooth muscles that attach to the lens by suspensory ligaments.
  b. produces the vitreous humor.
  c. is part of the iris of the eye.
  d. is part of the sclera.

_____ 7. The lens normally focuses light onto the
  a. optic disc.
  b. iris.
  c. macula lutea.
  d. aqueous humor.

_____ 8. Given the following structures:
  1. lens
  2. aqueous humor
  3. vitreous humor
  4. cornea

  Choose the arrangement that lists the structures in the order that light entering the eye would encounter them.
  a. 1, 2, 3, 4
  b. 1, 4, 2, 3
  c. 4, 1, 2, 3
  d. 4, 2, 1, 3
  e. 4, 3, 1, 2

_____ 9. Aqueous humor
  a. is the pigment responsible for the black color of the choroid.
  b. exits the eye through the canal of Schlemm.
  c. is produced by the iris.
  d. can cause cataracts if over produced.

_____ 10. Tears
  a. are released onto the surface of the eye near the medial corner of the eye.
  b. in excess are removed by the canal of Schlemm.
  c. in excess can cause a sty.
  d. can eventually end up in the nasal cavity.

11. Contraction of the smooth muscles in the ciliary body causes the
    a. lens to flatten.
    b. lens to become more spherical.
    c. pupil to constrict.
    d. pupil to dilate.

12. Given the following events:
    1. medial rectus contracts
    2. lateral rectus contracts
    3. pupils dilate
    4. pupils constrict
    5. lens of the eye flattens
    6. lens of the eye becomes more spherical

    Assume that you were looking at an object that was 30 feet away. If you suddenly looked at an object that was 1 foot away, which of the above events would occur?
    a. 1, 3, 6
    b. 1, 4, 5
    c. 1, 4, 6
    d. 2, 3, 6
    e. 2, 4, 5

13. Concerning the arrangement of neurons within the retina,
    a. photoreceptor cells synapse with bipolar cells.
    b. axons of ganglionic cells leave the retina to form the optic nerve.
    c. light must pass through a layer of ganglionic cells and a layer of bipolar cells to reach the photoreceptor cells.
    d. all of the above

14. Which of the following photoreceptor cells is correctly matched with its function?
    a. rods - vision in low light
    b. cones - visual acuity
    c. cones - color vision
    d. all of the above

15. Concerning dark adaptation,
    a. the amount of rhodopsin increases.
    b. the pupils constrict.
    c. it occurs more rapidly than light adaptation.
    d. all of the above

16. In the retina there are cones that are most sensitive to a particular color. Given the following list of colors:
    1. red
    2. yellow
    3. green
    4. blue

    Indicate which colors correspond to specific types of cones.
    a. 2, 3
    b. 3, 4
    c. 1, 2, 3
    d. 1, 3, 4
    e. 1, 2, 3, 4

17. Given the following areas of the retina:
    1. macula lutea
    2. fovea centralis
    3. optic disc
    4. periphery of the eye

    Choose the arrangement that lists the areas according to their cone density, starting with the area that has the highest density of cones.
    a. 1, 2, 3, 4
    b. 1, 3, 2, 4
    c. 2, 1, 4, 3
    d. 2, 4, 1, 3
    e. 3, 4, 1, 2

18. Concerning axons in the optic nerve from the right eye,
    a. they all go to the right occipital lobe.
    b. they all go to the right lobe of the thalamus.
    c. some go to the right occipital lobe, and some go to the left occipital lobe.
    d. some go to the right thalamic lobe, and some go to the left thalamic lobe.

19. A lesion that destroyed the left optic tract of a boy affected his eye by eliminating action potentials that would normally have been generated in response to light in
    a. the nasal retina of the left eye.
    b. the nasal retina of the right eye.
    c. the temporal retina of the left eye.
    d. a and c
    e. b and c

_____ 20. A person with an abnormally long eyeball (anterior to posterior) would be _____ and would wear _____ to correct his or her vision.
   a. nearsighted, concave lens
   b. nearsighted, convex lens
   c. farsighted, concave lens
   d. farsighted, convex lens

_____ 21. The following structures are found within or are a part of the external ear.
   a. oval window
   b. auditory tube
   c. ossicles
   d. auricle
   e. cochlear duct

_____ 22. Given the following ear bones:
   1. incus
   2. malleus
   3. stapes

   Choose the arrangement that lists the ear bones in order from the tympanic membrane to the inner ear.
   a. 1, 2, 3
   b. 1, 3, 2
   c. 2, 1, 3
   d. 2, 3, 1
   e. 3, 2, 1

_____ 23. Given the following structures:
   1. perilymph
   2. endolymph
   3. vestibular membrane
   4. basilar membrane

   Choose the arrangement that lists the structures in the order sound waves coming from the outside would encounter them.
   a. 1, 3, 2, 4
   b. 1, 4, 2, 3
   c. 2, 3, 1, 4
   d. 2, 4, 1, 3

_____ 24. The organ of Corti is found within the
   a. cochlear duct.
   b. scala vestibuli.
   c. scala tympani.
   d. vestibuli.

_____ 25. The _____ allows for release of excess pressure within the INNER ear.
   a. tectorial membrane
   b. basilar membrane
   c. vestibular membrane
   d. round window
   e. Eustachian tube

_____ 26. An increase in the loudness of sound occurs as a result of an increase in the
   a. frequency of the sound wave.
   b. amplitude of the sound wave.
   c. resonance of the sound wave.
   d. a and b

_____ 27. If a person's acoustic reflex were non-functional, how would the perception of sound be affected when that person was exposed to a loud noise?
   a. The noise would be louder than normal.
   b. The noise would be quieter than normal.
   c. The noise would be normal.

_____ 28. Interpretation of sounds of different frequencies is possible because of the ability of the _____ to vibrate at different frequencies and stimulate the _____.
   a. vestibular membrane, vestibular nerve
   b. vestibular membrane, organ of Corti
   c. basilar membrane, vestibular nerve
   d. basilar membrane, organ of Corti

_____ 29. Specialized receptor structure found within the utricle?
   a. macula
   b. crista ampullaris
   c. organ of Corti
   d. cupula

_____ 30. Damage to sensory structures in the semicircular canals would primarily affect the ability to detect
   a. linear acceleration.
   b. the position of the head relative to the ground.
   c. movement of the head in all directions.
   d. none of the above

Use a separate sheet of paper to complete this section.

1. At the battle of Bunker Hill, it could have been said, "Don't shoot until you see the _____ of their eyes!"

2. Crash McBang, a 20-year-old student, wears glasses when he reads. However, he never wears glasses when he drives. His optometrist seems to think that this practice is all right. What eye disorder does Crash probably have? What kind of lens are in his reading glasses?

3. A woman has lost the ability to see "out of the corner of both of her eyes" (i.e., loss of peripheral vision). What visual fields are affected? Assuming that the loss is due to nerve tract damage, where is the most likely place for that damage to have occurred?

4. Retinitis pigmentosa is a hereditary disorder that typically begins in childhood with the development of night blindness. As the disease progresses, the visual fields become concentrically smaller until eventually little useful vision is left. Given that the disorder involved the degeneration of rods and cones, explain the symptoms produced as the disease progresses.

5. Otto Harp notices that he can't hear music when it is played softly, but he can hear perfectly well when it is played loudly. Following his doctor's advice, he gets a hearing aid that solves his hearing problems. Do you think Otto's problem is in the middle or inner ear? How did the hearing aid help? Explain.

6. Compared to their normal position when a person is standing upright, if the hair cells in the macula were stretched (not bent), in what position would a person be relative to the ground? Explain.

7. As part of a game you are spun around and around in a clockwise direction. When you suddenly stop spinning it feels like you are spinning in a counterclockwise direction. Explain.

ANSWERS TO CHAPTER 16

WORD PARTS

1. olfactory; olfaction
2. gustatory
3. sclera
4. vitreous humor
5. retinal

6. opsin; rhodopsin; iodopsin
7. fovea
8. lacrimal
9. otolith
10. otolith

CONTENT LEARNING ACTIVITY

Olfaction

A. 1. Olfactory epithelium
2. Olfactory bulbs
3. Olfactory tracts
4. Olfactory vesicles
5. Basal cells

B. 1. Olfactory nerves
2. Mitral cells
3. Association neurons
4. Lateral olfactory area
5. Medial olfactory area
6. Intermediate olfactory area

Taste

A. 1. Circumvallate papillae
2. Fungiform papillae
3. Foliate papillae
4. Filiform papillae

B. 1. Taste buds
2. Gustatory (taste) cells

3. Gustatory hairs
4. Taste (gustatory) pores

C. 1. Facial nerve
2. Facial nerve
3. Glossopharyngeal nerve

Anatomy of the Eye

A. 1. Fibrous tunic
2. Vascular tunic
3. Nervous tunic
4. Sclera
5. Cornea

B. 1. Choroid
2. Ciliary body
3. Ciliary ring
4. Ciliary processes
5. Iris
6. Dilator pupillae

C. 1. Pigmented retina
2. Rods and cones

3. Macula lutea
4. Fovea centralis
5. Optic disc

D. 1. Cornea
2. Iris
3. Pupil
4. Lens
5. Suspensory ligaments
6. Ciliary body
7. Sclera
8. Choroid
9. Retina
10. Optic nerve

E. 1. Posterior chamber
    2. Aqueous humor
    3. Canal of Schlemm
    4. Vitreous humor

F. 1. Cuboidal epithelial
    2. Lens fibers
    3. Crystallines
    4. Capsule
    5. Suspensory ligaments

Accessory Structures

A. 1. Palpebrae
    2. Canthi
    3. Caruncle
    4. Tarsal plate
    5. Ciliary glands
    6. Sty
    7. Meibomian glands
    8. Chalazion
    9. Palpebral conjunctiva

B. 1. Lacrimal gland
    2. Lacrimal canal iculi
    3. Punctum

    4. Lacrimal papilla
    5. Lacrimal sac
    6. Nasolacrimal duct

C. 1. Lacrimal canal iculi
    2. Lacrimal sac
    3. Nasolacrimal duct
    4. Puncta
    5. Lacrimal gland

D. 1. Rectus muscles
    2. Oblique muscles
    3. Rectus muscles

Function of the Complete Eye

A. 1. Visible light
    2. Refraction
    3. Convex lens surface
    4. Focal point
    5. Focusing
    6. Reflection

B. 1. Suspensory ligament
    2. Emmetropia
    3. Ciliary muscles
    4. Accommodation
    5. Far point of vision

Structure and Function of the Retina

A. 1. Sensory retina
    2. Pigmented retina

B. 1. Rod cells
    2. Rhodopsin
    3. Retinal
    4. Bleaching
    5. Cone cells
    6. Iodopsin
    7. Fovea centralis

C. 1. Dark adaptation
    2. Light adaptation
    3. Dark adaptation
    4. Light adaptation

D. 1. Photoreceptor cells
    2. Ganglion cells
    3. Optic nerve
    4. Association neurons

E. 1. Bipolar
    2. Ganglion
    3. Spatial summation
    4. Increases
    5. Decreases

Neuronal Pathways

A.
1. Optic nerve
2. Optic chiasma
3. Optic tract
4. Thalamus
5. Superior colliculi
6. Visual radiations
7. Visual cortex

B.
1. Temporal visual field
2. Nasal visual field

Clinical Applications

A.
1. Myopia
2. Presbyopia
3. Astigmatism
4. Hyperopia
5. Presbyopia
6. Astigmatism

B.
1. Strabismus
2. Glaucoma
3. Colorblindness
4. Cataract
5. Diabetes
6. Trachoma

Auditory Structures and Their Functions

A.
1. Pinna
2. External auditory meatus
3. Cerumen
4. Tympanic membrane

B.
1. Oval window
2. Round window
3. Mastoid air cells
4. Auditory tube
5. Malleus
6. Incus

C.
1. External ear
2. Auricle
3. External auditory meatus
4. Malleus
5. Incus
6. Stapes
7. Auditory tube
8. Round window
9. Vestibule
10. Cochlea
11. Oval window
12. Semicircular canal
13. Inner ear
14. Middle ear
15. Tympanic membrane

D.
1. Bony labyrinth
2. Cochlea
3. Scala vestibuli
4. Scala tympani
5. Vestibular membrane

6. Basilar membrane
7. Cochlear duct
8. Modiolus

E.
1. Organ of Corti
2. Hair cells
3. Tectorial membrane
4. Cochlear ganglion
5. Cochlear nerve
6. Vestibulocochlear nerve

F.
1. Amplitude
2. Frequency
3. Tympanic membrane
4. Auditory ossicles
5. Attenuation reflex
6. Scala vestibuli
7. Endolymph
8. Basilar membrane
9. Tectorial membrane

G.
1. Helicotrema
2. Cochlear duct
3. Scala vestibuli
4. Oval window
5. Round window
6. Scala tympani
7. Basilar membrane
8. Organ of corti
9. Tectorial membrane
10. Vestibular membrane

Neuronal Pathways for Hearing

1. Medulla
2. Superior olivary nucleus
3. Inferior colliculi
4. Thalamus
5. Auditory cortex
6. Superior colliculus

Balance

1. Static labyrinth
2. Macula
3. Otoliths
4. Semicircular canals
5. Crista ampullaris
6. Cupula

Neuronal Pathways for Balance

1. Vestibular nucleus
2. Vestibular nuclear complex
3. Proprioceptive nerves
4. Nystagmus

Clinical Applications

1. Otosclerosis
2. Tinnitus
3. Otitis media

QUICK RECALL

1. Lateral olfactory area: conscious perception of smell
   Intermediate olfactory area: modulating smell
   Medial olfactory area: visceral and emotional responses to smell
2. Sour, salty, bitter, and sweet
3. Fibrous, vascular, and nervous tunics
4. Rods: high sensitivity to light, but lower visual acuity; shades of gray
   Cones: lower sensitivity to light, but greater visual acuity; color
5. Anterior compartment filled with aqueous humor and posterior compartment filled with vitreous humor
6. Photoreceptors (rods and cones), bipolar cells, and ganglion cells
7. Rods: rhodopsin; Cones: iodopsin
8. Red-sensitive , green-sensitive , and blue-sensitive cone cells
9. Malleus, incus, and stapes
10. Vestibule, cochlea, and semicircular canals. The vestibule and semicircular canals are involved primarily in balance, and the cochlea is involved in hearing.
11. Scala vestibuli: perilymph; Scala tympani: perilymph; and the cochlear duct, filled with endolymph
12. Static labyrinth and the kinetic labyrinth
13. Middle ear: auditory tube; Inner ear: round window

1. E. The olfactory neurons are bipolar neurons with projections called cilia. The cilia have receptors that react to airborne molecules that have dissolved in the fluid covering the olfactory epithelium. Action potentials generated in the olfactory neurons are propagated through the olfactory neuron axons (which form the olfactory nerve) to the olfactory bulb.

2. C. The axons from the intermediate area synapse with association neurons in the olfactory bulb and function to modify olfactory bulb input to the olfactory cortex. The lateral olfactory area is involved with conscious sensation of odors and the medial olfactory area with the emotional reactions to smell.

3. E. Taste cells are found within taste buds. The gustatory hairs of taste cells extend through openings called taste pores. When receptors react with substances dissolved in fluid, the taste cells release a neurotransmitter that stimulates action potentials in the neurons of cranial nerves (trigeminal, glossopharyngeal, and vagus). The neurons synapse with neurons that extend to the thalamus. Thalamic neurons then extend to the taste area of the cerebral cortex.

4. A. The four basic tastes are sour, salty, bitter, and sweet.

5. B. The fibrous tunic includes the sclera and cornea. The vascular tunic includes the choroid, ciliary body, and iris. The nervous tunic consists of the retina. The conjuctiva is a mucous membrane that covers the anterior sclera and inner surface of the eyelids.

6. A. The ciliary body, which is part of the choroid, consists of an outer ciliary ring (smooth muscles that attach to the lens by suspensory ligaments) and an inner group of ciliary processes that produce aqueous humor.

7. C. The macula lutea is located almost exactly at the posterior pole of the eye. A depression, the fovea centralis, is located in the center of the macula lutea. The fovea centralis is the most sensitive portion of the retina. The optic disc contains no photoreceptor cells and is also called the blind spot. The iris regulates the amount of light reaching the lens.

8. D. Light passes through the cornea, aqueous humor, lens, and vitreous humor.

9. B. The ciliary processes produce aqueous humor, which circulates through the anterior cavity of the eye and exits through the canal of Schlemm. The aqueous humor maintains intraocular pressure, provides nutrients, and refracts light. Overproduction of aqueous humor results in glaucoma.

10. D. The lacrimal glands are located in the superolateral corner of the orbit. Released tears move across the anterior surface of the eye due to gravity and blinking. The tears exit through the lacrimal canaliculi in the medial corner of the eye and eventually reach the nasal cavity.

11. B. When the ciliary muscles contract, tension on the suspensory ligaments is reduced, and the lens becomes more spherical.

12. C. The medial rectus would contract, directing the gaze more medially. The pupil would constrict, increasing the region around the object that is focused on the retina. The lens would become more spherical (accommodation).

13. D. Photoreceptor cells synapse with bipolar cells that synapse with ganglionic cells. The axons of the ganglionic cells form the optic nerve. Light passes through the ganglionic and bipolar layers to reach the photoreceptor cells, which are located next to the pigmented retina.

14. D. Rods are responsible for vision in low light. Cones are responsible for vision in bright light, color vision, and visual acuity (the fovea centralis has only cones)

15. A. During dark adaptation the production of rhodopsin occurs more rapidly than the break down of rhodopsin, and the amount of rhodopsin increases, making the rods more sensitive to low levels of light. This process occurs more slowly than light adaptation, which is a decrease in rhodopsin that occurs when the rate of breakdown of rhodopsin exceeds its rate of production. During dark adaptation the pupils dilate, and during light adaptation they constrict.

16. D. The three kinds of cones are blue sensitive, red sensitive, and green sensitive.

17. C. The fovea centralis has the highest density of cones (and has no rods). The macula lutea, in which the fovea centralis is located, has the next highest density of cones. From the macula lutea toward the periphery the number of cones decreases (and the number of rods increases). The optic disc has no cones or rods.

18. D. Some of the axons in the optic nerve go to the right thalamic lobe, and others cross to the left thalamic lobe through the optic chiasma. In the thalamus the axons synapse with thalamic neurons that extend to the visual cortex through the visual radiations.

19. E. Neurons that carry action potentials from the temporal portion of the retina do not decussate, whereas the neurons that carry action potentials from the nasal portion of the retina do decussate in the optic chiasma. Thus the left optic tract is composed of neurons that carry action potentials from the temporal retina of the left eye and the nasal retina of the right eye.

20. A. An abnormally long eyeball would result in images being focused anterior to the retina. This condition is nearsightedness and is corrected with a concave lens.

21. D. The external ear includes the auricle (pinna), external auditory meatus, and the tympanic membrane. The middle ear includes the ossicles, oval window, round window, and auditory tube. The cochlear duct is part of the inner ear.

22. C. The malleus is attached to the tympanic membrane. The incus connects the malleus to the stapes, which is attached to the oval window.

23. A. Movement of the stapes produces sound waves in the perilymph of the scala vestibuli. The waves cause the vestibular membrane to vibrate, which produces waves in the endolymph of the cochlear duct. Waves in the endolymph cause the basilar membrane to vibrate.

24. A. The cochlea is subdivided into three parts: the cochlear duct, the scala vestibuli, and the scala tympani. The organ of Corti is found within the cochlear duct. The vestibule is the entry way of the inner ear that connects to the middle ear through the oval window.

25. D. The round window is covered by a membrane. When pressure in the inner ear increases, the membrane bulges outward, relieving the pressure. The auditory tube equalizes pressure between the middle ear and the external environment.

26. B. The loudness of a sound is directly proportional to the amplitude of the sound wave, whereas the pitch of the sound is a function of the wave frequency.

27. A. When one is exposed to a loud noise, the acoustic reflex initiates contraction of the tensor typani and stapedius muscles. This causes the ossicles of the middle ear to become more rigid and dampen the amplitude of sound vibrations. The result is that loud sounds are perceived as being more quiet. The reflex may function to protect the inner ear from loud noises. Without the reflex a noise would be louder than normal.

28. D. High-pitched sounds cause the base of the basilar membrane to vibrate the most, whereas low-pitched sounds cause the apex to vibrate the most. Vibration of the basilar membrane causes hair cells in the organ of Corti to be bent. The hair cells induce action potentials in the cochlear nerve (the vestibular nerve is involved with balance).

29. A. The macula is found within the utricle. The cupula is a gelatinous mass that is part of the crista ampullaris. The crista ampullaris is part of the semicircular canals. The organ of Corti is found within the cochlea.

30. C. The semicircular canals detect the rate of movement of the head in all directions. The utricle and the saccule detect linear acceleration and the position of the head relative to the ground.

1. Sclera. The sclera forms the "whites of the eyes".

2. Farsightedness is characterized by normal vision at a distance but requires corrective concave lens to see close up because the near point of vision is greater than normal. Due to his young age, presbyopia is not likely.

3. Loss of peripheral vision "out of the corner of the eyes" affects the nasal visual fields. The most likely place for nerve tract damage is in the optic chiasma.

4. Early in its development the disease affects rods, causing night blindness. Also, the disease must start at the periphery of the retina and gradually extend toward the macula lutea, thus accounting for the concentric narrowing of vision.

5. Otto's hearing problem is in the middle ear and is caused by reduced transfer of sound from the typmanic membrane to the perilymph of the inner ear. The hearing aid amplifies sound and increases the movement of the tympanic membrane and thus the ear ossicles. The increased movement of the ossicles better transfers sound waves to the perilymph of the inner ear. A problem with the organ of Corti or nerve pathways is not indicated since he hears all frequencies of sound when they are loud enough.

6. If the hair cells of the macula were stretched (not bent), a person would be upside down. Gravity would act on the otoliths in the macula and cause the hair cells to stretch. If the hair cells were bent, it could indicate that the person was lying on their side.

7. When spinning in a clockwise direction, at first the endolymph in the semicircular canals lags behind the movement of the skull bones. Consequently, the hair cells embedded in the cupula are pushed in the opposite direction of the spin; this is interpreted by the brain as clockwise spin. Note that the perceived motion of spin (clockwise) is the opposite of the movement of the cupula (counterclockwise). As spinning continues, the endolymph catches up with the skull bones, and the cupula is no longer pushed over. When spinning stops, the skull bones immediately stop moving, but the endolymph in the semicircular canals continues to move, causing the hair cells embedded in the cupula to bend in the direction of the spin. The clockwise movement of the cupula is interpreted as counterclockwise movement, even though no such movement of the head is actually occurring.

Functional Organization Of The Endocrine System

FOCUS: Endocrine tissues internally produce hormones that are released into the blood, where they are carried to target tissues and produce a response. Hormones are either proteins, polypeptides, amino acid derivatives, or lipids. The release of hormones can be regulated by nonhormone substances (e.g., glucose, calcium, sodium), by other hormones, or by the nervous system. Some hormones are released at relatively constant rates, others are released suddenly in response to certain stimuli, and a few increase or decrease in a cyclic fashion. Once in the blood, hormones are transported dissolved in plasma or bound to plasma proteins. Hormones are eliminated from the blood primarily by excretion into urine or bile and by enzymatic degradation. Hormones bind to protein or glycoprotein receptors in their target tissues. Each hormone has a specific receptor, and target tissues have more than one kind of receptor. In some cases the target tissue response to the hormone is constant; in others it may decrease due to a decrease in the number of receptors (down regulation) or increase due to an increase in the number of receptors (up regulation). Hormone receptors can be classified as membrane-bound or intracellular receptors. When a hormone binds to a membrane-bound receptor, membrane permeability may change, or a second messenger (e.g., cyclic AMP) may be produced. The change in membrane permeability or the second messenger are responsible for activating the response of the target tissue to the hormone. When a hormone binds to an intracellular receptor, genes are activated, and enzymes are synthesized, producing the response of the target tissue.

WORD PARTS

Give an example of a new vocabulary word that contains each word part.

WORD PART	MEANING	EXAMPLE
endo-	within	1. _____
hormon-	to set into motion	2. _____

WORD PART	MEANING	EXAMPLE
modul-	measure	3. _____
para-	near	4. _____
conjug-	joined together	5. _____
glyco-	sweet	6. _____
poly-	many	7. _____
anti-	against	8. _____
diure-	urinate	9. _____
neuro-	nerve	10. _____

— **CONTENT LEARNING ACTIVITY** —

General Characteristics of the Endocrine System

"*Endocrine glands secrete their products internally and influence tissues that are separated by***"** *some distance from the endocrine glands.*

A. Match these terms with the correct statement or definition:

Amplitude-modulated signals Frequency-modulated signals
Endocrine system   Hormones

_____ 1. Composed of glands that secrete their products into the circulatory system.

_____ 2. The secretory products of endocrine glands.

_____ 3. Communication between cells that involves increases or decreases in the concentration of hormones in the body fluids.

_____ 4. The method of communication between neurons and their effectors; an all-or-none response.

B. Match these chemical messengers with their correct location in the following table:

Hormone
Neurohormone
Neuromodulator

Neurotransmitter (neurohumor)
Parahormone
Pheromone

INTERCELLULAR CHEMICAL MESSENGER	PRODUCED BY	FUNCTION	EXAMPLE
1._____	Specialized cells	Travels in blood; influences specific activities	Thyroxine
2._____	Neurons	Like hormones	Oxytocin, ADH
3._____	Neurons	Released from pre-synaptic terminals; affects postsynaptic terminals	Acetylcholine, epinephrine
4._____	Neurons	Released from pre-synaptic terminals; alters sensitivity of postsynaptic neurons to neurotransmitters	Prostaglandins, endorphins
5._____	Wide variety of tissue	Secreted into tissue space; localized effect	Prostaglandins, histamine

☞ Hormones and neurohormones are either proteins, polypeptides, amino acid derivatives, or lipids.

Control of Secretion Rate

"Most hormones are not secreted at a constant rate; most endocrine glands increase and decrease their secretory activity dramatically over time."

Using the terms provided, complete the following statements:

Hormone   Nonhormone
Neural

1. _____

2. _____

3. _____

There are three major patterns of regulation for hormones. An example is the effect glucose has on the secretion of insulin. This is a case of a _(1)_ influencing an endocrine gland. The release of antidiuretic hormone from the pituitary is a different situation. In this pattern the release of a hormone is regulated by _(2)_ control. A third pattern occurs when thyroid-stimulating hormone causes the thyroid to release thyroid hormones. In this case a _(3)_ controls the secretory activity of an endocrine gland.

Negative-feedback mechanisms play essential roles in maintaining hormone levels within normal concentration ranges. There are a few examples of positive-feedback mechanisms (e.g., oxytocin during delivery).

Transport and Distribution in the Body

"*Because hormones circulate in the blood, they are distributed throughout the body.***"**

Using the terms provided, complete the following statements:

Equilibrium   Plasma proteins
Free hormone   Target tissue

Hormones are dissolved in blood plasma and are transported either in a free form or bound to (1) . An (2) exists between the unbound hormone and those bound to plasma proteins. A large increase or decrease in plasma protein concentration can influence the concentration of (3) in the blood. In general, the amount of free hormone that reaches the (4) is directly correlated with the concentration of hormone in the blood.

1. _____

2. _____

3. _____

4. _____

Metabolism and Excretion

"*The destruction and elimination of hormones limits the length of time they are active.***"**

A. Match these factors with the correct statement or definition:

Active transport   Metabolism
Conjugation   Reversible binding
Excretion   Structural protection

_____ 1. Elimination of hormones from the circulatory system into the urine or bile.

_____ 2. In this process hormones are enzymatically degraded.

_____ 3. Hormones are made less active and eliminated by attaching compounds such as sulfates to them.

_____ 4. Method of prolonging the half-life of hormones by reacting with plasma proteins.

_____ 5. Half-life of hormones is prolonged by the carbohydrate components of glycoprotein hormones.

B. The length of time it takes for half a dose of a substance to be eliminated from the circulatory system is its half-life. Match the length of the half-life with the correct hormone characteristic:

Long half-life
Short half-life

_____ 1. Typical of water-soluble hormones (proteins, glycoproteins, and epinephrine).

_____ 2. These hormones generally regulate activities that have a rapid onset and a short duration.

_____ 3. The concentration of these hormones is maintained at a relatively constant level through time.

Interaction of Hormones with Their Target Tissue

66*Hormones bind to receptors in their target tissues and alter the rate at which certain activities occur.*99

Match these terms with the correct statement or definition:

Down regulation   Up regulation
Hormone receptors

_____ 1. Protein or glycoprotein molecules of specific three-dimensional shape that bind to a single type of hormone.

_____ 2. Decrease in the number of hormone receptors after exposure to certain hormones.

_____ 3. Increase in the number of hormone receptors after exposure to certain hormones.

☞ A hormone may bind to a number of different types of receptors. Different cell types can have different types of receptors.

Classes of Hormone Receptors

66*There are two major classes of hormone receptors.*99

A. Match these terms with the correct statement or definition:

Intracellular receptors
Membrane-bound receptors

_____ 1. Water-soluble or large-molecular-weight hormones bind to these.

_____ 2. Lipid-soluble hormones (e.g., steroids, thyroid hormones) bind to these.

_____ 3. When these are activated, a change in membrane permeability can result, or the activity of already existing enzymes are increased or decreased.

_____ 4. When these are activated, new enzymes (proteins) are produced.

343

B. Using the terms provided, complete the following statements:

Acceptor molecule   Phosphodiesterase
Cascade effect   Protein synthesis
Cyclic AMP   Second messenger
First messenger

The combination of a hormone with its membrane-bound
receptor may lead to the activation of an enzyme in the
membrane. Within the cell the activated enzyme catalyzes the
production of a chemical that diffuses throughout the cytoplasm,
binds to specific enzymes, and alters their activity. The hormone
acts as the _(1)_, which carries a signal to the cell membrane. The
chemical produced at the membrane carries the signal into the
cell and thus is called the _(2)_. The second messenger molecule
in many cells is _(3)_, and for each cell type a different set of
enzymes is stimulated. Cyclic AMP is broken down by the
enzyme _(4)_, which limits the length of time cAMP influences
activities in the cell. Hormones that stimulate synthesis of a
second messenger often produce a rapid response because, when
enzymes are activated, they in turn activate other enzymes. This
is called a _(5)_. Hormones may bind to intracellular receptors
inside the cell. The hormone-receptor combination diffuses into
the nucleus and binds to an _(6)_. As a result, DNA produces
mRNA, which moves into the cytoplasm and initiates _(7)_.

1. _____

2. _____

3. _____

4. _____

5. _____

6. _____

7. _____

1. List six types of chemical messengers produced by cells.

2. Explain the three mechanisms for regulation of hormones, and give an example of each type.

3. List three important patterns of secretion of hormones.

4. List four means by which hormones are eliminated from the circulatory system.

5. List two means by which the half-life of hormones is prolonged.

6. List the two major classes of hormone receptors, and list the types of hormone molecules that bind to each.

MASTERY LEARNING ACTIVITY

Place the letter corresponding to the correct answer in the space provided.

_____ 1. An endocrine gland
   a. lacks a duct.
   b. releases its hormone into the surrounding interstitial fluid.
   c. depends upon the blood for transport of its secretion.
   d. all of the above

_____ 2. When comparing the endocrine system to the nervous system, generally speaking, the endocrine system
   a. is faster acting than the nervous system.
   b. produces effects that are of a shorter duration.
   c. is regulated by less elaborate feedback mechanisms.
   d. uses amplitude modulated signals.

_____ 3. Given the following list of molecule types:
1. amino acid derivatives
2. fatty acid derivatives
3. polypeptides
4. proteins
5. steroids

Of the molecule types listed, which include hormones?
a. 1, 2, 3
b. 2, 3, 4
c. 1, 2, 3, 4
d. 2, 3, 4, 5
e. 1, 2, 3, 4, 5

_____ 4. The secretion of a hormone from an endocrine tissue is regulated by
a. other hormones.
b. nonhormone substances in the blood.
c. the nervous system
d. all of the above

_____ 5. Hormones are released into the blood
a. at relatively constant levels.
b. in large amounts in response to a stimulus.
c. in a cyclic fashion (increasing and decreasing in amounts in a regular fashion).
d. all of the above

_____ 6. Concerning the half-life of hormones,
a. lipid-soluble hormones generally have a longer half-life.
b. hormones with shorter half-lives regulate body action more precisely than hormones with longer half-lives.
c. hormones with a longer half-life are maintained at more constant levels in the blood.
d. all of the above

_____ 7. Given the following observations:
1. A hormone will affect only a specific tissue (i.e., a hormone will not affect all tissues).
2. A tissue can respond to more than one hormone.
3. Some tissues respond rapidly to a hormone while others take many hours to respond.

Which of the observations can be explained by hormone receptors?
a. 1
b. 1, 2
c. 2, 3
d. 1, 3
e. 1, 2, 3

_____ 8. Down regulation
a. results in a decrease in the number of receptors in the target tissue.
b. produces an increase in the sensitivity of the target tissue to the hormone.
c. is found in target tissues that respond to hormones that are maintained at constant levels.
d. all of the above

_____ 9. When a hormone binds to a membrane-bound receptor,
a. the receptor may change shape.
b. the permeability of the membrane may change.
c. enzyme activity may be altered.
d. all of the above

_____ 10. Given the following events:
1. activation of cyclic AMP
2. activation of genes
3. increase of enzyme activity

Which of the events occurs when a hormone binds to an intracellular hormone receptor?
a. 1
b. 1, 2
c. 2, 3
d. 1, 2, 3

Use a separate sheet of paper to complete this section.

1. A diabetic hears about a method of birth control that uses a skin patch containing estrogen. The estrogen diffuses from the patch through the skin, ensuring a steady release of small amounts of estrogen into the blood. The estrogen then inhibits the development of the ovum (see Chapter 28). The diabetic wonders if such a system could be used to administer insulin. What would you tell him?

2. Assuming that negative feedback is the major means by which a hormone's secretion rate is controlled and that the hormone causes the concentration of a substance called X to decrease in the blood, predict the effect on the rate of hormone secretion if some abnormal condition causes the levels of X in the blood to remain higher than normal.

3. Suppose the following experimental data were collected: (1) a few minutes after exposure to a hormone the target tissue began producing a secretion, and (2) after several hours of continual exposure to the hormone the production of secretion by the target tissue decreased. What conclusions can you come to about the regulatory mechanisms that control the target tissue's secretion? Would the response of the target tissue be mediated by a second messenger mechanism or by an intracellular receptor mechanism? Explain.

4. Given that you could measure any two intracellular substances, which two would you choose if you wanted to determine if the response of a cell to a hormone was mediated by the second messenger mechanism or by the intracellular receptor mechanism?

5. Antidiuretic hormone decreases the amount of urine produced by the kidneys. What would happen to urine production if the conjugation of antidiuretic hormone in the liver were impaired?

ANSWERS TO CHAPTER 17

WORD PARTS

1. endocrine
2. hormone
3. modulated
4. parahormone
5. conjugation

6. glycoprotein
7. polypeptide
8. antidiuretic
9. antidiuretic
10. neurohormone; neuron

CONTENT LEARNING ACTIVITY

General Characteristics of the Endocrine System

A. 1. Endocrine system
    2. Hormones
    3. Amplitude-modulated signals
    4. Frequency-modulated signals

B. 1. Hormone
    2. Neurohormone
    3. Neurotransmitter (neurohumor)
    4. Neuromodulator
    5. Parahormone

Control of Secretion Rate

1. Nonhoromone
2. Neural

3. Hormone

Transport and Distribution in the Body

1. Plasma proteins
2. Equilibrium

3. Free hormone
4. Target tissue

Metabolism and Excretion

A. 1. Excretion
    2. Metabolism
    3. Conjugation
    4. Reversible binding
    5. Structural protection

B. 1. Short half-life
    2. Short half-life
    3. Long half-life

Interaction of Hormones with Their Target Organs

1. Hormone receptors
2. Down regulation

3. Up regulation

Classes of Hormone Receptors

A. 1. Membrane-bound receptors
   2. Intracellular receptors
   3. Membrane-bound receptors
   4. Intracellular receptors

   4. Phosphodiesterase
   5. Cascade effect
   6. Acceptor molecule
   7. Protein synthesis

B. 1. First messenger
   2. Second messenger
   3. Cyclic AMP

QUICK RECALL

1. Hormones, neurohormones, neurotransmitters, neuromodulators, parahormones, and pheromones
2. Regulation by a nonhormone (glucose); by neural control (ADH); or by a hormone (thyroid hormone)
3. Relatively constant; change suddenly in response to stimuli; and change occurring in cycles

4. Excretion, enzymatic degradation, conjugation, and active transport
5. Reversible binding and structural protection
6. Membrane-bound receptors: water-soluble or large molecular weight hormones
   Intracellular receptors: lipid-soluble hormones

MASTERY LEARNING ACTIVITY

1. D. An endocrine gland has no ducts, releases its secretions into the surrounding interstitial fluid, and depends on the circulatory system to transport the secreted hormone to its target tissue.

2. D. The endocrine system uses amplitude-modulated signals (increase or decrease in hormone concentration), whereas the nervous system uses frequency-modulated signals (the number of action potentials produced). The endocrine system also tends to be slower acting than the nervous system but produces longer lasting effects. The endocrine sytem also has more elaborate feedback mechanisms.

3. E. s correct. Hormones can be classified into a "protein" group (i.e., amino acid derivatives, polypeptides, proteins, and glycoproteins) and a lipid group (i.e., steroids and fatty acid derivatives).

4. D. Other hormones, nonhormone substances, and the nervous system can regulate hormone secretion.

5. D. Some hormones are released at a constant rate (i.e., hormones involved with long-term maintenance), some are released in large amounts in response to a stimulus (allowing rapid adjustment to changing conditions), and some are released in a cyclic fashion (allowing periodic changes).

6. D. Lipid-soluble hormones (lipid hormones, thyroid hormones) bind to plasma proteins, increasing their half-life and contributing to the maintenance of constant levels of lipid-soluble hormones in the blood. Water-soluble hormones (e.g., protein hormones and epinephrine) are rapidly degraded by enzymes and have a short half-life. Hormones with a long half-life are maintained at constant levels in the blood whereas hormones with a short half-life regulate activities that have a rapid onset and a short duration. Hormones with a short half-life allow precise regulation of body actions because the level of the hormones can rapidly be adjusted.

7. E. A hormone receptor is specific for a given hormone, and only tissues with that receptor can respond to the hormone. Since a tissue can have more than one kind of receptor, a tissue can respond to more than one kind of hormone. Membrane-bound receptors produce rapid responses in tissues, whereas intracellular receptors may take hours to respond.

8. A. Down regulation results in a decrease in the number of receptors in the target tissue, due to decreased receptor synthesis and/or increased receptor degradation. The decrease in receptors decreases the sensitivity of the target tissue to hormone. Down regulation is found in target tissue that respond to short-term increases in hormone concentration.

9. D. Membrane-bound receptors operate in different ways. In some cases the receptor changes shape, producing a permeability change in the membrane. Other receptors activate a second messenger, which leads to an alteration of enzyme activity.

10. C. The combination of a hormone with an intracellular receptor leads to activation of genes. The gene produces mRNA, which leaves the nucleus; at the ribosome the mRNA is involved in the production of enzymes (proteins). The enzyme activity that results is responsible for the response of the cell to the hormone.

 FINAL CHALLENGES

1. Insulin levels normally change in order to maintain normal blood sugar levels, despite periodic fluctuations in sugar intake. A constant supply of insulin from a skin patch would result in insulin levels that would be too low when blood sugar levels were high (after a meal) and would be too high when blood sugar levels were low (between meals).

2. Negative feedback mechanisms operate by returning values to normal levels (homeostasis). If substance X is higher than normal, the rate of hormone secretion should increase, because this would normally cause the level of substance X to decrease back toward a normal value.

3. One possibility is down regulation, in which exposure to the hormone resulted in a decrease in the number of hormone receptors and therefore a decreased responsiveness to the hormone. Another possiblity is that the secretion of the target tissue produced an effect elsewhere in the body such as the production of another hormone, and the other hormone then inhibited the target tissue.

Since the target tissue began producing its secretion within a few minutes after being exposed to the hormone, this suggests that the second-messenger mechanism activated enzymes already present in the target tissue. With the intracellular mechanism one would expect a delay before production began, followed by an increase in the amount of secretion produced.

4. An increase in cyclic AMP or other second messengers would indicate the second messenger mechanism, and an increase in mRNA would indicate the intracellular receptor mechanism.

5. Impairment of antidiuretic hormone conjugation in the liver would increase the half-life of the hormone, increasing the concentration of the hormone in the blood. Consequently, one would expect a lower-than-normal production of urine.

Endocrine Glands

FOCUS: The major endocrine structures are in the brain (i.e., hypothalamus, pituitary gland, and pineal body) or are specialized glands (i.e., thyroid gland, parathyroid glands, thymus gland, pancreas, adrenal glands, testes, or ovaries). However, other tissues (e.g., the stomach, small intestine, liver, kidneys, and placenta) also produce hormones. The hypothalamus plays a major role in regulating the secretions of the pituitary gland. Through the hypothalamohypophyseal portal system the anterior pituitary receives regulatory hormones from hypothalamic neurosecretory cells. In response to these regulatory hormones, the anterior pituitary releases its own hormones. The posterior pituitary receives and stores the hormones produced by hypothalamic neurosecretory cells. These hormones are released in response to action potentials from the hypothalamus. The pituitary hormones affect a number of different target tissues, including other endocrine glands, and they regulate a variety of functions. Growth hormone (GH), thyroid-stimulating hormone (TSH), and adrenocorticotropin hormone (ACTH) all affect metabolism: GH directly promotes growth and stimulates the production of somatomedins, which also affect growth; TSH stimulates the thyroid gland to produce thyroid hormones, which increase metabolic rate; and ACTH stimulates the adrenal cortex to release cortisol, which promotes fat and protein breakdown. A number of hormones from the pituitary are involved in reproduction: oxytocin increases uterine contractions during delivery and is responsible for milk release during lactation; prolactin stimulates milk production; follicle-stimulating hormone (FSH) and luteinizing hormone (LH) are necessary for the development of sperm or ova and they also regulate the production of male (testosterone) and female (estrogen and progesterone) sex hormones. Other pituitary hormones include antidiuretic hormone (ADH), which increases water reabsorption in the kidneys, and melanocyte-stimulating hormone (MSH), which increases skin pigmentation. Many important body functions are regulated by other endocrine glands. For example, calcium levels are regulated by parathyroid hormone from the parathyroid glands and by calcitonin from the parafollicular cells of the thryoid gland; blood sugar levels are maintained by insulin and glucagon from the pancreas; sodium, potassium, and hydrogen ion levels are regulated by aldosterone from the adrenal cortex; epinephrine from the adrenal medulla assists the sympathetic nervous system during times of physcial activity; thymosin from the thymus gland is necessary for the immune system to function properly; and melatonin from the pineal body may be involved with the onset of puberty.

WORD PARTS

Give an example of a new vocabulary word that contains each word part.

WORD PART	MEANING	EXAMPLE
thalam-	chamber	1. _____
physis	growth	2. _____
adeno-	gland	3. _____
trop-	change; turn	4. _____
gon(ado)	a seed	5. _____
troph-	nourish	6. _____
pars	a part	7. _____
lacto-	milk	8. _____
folli-	a bag	9. _____
algesia	pain	10. _____

CONTENT LEARNING ACTIVITY

Introduction

Match these terms with the correct parts of the diagram labeled in Figure 18-1:

Adrenal gland
Gastrointestinal tract
Ovary
Pancreas

Pituitary
Testis
Thymus
Thyroid

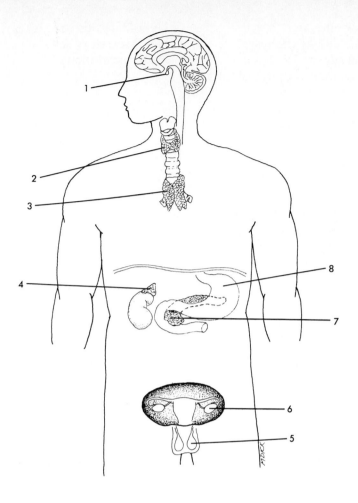

1. _____

2. _____

3. _____

4. _____

5. _____

6. _____

7. _____

8. _____

Figure 18-1

Pituitary Gland and Hypothalamus

❝*The hypothalamus and the pituitary are the major sites where the two major regulatory systems in the body interact.*❞

A. Match these terms with the correct statement or definition:

Action potentials
Adenohypophysis
Hypothalamus
Hypothalamohypophyseal
  portal system

Hypothalamohypophyseal tract
Neurohormones
Neurohypophysis

_____ 1. Also called the posterior pituitary.

_____ 2. Develops from the roof of the mouth.

_____ 3. Communicates between the hypothalamus and the adenohypophysis.

_____ 4. Neurohormones secreted from the neurohypophysis are produced here.

_____ 5. Hormones secreted from the adenohypophysis are produced here.

_____ 6. These regulate the release of hormones from the neurohypophysis.

_____ 7. These regulate the release of hormones from the adenohypophysis.

B. Match these terms with the correct parts of the diagram labeled in Figure 18-2:

Anterior pituitary
Hypothalamohypophyseal portal system
Hypothalamohypophyseal tract
Hypothalamus
Neurosecretory cells
Posterior pituitary

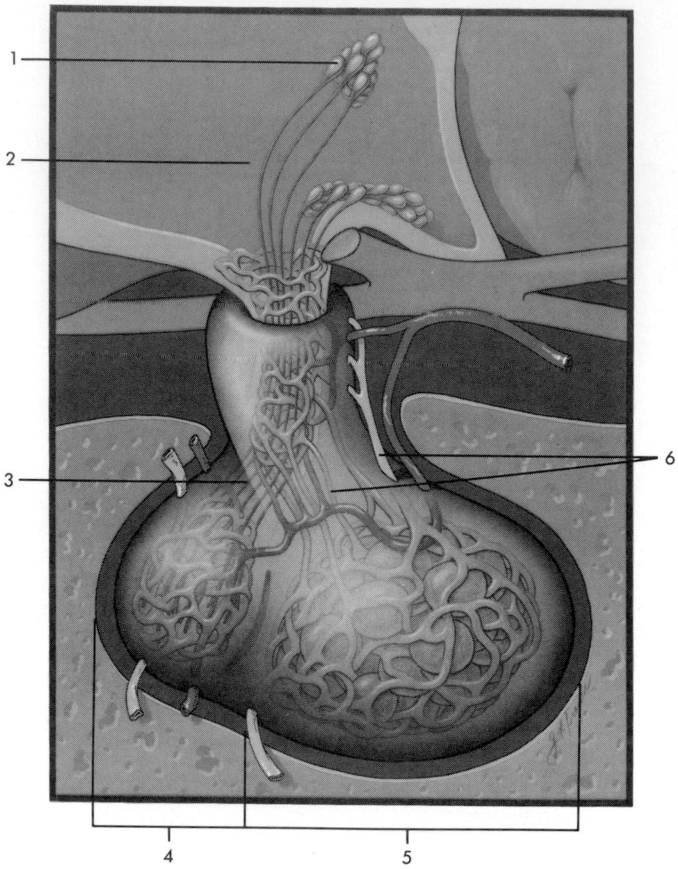

1. _____

2. _____

3. _____

4. _____

5. _____

6. _____

Figure 18-2

Neurohypophysis

66 *The neurohypophysis is directly connected to, and continuous with, the brain.* **99**

A. Match these terms with the correct statement as it pertains to the effects of increased ADH:

Decreases
Increases

_____ 1. Effect of ADH on urine volume.

_____ 2. Effect of ADH on blood osmolality.

_____ 3. Effect of ADH on blood volume.

B. Match these terms with the correct statement:

Blood osmolality
Blood volume

_____ 1. Osmoreceptors in the hypothalamus respond to this factor.

_____ 2. Sensory receptors for blood pressure indirectly measure this factor.

_____ 3. Increase in this factor results in an increase of ADH secretion.

_____ 4. Increase in this factor results in a decrease of ADH secretion.

C. Match these terms with the correct symptom:

Hypersecretion of ADH
Hyposecretion of ADH

_____ 1. Production of a large amount of dilute urine; diabetes insipidus.

_____ 2. Production of small quantities of very concentrated urine.

D. Match these terms with the correct statement:

Decreases
Increases

_____ 1. Effect of oxytocin on smooth-muscle contraction in the uterus.

_____ 2. Effect of oxytocin on milk ejection in lactating females.

E. Match these terms with the correct statement:

Decreases oxytocin secretion
Increases oxytocin secretion

_____ 1. Effect of stretch of uterus on oxytocin secretion.

_____ 2. Effect of nursing stimulation of nipples on oxytocin secretion.

Adenohypophysis

❝_Adenohypophyseal hormones are influenced by neurohormones from the hypothalamus._**❞**

A. Match these terms with the correct statement as it pertains to the effect of increased growth hormone:

Decreases
Increases

_____ 1. Effect of GH on growth and metabolic rate.

_____ 2. Effect of GH on amino acid uptake and protein synthesis by cells.

_____ 3. Effect of GH on the rate of breakdown of lipids.

_____ 4. Effect of GH on blood glucose levels.

_____ 5. Effect of GH on somatomedin production by the liver and skeletal muscle.

B. Match these terms with the correct statement:

Decreases GH secretion
Increases GH secretion

_____ 1. Effect of low blood glucose levels.

_____ 2. Effect of stress.

_____ 3. Effect of GH-RH from the hypothalamus.

_____ 4. Effect of GH-IH (somatostatin) from the hypothalamus.

C. Match these terms with the correct statement:

Hypersecretion of GH
Hyposecretion of GH

_____ 1. In children this produces dwarfism.

_____ 2. In children this produces giantism.

_____ 3. In adults this produces acromegaly.

D. Match these terms with the correct statement or definition:

ACTH   MSH
Beta-endorphin   Proopiomelanocortin
Lipotropins

_____ 1. Large precursor molecule that produces ACTH and others.

_____ 2. Has the same effect as opiate drugs.

_____ 3. Stimulates melanocytes in the skin to increase melanin production.

_____

E. Match these terms with the correct statement or definition:

FSH and LH   PRH
GnRH   Prolactin
PIH

_____ 1. Secreted from the adenohypophysis; regulates gamete and reproductive hormone production.

_____ 2. Neurohormone that stimulates the secretion of LH and FSH.

_____ 3. Responsible for milk production.

_____ 4. Neurohormone that stimulates prolactin production.

The Thyroid Gland

❝The thyroid gland is composed of two lobes connected by a narrow band of thyroid tissue called the isthmus.**❞**

A. Match these terms with the correct statement or definition:

Calcitonin   Thyroglobulin
Follicle   Thyroid hormones
Parafollicular cells

_____ 1. Small spheres of cuboidal epithelium in the thyroid which are filled with thyroglobulin.

_____ 2. Scattered cells between follicles in the thyroid.

_____ 3. Product of parafollicular cells.

B. Match these terms with the correct statement or definition:

I⁻ ions   Triiodothyronine
Thyroglobulins   Tyrosine
Tetraiodothyronine

_____ 1. Large proteins synthesized in thyroid follicles.

_____ 2. Amino acid used in the synthesis of thyroid hormones.

_____ 3. Actively absorbed into thyroid follicles; oxidized and bound to tyrosines.

_____ 4. Also called thyroxine.

C. Match these terms with the correct statement:

Albumin and prealbumin
Thyroxin-binding globulin

_____ 1. Transports most thyroid hormones; increases the half-life of thyroid hormones.

D. Match these terms with the correct statement as it pertains to the effects of increased thyroid hormones:

Decreases
Increases

_____ 1. Effect on glucose, fat, and protein metabolism.

_____ 2. Effect on body temperature.

E. Match these terms with the correct statement:

Decreases
Increases

_____ 1. Effect of stress or exposure to cold on TRH.

_____ 2. Effect of fasting on TRH.

_____ 3. Effect of increase of TRH on TSH.

_____ 4. Effect of increase in TSH on T3 and T4.

_____ 5. Effect of increase in T3 and T4 on TRH.

_____ 6. Effect of increase of T3 and T4 on TSH.

F. Match these terms with the correct symptom or condition:

Hypersecretion of thyroid hormone
Hyposecretion of thyroid hormone

_____ 1. Increased metabolic rate, weight loss, sweating.

_____ 2. Hyperactivity, rapid heart rate, exophthalmos.

_____ 3. Weight gain; reduced appetite; dry, cold skin.

_____ 4. Myxedema, decreased iodide uptake, cold intolerance.

_____ 5. Iodine deficiency with goiter.

_____ 6. Cretinism.

_____ 7. Grave's disease.

_____ 8. Thyroid storm.

G. Match these terms with the
   correct statement:

Decreases
Increases

_____ 1. Effect of calcitonin on breakdown of bone by osteoclasts.

_____ 2. Effect of calcitonin on levels of calcium and phosphate in blood.

_____ 3. Effect of increased calcium level of blood on calcitonin secretion.

The Parathyroid Glands

❝_The parathyroid glands are usually embedded in the posterior portion of each lobe of the thyroid gland._**❞**

A. Match these terms with the
   correct statement as it pertains
   to the effects of increased
   parathyroid hormone:

Decreases
Increases

_____ 1. Effect of PTH on osteoclast activity in bone.

_____ 2. Effect of PTH on calcium reabsorption in the kidneys.

_____ 3. Effect of PTH on the formation of active vitamin D in the kidneys,
                            which increases the rate of calcium and phosphate absorption in the
                            small intestine.

_____ 4. Effect on blood calcium levels.

_____ 5. Effect of PTH on blood phosphate levels.

B. Match these terms with the
   correct statement:

Decreases PTH secretion
Increases PTH secretion

_____ 1. Effect of low blood calcium levels on PTH secretion.

C. Match these terms with the
   correct symptom:

Hypersecretion of PTH
Hyposecretion of PTH

_____ 1. Kidney stones, osteomalacia, muscular weakness, constipation.

_____ 2. Increased muscular excitability, tachycardia, muscle tetany, diarrhea.

_____ 3. Increased cell permeability to sodium ions, depolarization of
                            cell membrane.

Adrenal Glands

❝_The adrenal glands are near the superior pole of each kidney._**❞**

Match these terms with the
correct statement or definition:

Adrenal cortex
Adrenal medulla

_____ 1. Inner portion of adrenal glands.

_____ 2. Derived from neural crest cells - part of the sympathetic nervous system.

_____ 3. Contains the zona glomerulosa, zona fasciculata, and zona reticularis
                            layers.

Adrenal Medulla

"*The adrenal medulla secretes two major hormones.***"**

A. Match these terms with the correct statement:

Epinephrine   Norepinephrine
Neurohormone

_____ 1. Adrenal medullary hormone secreted in larger quantity.

_____ 2. General category of hormones produced by the adrenal medulla.

B. Match these terms with the correct statement as it pertains to the effects of increased adrenal medullary hormones:

Decreases
Increases

_____ 1. Effect on heart rate and the force of contraction of the heart.

_____ 2. Effect on blood glucose level.

_____ 3. Effect on blood flow to the skin, kidneys, and digestive system.

_____ 4. Effect on blood flow to heart and skeletal muscle.

C. Match these terms with the correct statement:

Decreases secretion of adrenal medullary hormones
Increases secretion of adrenal medullary hormones

_____ 1. Effect of emotional excitement, stress, exercise, or injury.

_____ 2. Effect of stimulation of sympathetic neurons.

_____ 3. Effect of low blood glucose levels.

D. Match these terms with the correct symptom or definition:

Hypersecretion of adrenal medullary hormones
Hyposecretion of adrenal medullary hormones

_____ 1. Also called pheochromocytoma.

_____ 2. Hypertension.

_____ 3. Pallor and sweating.

Adrenal Cortex

"*The adrenal cortex secretes three hormone types.***"**

A. Match these terms with the correct statement as it pertains to the adrenal cortex:

Adrenal androgens
Glucocorticoids
Mineralocorticoids

_____ 1. Example is aldosterone.

_____ 2. Example is cortisol.

_____ 3. Converted into testosterone.

B. Match these terms with the correct statement as it pertains to the effects of increased aldosterone:

Decreases
Increases

_____ 1. Effect on sodium ion concentration in the blood.

_____ 2. Effect on potassium ion concentration in the blood.

_____ 3. Effect on hydrogen ion concentration in the blood.

C. Match these terms with the correct statement:

Hypersecretion of aldosterone
Hyposecretion of aldosterone

_____ 1. Hypernatremia and high blood pressure.

_____ 2. Alkalosis.

_____ 3. Hyperkalemia and skeletal muscle tetany.

D. Match these terms with the correct statement as it pertains to the effects of increased cortisol:

Decreases
Increases

_____ 1. Effect on fat and protein breakdown.

_____ 2. Effect on glucose and amino acid uptake by skeletal muscle.

_____ 3. Effect on the synthesis of glucose from amino acids.

_____ 4. Effect on blood glucose levels.

_____ 5. Effect on the intensity of the inflammatory response.

E. Match these terms with the correct statement or definition:

ACTH   Hypoglycemia or stress
Cortisol   Hypothalamus
CRH

_____ 1. Location for CRH production.

_____ 2. Production of ACTH is stimulated by this neurohormone.

_____ 3. Hormone that stimulates cortisol production.

_____ 4. Two hormones that inhibit CRH secretion.

_____

_____ 5. Hormone that inhibits ACTH production.

_____ 6. External factors that stimulate CRH production.

F. Match these term with each symptom:

Hypersecretion
Hyposecretion

_____ 1. Hyperglycemia leading to diabetes mellitus.

_____ 2. Osteoporosis, muscle atrophy, and weakness.

_____ 3. Fat redistributed--moon face and buffalo hump.

_____ 4. Increased skin pigmentation.

☞ The immune system is depressed in both hyposecretion and hypersecretion of cortisol.

G. Match these conditions with the correct disease:

Hyposecretion of aldosterone   Hypersecretion of androgens
Hyposecretion of cortisol   Hypersecretion of cortisol
Hypersecretion of aldosterone

_____ 1. Addison's disease.

_____

_____ 2. Aldosteronism.

_____ 3. Cushing's syndrome.

_____

H. Match these terms with the correct statement as it pertains to the effects of increased adrenal androgens:

Decreases
Increases

_____ 1. Effect on amount of pubic and axillary hair in women.

_____ 2. Effect on sex drive in women.

☞ In males the effects of adrenal androgens are negligible in comparison to the testosterone produced by the testes.

The Pancreas and Pancreatic Hormones

❝_The pancreas is both an exocrine gland and an endocrine gland._**❞**

A. Match these terms with the correct statement as it pertains to the pancreas:

Alpha cells   Ducts and acini
Beta cells   Pancreatic islets
Delta cells

_____ 1. Constitute the exocrine portion of the pancreas.

_____ 2. Islet cells that secrete glucagon.

_____ 3. Islet cells that secrete insulin.

B. Match these terms with the
correct statement as it pertains
to the effects of increased insulin
and glucagon:

Decreases
Increases

_____ 1. Effect of insulin on the uptake and use of glucose and amino acids in
muscle cells.

_____ 2. Effect of insulin on glycogen and fat synthesis.

_____ 3. Effect of insulin on blood sugar levels.

_____ 4. Effect of glucagon on breakdown of liver glycogen to glucose.

_____ 5. Effect of glucagon on glucose synthesis from amino acids and fats.

_____ 6. Effect of glucagon on blood sugar levels.

_____ 7. Effect of glucagon on fat breakdown and the production of ketones.

☞ Insulin has little direct effect on the nervous system, except to increase the uptake of glucose by
the satiety (hunger) center.

C. Match these terms with the
correct statement:

Decreases insulin secretion
Increases insulin secretion

_____ 1. Hyperglycemia, certain amino acids.

_____ 2. Parasympathetic innervation.

_____ 3. Gastrointestinal hormones (e.g., gastrin, secretin, and cholecystokinin).

D. Match these terms with the
correct statement:

Decreases glucagon secretion
Increases glucagon secretion

_____ 1. Hypoglycemia, certain amino acids.

_____ 2. Sympathetic innervation.

E. Match these terms with the
correct symptoms:

Diabetes mellitus
Insulin shock

_____ 1. Polyuria, polydipsia, and polyphagia.

_____ 2. Glucosuria and hyperglycemia.

_____ 3. Headache, drowsiness, fatigue, convulsive seizures, and death.

_____ 4. Sweating, pale skin, and tachycardia.

_____ 5. Ketonuria, acetone breath, and acidosis.

_____ 6. Atherosclerosis and peripheral vascular disease.

Hormonal Regulation of Nutrients

66 Several hormones function together to regulate blood nutrient levels. 99

A. Match these terms with the
    correct statement for conditions
    immediately after a meal:

Decrease(s)
Increase(s)

_____ 1. Levels of glucagon, cortisol, growth hormone and epinephrine.

_____ 2. Insulin secretion.

_____ 3. Uptake of glucose, amino acids, and fats.

_____ 4. Glucose converted to glycogen.

B. Match these terms with the
    correct statement for conditions
    two hours after a meal:

Decrease(s)
Increase(s)

_____ 1. Levels of glucagon, cortisol, growth hormone, and epinephrine.

_____ 2. Insulin secretion.

_____ 3. Uptake of glucose.

_____ 4. Glycogen converted to glucose.

_____ 5. Fat and protein metabolism for most tissues.

C. Match these terms with the
    correct statement for conditions
    during exercise:

Decreases
Increases

_____ 1. Sympathetic nervous system stimulation.

_____ 2. Release of epinephrine.

_____ 3. Release of insulin.

_____ 4. Fatty acids, triglycerides, and ketones in blood.

_____ 5. Fat and glycogen metabolism in skeletal muscle.

Hormones of the Pineal Body, Thymus Gland, and Others

66 Hormones are produced at several other locations in the body. 99

Match these terms with the   Melatonin   Thymosin
correct statement or definition: Pineal body

_____ 1. Endocrine gland in the epithalamus that secretes hormones that inhibit reproductive function.

_____ 2. One secretion of the pineal body; production of this hormone decreases as day length increases.

_____ 3. Hormone produced by the thymus; affects the immune system.

363

Hormonelike Substances

66 *Several substances have some characteristics in common with hormones and some not in common.* 99

Match these terms with the
correct statement or definition:

Endorphins and enkephalins
Prostaglandins

_____ 1. Fatty acids that help regulate uterine contractions.

_____ 2. Substances that initiate some symptoms of inflammation and pain.

_____ 3. Small polypeptides that bind to the same receptors as morphine.

QUICK RECALL

A. Match these endocrine glands with the correct hormone it secretes:

Adenohypophysis
Adrenal cortex
Adrenal medulla
Hypothalamus
Neurohypophysis
Ovaries
Pancreas (alpha cells)
Pancreas (beta cells)
Pancreas (delta cells)
Parathyroid glands
Pineal body
Thymus gland
Thyroid gland (follicle cells)
Thyroid gland (parafollicular cells)
Testes

_____ 1. ACTH.

_____ 2. ADH.

_____ 3. Adrenal androgens.

_____ 4. Aldosterone.

_____ 5. Calcitonin.

_____ 6. Cortisol.

_____ 7. CRH.

_____ 8. Epinephrine.

_____ 9. Estrogen.

_____ 10. FSH.

_____ 11. GH.

_____ 12. GH-RH and GH-IH.

_____ 13. Glucagon.

_____ 14. GnRH.

_____ 15. Insulin.

_____ 16. LH.

_____ 17. Melatonin.

_____ 18. MSH.

_____ 19. Norepinephrine.

_____ 20. Oxytocin.

_____ 21. Parathyroid hormone.

_____ 22. Progesterone.

_____ 23. Prolactin.

_____ 24. PRH and PIH.

_____ 25. Testosterone.

_____ 26. Tetraiodothyronine and
                         triiodothyronine.

_____ 27. Thymosin.

_____ 28. TRH.

_____ 29. TSH.

B. Match these target tissues with the correct hormone that affects it:

Adipose tissue, liver, and skeletal muscle
Adrenal cortex
Blood vessels
Bone
Immune tissues
Kidneys
Liver
Mammary gland
Most tissues of the body
Ovaries or testes
Skin
Thyroid gland
Uterus

_____ 1. ADH.

_____ 2. Oxytocin (two).

_____

_____ 3. GH.

_____ 4. TSH.

_____ 5. ACTH.

_____ 6. MSH.

_____ 7. LH and FSH.

_____ 8. Prolactin.

_____ 9. Thyroid hormones.

_____ 10. Calcitonin.

_____ 11. PTH (two).

_____

_____ 12. Epinephrine (two).

_____

_____ 13. Aldosterone.

_____ 14. Glucocorticoids.

_____ 15. Insulin.

_____ 16. Glucagon.

_____ 17. Testosterone, progesterone, and
estrogen.

_____ 18. Thymosin.

C. Simmond's disease results from a tumor in the pituitary gland or from a
lack of blood supply to the pituitary. Symptoms are similar to those that
occur when the pituitary gland is removed. What effects (increase, decrease,
or no effect) would Simmond's disease have on the following variables?

_____ 1. Metabolic rate.

_____ 2. Height, if the condition developed
in a child.

_____ 3. Ability to deal with stress.

_____ 4. Blood sugar levels.

_____ 5. Blood calcium levels.

_____ 6. Blood sodium levels.

_____ 7. Blood pH.

_____ 8. Milk production.

_____ 9. Ability to produce sperm or egg.

MASTERY LEARNING ACTIVITY

Place the letter corresponding to the correct answer in the space provided.

_____ 1. The pituitary gland
a. is derived from the brain.
b. is derived from the roof of the
mouth.
c. is divided into two parts.
d. all of the above

_____ 2. The hypothalamohypophyseal portal
system
a. contains one capillary system.
b. carries hormones from the
adenohypophysis to the body.

c. carries hormones from the
neurohypophysis to the body.
d. carries hormones from the
hypothalamus to the
adenohypophysis.
e. carries hormones from the
hypothalamus to the
neurohypopyhsis.

_____ 3. Hormones secreted from the neurohypophysis
  a. are produced in the hypothalamus.
  b. are transported to the neurohypophysis within axons.
  c. include ADH and oxytocin.
  d. all of the above

_____ 4. Which of the following is a consequence of an elevated blood osmolality?
  a. decreased prolactin secretion
  b. increased prolactin secretion
  c. decreased ADH secretion
  d. increased ADH secretion

_____ 5. Oxytocin is responsible for
  a. preventing milk release from the mammary glands.
  b. preventing goiter.
  c. causing contraction of the uterus.
  d. maintaining normal calcium levels.

_____ 6. Growth hormone
  a. increases the usage of glucose.
  b. increases the breakdown of lipids.
  c. decreases the synthesis of proteins.
  d. all of the above

_____ 7. Hypersecretion of growth hormone
  a. results in giantism if it occurs in children.
  b. causes acromegaly in adults.
  c. increases the probability that one will develop diabetes mellitus.
  d. a and b
  e. all of the above

_____ 8. LH and FSH
  a. are produced in the hypothalamus.
  b. production is increased by TSH.
  c. promote the production of gametes and reproductive hormones.
  d. inhibit the production of prolactin.

_____ 9. Thyroid hormones
  a. require iodine for their production.
  b. are made from the amino acid tyrosine.
  c. are transported in the blood bound to thyroxin-binding globulin.
  d. all of the above

_____ 10. Which of the following symptoms are associated with hyposecretion of the thyroid gland?
  a. hypertension
  b. nervousness
  c. diarrhea

  d. weight loss with a normal or increased food intake
  e. decreased metabolic rate

_____ 11. Which of the following would probably occur if a normal person received an injection of thyroid hormone?
  a. The secretion rate of TSH would decline.
  b. The person would develop symptoms of hypothyroidism.
  c. The person would develop hypercalcemia.
  d. The person would secrete more thyroid-releasing hormone.

_____ 12. Which of the following would result in response to a thyroidectomy (removal of the thyroid gland)?
  a. increased calcitonin secretion
  b. increased T3 and T4 secretion
  c. decreased TRH secretion
  d. increased TSH secretion

_____ 13. Choose the statement below that most accurately predicts the long-term effect of a substance that prevents active transport of iodide by the thyroid gland.
  a. Large amounts of thyroid hormone would accumulate within the thyroid follicles, and little would be released.
  b. The person would exhibit symptoms of hypothyroidism.
  c. The anterior pituitary would secrete smaller amounts of TSH.
  d. The circulating levels of T3 and T4 would be increased.

_____ 14. A patient exhibited symptoms of hyperthyroidism. She had elevated blood levels of T3 and T4 and reduced TSH levels. There was no evidence of a tumor within the thyroid gland. Which of the following explanations are consistent with the data?
  a. She experienced hypersecretion of TRH.
  b. Neither the hypothalamus nor the pituitary was sensitive to the negative feedback effect of T3 and T4.
  c. She had an elevated antibody in her blood that is similar in structure to TSH.
  d. She suffered from anterior pituitary hypofunction.

_____ 15. Calcitonin
   a. is produced by the parathyroid
      glands.
   b. levels increase when blood calcium
      levels decrease.
   c. causes blood calcium levels to
      decrease.
   d. insufficiency results in weak bones
      and tetany.

_____ 16. Parathyroid hormone secretion
      increases in response to
   a. a decrease in blood calcium levels.
   b. increased production of parathyroid-
      stimulating hormone from the
      anterior pituitary.
   c. increased secretion of parathyroid-
      releasing hormone from the
      hypothalamus.
   d. all of the above

_____ 17. If parathyroid hormone levels increase,
      which of the following would be
      expected?
   a. Osteoclast activity is increased.
   b. Calcium absorption from the small
      intestine is inhibited.
   c. Calcium reabsorption from urine is
      inhibited.
   d. Less active vitamin D would be
      formed in the kidneys.

_____ 18. Choose the response below that is con-
      sistent with the development of a tumor
      that destroys the parathyroid glands.
   a. The person may develop tetany.
   b. The person will be highly lethargic.
   c. The person's bones will break more
      easily than normal.
   d. Increased blood levels of calcium
      will result.

_____ 19. The adrenal medulla
   a. is formed from a modified portion of
      the sympathetic nervous system.
   b. has epinephrine as its major
      secretory product.
   c. increases its secretions during
      exercise.
   d. all of the above

_____ 20. Pheochromocytoma is a condition in
      which a benign tumor results in hyper-
      secretion of the adrenal medulla. The
      symptoms that one would expect
      include
   a. hypotension.
   b. bradycardia.

   c. pallor.
   d. lethargy.
   e. a and b

_____ 21. The hormone secreted from the adrenal
      cortex is
   a. aldosterone.
   b. cortisol.
   c. androgen.
   d. a and b
   e. all of the above

_____ 22. Which of the following would occur as
      a result of a decrease in aldosterone
      secretion?
   a. Blood potassium levels decrease, and
      cells become more excitable.
   b. Blood potassium levels increase, and
      cells become less excitable.
   c. Blood hydrogen levels decrease, and
      acidosis results.
   d. Blood hydrogen levels increase, and
      alkalosis results.
   e. Blood sodium levels decrease, and
      blood volume decreases.

_____ 23. Glucocorticoids (cortisol)
   a. increase the breakdown of fats.
   b. increase the breakdown of proteins.
   c. increase blood sugar levels.
   d. decrease inflammation.
   e. all of the above

_____ 24. The release of cortisol from the adrenal
      cortex is regulated by other hormones.
      Which of the following hormones is
      correctly matched with its origin and
      function?
   a. CRH: secreted by the hypothalamus
      and stimulates the adrenal cortex to
      secrete cortisol
   b. CRH: secreted by the anterior
      pituitary and stimulates the adrenal
      cortex to secrete cortisol
   c. ACTH: secreted by the
      hypothalamus and stimulates the
      adrenal cortex to secrete cortisol
   d. ACTH: secreted by the anterior
      pituitary and stimulates the adrenal
      cortex to secrete cortisol

_____ 25. Which of the following would be
      expected in Cushing's syndrome?
   a. loss of hair in women
   b. deposition of fat in the face, neck,
      and abdomen
   c. low blood glucose
   d. low blood pressure

369

26. Within the pancreas, the islets of Langerhans produce
    a. insulin.
    b. glucagon.
    c. digestive enzymes.
    d. a and b
    e. all of the above

27. Insulin
    a. increases the uptake of glucose by its target tissues.
    b. increases the uptake of amino acids by its target tissues.
    c. increases glycogen synthesis in the liver and in skeletal muscle.
    d. a and b
    e. all of the above

28. Which of the following tissues is least affected by insulin?
    a. adipose tissue
    b. heart
    c. skeletal muscle
    d. brain

29. Glucagon
    a. primarily affects the liver.
    b. causes blood sugar levels to decrease.
    c. decreases fat metabolism.
    d. all of the above

30. Concerning ketones,
    a. they are a byproduct of the breakdown of fats.
    b. excess production of ketones can produce acidosis.
    c. ketones can be used as a source of energy.
    d. all of the above

31. When blood sugar levels increase,
    a. insulin and glucagon secretion increase.
    b. insulin and glucagon secretion decrease.
    c. insulin secretion increases, and glucagon secretion decreases.
    d. insulin secretion decreases, and glucagon secretion increases.

32. If a person that has diabetes mellitus forgot to take an insulin injection, symptoms that may soon appear include
    a. acidosis.
    b. hyperglycemia.
    c. increased urine production.
    d. all of the above

33. Melatonin
    a. is produced by the posterior pituitary.
    b. production increases as day length increases.
    c. inhibits development of the reproductive system.
    d. all of the above

34. A researcher noticed that following a heavy meal in the evening, rich in carbohydrates, he would be hungrier than usual in the morning. He decided to do an experiment to determine the cause for the condition. In the first experiment, several volunteers fasted for 24 hours. Then they consumed 1 g of glucose per kilogram of body weight (time 0 hours in the graph below). For the next four hours blood glucose levels were measured.

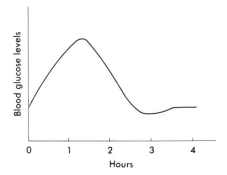

In the second experiment three different groups were used. Each group fasted for 24 hours before the experiment. Group A then consumed 1 g of glucose per kilogram of body weight. Group B consumed a heavy meal rich in carbohydrates. Group C continued to fast. Measurement of blood insulin levels was made immediately following each treatment (time 0 in the graph below) and at 1 and 3 hours.

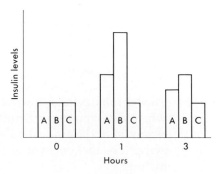

On the basis of the data, the investigator concluded that

a. large amounts of insulin are secreted in response to food consumption, and the insulin levels remain high in the blood long enough for blood levels of glucose to decrease below normal, thus creating the hunger sensation.

b. blood glucose levels become elevated following food consumption and remain elevated for several hours due to the large amounts of insulin secreted, thus creating the hunger sensation.

c. the ability for the pancreas to produce insulin is limited to a 2 to 3 hour period. Thereafter only small amounts are secreted, thus resulting in the sensation of hunger.

d. the hunger sensation cannot be related to either blood glucose levels or insulin levels.

_____ 35. Which of the following hormones, produced by most tissues of the body, can promote inflammation?

a. endorphin
b. enkephalin
c. thymosin
d. prostaglandin

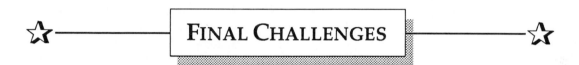

FINAL CHALLENGES

Use a separate sheet of paper to complete this section

1. If there is insufficient dietary intake of iodide, goiter can develop. Would the levels of T3, T4, TSH, and TRH be higher or lower than normal? Explain.

2. During pregnancy and/or lactation, parathyroid hormone levels may be elevated. Explain why this would occur and give a reason why it might be harmful.

3. A patient has pheochromocytoma. Would you expect the pupils to be dilated or constricted? Explain.

4. One of the symptoms of Addison's disease is increased skin pigmentation due to high levels of ACTH. Explain why ACTH levels are high.

5. A young boy (6 years old) exhibited marked and rapid development of sexual characteristics. On examination his testicles were not found to be larger than normal, but his plasma testosterone levels were elevated. As a mental exercise a student nurse decided that she would propose a cure. She considered the symptoms and decided on surgery to remove an adrenal tumor. Explain why you agree or disagree with her diagnosis.

6. The glucose tolerance test determines the ability of a person to dispose of a standard dose of glucose. If the rate of removal of glucose from the blood is significantly slower than normal, a diagnosis of diabetes mellitus is made. However, the diagnosis can be wrong. One explanation for these false tests is patient anxiety over the test. Explain how this could produce a false test result.

7. Would you expect a person who was blind from birth to enter puberty at an earlier or later age than a normally sighted person? Explain.

ANSWERS TO CHAPTER 18

WORD PARTS

1. thalamus; hypothalamus; subthalamus; epithalamus
2. hypopophysis; adenohypophysis; neurohypophysis
3. adenohypophysis
4. gonadotropin; corticotropin; tropic; tropin
5. gonadotropin
6. somatotrophic; somatotrophs
7. pars distalis; pars tuberalis; pars intermedia
8. prolactin
9. follicle; follicular; parafollicular
10. analgesia; analgesic

CONTENT LEARNING ACTIVITY

Introduction

1. Pituitary gland
2. Thyroid gland
3. Thymus
4. Adrenal gland
5. Testis
6. Ovary
7. Pancreas
8. Gastrointestinal tract

Pituitary Gland and Hypothalamus

A.
1. Neurohypophysis
2. Adenohypophysis
3. Hypothalamohypophyseal portal system
4. Hypothalamus
5. Adenohypophysis
6. Action potentials
7. Neurohormones

B.
1. Neurosecretory cells
2. Hypothalamus
3. Hypothalamohypophyseal tract
4. Posterior pituitary
5. Anterior pituitary
6. Hypothalamohypophyseal portal system

Neurohypophysis

A.
1. Decreases
2. Decreases
3. Increases

B.
1. Blood osmolality
2. Blood volume
3. Blood osmolality
4. Blood volume

C.
1. Hyposecretion of ADH
2. Hypersecretion of ADH

D.
1. Increases
2. Increases

E.
1. Increases oxytocin secretion
2. Increases oxytocin secretion

Adenohypophysis

A. 1. Increases
   2. Increases
   3. Increases
   4. Increases
   5. Decreases

B. 1. Increases GH secretion
   2. Increases GH secretion
   3. Increases GH secretion
   4. Decreases GH secretion

C. 1. Hyposecretion of GH
   2. Hypersecretion of GH
   3. Hypersecretion of GH

D. 1. Proopiomelanocortin
   2. Beta-endorphin
   3. MSH; ACTH

E. 1. FSH and LH
   2. GnRH
   3. Prolactin
   4. PRH

The Thyroid Gland

A. 1. Follicle
   2. Parafollicular cells
   3. Calcitonin

B. 1. Thyroglobulins
   2. Tyrosine
   3. I- ions
   4. Tetraiodothyronine

C. 1. Thyroxine-binding globulin

D. 1. Increases
   2. Increases

E. 1. Increases
   2. Decreases
   3. Increases
   4. Increases

   5. Decreases
   6. Decreases

F. 1. Hypersecretion of thyroid hormones
   2. Hypersecretion of thyroid hormones
   3. Hyposecretion of thyroid hormones
   4. Hyposecretion of thyroid hormones
   5. Hyposecretion of thyroid hormones
   6. Hyposecretion of thyroid hormones
   7. Hypersecretion of thyroid hormones
   8. Hypersecretion of thyroid hormones

H. 1. Decreases
   2. Decreases
   3. Increases

Parathyroid Glands

A. 1. Increases
   2. Increases
   3. Increases
   4. Increases
   5. Decreases

B. 1. Increases PTH secretion

C. 1. Hypersecretion of PTH
   2. Hyposecretion of PTH
   3. Hyposecretion of PTH

Adrenal Glands

1. Adrenal medulla
2. Adrenal medulla

3. Adrenal cortex

Adrenal Medulla

A. 1. Epinephrine
   2. Neurohormone

B. 1. Increases
   2. Increases
   3. Decreases
   4. Increases

C. 1. Increases secretion of adrenal medullary
      hormones

2. Increases secretion of adrenal medullary
   hormones
3. Increases secretion of adrenal medullary
   hormones

D. 1. Hypersecretion of adrenal medullary
      hormones
   2. Hypersecretion of adrenal medullary
      hormones
   3. Hypersecretion of adrenal medullary
      hormones

Adrenal Cortex

A. 1. Mineralocorticoids
   2. Glucocorticoids
   3. Adrenal androgens

B. 1. Increases
   2. Decreases
   3. Increases

C. 1. Hypersecretion of aldosterone
   2. Hypersecretion of aldosterone
   3. Hyposecretion of aldosterone

D. 1. Increases
   2. Decreases
   3. Increases
   4. Increases
   5. Decreases

E. 1. Hypothalamus
   2. CRH
   3. ACTH
   4. ACTH; cortisol
   5. Cortisol
   6. Hypoglycemia or stress

F. 1. Hypersecretion of cortisol
   2. Hypersecretion of cortisol
   3. Hypersecretion of cortisol
   4. Hyposecretion of cortisol

G. 1. Hyposecretion of aldosterone and cortisol
   2. Hypersecretion of aldosterone
   3. Hypersecretion of cortisol and androgens

H. 1. Increases
   2. Increases

The Pancreas and Pancreatic Hormones

A. 1. Ducts and acini
   2. Alpha cells
   3. Beta cells

B. 1. Increases
   2. Increases
   3. Decreases
   4. Increases
   5. Increases
   6. Increases
   7. Increases

C. 1. Increases insulin secretion
   2. Increases insulin secretion
   3. Increases insulin secretion

D. 1. Increases glucagon secretion
   2. Increases glucagon secretion

E. 1. Diabetes mellitus
   2. Diabetes mellitus
   3. Insulin shock
   4. Insulin shock
   5. Diabetes mellitus
   6. Diabetes mellitus

Hormonal Regulation of Nutrients

A. 1. Decreases
   2. Increases
   3. Increases
   4. Increases

B. 1. Increases
   2. Decreases
   3. Decreases

4. Increases
5. Increases

C. 1. Increases
   2. Increases
   3. Decreases
   4. Increases
   5. Increases

Hormones of the Pineal Body, Thymus Gland, and Others

1. Pineal body
2. Melatonin

3. Thymosin

Hormonelike Substances

1. Prostaglandins
2. Prostaglandins

3. Endorphins and enkephalins

QUICK RECALL

A. 1. Adenohypophysis
   2. Neurohypophysis
   3. Adrenal cortex
   4. Adrenal cortex
   5. Thyroid gland (parafollicular cells)
   6. Adrenal cortex
   7. Hypothalamus
   8. Adrenal medulla
   9. Ovaries
   10. Adenohypophysis
   11. Adenohypophysis
   12. Hypothalmus
   13. Pancreas (alpha cells)
   14. Hypothalamus
   15. Pancreas (beta cells)
   16. Adenohypophysis
   17. Pineal body
   18. Adenohypophysis
   19. Adrenal medulla
   20. Neurohypophysis
   21. Parathyroid glands
   22. Ovaries
   23. Adenohypophysis
   24. Hypothalamus
   25. Testes
   26. Thyroid gland (follicular cells)
   27. Thymus gland
   28. Hypothalamus
   29. Adenohypophysis

B. 1. Kidney
   2. Uterus; mammary gland
   3. Most tissues of the body
   4. Thyroid gland
   5. Adrenal cortex
   6. Skin
   7. Ovaries and testes
   8. Mammary glands
   9. Most tissues of the body
   10. Bone
   11. Bone; kidney
   12. Blood vessels and heart; adipose tissue, liver, and skeletal muscle
   13. Kidneys
   14. Adipose tissue, liver, and skeletal muscle
   15. Adipose tissue, liver, and skeletal muscle
   16. Liver
   17. Most tissues of the body
   18. Immune cells

C. 1. Decrease
   2. Decrease
   3. Decrease
   4. Decrease
   5. No effect
   6. Decrease
   7. Decrease
   8. Decrease
   9. Decrease

1. D. The pituitary gland can be divided into the anterior pituitary (adenohypophysis) and the posterior pituitary (neurohypophysis). The anterior pituitary is derived from the roof of the mouth, and the posterior pituitary is derived from the brain.

2. D. The hypothalamohypophyseal portal system orginates as a capillary bed within the hypothalamus and extends to the adenohypophysis, where it terminates as a capillary bed. It is responsible for carrying hormones from the hypothalamus to the adenohypophysis.

3. D. Hormones such as ADH and oxytocin are produced by neurosecretory cells in the hypothalamus. The hormones are transported within the axons of the neurosecretory cells (through the hypothalamohypophyseal tract) to the neurohypophysis. The hormones are stored in the ends of the axons, and they are secreted in response to action potentials.

4. D. Increased blood osmolality increases the frequency of action potentials produced by osmoreceptors within the hypothalamus. In response, an increased amount of ADH is secreted to initiate water conservation in the kidneys. Increased prolactin results in production of milk.

5. C. Oxytocin causes uterine contractions. It is released during parturition (delivery) and helps stimulate the uterus to contract. Oxytocin also causes the myoepithelial cells that surround the alveoli of the mammary gland to contract. Consequently, in the lactating female oxytocin assists in "ejecting" milk from the mammary gland.

6. B. Growth hormone spares glucose usage, increases the breakdown of lipids, and increases the synthesis of proteins.

7. E. Hypersecretion of growth hormone increases bone growth in length, if it occurs before the epiphyseal plates are ossified (in children), and in width. In adults the epiphyseal plates are ossified so the bone growth in width is most obvious (acromegaly). Since growth hormone increases blood glucose levels, those that suffer from hypersecretion of growth hormone are likely to develop diabetes mellitus.

8. C. Produced in the adenohypophysis, LH and FSH promote the production of gametes (eggs and sperm) and reproductive hormones (testosterone in the male, estrogen and progesterone in the female). GnRH from the hypothalamus stimulates the release of LH and FSH. Prolactin is regulated by PRH and PIH from the hypothalmus.

9. D. Iodine binds with the amino acid tyrosine to form the thyroid hormones T3 and T4. The thyroid hormones are stored within the thyroid follicles as components of thyroglobulin. When released from the follicles, thyroid hormones are transported in the blood bound to thyroxin-binding globulin (TBG), or albumin.

10. E. Hypotension, lethargy, sluggishness, constipation, weight gain, and decreased metabolic rate are symptoms of hypothroidism (hyposecretion of the thyroid gland).

11. A. An injection of thyroid hormone (T3 and T4) would result in an elevated blood level of these hormones. It would mimic hyperthyroidism. The elevated T3 and T4 levels in the blood would act on both the hypothalamus and the anterior pituitary to decrease TRH and TSH secretion, respectively.

12. D. Removal of the thyroid gland would eliminate T3 and T4 secretion, since they are the thyroid hormones. In response to the decreased T3 and T4 secretion, both TRH and TSH would be released in larger amounts. Calcitonin is secreted by the parafollicular cells of the thyroid gland. Its removal would eliminate calcitonin secretion.

13. B. The reduced iodine uptake results in decreased T3 and T4 synthesis. The T3 and T4 stored within the thyroid gland will be depleted, and the T3 and T4 levels in the blood will decrease. The reduced blood levels of T3 and T4 will result in the development of hypothyroidism and also remove a negative feedback effect on TSH secretion, resulting in increased TSH secretion.

14. C. Elevated T3 and T4 levels are normally the result of elevated TSH levels in the blood. However, in this case the T3 and T4 levels were

elevated, whereas the TSH levels were suppressed. The only response that is consistent with the data is that there is another substance (an antibody) that has a TSH-like effect on the thyroid gland (as in Grave's disease).

15. C. Calcitonin causes blood calcium levels to decrease. It is produced by the parafollicular cells of the thyroid gland, it increases in response to increased blood calcium levels, and there is no known pathology associated with lack of calcitonin secretion.

16. A. Parathyroid hormone secretion is mediated through changes in blood calcium levels.

17. A. Increased parathyroid hormone levels result in increased blood calcium levels. Parathyroid hormone acts in several ways to increase blood calcium levels: it increases osteoclast activity, which increases bone breakdown and release of calcium into the blood; it promotes the production of active vitamin D, which increases calcium absorption from the small intestine; and it increases calcium reabsorption from urine.

18. A. Degeneration of the parathyroid gland results in decreased parathyroid hormone (PTH) secretion. PTH increases calcium absorption in the gut, reabsorption in the kidney, and reabsorption from bone. Decreased PTH results in hypocalcemia. The only symptom listed that is consistent with reduced PTH secretion and reduced blood levels of calcium is tetany.

19. D. The medulla is a modified portion of the sympathetic nervous system. Its major secretory products are epinephrine and small amounts of norepinephrine, which are released during exercise, stress, injury, or emotional excitement or in response to low blood glucose levels.

20. C. Hypersecretion of the adrenal medulla would result in large amounts of epinephrine released into the circulatory system, causing cutaneous vasoconstriction and pallor. Epinephrine would increase heart rate, stroke volume, and vasoconstriction in the viscera and skin. Hypertension is therefore consistent with pheochromocytoma. Epinephrine also tends to induce hyperexcitability rather than lethargy.

21. E. The zona glomerulosa of the adrenal cortex secretes aldosterone, the zona fasciculata secretes cortisol, and the zona reticularis secretes androgens.

22. E. A decrease in aldosterone results in a decrease in blood sodium levels and therefore blood volume. Aldosterone decrease also causes blood potassium levels to increase, resulting in more excitable cells (hypopolarization of cell membranes), and blood hydrogen levels to increase resulting in acidosis.

23. E. Glucocorticoids increase the breakdown of fats and proteins and increase blood sugar levels. They also decrease the intensity of the inflammatory response.

24. D. ACTH is secreted by the anterior pituitary and stimulates the adrenal cortex to secrete cortisol. CRH is secreted by the hypothalamus and stimulates the secretion of ACTH from the anterior pituitary.

25. B. Cushing's syndrome results from hypersecretion of cortisol and androgens. The cortisol causes fat to be redistributed to the face (moon face), neck (buffalo hump), and abdomen. It also causes hyperglycemia and depresses the immune system. The increased androgens cause hirsutism (excessive facial and body hair). Low blood pressure and low blood glucose are associated with decreased aldosterone levels (Addison's disease).

26. D. Cells within the islets of Langerhans produce insulin and glucagon. The beta-cells produce insulin, and the alpha-cells produce glucagon. The exocrine portion of the pancreas produces digestive enzymes.

27. E. Insulin promotes the uptake of glucose and amino acids, both of which can be used as an energy source. In the liver and in skeletal muscle the glucose is stored as glycogen. In adipose tissue glucose is converted into fat. The amino acids can be used to synthesize proteins or glucose.

28. D. Insulin affects brain tissue less than adipose, heart, or skeletal muscle. Glucose enters the brain cells without the presence of insulin. An exception is the satiety (hunger) center within the hypothalamus of the brain.

29. A. Glucagon primarily affects the liver, causing the breakdown of glycogen to glucose and the synthesis of glucose from amino acids and fats. Consequently, blood sugar levels increase. Glucagon also promote fat metabolism.

30. D. When blood levels decrease, insulin levels drop, and glucagon levels increase. One result is a switch to the use of fats or a source of energy. One product produced during fat metabolism is ketones, which can be used by most tissue for energy. Ketones are also acidic and can produce acidosis.

31. C. Both insulin and glucagon secretions are regulated by blood sugar levels, but in the opposite direction.

32. D. A lack of insulin would be the result of a diabetic forgetting to take an insulin injection. Symptoms consistent with hyposecretion of insulin would develop such as acidosis, hyperglycemia, increased urine production, and others.

33. C. Melatonin inhibits the development of the reproductive system in some animals and may be involved with the onset of puberty in humans. Melatonin is produced by the pineal body, and its production increases as day length decreases.

34. A. The data in the first experiment indicate that, following glucose (or food) intake, the blood glucose levels increase and remains elevated for roughly 2 hours. Then, blood glucose falls and remain less than normal for several hours. From the second experiment, it can be seen that the insulin levels are above normal following ingestion of food and, further, that they are correlated with the depressed blood glucose levels after 3 hours. That effect is exaggerated in those who consumed a large meal. Thus the data indicate that following a heavy meal a large amount of insulin is secreted. Since the insulin levels are elevated and remain elevated after 3 hours, one would expect the blood levels of glucose to decrease rapidly to below normal levels. Low blood glucose levels are known to stimulate the "hunger" center in the brain.

35. D. Prostaglandins are "local" hormones produced by most tissues. They have a variety of effects, including the promotion of inflammation, fever, pain, and uterine contractions. The endorphin and enkephalin reduce pain within the central nervous system. Thymosin is produced by the thymus gland and is involved with immunity.

 FINAL CHALLENGES

1. Without sufficient iodide thyroid hormones are not synthesized, resulting in low blood levels of T3 and T4. Without the negative feedback effects of T3 and T4, TSH and TRH levels would be elevated.

2. During pregnancy the fetus withdraws calcium from the mother's blood and during lactation calcium is lost with the milk. In both cases, lower blood calcium levels stimulate parathyroid hormone release. Consequently, calcium is released from the bones or is absorbed more efficiently from the small intestine and urine, helping to maintain blood calcium levels. The removal of calcium from the bones could result in weakened bones.

3. Pheochromocytoma results in overproduction of epinephrine and norepinephrine by the adrenal medulla. These chemicals are released into the blood and produce the same effects as sympathetic nervous system stimulation (see Chapter 15). Consequently one would expect dilated pupils.

4. Addison's disease is due to hyposecretion of aldosterone and cortisol. With low levels of cortisol, the adenohpophysis is not inhibited, and ACTH secretion increases.

5. Rapid sexual development in a prepubertal boy is indicative of hypersecretion of sex steroids. The two most likely tissues are the testes and the adrenal glands. Since the testes are of normal size and there is no indication of abnormal testes function, the most logical tissue to suspect would be the adrenal glands. It is possible that removal of an adrenal tumor would cure the boy.

6. Anxiety could result in sympathetic system activity. The sympathetic system inhibits insulin secretion and can therefore bias the test results. In addition, sympathetic activity could result in epinephrine release. Epinephrine causes an increase in blood sugar levels that can also bias the test results.

7. One might expect a person blind from birth to enter puberty at a later age than a normally sighted person. A decrease in melatonin production may be involved with the onset of puberty. In a blind person melatonin may be produced at higher than normal levels, since melatonin production increases in the dark.

Cardiovascular System: Blood

FOCUS: Blood consists of plasma and formed elements. The plasma is 92% water with dissolved or suspended molecules, including albumin, globulins, and fibrinogen. The formed elements include erythrocytes, leukocytes, and platelets. Erythrocytes contain hemoglobin, which can transport oxygen and carbon dioxide, and carbonic anhydrase, which is involved in carbon dioxide transport. Erythrocyte production is stimulated by renal erythropoietic factor released from the kidneys when blood oxygen levels decrease. Erythrocytes can be typed according to their surface antigens into the ABO and Rh blood groups. Leukocytes protect the body against microorgansisms and remove dead cells and debris from the body. The different leukocytes include neutrophils, eosinophils, basophils, lymphocytes, and monocytes. Platelets prevent bleeding by forming platelet plugs and by producing factors involved in clotting. Blood clotting can be divided into an extrinsic (initiated by chemicals outside the blood) and intrinsic pathway (initiated by platelets and plasma coagulation factors). Once initiated, both pathways cause the activation of thrombin, which converts fibrinogen into fibrin (the clot). Overproduction of clots is prevented by antithrombin and heparin, and clots are dissolved by plasmin.

WORD PARTS

Give an example of a new vocabulary word that contains each word part.

WORD PART	MEANING	EXAMPLE
erythro-	red	1. _____
leuko-	white	2. _____
thrombo-	clot	3. _____
poie-	to make or produce	4. _____
hemo-	blood	5. _____

WORD PART	MEANING	EXAMPLE
poly-	many	6. _____
-chrom-	color	7. _____
-morpho-	shape	8. _____
-emia	blood	9. _____
-phil-	loving	10. _____

CONTENT LEARNING ACTIVITY

Introduction

66_Blood travels to and from the tissues of the body and plays an important role in homeostasis._**99**

Using the terms provided complete the following statements:

Carbon dioxide and waste products
Enzymes
Fluid loss

Hormones
Nutrients and Oxygen
Protection
Regulation

Blood has many functions in the body. It transports _(1)_ to cells to provide the fuel for cellular respiration. _(2)_ are transported away from cells by the blood, and _(3)_ are transported from endocrine glands to their target tissues. _(4)_ , which are important for catalyzing reactions in cells, are transported to certain tissues. The blood also has the important function of _(5)_ of the body from bacteria, foreign substances, and transformed cells, as well as the _(6)_ of temperature, fluids, electrolytes, and pH. In addition to these functions, blood also coagulates to prevent excessive _(7)_ .

1. _____

2. _____

3. _____

4. _____

5. _____

6. _____

7. _____

Blood Components

66_Blood, consisting of cells and cell fragments surrounded by a liquid intercellular matrix,_**99** _is classified as a connective tissue._

Match these terms with the correct statement or definition:

Formed elements
Hematocrit
Plasma

Platelets
Red blood cells
White blood cells

_____ 1. Collectively, the cells and cell fragments in blood.

_____ 2. Fluid matrix of blood.

380

_____ 3. Cells that constitute most of the formed elements.

_____ 4. Cell fragments that are part of the formed elements.

_____ 5. Percent of total blood volume composed of formed elements; normal range is 38 to 48% in females and 44 to 54% in males.

Plasma

66_The liquid matrix of the blood is called plasma._**99**

Match these terms with the correct statement or definition:

Colloidal solution   Serum
Plasma   Solutes
Plasma proteins   Water

_____ 1. Pale yellow fluid that accounts for more than half the blood volume.

_____ 2. Fine particles suspended in a liquid and resistant to sedimentation or filtration.

_____ 3. Albumin, globulins, and fibrinogen.

_____ 4. Plasma with the clot-producing proteins removed.

_____ 5. The major (over 90%) component of plasma.

Production of Formed Elements

66_All the formed elements of the blood are derived from a single population of stem cells._**99**

Match these terms with the correct statement or definition:

Hematopoiesis   Monoblasts
Hemocytoblasts   Myeloblasts
Lymphoblasts   Proerythroblasts
Megakaryoblasts

_____ 1. Blood cell production process that takes place in red bone marrow and lymphoid tissue after birth.

_____ 2. Stem cells that give rise to all the formed elements.

_____ 3. Cells from which granulocytes develop.

_____ 4. Cells from which lymphocytes develop.

_____ 5. Cells from which platelets develop.

☞ In adults, the hematopoietic red marrow is confined to the skull, ribs, sternum, vertebrae, pelvis, proximal femur, and proximal humerus.

Erythrocytes

"Erythrocytes are the most numerous of the formed elements in the blood....."

A. Using the terms provided, complete the following statements:

1. _____

2. _____

3. _____

Biconcave   Nuclei
Hemoglobin

Normal erythrocytes are _(1)_ disks that lose their _(2)_ and nearly all of their organelles during maturation. The main component of the erythrocyte is the pigmented protein _(3)_, which accounts for its red color.

B. Match these terms with the correct statement or definition:

Bicarbonate ion   Carbonic anhydrase
Carbonic acid   Hemolysis

_____ 1. Process in which erythrocytes rupture and hemoglobin is released.

_____ 2. Enzyme that catalyzes the reaction between carbon dioxide and water.

_____ 3. Product of the reaction between carbon dioxide and water.

_____ 4. Major form of carbon dioxide transported in the blood.

☞ The primary functions of erythrocytes are to transport oxygen from the lungs to the various tissues of the body and to transport carbon dioxide from the tissues to the lungs.

C. Match these terms with the correct statement or definition:

Carbaminohemoglobin   Heme
Deoxyhemoglobin   Iron
   (reduced hemoglobin)   Oxyhemoglobin
Globin

_____ 1. One of four protein chains in the hemoblobin molecule.

_____ 2. Red pigment molecule in the form of a porphyrin ring.

_____ 3. Element in the center of each heme molecule.

_____ 4. Hemoglobin that has oxygen associated with each heme group.

_____ 5. Hemoglobin with a darker red color.

_____ 6. Hemoglobin with carbon dioxide attached to amino groups of the globin molecule.

D. Using the terms provided, complete the following statements:

Anemia   Less
Iron   More

Embryonic and fetal globins are _(1)_ effective at binding oxygen than is adult globin. Abnormal globins are _(2)_ effective at attracting oxygen than is normal globin, and may result in _(3)_. _(4)_ is necessary for the normal function of hemoglobin. Dietary iron is absorbed into the circulation in the upper part of the digestive tract. If the iron content of the body is high, _(5)_ iron is absorbed, and, as the content drops, _(6)_ iron is absorbed. Iron deficiency can also result in _(7)_.

1. _____

2. _____

3. _____

4. _____

5. _____

6. _____

7. _____

E. Using the terms provided, complete the following statements:

Erythrocytes   Polychromatic erythroblasts
Erythropoiesis   Reticulocytes
Proerythroblasts

The process by which new erythrocytes are produced is called _(1)_. _(2)_, the cells from which erythrocytes develop, are derived from hemocytoblasts. After several mitotic divisions, proerythroblasts become basophilic erythroblasts, which continue to undergo mitosis and begin to produce hemoglobin. These cells then develop into _(3)_, which have almost a complete complement of hemoglobin. These cells then lose their nuclei by a process of extrusion, after which the cells are called _(4)_. These cells are released from red bone marrow into the circulating blood, and within 1 or 2 days lose their endoplasmic reticulum and become _(5)_.

1. _____

2. _____

3. _____

4. _____

5. _____

F. Match these terms with the correct statement or definition:

Cofactor   Oxygen
Erythropoietin   Renal erythropoietic factor

_____ 1. Function of iron, vitamin B12, and folic acid in erythropoiesis.

_____ 2. Humoral factor that stimulates erythropoiesis.

_____ 3. Enzyme produced by the kidneys that converts an inactive molecule to erythropoietin.

_____ 4. Production of renal erythropoietic factor increases when the levels of this substance decrease.

G. Match these terms with the correct statement or definition:

Bilirubin   Macrophages
Biliverdin

_____ 1. Cells that remove old, damaged, or defective erythrocytes from the blood.

_____ 2. First breakdown product of heme groups.

_____ 3. Breakdown product of heme groups that is released into the blood and is excreted by the liver in bile.

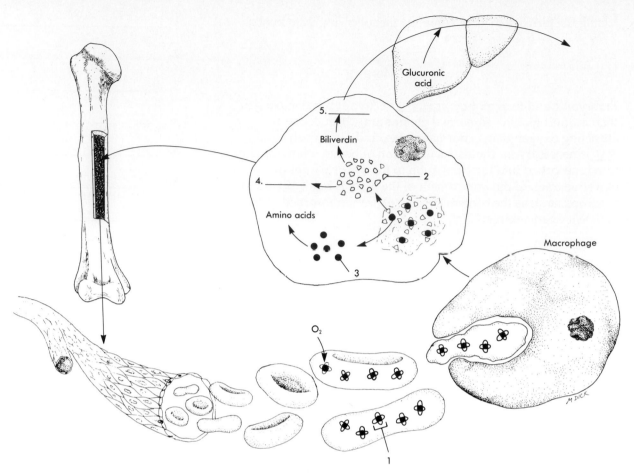

Figure 19-1

H. Match these terms with the correct parts of the diagram labeled in Figure 19-1:

Bilirubin
Globin
Heme

Hemoglobin
Iron

1. _____ 3. _____ 5. _____

2. _____ 4. _____

I. Match these terms with the correct statement or definition:

Antibodies
Antigens

Transfusion

_____ 1. Process of replacing lost blood.

_____ 2. Substances recognized by the immune system.

_____ 3. Proteins in the plasma of the blood that react with antigens on blood cells. The reaction results in the clumping or lysing of blood cells.

J. Match these blood types with
    the correct statement:

Type A blood   Type B blood
Type AB blood   Type O blood

_____ 1. This type blood could be given to a person with any blood type.

_____ 2. A person with this type of blood could receive blood from anyone.

_____ 3. These blood types would have type A antibodies.

_____

_____ 4. This blood type is the most common throughout the world.

_____ 5. In this blood type, erythocytes would contain both A and B antigens.

K. Match these blood types with
    the correct statement:

Rh-negative
Rh-positive

_____ 1. Blood type with no Rh antigens.

_____ 2. A fetus with this type of blood could develop erythroblastosis fetalis.

_____ 3. A woman with this type of blood will never have a baby who develops
                           erythroblastosis fetalis.

☞ Erythroblastosis fetalis is a disorder that occurs when maternal anti-Rh antibodies cross
    the placenta in large numbers and lyse fetal erythrocytes.

Leukocytes

❝*Leukocytes, or white blood cells, are nucleated blood cells that lack hemoglobin.***❞**

A. Match these terms with the
    correct statement or definition:

Agranulocytes   Granulocytes
Chemotaxis   Phagocytosis
Diapedesis   Pus
Differential count

_____ 1. Process by which leukocytes leave the circulation.

_____ 2. Process by which leukocytes are attracted to foreign material or dead
                           cells.

_____ 3. Accumulation of dead leukocytes, fluid, and debris.

_____ 4. Leukocytes containing large cytoplasmic granules; includes
                           neutrophils, eosinophils, and basophils.

_____ 5. Leukocytes with very small cytoplasmic granules; includes
                           lymphocytes and monocytes.

_____ 6. Evaluation of the relative proportions of various leukocyte types.

Blood Disorders

" *Blood disorders may seriously affect the function of the circulatory system.* **"**

A. Match these types of anemia with the correct description:

Aplastic anemia Hypochromic anemia
Hemolytic anemia Iron deficiency anemia
Hemorrhagic anemia Pernicious anemia

_____ 1. Deficiency of erythrocytes caused by the loss of large quantities of blood.

_____ 2. Insufficient erythrocyte production caused by abnormal red bone marrow or destruction of the red bone marrow.

_____ 3. Caused by insufficient vitamin B12.

_____ 4. Anemia in which less than normal amounts of hemoglobin are in the erythrocytes.

_____ 5. Disorders such as thalassemia and sickle cell anemia, in which red blood cells rupture or are destroyed at an excessive rate.

B. Match these blood disorders with the correct description:

Hemophilia Leukocytosis
Leukemia Polycythemia
Leukopenia Thombocytopenia

_____ 1. Condition characterized by an overabundance of erythrocytes.

_____ 2. Type of cancer in which abnormal production of one or more leukocytes occurs.

_____ 3. Genetic disorder in which coagulation factors are abnormal or absent.

_____ 4. Elevation of leukocyte numbers in the blood due to causes such as bacterial or parasitic infections.

_____ 5. Decrease in leukocyte numbers due to such things as viral infection, radiation or drug therapy, or liver sclerosis.

QUICK RECALL

1. List seven important functions of the circulatory system.

2. Name the parts of a hemoglobin molecule, give the function of each part, and state the fate of each part when hemoglobin is broken down.

3. List the events that lead to increased red blood cell production when blood oxygen levels decrease.

4. List the three types of granulocytes and give a function of each.

5. List the two types of agranulocytes and give a function of each.

6. Give two ways that platelets prevent blood loss.

7. For stage 1 of the extrinsic and intrinsic clotting mechanism, list the starting and ending chemicals.

8. For stages 2 and 3 of the clotting mechanism, list the chemical reactions that occur.

9. List the chemicals that prevent clot formation or dissolve clots.

MASTERY LEARNING ACTIVITY

Place the letter corresponding to the correct answer in the space provided.

_____ 1. Which if the following is a function of blood?
a. prevents fluid loss
b. transport of hormones
c. carries oxygen to cells
d. involved in regulation of body temperature
e. all of the above

_____ 2. Which of the following is NOT a component of plasma?
a. albumin
b. fibrinogen

c. platelets
d. water
e. glucose

_____ 3. The stem cells that give rise to all the formed elements are
a. hemocytoblasts.
b. lymphoblasts.
c. megakaryoblasts.
d. monoblasts.
e. myeloblasts.

_____ 4. Erythrocytes
   a. are the least numerous formed
      element in the blood.
   b. are cylindrically shaped cells.
   c. are produced in yellow marrow.
   d. do not have a nucleus.
   e. all of the above

_____ 5. Which of the following components of
      an erthyrocyte is correctly matched
      with its function?
   a. heme group of hemoglobin - oxygen
      transport
   b. globin portion of hemoglobin -
      carbon dioxide transport
   c. carbonic anhydrase - carbon dioxide
      transport
   d. all of the above

_____ 6. Which of the following constitutents of
      blood is used as a sensitive indicator of
      increased hematopoiesis?
   a. erythrocytes
   b. reticulocytes
   c. neutrophils
   d. platelets
   e. plasma

_____ 7. Which of the components of
      hemoblobin is correctly matched with
      its fate following the destruction of an
      erthrocyte?
   a. heme: reused to form new
      hemoglobin molecule
   b. globin: broken down into amino
      acids
   c. iron: mostly secreted in bile
   d. all of the above

_____ 8. Erythropoietin
   a. is produced mainly by the kidneys.
   b. inhibits the production of
      erythrocytes.
   c. activation increases when blood
      oxygen decreases.
   d. activation is inhibited by
      testosterone.

_____ 9. Which of the following changes would
      occur in the blood in response to the
      initiation of a vigorous exercise
      program?
   a. increased renal erythropoietic
      secretion
   b. increased concentration of
      reticulocytes

   c. increased bilirubin formation
   d. a and b
   e. all of the above

_____ 10. If you lived near the coast and you were
      training for a track meet in Denver, you
      would want to spend a few weeks
      before the meet training at
   a. sea level.
   b. an altitude similar to Denver's.
   c. a facility with a hyperbaric chamber.
   d. it doesn't matter

_____ 11. A person with type A blood
   a. has type A antibodies.
   b. has type B antigcns.
   c. will have a transfusion reaction if
      given type B blood.
   d. all of the above

_____ 12. Rh negative mothers that receive
      RhoGam injections are given that
      injection
   a. to initiate the synthesis of anti-Rh
      positive antibodies in the mother.
   b. to initiate anti-Rh positive antibody
      production in the baby.
   c. to prevent the mother from
      producing anti-Rh positive
      antibodies.
   d. to prevent the baby from producing
      anti-Rh positive antibodies.

_____ 13. The type of blood cells that function to
      inhibit inflammation are
   a. eosinophils.
   b. basophils.
   c. neutrophils.
   d. monocytes.

_____ 14. The most numerous type of leukocytes,
      whose primary function is
      phagocytosis, are
   a. macrophages.
   b. lymphocytes.
   c. neutrophils.
   d. thrombocytes.

_____ 15. Monocytes
   a. are the smallest sized leukocytes.
   b. increase in number in chronic
      infections.
   c. give rise to neutrophils.
   d. all of the above

_____ 16. An elevated neutrophil count is usually indicative of
a. an allergic reaction.
b. a bacterial infection.
c. a viral infection.
d. a parasitic infection.
e. increased antibody production.

_____ 17. A constituent of blood plasma that forms the network of fibers in a clot is
a. fibrinogen.
b. thromboplastin.
c. platelets.
d. thrombin.

_____ 18. The intrinsic pathway for clot formation begins with
a. tissue thromboplastin.
b. prothrombin activator.
c. factor X (Stuart factor).
d. factor XII (Hageman factor).

_____ 19. Given the following chemicals:
1. activated factor X
2. fibrinogen
3. prothrombin activator
4. thrombin

Choose the arrangement that lists the chemicals in the order they would be used during clot formation.
a. 1, 3, 4, 2
b. 2, 3, 4, 1
c. 3, 4, 1, 2
d. 3, 4, 2, 1

_____ 20. The chemical that is involved in the breakdown of a clot (fibrinolysis) is
a. fibrinogen.
b. antithrombin.
c. heparin.
d. plasmin.

☆ ——————— FINAL CHALLENGES ——————— ☆

Use a separate sheet of paper to complete this section.

1. Explain the effect of a decreased intake of iron on hematocrit.

2. Patients with advanced kidney diseases that impair kidney function often become anemic. On the other hand, patients with kidney tumors sometimes develop polycythemia. Can you explain these symptoms? (Hint: Tumors often cause overactivity of the tissue affected.)

3. If the plasma from a person suffering from hemolytic anemia were injected into a normal person, what would be the effect on erythrocyte production in the normal person? Explain.

4. Cigarette smoke produces carbon monoxide. If a nonsmoker smoked a pack of cigarettes a day for a few weeks, what would happen to his reticulocyte count? Explain.

5. A young child has periodic episodes of difficulty in breathing. Because of the patient's history, the physician suspects that the attacks are brought on by a stress-anxiety reaction. However, asthma (an allergic reaction) is also a legitimate possibility. The physician orders a complete blood count. What information in the complete blood count could be used to decide between the two diagnoses?

6. How could you distinguish between excessive bleeding due to a vitamin K deficiency and thrombocytopenia?

7. During pregnancy the developing fetus must manufacture many new red blood cells. What precautions should the mother take with her diet to prevent the development of anemia in herself and the fetus?

ANSWERS TO CHAPTER 19

WORD PARTS

1. erythrocyte; proerythroblast; erythropoietin
2. leukocyte; leukocytosis; leukemia
3. thrombocyte; thrombocytopenia; thrombus
4. erythropoietin; renal erythropoietic factor; hematopoiesis
5. hematocrit; hematopoiesis; hemocytoblasts; hemolysis; hemoglobin
6. polychromatic; polymorphonuclear; polycythemia
7. polychromatic
8. polymorphonuclear
9. anemia; polycythemia; leukemia
10. basophil; eosinophil

CONTENT LEARNING ACTIVITY

Introduction

1. Nutrients and oxygen
2. Carbon dioxide and waste products
3. Hormones
4. Enzymes
5. Protection
6. Regulation
7. Fluid loss

Blood Components

1. Formed elements
2. Plasma
3. Red blood cells
4. Platelets
5. Hematocrit

Plasma

1. Plasma
2. Colloidal solution
3. Plasma proteins
4. Serum
5. Water

Production of Formed Elements

1. Hematopoiesis
2. Hemocytoblasts
3. Myeloblasts
4. Lymphoblasts
5. Megakaryoblasts

Erythrocytes

A. 1. Biconcave
   2. Nuclei
   3. Hemoglobin

B. 1. Hemolysis
   2. Carbonic anhydrase
   3. Carbonic acid
   4. Bicarbonate ion

C. 1. Globin
   2. Heme
   3. Iron
   4. Oxyhemoglobin
   5. Deoxyhemoglobin (reduced hemoglobin)
   6. Carbaminohemoglobin

D. 1. More
   2. Less
   3. Anemia
   4. Iron
   5. Less
   6. More
   7. Anemia

E. 1. Erythropoiesis
   2. Proerythroblasts
   3. Polychromatic erythroblasts
   4. Reticulocytes
   5. Erythrocytes

F. 1. Cofactor
   2. Erythropoietin
   3. Renal erythropoietic factor
   4. Oxygen

G. 1. Macrophages
   2. Biliverdin
   3. Bilirubin

H. 1. Hemoglobin
   2. Heme
   3. Globin
   4. Iron
   5. Bilirubin

I. 1. Transfusion
   2. Antigens
   3. Antibodies

J. 1. Type O blood
   2. Type AB blood
   3. Type B blood; type O blood
   4. Type O blood
   5. Type AB blood

K. 1. Rh-negative
   2. Rh-positive
   3. Rh-positive

Leukocytes

A. 1. Diapedesis
   2. Chemotaxis
   3. Pus
   4. Granulocytes
   5. Agranulocytes
   6. Differential count

B. 1. Polymorphonuclear neutrophils
   2. Phagocytosis
   3. Lysozyme

C. 1. Eosinophils
   2. Eosinophils
   3. Eosinophils
   4. Histamine

D. 1. Lymphocyte
   2. Monocyte
   3. Macrophage
   4. Lymphocyte

Platelets

1. Platelet plug
2. Prostaglandin

3. Serotonin and thrombaxane
4. Thrombocytopenia

Homeostasis

A. 1. Coagulation
    2. Vitamin K
    3. Fibrinogen
    4. Thrombin

B. 1. Stage 1 (extrinisc)
    2. Stage 1 (intrinsic)
    3. Stage 1 (extrinsic); Stage 1 (intrinsic)
    4. Stage 2
    5. Stage 3

C. 1. Anticoagulant
    2. Heparin
    3. Embolus
    4. Antithrombin
    5. Clot retraction
    6. Fibrinolysis
    7. Plasmin

Blood Disorders

A. 1. Hemorrhagic anemia
    2. Aplastic anemia
    3. Pernicious anemia
    4. Hypochromic anemia
    5. Hemolytic anemia

B. 1. Polycythemia
    2. Leukemia
    3. Hemophilia
    4. Leukocytosis
    5. Leukopenia

QUICK RECALL

1. Transports gases, nutrients, waste products, and hormones; protects against disease; regulates temperature, fluid, and electrolytes
2. Globin: transports carbon dioxide; broken down to amino acids, that are recycled; Heme: transports oxygen; broken down to biliverdin, then to bilirubin, which is carried in the plasma to the liver, where it is incorporated into bile; iron is recycled
3. Decreased oxygen - renal erythropoietic factor produced from kidney - erythropoietin activated - erythropoiesis stimulated
4. Neutrophils: phagocytize foreign matter, secrete lysozyme; eosinophils: reduce inflammatory response; basophils: release histamine and heparin for inflammatory or allergic response.

5. Lymphocytes: immunity, including antibody production; monocytes: become macrophages
6. Formation of platelet plugs and formation of clots
7. Stage 1 extrinsic: tissue thromboplastin to Stuart Factor (Factor X)
   Stage 1 intrinsic: Hageman factor (Factor XII) to Stuart Factor (Factor X)
8. Stage 2: Prothrombin to thrombin; Stage 3: Fibrinogen to fibrin
9. Heparin and antithrombin prevent clots; plasmin dissolves clots

1. E. Blood performs many important functions such as transport (oxygen, carbon dioxide, nutrients, waste products, hormones, enzymes), protection against bacteria and other foreign substances, coagulation (prevents fluid loss), temperature regulation, and pH regulation.

2. C. Albumin, fibrinogen, water, and glucose are all important components of plasma. Platelets are formed elements and are not part of the plasma.

3. A. Hemocytoblasts give rise to all the formed elements. Hemocytoblasts give rise to lymphoblasts (from which lymphocytes develop), megakaryoblasts (from which platelets develop), monoblasts (from which monocytes develop), myeloblasts (from which granulocytes develop), and proerythroblasts (from which erythrocytes develop).

4. D. Erythrocytes do not have a nucleus. They are the most numerous formed elements, they are biconcave disks, and they are produced in red bone marrow.

5. D. Heme groups are involved with oxygen transport, and the globins with carbon dioxide transport. Carbonic anhydrase catalyzes the reaction between carbon dioxide and water to form carbonic acid, which dissociates to form bicarbonate and hydrogen ions.

6. B. Reticulocytes are immature red blood cells that have just been released into the blood. An increase in the number of reticulocytes would indicate an increased rate of hemopoiesis.

7. B. Globin is broken down into its component amino acids, most of which are used in the production of other proteins. Iron is recycled and used in the production of new hemoglobin. Heme is converted to biliverdin and then to bilirubin, which is secreted in bile.

8. C. Erythropoietin is produced by the liver and released into the blood in an inactive form. When blood oxygen levels decrease, the kidneys produce renal erythropoietic factor, which activates erythropoietin. The activated erythropoietin stimulates erythrocyte production. Testosterone stimulates renal erythropoietic factor and therefore stimulates erythropoietin activation.

9. E. Vigorous exercise reduces blood flow to the kidney (see Chapter 15); therefore there is reduced oxygen delivery. The kidneys release more renal erythropoietic factor, and more erythropoietin is activated. This results in increased hematopoiesis, and reticulocyte count increases. With increased exercises one might expect increased damage to erythrocytes and increased bilirubin formation as hemoglobin was broken down.

10. B. At higher altitudes the blood is less able to pick up oxygen from the air. The resulting decrease in blood oxygen stimulates hemopoiesis. Since the oxygen-carrying capacity of blood increases with an increased erythrocyte count, it is an advantage for a person who is going to compete at a high altitude to be exposed to that altitude long enough to acclimate.

11. C. A person with type A blood has type A antigens and type B antibodies; one with type B blood has type B antigens and type A antibodies. If the type A person receives type B blood, the A antibodies in the donated blood will bind to the A antigen of the person, and a transfusion reaction will occur. Also, the person's type B antibodies can bind to the type B antigens of the donated blood.

12. C. The RhoGam injection contains anti-Rh positive antibodies that remove fetal Rh positive-antigens from the maternal circulation before the mother's immune system recognizes their presence and begins to produce Rh positive antibodies.

13. A. Eosinophils produce enzymes that destroy inflammatory chemicals such as histamine. Basophils promote inflammation through the release of histamines.

14. C. Neutrophils are the most numerous leukocytes (60% to 70%); their primary function is phagocytosis. Macrophages are also phagocytic but are derived from monocytes that make up 2% to 8% of leukocytes. Thrombocytes or platelets are much more numerous than leukocytes, but they are not considered leukocytes and they are involved with clot formation, not phagocytosis.

15. B. Monocyte numbers typically increase during chronic infection. Monocytes are the largest of the leukocytes (lymphocytes are the smallest), and they give rise to macrophages.

16. B. Neutrophils are capable of phagocytizing bacteria, but not viruses. Neutrophil numbers often increase dramatically in response to bacterial infections. Eosinophil numbers increase in response to allergic reactions and certain parasitic infections. Lymphocytes are responsible for antibody production.

17. A. During the process of clotting fibrinogen gives rise to fibrin, which undergoes polymerization to form the network of fibers that make up a clot.

18. D. When factor XII comes into contact with damaged blood vessels, it becomes activated, and the intrinsic pathway begins. Tissue thromboplastin is released by damaged blood vessels and initiates the extrinsic pathway.

19. A. Through the extrinsic or intrinsic pathway factor X is activated. It combines with Ca^{2+} ions, factor V, and phospholipids to form prothrombin activator. Prothrombin activator converts prothrombin to thrombin, and the thrombin converts fibrinogen to fibrin.

20. D. Plasmin breaks down fibrin. Fibrinogen is converted into fibrin during clot formation. Antithrombin and heparin are anticoagulants that prevent clot formation.

 FINAL CHALLENGES

1. Inadequate iron in the diet results in decreased hemoglobin production. Since hemoglobin normally makes up about one-third of the volume of an erythrocyte, the erythrocytes would be smaller than normal. Hematocrit is the percent of the total blood volume that is formed elements, the bulk of which is erythrocytes. With decreased erythrocyte size, the hematocrit would be decreased.

2. Impairment of kidney function could result in decreased renal erythropoietic factor production. Thus, hemopoiesis would decrease and anemia would result. If a tumor caused overproduction of renal erythropoietin factor, polycythemia could result.

3. In hemolytic anemia erythrocytes are rapidly destroyed. Renal erythropoietin factor is released from the kidneys into the plasma in an attempt to stimulate the production of erythrocytes to replace those lost. If this plasma were injected into a normal person, one would expect erythrocyte production to increase.

4. Carbon monoxide binds to the iron of hemoglobin to form carboxyhemoglobin, which does not transport oxygen. The decreased oxygen would stimulate erythropoiesis, and the reticulocyte count should be elevated over normal.

5. An elevated eosinophil count could indicate an allergic reaction (asthma) that produces inflammation. The eosinophils function to reduce the severity of the inflammation.

6. A possible solution would be to take a blood sample and count the number of platelets. Thrombocytopenia is characterized by a reduced platelet count.

7. The mother should include adequate amounts of vitamin B12 and folic acid (to ensure erythrocyte production), iron (to ensure hemoglobin production), and vitamin K (to ensure proper blood clotting).

Cardiovascular System: The Heart

FOCUS: The heart is surrounded by the pericardium, which anchors the heart in the mediastinum, prevents overdistension of the heart, and protects the heart against friction. The heart has two atria, which receive blood from the body and the lungs, and two ventricles, which pump blood to the body and lungs. Atrioventricular valves and semilunar valves ensure one-way flow of blood through the heart, and heart sounds are produced as these valves close. The conducting system of the heart produces action potentials in the SA node that initiate contraction of the atria. Action potentials are delayed in the AV node, allowing time for the atria to contract and move blood into the ventricles. Then the ventricles contract, starting at the apex of the heart. Cardiac muscle cells are autorhythmic and have a prolonged depolarization (plateau phase), which extends the refractory period and prevents tetanus. The electrocardiogram measures the electrical activity of the heart, i.e., the electrical changes produced due to cardiac muscle cell action potentials. The action potentials result in the cardiac cycle, a repetitive sequence of systole (contraction) and diastole (relaxation). During systole, pressure builds within heart chambers and blood is ejected, whereas during diastole pressure in heart chambers decreases, and blood flows into the chambers. The rate and stroke volume (amount of blood ejected per beat) of the heart are regulated. Starling's law of the heart states that stroke volume is equal to venous return. The parasympathetic system inhibits heart rate whereas the sympathetic system, epinephrine, and norepinephrine increase heart rate and stroke volume. The baroreceptor reflex detects changes in blood pressure and causes alterations of heart rate and stroke volume that result in the return of blood pressure to normal levels.

WORD PARTS

Give an example of a new vocabulary word that contains each word part.

WORD PART	MEANING	EXAMPLE
brady-	slow	1. _____
tachy-	fast	2. _____

WORD PART	MEANING	EXAMPLE
diastol-	stand apart; relax	3. _____
systol-	stand together; contract	4. _____
ectop-	displaced	5. _____
iso-	equal	6. _____
metr-	measure	7. _____
dia-	apart	8. _____
-stasis	standing	9. _____
baro-	pressure	10. _____

CONTENT LEARNING ACTIVITY

Size, Form, and Location of the Heart

"The adult heart has the shape of a blunt cone and is about the size of a closed fist.**"**

Match these terms with the correct statement or definition:

Apex
Base

Pericardial cavity

_____ 1. Space within the mediastinum that contains the heart.

_____ 2. End of the heart where veins enter and arteries exit; superior part of heart.

Anatomy of the Heart

"The heart is a muscular pump consisting of four chambers.**"**

A. Match these terms with the correct statement or definition:

Fibrous pericardium
Parietal pericardium
Pericardial fluid

Pericardium
Serous pericardium
Visceral pericardium (epicardium)

_____ 1. Double-layered sac that surrounds the heart.

_____ 2. Tough outer layer of the pericardium.

_____ 3. Portion of the serous pericardium that lines the fibrous pericardium.

_____ 4. Fluid that fills the pericardial cavity.

The serous pericardium and pericardial fluid reduce friction as the heart moves within the pericardium.

B. Match these vessels with the correct description:

Auricles   Coronary sulcus
Coronary artery   Cardiac vein
Coronary sinus   Interventricular sulcus

_____ 1. Vessel that arises from the aorta and carries blood to the wall of the heart.

_____ 2. Large vessel that carries blood from the walls of the heart to the right atrium.

_____ 3. Small vessels that empty into the right atrium.

_____ 4. Flaplike extensions of the atria.

_____ 5. Large groove that separates the atria and ventricles, and runs obliquely around the heart.

C. Match these terms with the correct statement or definition:

Atrioventricular canal   Interatrial septum
Foramen ovale   Interventricular septum
Fossa ovalis

_____ 1. Wall separating the right and left atria.

_____ 2. Oval depression on the right side of the interatrial septum.

_____ 3. Opening between the right and left atria in embryonic and fetal stages of development.

_____ 4. Opening between an atrium and ventricle.

D. Match these terms with the correct statement or definition:

Atrioventricular valve   Semilunar valve
Bicuspid (mitral) valve   Tricuspid valve
Chordae tendinae

_____ 1. General term for a one-way valve between the atrium and ventricle.

_____ 2. Atrioventricular valve between the right atrium and right ventricle.

_____ 3. Connective tissue strings between papillary muscles and atrioventricular valves.

_____ 4. One-way valve in the aorta or pulmonary trunk; consists of pocketlike cusps.

 The papillary muscles and the chordae tendinae function to prevent the atrioventricular valves from opening back into the atria.

E. Match these terms with the correct parts of the diagram labeled in Figure 20-1:

Aorta
Aortic semilunar valve
Bicuspid (mitral) valve
Chordae tendinae
Interventricular septum
Left atrium
Left ventricle
Papillary muscles

Pulmonary semilunar valve
Pulmonary vein
Pulmonary trunk
Right atrium
Right ventricle
Superior vena cava
Tricuspid valve

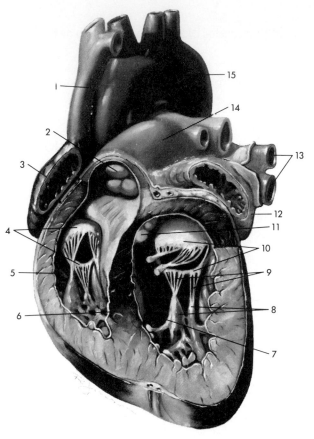

Figure 20-1

1. _____

2. _____

3. _____

4. _____

5. _____

6. _____

7. _____

8. _____

9. _____

10. _____

11. _____

12. _____

13. _____

14. _____

15. _____

Route of Blood Flow Through the Heart

66 *Blood flow through the heart occurs simultaneously in both right and left sides.* **99**

Arrange the following terms in sequential order as blood returns to the heart from the body and is pumped through the heart:

Aorta
Aortic semilunar valve
Bicuspid (mitral) valve
Left atrium
Left ventricle

Lungs
Pulmonary semilunar valve
Pulmonary trunk
Pulmonary veins
Right atrium

Right ventricle
Tricuspid valve
Vena cavae

1. _____   6. _____   11. _____

2. _____   7. _____   12. _____

3. _____   8. _____   13. _____

4. _____   9. _____

5. _____   10. _____

Histology

❝The heart contains cardiac muscle, epithelium, and connective tissue.❞

A. Match these terms with the correct statement or definition:

Crista terminalis
Endocardium
Epicardium
Musculi pectinati

Myocardium
Skeleton of the heart
Trabeculae carneae

_____ 1. Connective tissue plate that electrically isolates the atria from the ventricles; forms the fibrous rings.

_____ 2. Thin serous membrane that forms the outer smooth surface of the heart; visceral pericardium.

_____ 3. Layer of the heart responsible for the ability of the heart to contract.

_____ 4. Muscular ridges in the auricles and right atrial wall.

_____ 5. Ridges and columns on the interior walls of the ventricles.

B. Using the terms provided, complete the following statements:

Diad
Intercalated disks
Lactic acid
Oxygen debt

Sarcomeres
Sarcoplasmic reticulum
Transverse tubules

1. _____

2. _____

3. _____

4. _____

5. _____

6. _____

7. _____

Like skeletal muscle, cardiac muscle contains actin and myosin organized to form (1) . Cardiac muscle also has a smooth (2) that is loosely associated with membranes of the (3) . This loose association, which is called a (4) , is partly responsible for the slow onset of contraction in cardiac muscle. Fatty acids and (5) are used to produce ATPs that provide energy for cardiac muscle contraction. The production of energy in cardiac muscles depends on oxygen, and cardiac muscle cannot develop a significant (6) . Cardiac muscle cells are bound end-to-end and laterally by specialized cell-to-cell contacts called (7) , which hold the cells together and function as areas of low electrical resistance between the cells, allowing cardiac muscle cells to function as a single unit.

C. Match these terms with the correct statement or definition:

Apex
AV bundle
AV node

Bundle branches
Purkinje fibers
SA node

_____ 1. Modified cardiac muscle cells that delay action potentials between the atria and the AV bundle.

_____ 2. Conducting cells that arise from the AV node.

_____ 3. Right and left subdivisions of the AV bundle.

_____ 4. Inferior, terminal branches of the bundle branches, composed of large-diameter cardiac muscle fibers.

_____ 5. Part of the heart where ventricular contraction begins.

D. Match these terms with the correct parts of the diagram labeled in Figure 20-2:

AV bundle
AV node
Bundle branches
Purkinje fibers
SA node

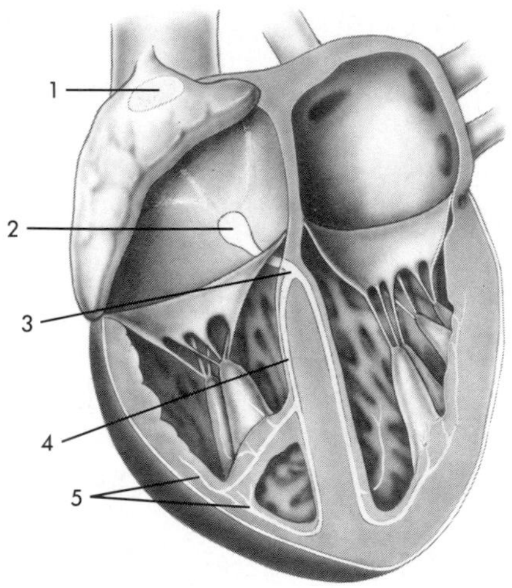

Figure 20-2

1. _____

2. _____

3. _____

4. _____

5. _____

Electrical Properties

"Cardiac muscle cells have many characteristics in common with other electrically excitable cells."

A. Match these terms with the correct statement or definition:

Absolute refractory period
Autorhythmic
Ectopic pacemakers (foci)

Plateau phase
Relative refractory period
Resting membrane potential

_____ 1. Condition necessary for electrically excitable cells to produce an action potential.

_____ 2. Prolonged period of depolarization in cardiac muscle cells; prevents tetanic contractions.

_____ 3. Locations away from the SA node that initiate heartbeats.

_____ 4. Period during which cardiac muscle is insensitive to stimulation.

B. Match these terms with the
    correct statement or definition:

Cardiac muscle
Skeletal muscle

_____ 1. Smaller-diameter, shorter, branching cells.

_____ 2. In this type of muscle multinucleate cells are present.

_____ 3. Depolarization phase of the action potential is due to sodium and
                            calcium.

_____ 4. Cells that generate spontaneous action potentials.

_____ 5. Has faster depolarization and a shorter refractory period.

C. Match these terms with the
    correct statement or definition:

Electrocardiogram (ECG)   QRS complex
P-Q (P-R) interval   Q-T interval
P wave   T wave

_____ 1. Summated record of all action potentials transmitted through the heart
                            during a given time period.

_____ 2. Record of action potentials that cause depolarization of the atrial
                            myocardium.

_____ 3. Record of action potentials from ventricular depolarization.

_____ 4. Record of repolarization of the ventricles.

_____ 5. Approximate time of ventricular contraction.

D. Match these terms with the correct location on the diagram in Figure 20-3:

P-Q interval
P wave
QRS complex
Q-T interval
T wave

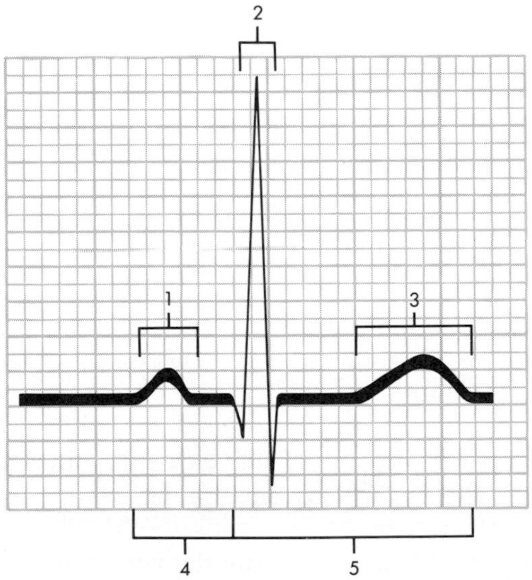

1. _____

2. _____

3. _____

4. _____

5. _____

Figure 20-3

403

E. Match these terms with the correct symptoms:

Atrial fibrillation
Bradycardia
Sinus arhythmia

Tachycardia
Ventricular fibrillation

_____ 1. Condition in which heart rate is in excess of 100 beats/minute.

_____ 2. Condition in which no P waves are seen on the ECG.

_____ 3. Condition in which no Q-T complexes are seen on the ECG.

_____ 4. Condition in which heart rate is less than 60 beats/min.

Cardiac Cycle

66 *The cardiac cycle is the repetitive pumping process that begins with cardiac muscle* 99
contraction and ends with the beginning of the next contraction.

A. Match these terms with the correct statement or definition:

Diastole
Final one third of ventricular
    diastole
First one third of ventricular
    diastole

Second one third of ventricular
    diastole

_____ 1. Contraction of the ventricular myocardium.

_____ 2. 70% of ventricular filling (due to low pressure in the ventricles).

_____ 3. Diastasis (little ventricular filling).

_____ 4. 30% of ventricular filling (due to atrial contraction).

B. Match these waves of the left atrial pressure curve with the correct statement or definition:

a wave
c wave
v wave

_____ 1. Result of atrial contraction.

_____ 2. Result of a slight increase in atrial pressure due to ventricular systole.

_____ 3. Result of continuous blood flow into the atria.

C. Match these time periods with the correct statement or definition:

Ejection
Isometric contraction

Isometric relaxation
Protodiastole

_____ 1. Period between atrioventricular valve closure and semilunar valve opening, when there is no movement of blood out of the ventricles.

_____ 2. Period of time that blood flows from the ventricles.

_____ 3. Last part of ventricular systole, when ventricular volume becomes very low and almost no blood flows from the ventricles.

_____ 4. Period between semilunar valve closure and atrioventricular valve opening, when no blood flows from the atria into the ventricles.

D. Match these terms with the
   correct statement or definition:

Blood pressure
Cardiac output
Cardiac reserve
End-diastolic volume

End-systolic volume
Incisura (aortic notch)
Stroke volume

_____ 1. Volume of blood in the ventricles when they are filled.

_____ 2. Volume of blood pumped during each cardiac cycle.

_____ 3. Total amount of blood pumped per minute.

_____ 4. The force responsible for blood movement in vessels (proportional to
                            cardiac output times the peripheral resistance).

_____ 5. Increase in aortic pressure when the semilunar valve closes and blood
                            flows back toward the ventricle from the aorta.

E. Match these terms with the
   correct statement or definition:

First heart sound
Second heart sound

Third heart sound

_____ 1. The vibrations associated with the semilunar valves closing.

_____ 2. The vibrations associated with the atrioventricular valves closing.

_____ 3. The sound of blood flowing in a turbulent fashion into the ventricles.

_____ 4. The low-pitched "lubb" sound.

F. Match these terms with the correct location on Graph A in Figure 20-4:

1. _____

Aortic valve closes   Semilunar valve closes
Aortic valve opens   Semilunar valve opens

2. _____

3. _____

Match these terms with the correct location on Graph B in Figure 20-4:

4. _____

Ejection   Isometric relaxation
End diastolic volume   Protodiastole
End systolic volume   Stroke volume
Isometric contraction

5. _____

6. _____

The difference between 5 and 6.:

7. _____

8. _____

9. _____

10. _____

11. _____

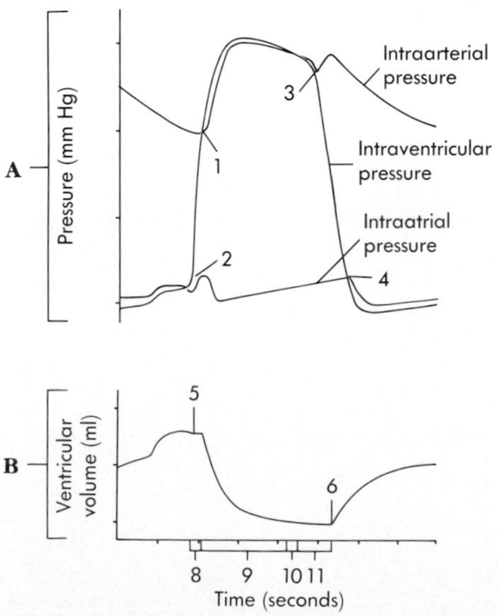

Figure 20-4

405

G. Match these terms with the
   correct statement or definition:

Incompetent valve   Stenosed valve
Murmur

_____ 1. Abnormal heart sound.

_____ 2. Valve that leaks significantly.

_____ 3. Valve with an abnormally narrow opening.

Regulation of the Heart

“_The amount of blood pumped by the heart may vary dramatically depending on a variety of conditions._**”**

A. Match these terms with the
   correct statement or definition:

Extrinisic regulation
Intrinisic regulation

_____ 1. Regulation of the heart due to its normal functional characteristics.

_____ 2. Regulation of the heart due to neural or hormonal control.

_____ 3. Regulation that involves venous return and Starling's law of the heart.

B. Match these terms with the
   correct statement or definition:

Decrease   No effect
Increase

_____ 1. Effect of decreased venous return on cardiac output.

_____ 2. Effect of stretching the right atrial wall on heart rate.

_____ 3. Effect of stimulating the vagus nerve on heart rate.

_____ 4. Effect of sympathetic stimulation on heart rate and the force of heart muscle contraction.

_____ 5. Effect on stroke volume if heart rate becomes too great.

_____ 6. Effect of norepinephrine on heart rate and the force of heart muscle contraction.

Heart and Homeostasis

66The heart's pumping efficiency plays an important role in the maintenance of homeostasis.99

A. Match these terms with the
correct statement or definition:

Baroreceptors
Cardioregulatory center

_____ 1. Sensory receptors that measure blood pressure, located in the walls of
the aorta and internal carotid arteries.

_____ 2. Area of the medulla oblongata that integrates sensory information and
sends efferent impulses to the heart.

B. Match these terms with the
correct statement or definition:

Decreases
Increases

_____ 1. Effect on heart rate when arterial blood pressure increases.

_____ 2. Effect on heart rate of an increase in blood pH and a decrease of blood
carbon dioxide level.

_____ 3. Effect on heart rate of a decrease in blood oxygen level.

_____ 4. Effect on heart rate and stroke volume of an excess of potassium ions.

_____ 5. Effect on the force of contraction of an increase in extracellular calcium
ions.

1. List the parts of the serous pericardium and describe their function.

2. Name the veins that enter the right and left atria.

3. Name the four valves that regulate the direction of blood flow in the heart, and give their location.

4. Give the two nodes of the conducting system of the heart and their functions.

5. List four histological (structural) differences between skeletal and cardiac muscle.

6. List four differences between skeletal and cardiac muscle in regard to action potentials.

7. State the cause of the P wave, the QRS complex, and the T wave of the ECG. Name the contraction events associated with each wave.

8. List the two major normal heart sounds, and give the reason for each.

9. Give two intrinsic regulatory mechanisms that cause cardiac output to equal venous return.

10. List the effects of sympathetic and parasympathetic stimulation of the heart.

11. List the locations where the nervous system detects changes in blood pressure, carbon dioxide, pH, and oxygen that affect the heart.

MASTERY LEARNING ACTIVITY

Place the letter corresponding to the correct answer in the space provided.

_____ 1. The fibrous pericardium
a. is in contact with the heart.
b. is a serous membrane.
c. is also known as the epicardium.
d. forms the outer layer of the pericardium (pericardial sac).

_____ 2. Which of the following structures carry blood to the right atrium?
a. coronary sinus
b. superior vena cava
c. inferior vena cava
d. all of the above

_____ 3. The valve located between the right atrium and the right ventricle is the
a. aortic semilunar valve.
b. pulmonary semilunar valve.
c. tricuspid valve.
d. bicuspid (mitral) valve.

_____ 4. The papillary muscles
a. are attached to the chordae tendinae.
b. are found in the atria.
c. contract to close the foramen ovale.
d. are attached to the semilunar valves.

_____ 5. Given the following blood vessels:
1. aorta
2. inferior vena cava
3. pulmonary trunk
4. pulmonary vein

Choose the arrangement that lists the vessels in the order a red blood cell would encounter them.
a. 1, 3, 4, 2
b. 2, 3, 4, 1
c. 2, 4, 3, 1
d. 3, 2, 1, 4
e. 3, 4, 1, 2

_____ 6. The skeleton of the heart
a. electrically insulates the atria from the ventricles.
b. provides a rigid site of attachment for the cardiac muscle.

c. functions to reinforce or support the valve openings.
d. all of the above

_____ 7. The bulk of the heart wall consists of the
a. epicardium.
b. pericardium.
c. myocardium.
d. endocardium.

_____ 8. Muscular ridges on the interior surface of the auricles are called
a. trabeculae carneae.
b. crista terminalis.
c. musculi pectinati.
d. endocardium.

_____ 9. Cardiac muscle has
a. sarcomeres.
b. a sarcoplasmic reticulum.
c. transverse tubules.
d. all of the above

_____ 10. Electrical impulses are conducted from one cardiac muscle cell to another
a. due to intercalated disks.
b. because of a special cardiac nervous system.
c. due to the large voltage of the action potentials.
d. because of the plateau phase of the action potential.

_____ 11. Given the following structures of the conduction system of the heart:
1. atrioventricular bundle
2. AV node
3. bundle branches
4. Purkinje fibers
5. SA node

Choose the arrangement that lists the structures in the order an action potential would pass through them.
a. 2, 5, 1, 3, 4
b. 2, 5, 3, 1, 4
c. 2, 5, 4, 1, 3
d. 5, 2, 1, 3, 4
e. 5, 2, 4, 3, 1

_____ 12. During the transmission of action potentials through the conducting system of the heart, there is a temporary delay of transmission at the
  a. bundle branches.
  b. Purkinje fibers.
  c. AV node.
  d. SA node.

_____ 13. Purkinje fibers
  a. are specialized cardiac muscle cells.
  b. conduct impulses much more slowly than ordinary cardiac muscle.
  c. conduct action potentials through the atria.
  d. all of the above

_____ 14. The apex of the heart is
  a. located superior to the base of the heart.
  b. where the major blood vessels exit the heart.
  c. where ventricular contraction begins.
  d. all of the above

_____ 15. If the SA-node is damaged and becomes nonfunctional, which of the following is most likely?
  a. The heart will stop.
  b. The ventricles will contract with a greater frequency than the atria.
  c. Another portion of the heart, possibly within the atria, will become the pacemaker.
  d. The heart will enter fibrillation.
  e. Tachycardia will develop.

_____ 16. Cardiac muscle cells will not tetanize because
  a. the nervous stimulation is not sufficiently strong.
  b. the SA node is the pacemaker.
  c. of the intercalated disks.
  d. it has a long refractory period.

_____ 17. A T wave represents
  a. depolarization of the ventricles.
  b. repolarization of the ventricles.
  c. depolarization of the atria.
  d. repolarization of the atria.

_____ 18. Which of the following conditions observed in an electrocardiogram would suggest that the AV node is not conducting action potentials?
  a. complete lack of the P wave
  b. complete lack of the QRS complex
  c. more QRS complexes than P waves
  d. a prolonged P-R interval
  e. complete asynchrony between the P wave and QRS complexes

_____ 19. The greatest amount of ventricular filling occurs during
  a. the first third of diastole.
  b. the second third of diastole.
  c. the last third of diastole.
  d. ventricular systole.

_____ 20. While the semilunar valves are open during a normal cardiac cycle,
  a. the pressure in the left ventricle is greater than the pressure in the aorta.
  b. the pressure in the left ventricle is less than the pressure in the aorta.
  c. the pressure in the left ventricle is exactly the same as the pressure in the aorta.
  d. the pressure in the left ventricle is the same as the pressure in the left atrium.

_____ 21. The pressure within the left ventricle fluctuates between
  a. 120 to 80 mm Hg
  b. 120 to 0 mm Hg
  c. 80 to 0 mm Hg
  d. 20 to 0 mm Hg

_____ 22. Blood neither flows into nor out of the ventricles during
  a. the period of isometric contraction.
  b. the period of isometric relaxation.
  c. diastasis.
  d. protodiastole.
  e. a and b

_____ 23. Stroke volume
  a. is the amount of blood pumped by the heart per minute.
  b. is the difference between end-diastolic and end-systolic volume.
  c. is the difference between the amount of blood pumped at rest and that pumped at maximum output.
  d. is the amount of blood pumped from the atria into the ventricles.

24. Cardiac output is defined as
   a. blood pressure times peripheral resistance.
   b. peripheral resistance times heart rate.
   c. heart rate times stroke volume.
   d. stroke volume times blood pressure.

25. Pressure in the aorta is at its lowest
   a. at the time of the first heart sound.
   b. at the time of the second heart sound.
   c. just before the AV valves open.
   d. just before the semilunar valves open.

26. Just after the aortic notch on the aortic pressure curve,
   a. the pressure in the aorta is greater than the pressure in the ventricle.
   b. the pressure in the ventricle is greater than the pressure in the aorta.
   c. the pressure in the left atrium is greater than the pressure in the ventricle.
   d. the pressure in the left atrium is greater than the pressure in the aorta.
   e. blood flow in the aorta has stopped.

27. The "lubb" sound (first heart sound) of the heart is caused by the
   a. closing of the AV valves.
   b. closing of the semilunar valves.
   c. blood rushing out of the ventricles.
   d. filling of the ventricles.
   e. ventricular contraction.

28. Increased venous return results in increased
   a. stroke volume.
   b. heart rate.
   c. cardiac output.
   d. all of the above

29. Parasympathetic nerve fibers are carried in the _____ nerves and release _____ at the heart.
   a. cardiac, acetylcholine
   b. cardiac, norepinephrine
   c. vagus, acetylcholine
   d. vagus, norepinephrine

30. Increased parasympathetic stimulation to the heart will
   a. increase the force of ventricular contraction.
   b. increase the rate of depolarization in the SA node.
   c. decrease the heart rate.
   d. increase cardiac output.

31. Sympathetic stimulation of the heart increases which of the parameters listed below?
   a. heart rate
   b. stroke volume
   c. force of ventricular contraction
   d. all of the above

32. Epinephrine released from the adrenal medulla
   a. increases the rate of heart contractions.
   b. decreases the force of heart contractions.
   c. produces the same effect as parasympathetic stimulation of the heart.
   d. a and c
   e. b and c

33. Through the baroreceptor reflex, when normal arterial blood pressure decreases, you would expect
   a. heart rate to decrease.
   b. stroke volume to decrease.
   c. blood pressure to return to normal.
   d. all of the above

34. A decrease in blood pH and an increase in blood carbon dioxide levels results in
   a. increased heart rate.
   b. increased stroke volume.
   c. increased sympathetic stimulation of the heart.
   d. all of the above

35. An increase in extracellular potassium levels could cause
   a. an increase in stroke volume.
   b. an increase in the force of contraction.
   c. a decrease in heart rate.
   d. a and b

411

Use a separate sheet of paper to complete this section.

1. Skeletal muscle exhibits graded responses due to multiple motor unit summation and multiple wave summation. Explain why the heart does not do this.

2. A friend tells you that her son had an ECG and it revealed that he had a slight heart murmur. Should you be convinced that he has a heart murmur? Explain.

3. Predict the effect on heart rate if the vagus nerves to the heart were cut.

4. Predict the effect on Starling's law of the heart if the vagus nerves to the heart were cut.

5. Predict the effect on heart rate if the glossopharyngeal nerves to the heart were cut.

6. An experiment on a dog was performed in which the mean arterial blood pressure was monitored before and after the common carotid arteries were clamped (at time A). The results are graphed below:

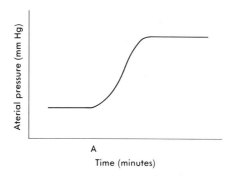

Explain the change in mean arterial blood pressure (hint: baroreceptors are located in the internal carotid arteries, which are superior to the site of clamping of the common carotid arteries).

7. What would happen to blood pressure following the ingestion of a large amount of isosmotic fluid? Explain.

8. During hemorrhagic shock (due to loss of blood) the blood pressure may fall dramatically, although the heart rate is elevated. Explain why the blood pressure falls despite the increase in heart rate.

9. During defecation it is not uncommon to "strain" by holding the breath and compressing the thoracic and abdominal muscles. Assume that the increased pressure in the thoracic cavity compresses the vena cavae. What would happen to blood pressure? Describe the compensatory mechanisms that would be activated to correct the change in blood pressure. Next, assume that the increased thoracic pressure increases the pressure in the aortic arch. What would happen to blood pressure (hint: sensory receptors).

10. A patient exhibited the following symptoms: chest pain, rapid pulse, a greater frequency of P waves than QRS complexes, and disappearance of these symptoms following the administration of a beta adrenergic blocking agent (inhibited the action of norepinephrine on the heart). Is the patient suffering from a myocardial infarct, heart block, or atrial tachycardia? Explain.

ANSWERS TO CHAPTER 20

WORD PARTS

1. bradycardia
2. tachycardia
3. diastole; diastolic; protodiastole
4. systole; systolic
5. ectopic

6. isometric
7. isometric
8. diastasis
9. diastasis
10. baroreceptors

CONTENT LEARNING ACTIVITY

Size, Form and Location of the Heart

1. Pericardial cavity

2. Base

Anatomy of the Heart

A. 1. Pericardium
2. Fibrous pericardium
3. Parietal pericardium
4. Pericardial fluid

B. 1. Coronary artery
2. Coronary sinus
3. Cardiac vein
4. Auricles
5. Coronary sulcus

C. 1. Interatrial septum
2. Fossa ovalis
3. Foramen ovale
4. Atrioventricular canal

D. 1. Atrioventricular valve
2. Tricuspid valve
3. Chordae tendinae
4. Semilunar valve

E. 1. Superior vena cava
2. Pulmonary semilunar valve
3. Right atrium
4. Tricuspid valve
5. Right ventricle
6. Interventricular septum
7. Left ventricle
8. Papillary muscles
9. Chordae tendinae
10. Bicuspid (mitral) valve
11. Aortic semilunar valve
12. Left atrium
13. Pulmonary veins
14. Pulmonary trunk
15. Aorta

Route of Blood Flow Through the Heart

1. Vena cavae
2. Right atrium
3. Tricuspid valve
4. Right ventricle
5. Pulmonary semilunar valve
6. Pulmonary trunk
7. Lungs

8. Pulmonary veins
9. Left atrium
10. Bicuspid (mitral) valve
11. Left ventricle
12. Aortic semilunar valve
13. Aorta

Histology

A. 1. Skeleton of the heart
   2. Epicardium
   3. Myocardium
   4. Musculi pectinati
   5. Trabeculae carneae

B. 1. Sarcomeres
   2. Sarcoplasmic reticulum
   3. Transverse tubules
   4. Diad
   5. ATP
   6. Oxygen debt
   7. Intercalated disks

C. 1. AV node
   2. AV bundle
   3. Bundle branches
   4. Purkinje fibers
   5. Apex

D. 1. SA node
   2. AV node
   3. Bundle of His
   4. Bundle branches
   5. Purkinje fibers

Electrical Properties

A. 1. Resting membrane potential
   2. Plateau phase
   3. Ectopic pacemakers (foci)
   4. Absolute refractory period

B. 1. Cardiac muscle
   2. Skeletal muscle
   3. Cardiac muscle
   4. Cardiac muscle
   5. Skeletal muscle

C. 1. Electrocardiogram (ECG)
   2. P wave
   3. QRS complex

   4. T wave
   5. Q-T interval

D. 1. P wave
   2. QRS complex
   3. T wave
   4. P-Q interval
   5. Q-T interval

E. 1. Tachycardia
   2. Atrial fibrillation
   3. Ventricular fibrillation
   4. Bradycardia

Cardiac Cycle

A. 1. Systole
   2. First one third of ventricular diastole
   3. Second one third of ventricular diastole
   4. Final one third of ventricular diastole

B. 1. a wave
   2. c wave
   3. v wave

C. 1. Isometric contraction
   2. Ejection
   3. Protodiastole
   4. Isometric relaxation

D. 1. End-diastolic volume
   2. Stroke volume
   3. Cardiac output
   4. Blood pressure
   5. Incisura (aortic notch)

E. 1. Second heart sound
   2. First heart sound
   3. Third heart sound
   4. First heart sound

F. 1. Semilunar valves open
   2. Atrioventricular valves close
   3. Semilunar valves close
   4. Atrioventricular valves open
   5. End diastolic volume
   6. End systolic volume
   7. Stroke volume
   8. Isometric contraction
   9. Ejection
   10. Isometric relaxation
   11. Protodiastole

G. 1. Murmur
   2. Incompetent valve
   3. Stenosed valve

Regulation of the Heart

A. 1. Intrinsic regulation
   2. Extrinsic regulation
   3. Intrinsic regulation

B. 1. Decrease
   2. Increase
   3. Decrease
   4. Increase
   5. Decrease
   6. Increase

Heart and Homeostasis

A. 1. Baroreceptors
   2. Cardioregulatory center

B. 1. Decreases
   2. Decreases
   3. Increases
   4. Decreases
   5. Increases

QUICK RECALL

1. Parietal pericardium, visceral pericardium, and pericardial cavity - reduce friction.
2. Right atrium - inferior and superior vena cava, cardiac veins, and coronary sinus; left atrium - four pulmonary veins.
3. Tricuspid valve - between right atrium and right ventricle;bicuspid (mitral) valve - between left atrium and left ventricle; aortic semilunar valve - in the aorta; pulmonary semilunar valve - in the pulmonary trunk.
4. SA node - pacemaker of the heart; AV node - slows action potentials, allowing atria to contract and move blood into ventricles.
5. Cardiac muscle - one nucleus, branches, no dilated cisternae, diads, intercalated disks; skeletal muscle - multinucleate, unbranched, dilated cisternae, triads.
6. Depolarization due to sodium and calcium in cardiac muscle (sodium in skeletal muscle); rate of action potential propagation is slower in cardiac muscle; action potentials are propagated from cell to cell in cardiac muscle but not in skeletal muscle; cardiac muscle has spontaneous generation of action potentials whereas skeletal muscle action potentials result from nervous system stimulation; and cardiac muscle has a prolonged depolarization (plateau) phase.
7. P wave: atrial depolarization, atrial systole; QRS complex: ventricular depolarization, ventricular systole; T wave: ventricular repolarization, ventricular diastole. 8. First heart sound: closing of atrioventricular valves; second heart sound: closing of semilunar valves.
9. Starling's law of the heart is "cardiac output = venous return". Venous return stretches SA node and increases heart rate.
10. Parasympathetic stimulation - decreased heart rate and small decrease in stroke volume; sympathetic stimulation - increased heart rate and stroke volume.
11. Baroreceptors in aorta and carotid arteries monitor blood pressure; chemoreceptors in medulla monitor carbon dioxide and pH changes; chemoreceptors near carotid arteries monitor oxygen changes.

MASTERY LEARNING ACTIVITY

1. D. The pericardium consists of an outer fibrous pericardium and an inner serous pericardium (a serous membrane). The serous pericardium consists of a parietal pericardium, which lines the fibrous pericardium, and a visceral pericardium (also called the epicardium), which lines the heart.

2. D. The coronary sinus, superior vena cava, and inferior vena cava all carry blood to the right atrium. In addition, a number of cardiac veins from the anterior surface of the heart empty into the right atrium.

3. C. The tricuspid valve is between the right atrium and the right ventricle. The bicuspid valve is between the left atrium and the left ventricle. The aortic semilunar valve is between the left ventricle and the aorta. The pulmonary semilunar valve is between the right ventricle and the pulmonary trunk.

4. A. The papillary muscles (located in the ventricles) are attached to the atrioventricular valves. Contraction of the papillary muscles helps to prevent the valves from opening back into the atria. The foramen ovale is an opening between the right and left atria during development. It becomes the fossa ovalis.

5. B. The red blood cell would enter the right atrium through the inferior vena cava and pass into the right ventricle. The red blood cell would exit the right ventricle through the pulmonary trunk, pass through the lungs, and return to the left atrium through the pulmonary vein. From the left atrium, the red blood cell would enter the left ventricle and leave the heart via the aorta.

6. D. The skeleton of the heart is composed of fibrous connective tissue. It serves as electrical insulation between the atria and ventricles, provides a site of attachment for cardiac muscles, and forms a support for the valves of the heart (fibrous rings).

7. C. The myocardium is the middle layer of the heart wall, is composed of cardiac muscle, and comprises the bulk of the heart.

8. C. The musculi pectinati are muscular ridges in the auricles and a portion of the right atrial wall. The crista terminalis separates the musculi pectinati of the right atrial wall from the smooth portion of the right atrial wall. The trabeculae carneae are modified ridges in the interior ventricular walls. The endocardium is not muscle.

9. D. Cardiac muscle has many similarities with skeletal muscle. Cardiac muscle has actin and myosin myofilaments arranged to form sarcomeres. The sarcoplasmic reticulum and transverse tubules in cardiac muscle forms structures called diads.

10. A. The intimate connection between cardiac muscle cells at the intercalated disk constitutes an area of low electrical resistance. Action potentials are propagated from one muscle cell to the adjacent muscle cell, resulting in the function of cardiac muscle cells as a single unit.

11. D. The SA node generates an action potential that passes to the AV node. From the AV node the action potential passes through the atrioventricular bundle, the bundle branches, and the Purkinje fibers.

12. C. The delay in the AV node allows the atria to contract before the ventricles.

13. A. Purkinje fibers are large-diameter cardiac muscle cells specialized to conduct action potentials rapidly. They are found in the ventricles.

14. C. Ventricular contraction begins at the apex of the heart. The result is that blood is pushed superiorly toward the base of the heart where the major blood vessels exit.

15. C. If the SA-node becomes nonfunctional, it is unlikely that the heart will either stop or fibrillate, although there is a chance that either condition may occur. Since heart muscle cells are autorhythmic, another region of the heart will probably become the pacemaker, probably with in the atria.

16. D. Cardiac muscles cells do not tetanize because the action potential has a refractory period that is sufficiently long to allow contraction and relaxation before another action potential can be initiated. The plateau phase of the action potential in which depolarization is prolonged is responsible for the prolonged refractory period.

17. B. The T wave represents repolarization of the ventricles. The QRS complex represents depolarization of the ventricles; and the P wave, depolarization of the atria.

18. E. No conduction of action potentials through the AV node is called a complete heart block. This prevents the conduction of action potentials from the atria to the ventricles, and a pacemaker develops in the ventricle. The result is complete asynchrony or a complete lack of a relationship between atrial (P wave) and ventricular depolarization (QRS complex). In addition, the frequency of the P waves would be greater than the frequency of the QRS complexes.

19. A. During the first one third of diastole the atrioventricular valves open, and blood flows into the ventricles under the influence of the small pressure differential between the atria and the ventricles. About 70% of ventricular filling occurs during that period. The remaining 30% of ventricular filling occurs during the last one third of diastole due to atrial contraction.

20. A. The semilunar valves are open during ventricular systole, and blood is flowing from the left ventricle to the aorta. Therefore the pressure in the left ventricle must be higher than the pressure in the aorta.

21. B. Left ventricular pressure reaches 120 mm Hg (normal systolic pressure) during ventricular systole (contraction). During diastole (relaxation) the pressure in the ventricle falls to nearly zero, whereas the pressure within the aorta drops to 80 mm Hg.

22. E. Blood does not leave the ventricles during the period of isometric contraction and does not enter the ventricles during the period of isometric relaxation. During diastasis (the second third of ventricular diastole) a small amount of blood enters the ventricles, and during protodiastole (the last part of ventricular systole) a small amount of blood leaves the ventricles.

23. B. Stroke volume is the difference between end-diastolic volume (the amount of blood in a filled ventricle) and end-systolic volume (the amount of blood in a ventricle after contraction has occurred). Therefore stroke volume is the amount of blood ejected per beat of the heart. Cardiac output is the amount of blood pumped by the heart per minute. Cardiac reserve is the difference between the amount of blood pumped at rest and at maximum output.

24. C. Cardiac output is equal to heart rate times stroke volume.

25. D. Following ejection of blood into the aorta, pressure drops until the next ejection of blood. Since the next ejection of blood occurs when the semilunar valves open, the lowest pressure in the aorta occurs just before the semilunar valves open.

26. A. The aortic notch represents closure of the aortic semilunar valve. That valve closes when the pressure within the aorta exceeds the pressure within the ventricle.

27. A. The first heart sound is caused by closure of the AV valves, and the second heart sound ("dupp") is caused by closure of the semilunar valves. A third heart sound can sometimes be heard. It is due to turbulent flow of blood into the ventricles.

28. D. Increased venous return results in increased stroke volume (Starling's law of the heart) and increased heart rate (due to stretching of the SA node). The increased stroke volume and heart rate results in increased cardiac output.

29. C. The vagus nerves carry parasympathetic fibers to the heart where the postganglionic parasympathetic neurons release acetylcholine. The cardiac nerves carry sympathetic fibers to the heart where the postganglionic sympathetic neurons release norepinephrine.

30. C. Increased parasympathetic stimulation decreases heart rate by decreasing the rate of depolarization of the SA node. Parasympathetic stimulation has little inhibitory effect on the force of ventricular contraction. Since heart rate decreases, an increase in cardiac output does not occur.

31. D. Sympathetic stimulation of the heart results in an increased heart rate and force of ventricular contraction. Stroke volume increases due to the increased force of ventricular contraction.

32. A. Epinephrine produces the same effects as sympathetic stimulation of the heart, i.e., increased rate and force of contraction of the heart.

33. C. When blood pressure decreases, the baro-receptor reflex causes an increase in heart rate and stroke volume (due to increased sympathetic stimulation of the heart). Consequently, blood pressure increases (returns to normal).

34. D. When pH decreases and carbon dioxide levels increase, chemoreceptors detect the change and reflexes are activated that increase sympathetic stimulation of the heart. Consequently, heart rate and stroke volume increase.

35. C. Increased extracellular potassium results in a decreased heart rate and stroke volume.

1. Since cardiac muscle cells are interconnected by intercalated disks, cardiac muscle acts as a single unit and does not exhibit multiple motor unit summation (i.e., recruitment of motor units). Since cardiac muscle action potentials have a prolonged plateau phase, cardiac muscle relaxes before the next contraction and does not exhibit multiple wave summation.

2. An ECG measures the electrical activity of the heart and would not indicate a slight heart murmur. Heart murmurs are detected by listening to the heart sounds. The boy may have a heart murmur, but the mother does not understand the basis for making such a diagnosis.

3. The vagus nerves carry parasympathetic fibers to the heart, which inhibit heart rate. This is called vagal tone. Cutting the vagus nerves would remove this inhibition, and heart rate would increase. This is called vagal escape.

4. Since Starling's law of the heart is an intrinsic regulatory mechanism not dependent on the nervous system, cutting the vagus nerves should have no effect on Starling's law. Therefore venous return will still equal stroke volume.

5. The glossopharyngeal nerves and the vagus nerves contain afferent neurons from baroreceptors. Cutting the glossopharyngeal nerves would reduce the number of afferent impulses sent from the baroreceptors, the same effect produced by a decrease in blood pressure. Consequently, one would expect heart rate to increase.

6. When both common carotid arteries are clamped, the blood pressure within the internal carotid arteries drops dramatically. The decreased blood pressure is detected by the baroreceptors, and the baroreceptor reflex causes an increase in heart rate and stroke volume. The resulting increase in cardiac output causes the increase in blood pressure.

7. The ingested fluid would increase blood volume and therefore venous return. As venous return increased, cardiac output would increase (Starling's law of the heart), producing an increase in blood pressure. The increased blood pressure would be detected and by means of the baroreceptor reflex heart rate would decrease, returning blood pressure to normal.

8. Venous return declines markedly in hemorrhagic shock due to the loss of blood volume. With decreased venous return, stroke volume decreases (Starling's law of the heart). The decreased stroke volume results in a decreased cardiac output, which produces a decreased blood pressure. In response to the decreased blood pressure, the baroreceptor reflex causes an increase in heart rate in an attempt to restore normal blood pressure. However, with inadequate venous return the increased heart rate is not able to restore normal blood pressure.

9. This act of holding the breath is called the Valsalva manuever. Increased intrathoracic pressure produced by straining causes compression of the venae cavae leading to a decreased venous return to the heart, a decreased stroke volume (Starling's law of the heart), and a decreased cardiac output. The decreased cardiac output results in a decreased mean arterial blood pressure, which, by means of the baroreceptor reflex, causes an increase in heart rate and stroke volume. The increased heart rate and stroke volume cause an increase in blood pressure. If one strains for too long it is possible to faint because of inadequate delivery of blood to the brain (i.e., the increase in blood pressure is not adequate). Increased intrathoracic pressure could also cause an increase in pressure in the aorta and by means of the baroreceptor reflex a decreased heart rate, stroke volume, and blood pressure. Although either an increase or a decrease in blood pressure is possible with the Valsalva manuever, usually there is an increase.

10. Atrial tachycardia results in a greater number of P waves than QRS complexes because the ventricles require a greater time to repolarize than the atria. The tachycardia results in inadequate blood pressure within the coronary blood vessels and increases energy requirements of the cardiac muscle. The result is anoxia of the cardiac tissue and chest pain. Decreasing sympathetic stimulation of the heart by administering beta-adrenergic blocking agents is often used as a treatment because these drugs slow the heart rate. One would not expect a rapid pulse with heart block or a greater frequency of P waves than QRS complexes with a myocardial infarct.

Cardiovascular System: Peripheral Circulation And Regulation

FOCUS: Blood from the heart flows through elastic arteries, muscular arteries, and arterioles to capillaries in the tissues. From the capillaries blood returns to the heart through venules, small veins, and large veins. Generally, blood vessels consist of endothelium (tunica intima), smooth muscle and elastic tissue (tunica media), and connective tissue (tunica adventitia). The pulmonary circulation transports blood from the heart to the lungs and back to the heart, whereas the systemic circulation carries blood from the heart to the body and back to the heart. Blood flow through the blood vessels is mostly laminar. Turbulent blood flow produces Korotkoff sounds which can be used to estimate aortic blood pressure. Blood pressure is responsible for the movement of blood (Poiseuille's law), for keeping blood vessels from collapsing (law of Laplace), for the volume of blood contained within the vessel (vascular compliance), and the movement of fluid out of capillaries. Pressure can be increased or decreased within vessels due to the effect of gravity. Blood flow through tissues is controlled by relaxation and contraction of precapillary sphincters (vasodilator and nutrient demand theories), and blood pressure is regulated by nervous control (e.g., baroreceptor reflexes, chemoreceptor reflexes, central nervous system ischemic response) and humoral control (epinephrine, renin-angiotensin-aldosterone, ADH, and atrial natriuretic factor). The baroreceptor reflexes are the most important short-term regulator of blood pressure, achieving changes in blood pressure by altering heart rate, stroke volume, and peripheral resistance. The kidneys are the most important long-term regulator of blood pressure, accomplished by controlling blood volume.

WORD PARTS

Give an example of a new vocabulary word that contains each word part.

WORD PART	MEANING	EXAMPLE
-sclero-	hard	1. _____
arteri-	an artery	2. _____
vaso-	vessel	3. _____
carot-	stupor; sleep	4. _____
-rhagia	breaking out	5. _____
celia-	the abdominal cavity	6. _____
jugu-	the throat	7. _____
hepato-	the liver	8. _____
ischem-	stopping blood	9. _____
auscult-	listen to	10. _____

CONTENT LEARNING ACTIVITY

General Features of Blood Vessel Structure

66 *Blood flows from the heart through elastic arteries, muscular arteries, and arterioles to capillaries,* 99
then to venules, small veins and large veins to return to the heart.

A. Match these terms with the correct statement or definition:

Adventitia   Endothelium
Basement membrane   Pericapillary cells

_____ 1. Layer of simple squamous epithelium lining all blood vessels.

_____ 2. Thin layer to which endothelial cells are attached.

_____ 3. Delicate layer of loose connective tissue surrounding the basement membrane of capillaries.

_____ 4. Scattered cells that lie between the basement membrane and endothelial cells.

B. Match these terms with the correct statement or definition:

Continuous capillaries
Fenestrated capillaries
Sinusoidal capillaries
Sinusoids
Venous sinuses

_____ 1. Capillaries with no gaps between endothelial cells; present in nervous and muscle tissue.

_____ 2. Capillaries with endothelial cells possessing numerous fenestrae; present in intestinal villi and glomeruli of the kidney.

_____ 3. Capillaries with large diameters, large fenestrae, and a less prominent basement membrane; present in endocrine glands.

_____ 4. Large-diameter sinusoidal capillaries; present in the liver and bone marrow.

C. Match these terms with the correct statement or definition:

Arterial capillaries
Metarterioles
Precapillary sphincters
Thoroughfare channels
Venous capillaries

_____ 1. End capillaries closest to the arterioles.

_____ 2. Channels, through which blood flow is relatively continuous, that extend from a metarteriole to a venule.

_____ 3. Arterioles with isolated smooth muscle cells along their walls.

_____ 4. Smooth muscle cells that regulate blood flow from the thoroughfare channel into capillaries.

D. Match these terms with the correct statement or definition:

Tunica adventitia
Tunica intima
Tunica media

_____ 1. Tunic closest to the lumen of blood vessels, consisting of endothelium, a basement membrane, lamina propria, and a layer of elastic fibers.

_____ 2. Middle layer of blood vessel walls, consisting of smooth muscle cells, and elastic and collagen fibers.

_____ 3. Outer layer of blood vessel walls, composed of connective tissue that varies from dense to loose, depending on the blood vessel.

E. Match these types of arteries with the correct statement or definition:

Arterioles
Elastic arteries
Medium-sized and small arteries

_____ 1. Largest-diameter arteries; often called conducting arteries.

_____ 2. Have relatively thick walls due to smooth muscle layers in the tunica media; often called distributing arteries.

_____ 3. Transport blood from the small arteries to the capillaries.

_____ 4. Capable of vasoconstriction and vasodilation.

F. Match these terms with the correct statement or definition:

Medium-sized and large veins   Venules
Small veins

_____ 1. Structure of these veins is similar to capillaries; the smallest are capable of nutrient exchange.

_____ 2. Smallest veins to have continuous layer of smooth muscle.

_____ 3. Most of the veins in this category have valves (folds in the tunica intima that prevent backflow of blood).

G. Match these terms with the correct statement or definition:

Arteriovenous anastomoses   Vasa vasorum
Glomus

_____ 1. Small vessels that supply blood to the walls of veins and arteries.

_____ 2. Arterioles that allow blood to flow directly into small veins without an intermediate capillary; function in temperature regulation.

_____ 3. Arteriovenous anastomosis that consists of arterioles arranged in a convoluted fashion.

H. Match these terms with the correct statement or definition:

Arteriosclerosis   Phlebitis
Atherosclerosis   Varicose veins

_____ 1. Dilated veins with incompetent valves.

_____ 2. Inflammation of the veins.

_____ 3. Degenerative changes in arteries that make them less elastic.

_____ 4. Deposition of a fatlike substance containing cholesterol in the walls of arteries to form plaques.

Pulmonary Circulation

> ❝The pulmonary vessels transport blood from the right ventricle through the lungs and back to the left atrium.❞

Match these terms with the correct statement or definition:

Pulmonary arteries   Pulmonary veins
Pulmonary trunk

_____ 1. Blood passes from the right ventricle directly into this vessel.

_____ 2. These vessels transport blood to each lung.

_____ 3. There are four of these vessels from the lungs, which enter the left atrium.

Systemic Circulation: Arteries

"Oxygenated blood passes from the left ventricle to the aorta and is distributed to all parts of the body.**"**

A. Match these parts of the aorta with the correct statement:

Abdominal aorta   Descending aorta
Aortic arch   Thoracic aorta
Ascending aorta

_____ 1. Portion of the aorta that gives rise to the right and left coronary arteries.

_____ 2. Three major branches of this part of the aorta are the brachiocephalic, the left common carotid, and the left subclavian arteries.

_____ 3. Longest part of the aorta, running from the aortic arch to the common iliac arteries.

_____ 4. Portion of the aorta between the aortic arch and the diaphragm.

B. Match these arteries with the correct parts of the diagram labeled in Figure 21-1:

Brachiocephalic artery
Left common carotid artery
Left subclavian artery
Left vertebral artery
Right common carotid artery
Right subclavian artery
Right vertebral artery

1. _____

2. _____

3. _____

4. _____

5. _____

6. _____

7. _____

Figure 21-1

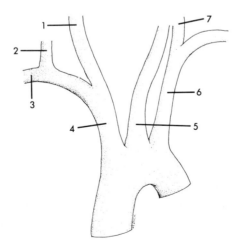

C. Match these arteries with the correct parts of the diagram labeled in Figure 21-2:

Basilar artery
Circle of Willis
External carotid artery

Internal carotid artery
Vertebral artery

1. _____

2. _____

3. _____

4. _____

5. _____

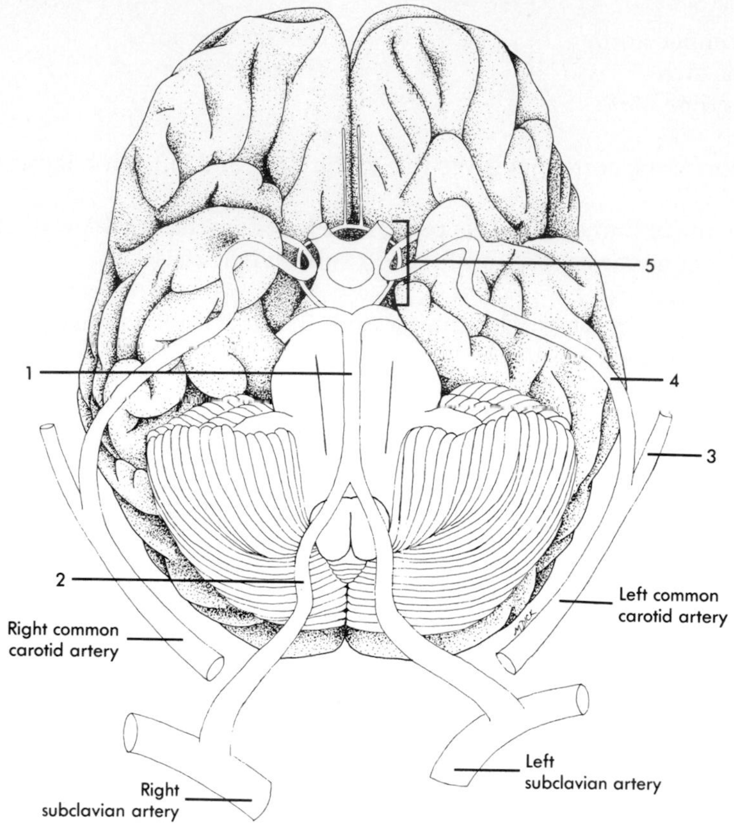

Figure 21-2

Labels on figure: 5, 4, 3, 1, 2, Right common carotid artery, Left common carotid artery, Left subclavian artery, Right subclavian artery

D. Match these arteries with the correct parts of the diagram labeled in Figure 21-3:

Axillary artery
Brachial artery
Digital artery
Palmar arches
Radial artery
Ulnar artery

1. _____

2. _____

3. _____

4. _____

5. _____

6. _____

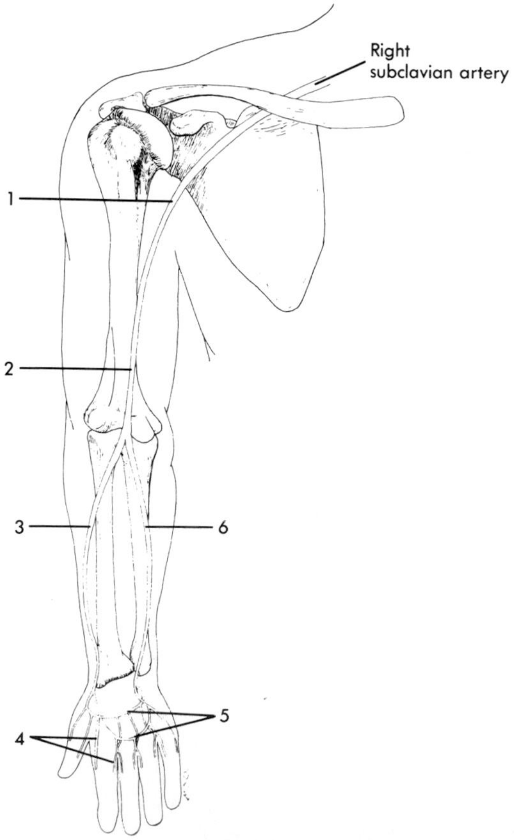

Right subclavian artery

Figure 21-3

Labels on figure: 1, 2, 3, 6, 4, 5

E. **M**atch these arteries with the correct parts of the diagram labeled in Figure 21-4:

Anterior intercostal artery
Internal thoracic artery
Posterior intercostal artery
Visceral arteries

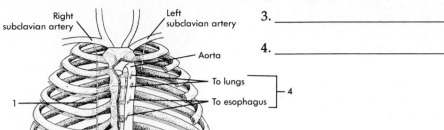

Figure 21-4

1. _____

2. _____

3. _____

4. _____

F. **M**atch these arteries with the correct parts of the diagram labeled in Figure 21-5:

Celiac trunk
Common hepatic artery
Common iliac artery
Inferior mesenteric artery
Left gastric artery
Gonadal artery
Renal artery
Splenic artery
Superior mesenteric artery
Suprarenal artery

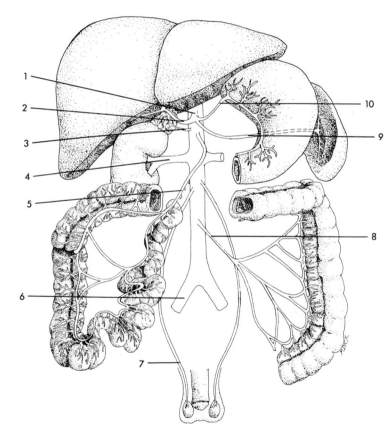

Figure 21-5

1. _____   5. _____   9. _____

2. _____   6. _____   10. _____

3. _____   7. _____

4. _____   8. _____

G. Match these arteries with the correct parts of the diagram labeled in Figure 21-6:

Anterior tibial artery
Digital arteries
Dorsalis pedis artery
Femoral artery
Lateral plantar artery
Medial plantar artery
Peroneal artery
Popliteal artery
Posterior tibial artery

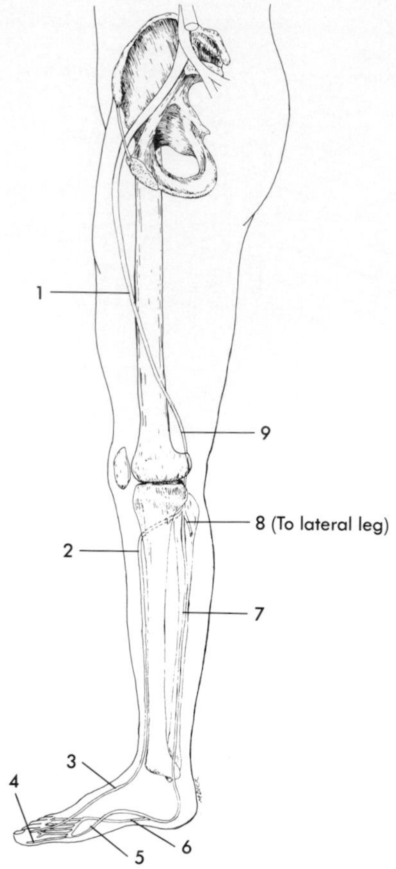

1. _____
2. _____
3. _____
4. _____
5. _____
6. _____
7. _____
8. _____
9. _____

Figure 21-6

Systemic Circulation: Veins

❝*Veins transport blood from the capillary beds toward the heart.*❞

A. Match these major veins with the correct description:

Coronary sinus
Inferior vena cava
Internal jugular vein
Superior vena cava
Venous sinuses

_____

_____

_____

_____

_____

1. Vein that returns blood from the walls of the heart.

2. Vein that returns blood from the head, neck, thorax and upper limbs to the heart.

3. Vein that returns blood from the abdomen, pelvis and lower limbs to the heart.

4. Spaces within the dura mater surrounding the brain; the superior sagittal sinus is an example.

5. Vein that drains blood from the venous sinuses of the brain.

B. Match these arteries with the correct parts of the diagram labeled in Figure 21-7:

Inferior vena cava
Pulmonary veins
Right brachiocephalic vein
Right external jugular vein
Right internal jugular vein
Right subclavian vein
Superior vena cava

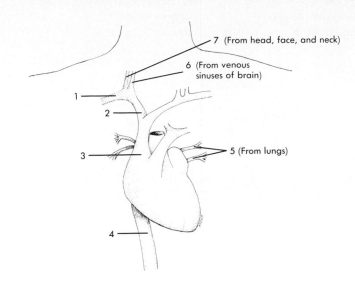

Figure 21-7

1. _____   4. _____   6. _____

2. _____   5. _____   7. _____

3. _____

C. Match these arteries with the correct parts of the diagram labeled in Figure 21-8:

Axillary vein
Basilic vein
Brachial veins
Cephalic vein
Median cubital vein
Venous arches

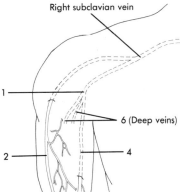

Figure 21-8

1. _____

2. _____

3. _____

4. _____

5. _____

6. _____

D. **Match these arteries with the correct parts of the diagram labeled in Figure 21-9:**

Accessory hemiazygos vein
Azygos vein
Hemiazygos vein

1. _____

2. _____

3. _____

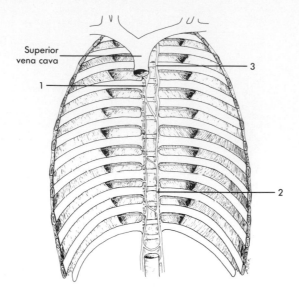

Superior
vena cava

Figure 21-9

E. **Match these arteries with the correct parts of the diagram labeled in Figure 21-10:**

Common iliac vein   Internal iliac vein
External iliac vein   Renal vein
Gonadal vein   Suprarenal vein
Hepatic veins

1. _____

2. _____

3. _____

4. _____

5. _____

6. _____

7. _____

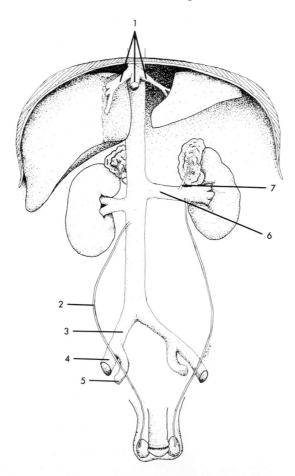

Figure 21-10

F. **M**atch these arteries with the correct parts of the diagram labeled in Figure 21-11:

Gastric vein
Hepatic portal vein
Hepatic veins
Inferior mesenteric vein

Inferior vena cava
Splenic vein
Superior mesenteric vein

1. _____

2. _____

3. _____

4. _____

5. _____

6. _____

7. _____

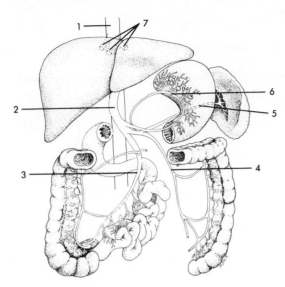

Figure 21-11

G. **M**atch these arteries with the correct parts of the diagram labeled in Figure 21-12:

Femoral vein
Great saphenous vein
Popliteal vein
Small saphenous vein

1. _____

2. _____

3. _____

4. _____

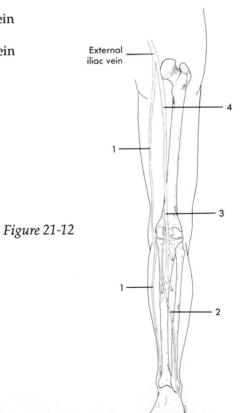

Figure 21-12

The Physics of Circulation

"*The physical characteristics of blood and the physical principles affecting the flow of liquids***"** *through vessels dramatically influence the circulation of blood.*

A. Match these terms with the correct statement or definition:

    Laminar flow   Viscosity
    Turbulent flow

_____   1. Measure of the resistance of a liquid to flow.

_____   2. Tendency for a fluid to flow through tubes as if the fluid is composed of concentric layers.

_____   3. Numerous small currents flowing crosswise or obliquely to the long axis of a vessel.

☞ As the hematocrit increases, the viscosity of blood increases logarithmically.

B. Match these terms with the correct statement or definition:

    Auscultatory   Korotkoff sounds
    Blood pressure   Sphygmomanometer

_____   1. Measure of the force blood exerts against the blood vessel walls.

_____   2. Device that uses an inflatable cuff to measure blood pressure.

_____   3. Method of determining blood pressure by listening to the sound of blood flowing through the arteries.

_____   4. Sound of turbulent blood flow through a blood vessel.

C. Using the terms provided, complete the following statements:

    Decreased   Increased

According to Poiseuille's Law, several factors affect resistance to blood flow. Flow is dramatically _(1)_ when the radius of a blood vessel is increased. Increased blood viscosity will cause _(2)_ flow, and decreased vessel length will result in _(3)_ flow. Also, if the pressure gradient in a vessel is decreased, flow will be _(4)_.

1. _____

2. _____

3. _____

4. _____

D. Match these terms with the correct statement or definition:

    Aneurysm   The law of Laplace
    Critical closing pressure   Vascular compliance

_____   1. Blood pressure below which a blood vessel will collapse.

_____   2. The force that stretches the vascular wall is proportional to the diameter of the vessel times the blood pressure.

_____   3. Bulge in a weakened blood vessel wall.

_____   4. Tendency for blood vessel volume to increase as blood pressure increases.

Veins have a much higher vascular compliance than arteries. Consequently, a small increase in pressure causes a large increase in venous volume, and the veins act as a storage area for blood.

E. Match these terms with the correct statement or definition:

Decreases
Increases

_____ 1. What happens to critical closing pressure when sympathetic stimulation of blood vessels increases.

_____ 2. What happens to the force acting on the wall of a blood vessel when the diameter of the vessel decreases (e.g., with sympathetic stimulation).

_____ 3. What happens to blood vessel volume following an increase in blood pressure.

Physiology of Systemic Circulation

66 *The anatomy of the circulatory system and the physics of blood flow participate in determining* **99** *the physiological characteristics of the circulatory system.*

A. Match these terms with the correct statement or definition:

Aorta   Capillaries
Arteries   Veins
Arterioles

_____ 1. The largest percentage of blood volume is contained in these vessels.

_____ 2. These vessels have the slowest velocity of blood flow, but the greatest cross-sectional area.

_____ 3. Blood flowing through this vessel has the greatest velocity and the greatest pressure.

_____ 4. These vessels have the highest resistance to flow.

_____ 5. These vessels have the lowest resistance to flow.

B. Using the terms provided, complete the following statements:

Decreased   Pulse pressure
Increased

The difference between the systolic and diastolic pressure is called the _(1)_. When stroke volume is increased, pulse pressure is _(2)_. For a given stroke volume, pulse pressure is _(3)_ when vascular compliance is decreased (e.g., with age). Pulse pressure produces a pressure wave that can be monitored as the pulse. Weak pulses usually indicate a(n) _(4)_ stroke volume or a(n) _(5)_ constriction of the arteries as a result of intense sympathetic stimulation.

1. _____

2. _____

3. _____

4. _____

5. _____

C. Match these terms with the correct statement or definition:

Diffusion   Lymphatic system
Edema   Osmotic pressure
Hydrostatic pressure

_____ 1. Major means by which nutrients and waste products are exchanged across capillary surfaces.

_____ 2. Physical force that moves fluid out of the capillary.

_____ 3. At the arteriolar end of the capillary, this force is greatest.

_____ 4. At the venous end of the capillary, this force is greatest.

_____ 5. System that picks up excess tissue fluid and returns it to the general circulation.

_____ 6. Swelling caused by excess tissue fluid accumulation.

D. Match these terms with the correct statement or definition:

Decreases
Increases

_____ 1. Effect on cardiac output when blood volume increases.

_____ 2. Effect on venous return to the heart when sympathetic stimulation increases venous tone.

_____ 3. Effect on venous return to the heart when exercising.

_____ 4. Effect on venous pressure in the legs when standing still; can result in edema.

Local Control of Blood Flow by the Tissues

"_In most tissues, blood flow to the tissues is proportional to the metabolic needs of the tissue._**"**

Using the terms provided, complete the following statements:

Autoregulation   Metarterioles and precapillary
Decrease   sphincters
Increase   Nutrient demand
Metabolism   Vasodilator substances

1. _____

2. _____

3. _____

4. _____

5. _____

6. _____

7. _____

Control of local blood flow occurs through the _(1)_ in capillary beds. Because there is little innervation of these structures, local factors regulate blood flow. As the rate of _(2)_ of a tissue increases, rate of blood flow through its capillaries increases. _(3)_ , including carbon dioxide, lactic acid, potassium ions and hydrogen ions, increase in extracellular fluid as the rate of metabolism increases. _(4)_ may also be important in regulating local blood flow; for instance, increased metabolism reduces the amount of oxygen and other nutrients in the tissues. In response to a(n) _(5)_ in vasodilator substances or a(n) _(6)_ in nutrients, metarterioles and precapillary sphincters dilate, resulting in an increased blood flow. _(7)_ refers to the local control mechanisms that determine blood flow to tissues, despite relatively large changes in systemic blood pressure.

Regulation of Local Circulation

66 *Nervous control of arterial blood pressure is important in minute-to-minute regulation, and during exercise or shock.* 99

Match these terms with the
correct statement or definition:

Acetylcholine   Sympathetic
Hypothalamus and cerebral cortex   Vasomotor center
Norepinephrine   Vasomotor tone

_____ 1. Most important division of the autonomic nervous system for nervous regulation of blood flow.

_____ 2. Area of the lower pons and upper medulla that controls sympathetic nerve impulses to blood vessels.

_____ 3. Areas of the brain that can inhibit or stimulate the vasomotor center.

_____ 4. Condition in which peripheral blood vessels are partially constricted.

_____ 5. Neurotransmitter for vasodilator fibers of blood vessels.

☞ Part of the vasomotor center is excitatory and is tonically active, and part of the vasomotor center is inhibitory and induces vasodilation when it is active.

Regulation of Mean Arterial Pressure

66 *A series of regulatory mechanisms exist that maintain the average systemic blood pressure* 99 *between 90 and 100 mm Hg.*

Match these terms with the
correct statement or definition:

Cardiac output   Mean blood pressure
Decrease   Peripheral resistance
Heart rate   Stroke volume
Increase

_____ 1. Equal to heart rate times stroke volume.

_____ 2. Volume of blood pumped by the heart during each contraction.

_____ 3. Equal to heart rate times stroke volume times peripheral resistance.

_____ 4. Resistance to the flow of blood in the blood vessels.

_____ 5. Effect on mean blood pressure when a decrease in heart rate occurs.

_____ 6. Effect on mean blood pressure when an increase in peripheral resistance occurs.

_____ 7. Effect on mean blood pressure when an increase in stroke volume occurs.

Nervous Regulation of Blood Pressure

"*Several mechanisms are involved in nervous regulation of blood pressure.*"

A. Match these types of short-term blood pressure regulation with the correct statement or definition:

Baroreceptor (pressoreceptor) reflex
Central nervous system ischemic response
Chemoreceptor reflex

_____ 1. Mechanism that involves sensory receptors sensitive to stretch; receptors are in the carotid sinus and aortic arch.

_____ 2. Mechanism that involves sensory receptors sensitive to decreased oxygen or increased carbon dioxide and hydrogen ion levels (i.e., decreased pH); receptors are in the carotid bodies and aortic bodies.

_____ 3. Mechanism that responds to increased carbon dioxide and hydrogen ions (i.e., decreased pH) in the medulla.

_____ 4. Mechanisms that usually are active only during emergency situations.

_____

_____ 5. Mechanism that is important in regulating blood pressure on a moment-to-moment basis.

B. Match these terms with the correct statements:

Decreased
Increased

Vasoconstriction
Vasodilation

_____ 1. Effect on heart rate and stroke volume produced by the baroreceptor reflexes with sudden decreases in arterial pressure.

_____ 2. Effect on blood vessels produced by the baroreceptor reflexes when blood pressure increases.

_____ 3. Effect on blood vessels produced by the chemoreceptor reflexes when arterial oxygen levels decrease markedly.

_____ 4. Effect on blood vessels produced by the central nervous system ischemic response when medullary carbon dioxide levels increase markedly.

Hormonal Regulation of Blood Pressure

"*Four important hormonal mechanisms control arterial pressure.*"

A. Match these types of hormonal control with the correct statement or definition:

Adrenal medullary mechanism
Atrial natriuretic mechanism
Renin-angiotensin-aldosterone mechanism
Vasopressin mechanism

_____ 1. Involves secretion of epinephrine and norepinephrine from the adrenal medulla.

_____ 2. Involves the release of an enzyme from the juxtaglomerular apparatuses in the kidneys.

_____   3. Involves the secretion of a hormone from the neurohypophysis.

_____   4. Involves the release of a polypeptide from cells in the atria of the heart.

B. Using the terms provided, complete the following statements:

Aldosterone   Decreased
Angiotensin II   Increased
Angiotensinogen   Renin

The kidneys release an enzyme called _(1)_ into the circulatory
system from specialized structures called juxtaglomerular
apparatuses. Renin acts on plasma proteins called _(2)_ to split a
fragment off one end. The fragment, called angiotensin I, has
two more amino acids cleaved from it to become _(3)_ .
Angiotensin II causes _(4)_ vasomotor tone in veins, which
functions to raise blood pressure. Angiotensin II also stimulates
_(5)_ secretion from the adrenal cortex. Aldosterone acts on the
kidneys resulting in _(6)_ urine production. Angiotensin II also
stimulates thirst, increases salt appetite, and stimulates ADH
secretion. Stimuli that increase renin secretion include _(7)_
blood pressure, _(8)_ potassium ion levels, and _(9)_ sodium ion
levels.

1. _____

2. _____

3. _____

4. _____

5. _____

6. _____

7. _____

8. _____

9. _____

☞ The renin-angiotensin-aldosterone mechanism is important in maintaining blood pressure
    under conditions of circulatory shock.

C. Match these arteries with the   Aldosterone   Angiotensin
    correct parts of the diagram   Angiotensin II   Renin
    labeled in Figure 21-13:

1. _____

2. _____

3. _____

4. _____

Adrenal cortex

(decreases urine volume;
conserves blood pressure)

1

Angiotensin I

(increases due
to a decrease
in blood
pressure)

Kidney

3

Potent
vasoconstrictor
substance
(causes an
increase in
blood pressure)

2   (produced in liver)

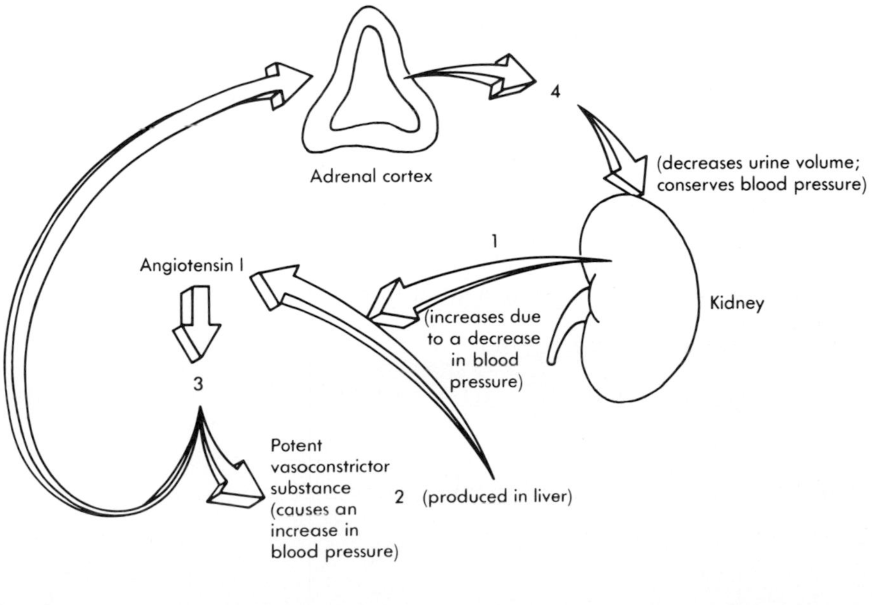

Figure 21-13

D. Match these terms with the
correct statements as they apply
to hormonal mechanisms that
control arterial pressure:

Decrease
Increase

_____ 1. Effect of epinephrine and norepinephrine on heart rate and stroke volume.

_____ 2. Effect of epinephrine on vasoconstriction (of skin and visceral blood vessels).

_____ 3. Effect of ADH on vasoconstriction.

_____ 4. Effect of ADH on urine production.

_____ 5. Effect of a decrease in blood pressure on ADH secretion.

_____ 6. Effect of atrial natriuretic factor on urine production.

_____ 7. Effect of increased atrial blood pressure on atrial natriuretic factor secretion.

The Fluid Shift Mechanism and the Stress-Relaxation Response

“_Two mechanisms in addition to the nervous and hormonal mechanisms help to regulate blood pressure._**”**

A. Match these terms with the
correct statement or definition:

Decrease   Increase
Fluid shift mechanism   Stress-relaxation response

_____ 1. When blood pressure increases, fluid moves from the blood vessels into the interstitial spaces; when blood pressure decreases, the opposite effect occurs.

_____ 2. When blood pressure declines, smooth muscle cells in blood vessel walls contract; when blood pressure increases, the smooth muscle cells relax.

_____ 3. Change in the amount of fluid moved out of capillaries when blood pressure decreases.

_____ 4. Effect on blood pressure when fluid moves out of tissues.

_____ 5. Change in contraction of smooth muscle in blood vessel walls when blood volume decreases.

_____ 6. Effect on blood pressure when smooth muscle in blood vessel walls contract.

Long-Term Regulation of Blood Pressure

“_Long-term regulation of blood pressure is achieved by the kidneys, which affect blood pressure by_ **”** _changing blood volume._

Match these terms with the
correct statement:

Decrease
Increase

436

_____ 1. Effect on urine production of an increase in blood pressure.

_____ 2. Effect on blood volume of an increase in urine production.

_____ 3. Effect on blood pressure of a decrease in blood volume.

☞ Urine production is regulated by several hormones. Increased blood volume results in increased atrial natriuretic hormone and decreased renin, angiotensin, aldosterone, and ADH. These changes cause an increased urine production and a reduction in blood volume.

QUICK RECALL

1. Name the types of blood vessels, starting and ending at the heart.

2. List the three types of capillaries.

3. Name the subdivisions of the aorta.

4. Name the three major arteries that branch from the aorta to supply the head and upper limbs.

5. List the three unpaired arteries and the four major paired arteries that branch from the abdominal aorta.

6. List three veins that return blood to the superior vena cava.

7. Name the veins from the brain and upper limbs that empty into the brachiocephalic veins.

8. Name the veins that join the inferior vena cava to return blood from the small intestine, the kidneys, and the lower limbs.

9. List the two major superficial veins of the upper limbs and the two major superficial veins of the lower limbs.

10. Name four factors in Poiseuille's Law that influence blood flow.

11. List two factors that influence the force acting on the wall of a blood vessel.

12. List the major force responsible for the movement of fluid out of capillaries and that responsible for the movement of fluid into capillaries.

13. List the three nervous mechanisms for short-term regulation of blood pressure.

14. List four hormonal mechanisms for control of arterial pressure.

15. List two mechanisms in addition to nervous and hormonal that help to regulate systemic blood pressure.

MASTERY LEARNING ACTIVITY

Place the letter corresponding to the correct answer in the space provided.

_____ 1. Given the following blood vessels:
1. arteriole
2. capillary
3. elastic artery
4. muscular artery
5. vein
6. venule

Choose the arrangement that lists the blood vessels in the order a red blood cell would pass through them as the red blood cell leaves the heart, travels to a tissue, and returns to the heart.

a. 3, 4, 2, 1, 5, 6
b. 3, 4, 1, 2, 6, 5
c. 4, 3, 1, 2, 5, 6
d. 4, 3, 2, 1, 6, 5

_____ 2. The endothelial cells of capillaries can have areas where the cytoplasm is absent and the cell membrane consists of a thin, porous diaphragm. Which of the following types of capillaries would have such an area?

a. fenestrated capillary
b. sinusoidal capillary
c. continuous capillary
d. a and b
e. all of the above

_____ 3. Given the following structures:
1. metarteriole
2. precapillary sphincter
3. thoroughfare channel

Choose the arrangement that lists the structures in the order a red blood cell would encounter them as it passes through a tissue.
a. 1, 2, 3
b. 1, 3, 2
c. 2, 1, 3
d. 2, 3, 1

_____ 4. The layer of a blood vessel that contains smooth muscle?
a. tunica intima
b. tunica media
c. tunica adventitia
d. tunica muscularis

_____ 5. In which of the following blood vessels would elastic fibers be present in the greatest amounts?
a. large arteries
b. medium-sized arteries
c. arterioles
d. venules
e. large veins

_____ 6. Comparing and contrasting arteries and veins,
a. veins have thicker walls.
b. veins have a greater amount of smooth muscle than arteries.
c. veins have a tunica media and arteries do not.
d. veins have valves and arteries do not.
e. all of the above

_____ 7. Which structure supplies the walls of blood vessels with blood?
a. venous shunt
b. tunic channels
c. arteriovenous anastamoses
d. vasa vasorum

_____ 8. Given the following blood vessels:
1. aorta
2. inferior vena cava
3. pulmonary arteries
4. pulmonary veins

Which of the vessels carries oxygen rich blood?
a. 1, 3
b. 1, 4
c. 2, 3
d. 2, 4

_____ 9. Given the following list of arteries:
1. basilar
2. common carotids
3. internal carotids
4. vertebral

Which of the arteries has a DIRECT connection with the circle of Willis?
a. 2, 1
b. 2, 4
c. 3, 1
d. 3, 4

_____ 10. Given the following vessels:
1. axillary artery
2. brachial artery
3. brachiocephalic artery
4. radial artery
5. subclavian artery

Choose the arrangement that lists the vessels in order going from the aorta to the right hand.
a. 2, 5, 4, 1
b. 5, 2, 1, 4
c. 5, 3, 1, 4, 2
d. 3, 5, 1, 2, 4

_____ 11. Artery most commonly used to take the pulse near the wrist?
a. basilar
b. brachial
c. cephalic
d. radial
e. ulnar

_____ 12. A major branch of the aorta that subdivides to supply the liver, stomach, and spleen?
a. celiac
b. common iliac
c. inferior mesenteric
d. superior mesenteric

_____ 13. Given the following arteries:
1. common iliac
2. external iliac
3. femoral
4. popliteal

Choose the arrangement that lists the arteries in order going from the aorta to the knee.
a. 1, 2, 3, 4
b. 1, 2, 4, 3
c. 2, 1, 3, 4
d. 2, 1, 4, 3

_____ 14. Given the following veins:
1. brachiocephalic
2. internal jugular
3. superior vena cava
4. venous sinus

Choose the arrangement that lists the veins in order going from the brain to the heart.
a. 2, 4, 1, 3
b. 2, 4, 3, 1
c. 4, 2, 1, 3
d. 4, 2, 3, 1

_____ 15. Blood returning from the arm to the subclavian vein would pass through which of the following veins?
a. cephalic
b. basilic to axillary
c. brachial to axillary
d. a and b
e. all of the above

_____ 16. Given the following veins:
1. accessory hemiazygous
2. azygous
3. hemiazygous
4. inferior vena cava
5. superior vena cava

Trace the route a red blood cell would take when returning from the LEFT INFERIOR anterior surface of the thorax to the heart.
a. 2, 4
b, 2, 5
c. 1, 2, 5
d. 3, 2, 4
e. 3, 2, 5

_____ 17. Given the following vessels:
1. inferior mesenteric vein
2. superior mesenteric vein
3. hepatic portal vein
4. hepatic vein

Choose the arrangement that lists the vessels in order going from the small intestine to the inferior vena cava.
a. 1, 3, 4
b. 1, 4, 3
c. 2, 3, 4
d. 2, 4, 3

_____ 18. Given the following list of veins:
1. small saphenous
2. great saphenous
3. peroneal
4. posterior tibial

Which are superficial veins?
a. 1, 2
b. 1, 3
c. 2, 3
d. 2, 4
e. 3, 4

_____ 19. The viscosity of blood
a. is about the same as water.
b. increases as hematocrit increases.
c. is mostly due to plasma.
d. all of the above

_____ 20. Korotkoff sounds
a. are caused by closing the AV valves.
b. are caused by closing of the semilunar valves.
c. are caused by turbulence in the arteries.
d. equals the pulse rate multiplied by two.

_____ 21. If one could increase any of the following factors that affect blood flow by twofold, which one would cause the greatest increase in blood flow?
a. blood viscosity
b. the pressure gradient
c. vessel radius
d. vessel length

_____ 22. According to the law of Laplace,
a. the force that stretches the wall of a blood vessel is proportional to the diameter of the vessel times the blood pressure .
b. as blood pressure decreases, the force acting on the wall of a blood vessel decreases.

c. as the diameter of a blood vessel increases, the force acting on the wall of the blood vessel increases.
d. all of the above

_____ 23. Vascular compliance is
a. greater in arteries than in veins.
b. increase in vessel volume divided by increase in vessel pressure.
c. the pressure at which blood vessels collapse.
d. all of the above

_____ 24. The total cross sectional area would be the greatest in which of the following?
a. elastic arteries
b. muscular arteries
c. arterioles
d. capillaries
e. veins

_____ 25. The resistance to blood flow is greatest in the
a. aorta.
b. arterioles.
c. capillaries.
d. venules.
e. veins.

_____ 26. Pulse pressure
a. is the difference between systolic and diastolic pressure.
b. increases when stroke volume increases.
c. increases as vascular compliance decreases.
d. all of the above

_____ 27. Concerning fluid movement at the capillary level,
a. the amount of fluid leaving the arterial end of a capillary is equal to the amount of fluid entering the venous end.
b. fluid enters the venous end of a capillary primarily due to a decrease in capillary hydrostatic pressure.
c. fluid tends to move into tissues at the arteriole end of a capillary because there is a greater concentration of proteins in the tissue than in the blood.
d. all of the above

_____ 28. Veins
a. increase their volume because of their large compliance.

b. decrease venous return to the heart when they vasodilate.
c. vasodilate due to increased sympathetic stimulation.
d. all of the above

_____ 29. Which of the following average pressure relationships is correctly matched in a standing person?
a. 0 mm Hg: pressure in the right atrium
b. 100 mm Hg: pressure in the aorta by the heart
c. 110 mm Hg: pressure in an arteriole in the foot
d. all of the above

_____ 30. Local direct control of blood flow through a tissue
a. maintains a relatively constant rate of flow despite large changes in arterial blood pressure.
b. is due to relaxation and contraction of the precapillary sphincters.
c. occurs in response to a buildup of carbon dioxide in the tissue.
d. occurs in response to a decrease of oxygen in the tissue.
e. all of the above

_____ 31. The sympathetic neurons that regulate blood vessels
a. are continually stimulated, accounting for vasomotor tone.
b. release norepinephrine, which causes vasoconstriction.
c. release acetylcholine, which causes vasodilation.
d. all of the above

_____ 32. An increase in mean arterial blood pressure may result from
a. an increase in peripheral resistance.
b. an increase in heart rate.
c. an increase in stroke volume.
d. all of the above

_____ 33. Through the baroreceptor reflex, if there was an increase in mean arterial blood pressure, the expected response would be
a. increased sympathetic nervous system activity.
b. a decrease in peripheral resistance.
c. stimulation of the vasomotor center.
d. vasoconstriction.

34. Through the chemoreceptor reflex, when oxygen levels markedly decrease in the blood, you would expect
    a. peripheral resistance to increase.
    b. mean arterial blood pressure to decrease.
    c. cardiace output to decrease.
    d. all of the above

35. Through the central nervous system ischemic response, as
    a. oxygen levels increase, vasoconstriction results.
    b. oxygen levels decrease, vasodilation results.
    c. carbon dioxide levels increase, vaso-constriction results.
    d. carbon dioxide levels decrease, vaso-constriction results.

36. When blood pressure is suddenly decreased a small amount (10 mm Hg), which of the following mechanisms would be activated to restore blood pressure to normal levels?
    a. chemoreceptor reflexes
    b. baroreceptor reflexes
    c. central nervous system ischemic response
    d. all of the above

37. A sudden release of epinephrine from the adrenal medulla would
    a. increase heart rate.
    b. increase stroke volume.
    c. cause vasoconstriction in visceral blood vessels.
    d. all of the above

38. When blood pressure decreases,
    a. renin secretion increases.
    b. angiotensin II formation decreases.
    c. aldosterone secretion decreases.
    d. all of the above

39. In response to an increase in blood pressure,
    a. ADH secretion decreases.
    b. the kidneys increase urine production.
    c. blood volume decreases.
    d. all of the above

40. In response to a decrease in blood pressure,
    a. more fluid than normal would enter tissues (fluid shift mechanism).
    b. smooth muscles in blood vessels would relax (stress-relaxation response).
    c. the kidneys would retain more salts and water than normal.
    d. all of the above

41. During spinal anesthesia a decrease in the frequency of action potentials carried along sympathetic nerves may occur, especially in the lower region of the spine. Choose the most appropriate consequence from the list provided.
    a. decreased systemic blood pressure
    b. markedly decreased heart rate
    c. increased systemic blood pressure
    d. marked cutaneous vasoconstriction
    e. a and b

42. A patient that suffers from a myocardial infarction that clearly compromises cardiac function will exhibit which of the following?
    a. increased sympathetic stimulation of the heart
    b. increased blood pressure
    c. peripheral vasodilation
    d. all of the above

43. Wilber Merocrine sat in a hot bath for 30 minutes. He then quickly got out of the hot bath and jumped into an ice bath. As a consequence, which of the following occurred?
    a. Wilber's blood pressure increased dramatically.
    b. Wilber's peripheral resistance increased.
    c. Wilber's heart rate increased dramatically.
    d. a and b
    e. all of the above

44. A patient was found to have severe arteriosclerosis of his renal arteries, which reduces renal blood pressure. Which of the following is consistent with that condition?
    a. hypotension
    b. hypertension
    c. decreased vasomotor tone
    d. exaggerated sympathetic stimulation of the heart
    e. a and c

_____ 45. During exercise the blood flow through skeletal muscle may increase up to twenty fold. However, the cardiac output doesn't increase that much. This occurs

a. due to vasoconstriction in the viscera.

b. due to vasoconstriction in the skin (at least temporarily).

c. due to vasodilation of skeletal muscle blood vessels.

d. a and b

e. all of the above

FINAL CHALLENGES

Use a separate sheet of paper to complete this section.

1. When cancer of the colon is discovered, a liver scan is often made to see if the patient has cancer of the liver. Explain why this is a reasonable test to make.

2. Why does vasoconstriciton of blood vessls increase resistance to blood flow? How can blood flow be maintained when resistance to blood flow increases?

3. In varicose veins the blood vessels become distended and enlarge. Once this process starts it becomes worse and worse. Explain why. Why does raising the feet help this condition?

4. Just prior to blast off an astronaut had her blood pressure measured while lying horizontal. Once into space, under weightless conditions (i.e., no gravity), the measurement was repeated. Would you expect the blood pressure in space to be higher, lower, or unchanged compared to the blood pressure on earth?

5. In cirrhosis of the liver there is often abdominal swelling. Explain (hint: cirrhosis impairs blood flow through the liver and impairs albumin production by the liver).

6. While donating blood a student nurse felt dizzy and faint. Explain why on the basis of the following: blood donation results in loss of blood; emotional reactions associated with donating blood could cause hyperventilation; and emotional reactions can decrease sympathetic and increase parasympathetic nervous system activity.

7. The student nurse in the preceding question was told to lie down and breathe into a paper bag. Explain why these therapies would help her.

8. Private U.P. Wright had to stand at attention in the hot sun for several hours. Eventually he fainted. Explain what happened.

9. Just as private U.P. Wright faints, his buddies realize what is happening. To save him from the embarassment of fainting they held him in a standing position. Explain what is wrong with this treatment. What would you suggest they do?

10. A patient has the following symptoms: below normal blood pressure and edema in the ankles, legs, and hands. Assume that the patient has had a myocardial infarct that has damaged one of the ventricles. On the basis of the symptoms, explain why you believe it is the left or the right ventricle that was damaged.

ANSWERS TO CHAPTER 21

1. arteriosclerosis; atherosclerosis
2. arteriosclerosis, artery, arteriole
3. vasoconstriction, vasodilation, vasa vasorum
4. carotid
5. hemorrhage, hemorrhagic, hemorrhagia
6. celiac
7. jugular
8. hepatic
9. ischemic
10. auscultatory

CONTENT LEARNING ACTIVITY

General Features of Blood Vessel Structure

A. 1. Endothelium
    2. Basement membrane
    3. Adventitia
    4. Precapillary cells

B. 1. Continuous capillaries
    2. Fenestrated capillaries
    3. Sinusoidal capillaries
    4. Sinusoids

C. 1. Arterial capillaries
    2. Thoroughfare channels
    3. Metarterioles
    4. Precapillary sphincters

D. 1. Tunica intima
    2. Tunica media
    3. Tunica adventitia

E. 1. Elastic arteries
    2. Medium-sized and small arteries
    3. Arterioles
    4. Medium-sized and small arteries
    5. Elastic arteries

F. 1. Venules
    2. Small veins
    3. Medium-sized and large veins

G. 1. Vasa vasorum
    2. Arteriovenous anastomoses
    3. Glomus

H. 1. Varicose veins
    2. Phlebitis
    3. Arteriosclerosis
    4. Atherosclerosis

Pulmonary Circulation

1. Pulmonary trunk
2. Pulmonary arteries

3. Pulmonary veins

Systemic Circulation: Arteries

A.
1. Ascending aorta
2. Aortic arch
3. Descending aorta
4. Thoracic aorta

B.
1. Right common carotid artery
2. Right vertebral artery
3. Right subclavian artery
4. Brachiocephalic artery
5. Left common carotid artery
6. Left subclavian artery
7. Left vertebral artery

C.
1. Basilar artery
2. Vertebral artery
3. External carotid artery
4. Internal carotid artery
5. Circle of Willis

D.
1. Axillary artery
2. Brachial artery
3. Radial artery
4. Digital artery
5. Palmar arches
6 Ulnar artery

E.
1. Internal thoracic artery
2. Anterior intercostal artery
3. Posterior intercostal artery
4. Visceral artery

F.
1. Celiac trunk
2. Common hepatic artery
3. Suprarenal artery
4. Renal artery
5. Superior mesenteric artery
6. Common iliac artery
7. Gonadal artery
8. Inferior mesenteric artery
9. Splenic artery
10. Left gastric artery

G.
1. Femoral artery
2. Anterior tibial artery
3. Dorsalis pedis artery
4. Digital arteries
5. Medial plantar artery
6. Lateral plantar artery
7. Posterior tibial artery
8. Peroneal artery
9. Popliteal artery

Systemic Circulation: Veins

A.
1. Coronary sinus
2. Superior vena cava
3. Inferior vena cava
4. Venous sinuses
5. Internal jugular vein

B.
1. Right subclavian vein
2. Right brachiocephalic vein
3. Superior vena cava
4. Inferior vena cava
5. Pulmonary veins
6. Internal jugular vein
7. External jugular vein

C.
1. Axillary vein
2. Cephalic vein
3. Venous arches
4. Basilic vein
5. Median cubital vein
6. Brachial veins

D.
1. Azygos vein
2. Hemiazygos vein
3. Accessory hemiazygos vein

E.
1. Hepatic veins
2. Gonadal vein
3. Common iliac vein
4. External iliac vein
5. Internal iliac vein
6. Renal vein
7. Suprarenal vein

F.
1. Inferior vena cava
2. Hepatic portal vein
3. Superior mesenteric vein
4. Inferior mesenteric vein
5. Splenic vein
6. Gastric vein
7. Hepatic veins

G.
1. Great saphenous vein
2. Small saphenous vein
3. Popliteal vein
4. Femoral vein

The Physics of Circulation

A. 1. Viscosity
   2. Laminar flow
   3. Turbulent flow

B. 1. Blood pressure
   2. Sphygmomanometer
   3. Auscultatory
   4. Korotkoff sounds

C. 1. Increased
   2. Decreased

   3. Increased
   4. Decreased

D. 1. Critical closing pressure
   2. The law of Laplace
   3. Aneurysm
   4. Vascular compliance

E. 1. Decreases
   2. Decreases
   3. Increases

Physiology of Systemic Circulation

A. 1. Veins
   2. Capillaries
   3. Aorta
   4. Arterioles
   5. Veins

B. 1. Pulse pressure
   2. Increased
   3. Increased
   4. Decreased
   5. Increased

C. 1. Diffusion
   2. Hydrostatic pressure
   3. Hydrostatic pressure
   4. Osmotic pressure
   5. Lymphatic system
   6. Edema

D. 1. Increases
   2. Increases
   3. Increases
   4. Increases

Local Control of Blood Flow by the Tissues

1. Metarterioles and precapillary sphincters
2. Metabolism
3. Vasodilator substances
4. Nutrient demand

5. Increase
6. Decrease
7. Autoregulation

Nervous Regulation of Local Circulation

1. Sympathetic
2. Vasomotor center
3. Hypothalamus and cerebral cortex

4. Vasomotor tone
5. Acetylcholine

Regulation of Mean Arterial Pressure

1. Cardiac output
2. Stroke volume
3. Mean blood pressure
4. Peripheral resistance

5. Decrease
6. Increase
7. Increase

Nervous Regulation of Blood Pressure

A. 1. Baroreceptor (pressoreceptor) reflex
2. Chemoreceptor reflex
3. Central nervous system ischemic response
4. Chemoreceptor reflex; central nervous system ischemic response
5. Baroreceptor (pressoreceptor) reflex

B. 1. Increased
2. Vasodilation
3. Vasoconstriction
4. Vasoconstriction

Hormonal Regulation of Blood Pressure

A. 1. Adrenal medullary mechanism
2. Renin-angiotensin-aldosterone mechanism
3. Vasopressin mechanism
4. Atrial natriuretic mechanism

B. 1. Renin
2. Angiotensinogen
3. Angiotenisn II
4. Increased
5. Aldosterone
6. Decreased
7. Decreased
8. Increased
9. Decreased

C. 1. Renin
2. Angiotensinogen
3. Angiotensin II
4. Aldosterone

D. 1. Increase
2. Increase
3. Increase
4. Decrease
5. Increase
6. Increase
7. Increase

The Fluid Shift Mechanism and the Stress-Relaxation Response

1. Fluid shift mechanism
2. Stress-relaxation response
3. Decrease

4. Decrease
5. Increase
6. Increase

Long-Term Regulation of Blood Pressure

1. Increase
2. Decrease

3. Decrease

QUICK RECALL

1. Heart, elastic arteries, muscular arteries, arterioles, capillaries, venules, small veins, large veins, heart
2. Continuous capillaries, fenestrated capillaries, and sinusoidal capillaries
3. Ascending aorta, aortic arch, descending aorta, thoracic aorta, abdominal aorta
4. Brachiocephalic artery, common carotid artery, and left subclavian artery

5. Unpaired: celiac trunk, superior mesenteric artery, inferior mesenteric artery; paired: renal arteries, suprarenal arteries, gonadal arteries, common iliac arteries
6. Left and right brachiocephalic veins and azygos vein
7. Internal jugular vein, external jugular vein, vertebral vein, and subclavian vein
8. Superior mesenteric vein, renal vein, and common iliac veins

447

9. Basilic and cephalic veins in the upper limbs, small and great saphenous veins in the lower limbs
10. Viscosity, diameter of blood vessel, length of blood vessel, and pressure gradient
11. Force = diameter X blood pressure (law of Laplace)
12. Hydrostatic pressure: out of blood vessels; osmosis: into blood vessels
13. Baroreceptor reflex, chemoreceptor reflex, and the central nervous system ischemic response
14. Adrenal medullary mechanism, renin-angiotensin-aldosterone mechanism, vasopressin mechanism, and atrial natriuretic hormone mechanism
15. Fluid-shift mechanism and stress-relaxation response

MASTERY LEARNING ACTIVITY

1. B. The red blood cell would pass through an elastic artery, muscular artery, arteriole, capillary, venule, and vein.

2. D. The areas described are fenestrae, which increase the permeability of the capillary. They are found in fenestrated and sinusoidal capillaries.

3. B. Blood flows from arterioles into metarterioles, from which thoroughfare channels extend to venules. Blood flow through the capillaries that branch off the thoroughfare channels is regulated by precapillary sphincters.

4. B. The tunica media consists of circularly arranged smooth muscles and variable amounts of elastic and collagen fibers. The tunica intima consists of endothelium, basement membrane, lamina propria, and an internal elastic membrane. The tunica adventitia is connective tissue. There is no tunica muscularis.

5. A. The large arteries such as the aorta contain large amounts of elastic connective tissue, which plays an important role in the maintenance of blood pressure during diastole.

6. D. Only veins have valves. Veins are thinner walled with less smooth muscle than arteries. Both veins and arteries have all three tunics.

7. D. Vasa vasorum are small blood vessels that penetrate the walls of blood vessels. Arteriovenous anastomses allow blood to pass from arteries to veins without passing through capillaries.

8. B. Oxygen rich blood returns from the lungs to the heart through the pulmonary veins, passes through the left side of the heart, and exits through the aorta.

9. C. The internal carotids connect to the circle of Willis after branching from the common carotids. The basilar artery is formed by the joining of the left and right vertebral arteries. The basilar then connects to the circle of Willis.

10. D. Blood going to the right hand passes from the aorta into the brachiocephalic artery and into the right subclavian artery. Blood going to the left hand passes from the aorta into the left subclavian artery. From the subclavian artery (either side) blood passes through the axillary artery, brachial artery, and radial artery to reach the hand.

11. D. The radial artery passes over the radial bone near the wrist, and the pulse can be easily taken by slightly compressing the artery against the bone.

12. A. The celiac artery branches to form the hepatic (to the liver), gastric (to the stomach), and the splenic (to the spleen) arteries.

13. A. The order is common iliac, external iliac, femoral and popliteal arteries.

14. C. Blood from the brain enters a number of venous sinuses that empty into the internal jugulars. The internal jugulars join the brachiocephalic veins, which combine to form the superior vena cava.

15. E. The cephalic and basilic veins are superficial veins of the arm and forearm. The cephalic vein empties into the subclavian vein. The basilic vein joins the axillary vein, which empties into the subclavian vein. The brachial vein is a deep vein that also joins the axillary vein.

16. E. Blood from the anterior surface of the thorax can return through three routes. Blood from the left inferior surface flows through the hemiazygous, azygous, and superior vena cava. Blood from the left superior surface flows through the accessory hemiazygous, azygous, and superior vena cava. Blood from the right surface flows through the azygous to the superior vena cava.

17. C. The superior mesenteric vein drains the small intestine and joins the hepatic portal vein, which enters the liver. Blood from the liver flows to the inferior vena cava through the hepatic veins.

18. A. The small and great saphenous veins are the major superficial veins of the lower limb. The anterior tibial, posterior tibial, and peroneal veins are the major deep veins of the leg.

19. B. Viscosity is a measure of the resistance of a liquid to flow. Water has viscosity of 1 and blood has a viscosity of 3 to 4.5. The viscosity of blood is mainly due to erythrocytes and not to plasma. Therefore, as hematocrit increases, the viscosity of blood increases.

20. C. Blood flow in the brachial artery, where blood pressure is determined by the ausculatory method, is normally laminar (streamlined). When blood pressure measurement is taken, the pressure cuff constricts the brachial artery, resulting in turbulent flow that produces Korotkoff sounds.

21. C. According to Poiseuille's law, blood flow is directly proportional to the blood vessel radius raised to the fourth power, directly proportional to the pressure gradient, inversely proportional to viscosity, and inversely proportional to vessel length. Changing the blood vessel radius by twofold would affect the blood flow more than an identical change in any of the other parameters.

22. D. The force acting on a vessel wall is proportional to the diameter of the vessel times the blood pressure. Consequently, as pressure decreases the force decreases, and as diameter increases the force increases.

23. B. Vascular compliance is the tendency for blood vessel volume to increase as the blood pressure increases. It is greater in veins than in arteries. Critical closing pressure is the pressure at which blood vessels collapse.

24. D. As the blood vessels become smaller in diameter, their number increases. Also, as the blood vessels become smaller in diameter the velocity of blood flow decreases. The total cross sectional area of the capillaries, which have the smallest diameter and blood flow velocity, must be the greatest in order to accommodate the blood pumped by the heart.

25. B. The greatest resistance to blood flow occurs in the arterioles, followed by the capillaries. The resistance to blood flow is relatively low in the aorta, venules, and veins.

26. D. Pulse pressure is the difference between systolic and diastolic pressure. It increases with increased stroke volume and decreased vascular compliance.

27. B. Fluid movement out of the capillary is due to capillary hydrostatic pressure and a slight negative pressure in the tissue. Fluid movement into the capillary is due to osmosis because the blood has a higher concentration of proteins than the tissue. At the arteriole end of the capillary, the forces moving fluid out exceeds the forces moving fluid in. At the venous end the situation is reversed, primarily due to a decrease in hydrostatic pressure. However, overall, more fluid moves out of the capillary than returns. The excess fluid in the tissue is removed by means of the lymphatic system.

28. A. Because of their large compliance a small change in venous blood pressure results in a large increase in venous volume. Sympathetic stimulation decreases venous compliance and volume, thus increasing venous return to the heart and cardiac output (Starling's law of the heart).

29. D. Blood pressure in the right atrium is about 0 mm Hg and in the aorta about 100 mm Hg. Due to the effect of gravity, the pressure in a standing person increases below the level of the heart (110 mm Hg in an arteriole of the foot) and decreases above the level of the heart (90 mm Hg in an artery of the head).

30. E. Local direct control is due to the relaxation and contraction of the precapillary sphincters and metarterioles. The effect is to produce a cyclic flow of blood (vasomotion) and to maintain a normal flow despite changes in arterial blood pressure (autoregulation). The precapillary sphincters relax in response to a buildup of carbon dioxide (vasodilator theory) or a decrease in oxygen (nutrient demand theory) in the tissue.

31. D. The vasomotor center tonically stimulates most blood vessels through sympathetic fibers. The result is a state of partial contraction of smooth muscle in blood vessels, which is called vasomotor tone. The sympathetic neurotransmitter responsible for vasoconstriction is norepinephrine. Some sympathetic fibers to skeletal muscle and the skin release acetylcholine, which causes vasodilation.

32. D. Mean blood pressure in the aorta is proportional to heart rate times stroke volume times peripheral resistance.

33. B. There would be inhibition of the vasomotor center resulting in decreased sympathetic nervous system activity. Consequently, blood vessels would vasodilate, and peripheral resistance would decrease.

34. A. The chemoreceptor reflex increases peripheral resistance by stimulating vasoconstriction. It also increases heart rate and stroke volume, i.e., cardiac output. The increases in peripheral resistance and cardiac output cause mean arterial blood pressure to increase.

35. C. Increased carbon dioxide or decreased pH levels directly stimulate the vasomotor center, resulting in vasoconstriction.

36. B. The baroreceptor reflexes are the most important for regulating small, sudden changes in blood pressure. The chemoreceptor reflexes and the central nervous system ischemic response are activated when blood pressure decrease significantly, i.e., emergency conditions.

37. D. Epinephrine increases heart rate and stroke volume. It also causes vasoconstriction of skin and visceral blood vessels and vasodilation of skeletal muscle blood vessels.

38. A. A decreased blood pressure is detected by the juxtaglomerular apparatuses of the kidneys, which increase their secretion of renin. The increased renin results in increased formation of angiotensin I, which increases the formation of angiotensin II. Angiotensin II causes vasoconstriction, which helps to raise blood pressure and stimulates increased aldosterone secretion. The aldosterone helps to increase blood volume, and therefore blood pressure, by decreasing urine production.

39. D. An increase in blood pressure is detected by baroreceptors, resulting in decreased ADH secretion. Consequently, the kidneys increase urine production, and blood volume decreases. The decreased blood volume helps to reduce blood pressure.

40. C. Following a decrease in blood pressure, the kidneys retain more salts and water than normal, resulting in an increase in blood volume and therefore blood pressure. A decrease in blood pressure results in less fluid movement into tissues (fluid shift mechanism) and contraction of blood vessel smooth muscle (stress-relaxation response).

41. A. Decreased sympathetic innervation to blood vessels in the lower region of the body results in decreased vasomotor tone. Therefore peripheral resistance and subsequently blood pressure decrease. The heart rate would probably increase in response to the baroreceptor reflex. Marked cutaneous vasoconstriction could not occur due to the effects of the anesthesia on vasomotor tone.

42. A. The decreased cardiac function results in increased sympathetic stimulation of the heart and systemic blood vessels due to the drop in blood pressure. The increased sympathetic stimulation results in an increased heart rate and peripheral vasoconstriction.

43. D. Wilber's peripheral resistance increased dramatically following his immersion in the cold water. Stimulation of cold receptors in the skin results in cutaneous vasoconstriction. As the peripheral resistance increased, Wilber's blood pressure increased markedly. His heart rate decreased due to the sudden increase in blood pressure (baroreceptor reflexes).

44. B. Renal arteriosclerosis causes decreased blood flow through the kidneys. Consequently the kidneys increase renin secretion. The renin causes the formation of angiotensin II, which causes vasoconstriction and significantly increases aldosterone secretion from the adrenal cortex. The aldosterone increases renal sodium chloride and water retention. The expanded blood volume results in hypertension.

45. E. During exercise the following events occur with respect to the peripheral vasculature: (1) visceral and cutaneous vasoconstriction and (2) skeletal muscle vasodilation. The result is that blood is shunted from the skin and viscera to the skeletal muscle. With the normal increase in blood pressure and the changes that occur in the peripheral vasculature the blood flow to skeletal muscle may increase by twenty fold.

 FINAL CHALLENGES

1. The abdominal and pelvic viscera drain through the hepatic portal system into the liver. A cancer in these viscera (e.g., the colon) can enter the blood and travel to the liver.

2. According to Poiseuille's law, resistance to blood flow is inversely proportional to the radius of the blood vessel raised to the fourth power. Therefore decreasing the radius greatly increases the resistance. As resistance increases, in order to maintain blood flow, pressure must increase.

3. According to the law of Laplace, the greater the diameter of a vessel, the greater the force acting on the vessel wall. Thus as the vein enlarges, the force causing it to enlarge also increases. The law of Laplace also states that the force acting on the vessel wall increases or decreases with the blood pressure within the vessel. Raising the feet helps because it reduces the hydrostatic pressure produced by gravity.

4. Since the astronaut's blood pressure on earth was taken while she was lying horizontal, there would be no hydrostatic effect due to gravity. Therefore there would be no difference between the measurement in space and on earth.

5. In the capillaries blood pressure forces fluid out of the capillaries and, due to the higher concentration of proteins in the blood than in the tissues, osmosis causes fluid to move into the capillaries. Reduced blood flow through the liver increases capillary blood pressure so more fluid leaves the capillaries. At the same time reduced blood proteins (albumin) result in less movement of fluid into the capillaries. The result is abdominal swelling.

6. Loss of blood due to the donation would cause a reduced blood volume, reduced venous return, and thus reduced cardiac output (Starling's law of the heart). Reduced cardiac output results in reduced blood pressure and inadequate delivery of blood to the brain. Emotional reactions could have caused hyperventilation and a "blow off" of carbon dioxide. The reduced blood carbon dioxide would directly inhibit the vasomotor center resulting in decreased peripheral resistance and thus decreased blood pressure. Emotional reactions through the limbic system could have caused decreased sympathetic activity (resulting in decreased peripheral resistance, decreased venous return, and decreased stroke volume) and increased parasympathetic activity (resulting in decreased heart rate). Consequently, cardiac output and blood pressure are reduced.

7. Since the veins are distensible (have a large compliance), when standing the hydrostatic pressure produced by gravity causes them to expand and the volume of the blood in the veins increases. Upon lying down, the hydrostatic pressure produced by gravity is removed, the venous volume becomes smaller, and venous return to the heart increases. Consequently, cardiac output increases (Starling's law of the heart), resulting in increased blood pressure and better delivery of blood to the brain. Breathing into the paper bag would raise carbon dioxide levels in the blood, stimulating the chemoreceptor reflex. This would increase heart rate, stroke volume, and peripheral resistance, restoring cardiac output and blood pressure. Raising blood carbon dioxide levels would also remove the inhibitory effect of low carbon dioxide on the vasomotor center.

8. Standing at attention in the hot sun reduces venous return to the heart for two reasons. First, loss of sweat results in a reduced blood volume. Second, hydrostatic pressure due to gravity causes a "pooling" of blood in the distensible veins. The reduced venous return results in reduced cardiac output, lowered blood pressure, and inadequate delivery of blood to the brain. In addition, in the hot sun blood vessels in the skin vasodilate as the body attempts to lose heat. The vasodilation lowers peripheral resistance and therefore blood pressure.

9. Holding him in a standing position maintains the hydrostatic pressure due to gravity. His friends should lay him horizontal to eliminate the effect of gravity on the blood. Raising the feet slightly above the head would also be helpful. When this is done venous volume will decrease, and venous return will increase. Thus cardiac output and blood pressure will increase, providing beter delivery of blood to the brain.

10. Damage to the right ventricle could result in inadequate pumping of blood by the right ventricle. Consequently, there would be a decreased delivery of blood to the left side of the heart, resulting in a decreased stroke volume. The decreased stroke volume would produce a decreased cardiac output and therefore a below-normal blood pressure. Since the right side of the heart does not pump blood effectively, there would be a buildup of blood in the venous system. This increases venous capillary pressure, less fluid returns from the tissues into the capillaries, and edema results.

Lymphatic System And Immunity

FOCUS: The lymphatic system performs three major functions: removal of excess fluid from tissues, absorption of fats from the digestive tract, and defense against microorganisms and other foreign substances. Fluids and fats are transported in lymph vessels, which converge to empty into the subclavian veins. Lymph organs include the tonsils, which form a protective ring of lymph nodules around the openings of the nasal and oral cavities; the lymph nodes, which filter lymph and produce lymphocytes; the spleen, which removes foreign substances and worn-out red blood cells from the blood; and the thymus, which processes lymphocytes. The nonspecific immune system includes mechanical mechanisms, chemicals (e.g., complement, interferon, lysozyme, histamine), phagocytic cells (neutrophils and macrophages), and cells involved with inflammation (basophils, mast cells, eosinophils). The specific immune system can recognize and remember specific antigens. Antibody-mediated immunity involves B cells and the production of antibodies, which activate mechanisms that result in the destruction of the antigen. Antibody-mediated immunity is most effective against extracellular antigens such as bacteria. Cell-mediated immunity involves T cells, which can lyse tumor or virus infected cells. T cells also produce lymphokines, which activate mechanisms that eliminate the antigen. Certain T cells, called T helper and T suppressor cells, regulate the activity of antibody-mediated and cell-mediated immunity.

WORD PARTS

Give an example of a new vocabulary word that contains each word part.

WORD PART	MEANING	EXAMPLE
germ-	embryo; sprout	1. _____
trab-	beam or timber	2. _____
reticul-	a network	3. _____

WORD PART	MEANING	EXAMPLE
anti-	against	4. _____
-gen	to bear	5. _____
epi-	upon	6. _____
-tope	place	7. _____
-kine	move	8. _____
-toxic	poison	9. _____
moti-	to move	10. _____

CONTENT LEARNING ACTIVITY

Introduction

"The lymphatic system includes lymph, lymphocytes, lymph vessels, lymph nodes,**"** tonsils, spleen and thymus.

Match these terms with the correct statement or definition:

Chyle   Lymph
Lacteal   Lymphocyte

_____ 1. Interstitial fluid that is returned to the circulatory system through lymph vessels.

_____ 2. Special lymph vessel in the small intestine that transports fats.

_____ 3. Lymph with a milky appearance because of its high fat content.

_____ 4. Leukocyte found in lymphatic tissue that is capable of destroying microorganisms and foreign substances.

Lymphatic System

"The lymphatic system, unlike the circulatory system, only carries fluid away from the tissues.**"**

Match these terms with the correct statement or definition:

Cisterna chyli   Lymph vessels
Lymph capillaries   Right lymphatic duct
Lymph nodes   Thoracic duct

_____ 1. Vessels that are responsible for collecting interstitial fluid.

_____ 2. Vessels that resemble small veins and contain one-way valves.

_____ 3. Largest lymph duct, which enters the left subclavian vein.

_____ 4. Enlargement of the thoracic duct in the superior abdominal cavity.

_____ 5. Lymph duct that drains the right thorax, right upper limb, and right side of the head and neck.

Lymph Organs

66 _Lymphatic tissue is a special type of reticular connective tissue._ **99**

A. Match these terms with the correct statement or definition:

Diffuse lymphatic tissue Palatine tonsils
Lingual tonsils Pharyngeal tonsils
Lymph nodules Peyer's patches

_____ 1. Lymphatic tissue that has no clear boundary and blends with surrounding tissues.

_____ 2. Lymphatic tissue that is organized into compact, spherical structures found in loose connective tissue of the digestive, respiratory, and urinary systems.

_____ 3. Aggregations of lymph nodules found in the small intestine, appendix, and tonsils.

_____ 4. Large, oval, lymphoid masses on each side of the posterior oral cavity.

_____ 5. Aggregations of lymphatic tissue near the internal opening of the nasal cavity; called adenoids when enlarged.

B. Match these terms with the correct statement or definition:

Afferent lymph vessel Lymph tissue
Capsule Medulla
Cortex Medullary cords
Efferent lymph vessel Reticular fibers
Germinal center Trabeculae
Lymph sinus

_____ 1. Extensions of the capsule that form an internal skeleton in lymph nodes.

_____ 2. Structures that extend from the capsule and trabeculae to form a fiber network throughout the lymph node.

_____ 3. Areas of the lymph node in which lymphocytes and macrophages are packed around the reticular fibers.

_____ 4. Open space in the lymph node through which reticular fibers extend.

_____ 5. Outer layer of lymph nodes, containing lymph nodules and sinuses.

_____ 6. Structures consisting of branching, irregular strands of diffuse lymphatic tissue; separated by sinuses.

_____ 7. Vessel that enters the lymph node.

_____ 8. Areas of rapid lymphocyte division in the lymph node.

C. Match these terms with the correct parts of the diagram labeled in Figure 22-1:

Afferent lymph vessel
Capsule
Efferent lymph vessel

Germinal center
Lymph sinus
Trabeculae

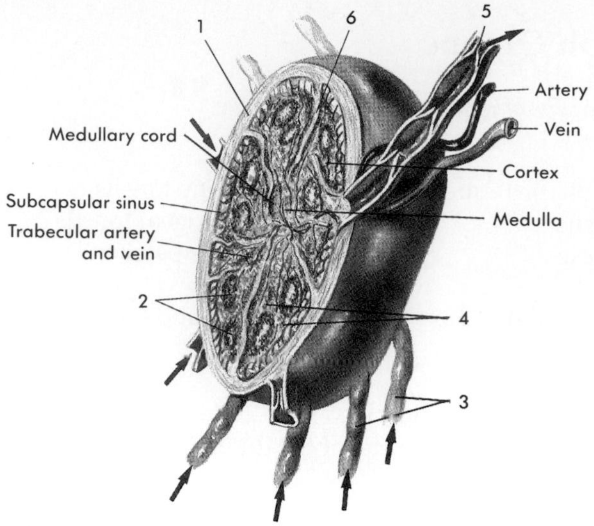

Medullary cord
Subcapsular sinus
Trabecular artery and vein

Artery
Vein
Cortex
Medulla

1. _____

2. _____

3. _____

4. _____

5. _____

6. _____

Figure 22-1

D. Match these terms with the correct statement as it applies to the spleen:

Capsule
Hilum
Red pulp

Trabeculae
Venous sinuses
White pulp

_____ 1. Fibrous outer covering of the spleen.

_____ 2. Dense accumulations of lymphocytes associated with the arterial supply in the spleen; the periarterial sheath and lymph nodules.

_____ 3. Network of reticular fibers filled with blood cells associated with the venous supply in the spleen.

_____ 4. Enlarged blood vessels that divide the red pulp of the spleen into pulp cords and unite to form veins.

E. Match these terms with the correct parts of the diagram labeled in Figure. 22-2:

Capsule
Lymph nodule
Periarterial sheath

Red pulp
Trabecula
White pulp

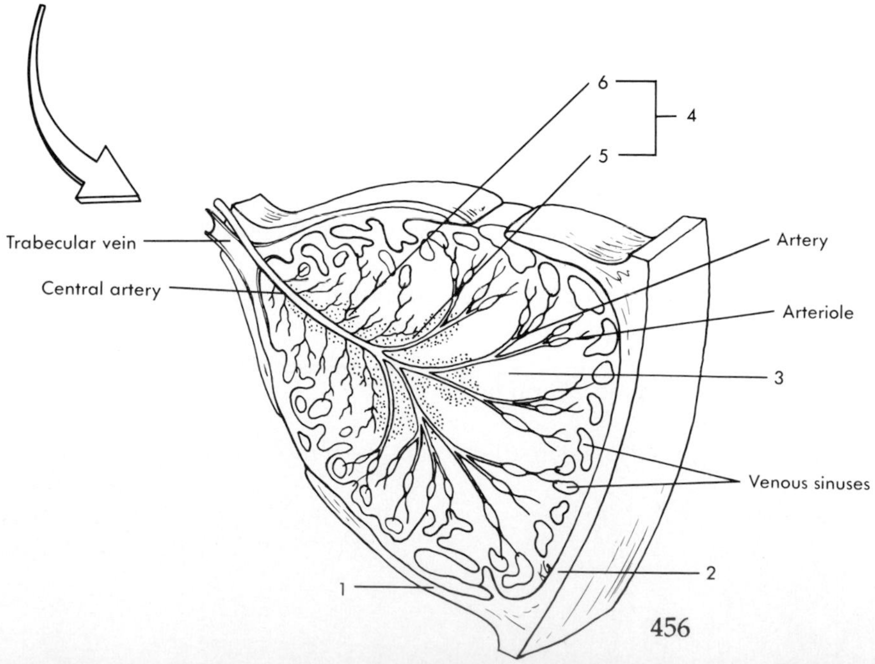

Trabecular vein
Central artery

Artery
Arteriole

Venous sinuses

1. _____

2. _____

3. _____

4. _____

5. _____

6. _____

Figure 22-2

456

F. Match these terms with the correct statement as it applies to the thymus:

Blood-thymic barrier   Lobule
Cortex   Medulla

_____ 1. Subdivisions of the thymus formed by the inward extension of trabeculae.

_____ 2. Outer portion of the thymus, which contains lymphocytes.

_____ 3. Relatively lymphocyte-free core of each lobule of the thymus.

_____ 4. Layer of reticular cells that prevents large molecules from leaving the capillaries in the thymus.

G. Match these lymph organs with the correct description or function:

Lymph node   Thymus
Spleen

_____ 1. Filters lymph and provides a source of lymphocytes; the only structure that has both afferent and efferent lymph vessels.

_____ 2. Detects and responds to foreign substances in the blood, destroys worn-out red blood cells, and acts as a reservoir for red blood cells.

_____ 3. Produces lymphocytes that move to other lymph tissue where they can respond to foreign substances.

_____ 4. Found primarily in the superior mediastinum.

Nonspecific Immune System

❝*In nonspecific immunity, each time the body is exposed to a substance the response is the same.***❞**

Using the terms provided, complete the following statements:

Cells   Mechanical mechanisms
Chemicals   Inflammation

Nonspecific immunity consists of several important components. The skin and mucous membrane are _(1)_ that prevent the entry of microorganisms or other foreign substances, whereas _(2)_ are substances that directly kill microorganisms or activate other mechanisms that result in the destruction of microorganisms. Phagocytosis and the production of chemicals is achieved by _(3)_ . The interaction of cells and chemicals results in the isolation and elimination of foreign substances in _(4)_ .

1. _____

2. _____

3. _____

4. _____

Chemicals

"A variety of chemicals are involved in nonspecific immunity."

A. Match these terms with the correct statement or definition:

Histamine   Mucus
Lysozyme   Sebum

_____ 1. Secretion in tears, saliva, nasal secretions, and sweat that lyse cells.

_____ 2. Secretion on membranes that traps microorganisms.

_____ 3. Chemical that promotes inflammation.

B. Using the terms provided, complete the following statements:

Alternate pathway   Complement cascade
Classical pathway   Properdin
Complement

_(1)_ is a group of at least 11 proteins that make up about 10% of the globulin portion of serum. These proteins become activated in the _(2)_ , a series of reactions in which each component of the series activates the next component. These reactions begin through one of two pathways. The _(3)_ is part of the specific immune system, whereas the _(4)_ is part of the nonspecific immune system. The alternate pathway is initiated when one of the complement proteins becomes spontaneously activated and combines with _(5)_ and certain foreign substances. Once activated, the complement promotes phagocytosis and inflammation.

1. _____
2. _____
3. _____
4. _____
5. _____

C. Using the terms provided, complete the following statements:

Antiviral   Nucleic acids
Defective   Reproduction
Does   Specific
Does not   Viral
Interferon

Interferon is a protein that protects the body against _(1)_ infection and perhaps some forms of cancer. When a virus infects a cell, viral _(2)_ take control of directing the activities of the cell. Viruses and other substances can also stimulate cells to produce interferon. Interferon _(3)_ protects the cell that produces it and _(4)_ act directly against viruses. Interferon binds to the surface of other cells and causes the production of _(5)_ proteins, which prevent the production of new viral nucleic acids and proteins. Interferon viral resistance is not _(6)_ ; the same interferon acts against many different viruses.

1. _____
2. _____
3. _____
4. _____
5. _____
6. _____

Cells

"Leukocytes and the cells derived from leukocytes are the most important cellular components of the immune system."

A. Match these terms with the correct statement or definition:

Chemotactic factors   Leukocytes
Chemotaxis   Phagocytosis

_____ 1. Parts of microbes or factors that are released by tissue cells and act as chemical signals to attract leukocytes.

_____ 2. Ability to detect and move toward chemotactic factors.

_____ 3. Endocytosis and destruction of particles by cells.

B. Match these terms with the correct statement or definition:

Macrophages
Neutrophils

_____ 1. Usually the first cells to enter an infected area; the primary constituent of pus.

_____ 2. Monocytes that leave the blood and enlarge and increase their numbers of lysosomes and mitochondria.

_____ 3. Responsible for most of the phagocytosis in the late stages of an infection.

_____ 4. Leukocytes that reside beneath free surfaces or in sinuses where they trap and destroy microbes.

☞ Monocytes and macrophages form the mononuclear phagocytic system.

C. Match these types of leukocytes with the correct function or description:

Basophils
Eosinophils
Mast cells

_____ 1. Motile leukocytes that become activated and secrete chemicals that promote inflammation.

_____ 2. Nonmotile leukocytes that are located in connective tissue and become activated to produce chemicals that promote inflammation.

_____ 3. Leukocytes that produce enzymes that inhibit inflammation.

Inflammatory Response

"The inflammatory response is a complex sequence of events."

A. Using the terms provided, complete the following statements:

Chemical mediators   Fibrin
Chemotactic   Vascular permeability
Complement and kinins

Most inflammatory responses are strikingly similar. The microbe itself or damage to tissues causes the release or activation of _(1)_ such as histamine, prostaglandins, leukotrienes, complement, and kinins. The mediators cause vasodilation, and this increased blood flow brings phagocytes and other leukocytes to the area. Some of the mediators are _(2)_ factors that stimulate phagocytes to leave the blood. Mediators also increase _(3)_ , allowing fibrin, complement, and kinins to enter the tissue. _(4)_ prevents the spread of infection by walling off the infected area. _(5)_ enhance the inflammatory response and attract additional phagocytes.

1. _____
2. _____
3. _____
4. _____
5. _____

B. Using the terms provided, complete the following statements:

Local inflammation   Systemic inflammation
Neutrophils   Vascular permeability
Pyrogens

An inflammatory response confined to a specific area of the body is a _(1)_ . An inflammatory response that occurs in many parts of the body is a _(2)_ . In addition to local symptoms, additional features may be present in systemic inflammation. These include production and release of large numbers of _(3)_ that promote phagocytosis, release of _(4)_ by microorganisms or leukocytes to produce fever, and a great increase in _(5)_ in severe cases, which may lead to shock and death.

1. _____
2. _____
3. _____
4. _____
5. _____

Specific Immune System

"Specific immunity involves the ability to recognize, respond to, and remember a particular substance."

A. Match these terms with the correct statement or definition:

Antigen   Hapten
Foreign antigen   Self antigen

_____ 1. General term for any substance that stimulates specific immune system responses.

_____ 2. Small molecule capable of combining with a larger molecule to stimulate a specific immune system response.

_____ 3. Antigen produced outside the body and introduced into the body.

_____ 4. Molecule produced by the body that stimulates a specific immune system response; stimulates autoimmune disease.

B. Match these terms with the correct statement or definition:

Antibody-mediated immunity   T effector cells
B cells   T helper cells
Cell-mediated immunity   T suppressor cells

_____ 1. Lymphocytes that produce antibodies.

_____ 2. Lymphocytes that produce the cell-mediated immune response.

_____ 3. Lymphocytes that increase the cell-mediated or antibody-mediated immune response.

_____ 4. Immunity produced by antibodies in the plasma; humoral immunity.

Origin and Development of Lymphocytes

" *All blood cells, including lymphocytes, are derived from stem cells in red bone marrow.* **"**

Match these terms with the correct statement or definition:

Clones   Stem cells
Red bone marrow   Thymus

_____ 1. Cells in the red bone marrow that give rise to all blood cells.

_____ 2. Location where pre-B cells are processed into B cells.

_____ 3. Location where hormones are produced for T cell maturation.

_____ 4. Small groups of identical lymphocytes produced during embryonic development.

☞ A specific clone can respond only to a particular antigen. Clones that can act against self-antigens are normally eliminated or suppressed.

Activation of Lymphocytes

" *To have a specific immune system response, lymphocytes must be activated by an antigen.* **"**

Match these terms with the correct statement or definition:

Antigen-binding receptors   Interleukin II
Antigenic determinants   Macrophages and dendritic cells
Interleukin I   T helper cells

_____ 1. Specific regions of a given antigen that activate a lymphocyte; an epitope.

_____ 2. Proteins on the surface of lymphocytes that combine with antigenic determinants.

_____ 3. Cells that accumulate and concentrate antigen.

_____ 4. Direct contact with macrophages (or dendritic cells) and this cell type may be necessary for activation of B cells.

_____ 5. Protein produced by activated T helper cells that can activate B cells.

Regulation of Lymphocyte Activity

" *Both the quantity and quality of the immune system's response are regulated.* **"**

Using the terms provided, complete the following statements:

Antibody-mediated   Tolerance
Antigens   T effector cells
Dampening   T suppressor cells

One way the specific immune system is inhibited is through the activity of (1) . Exposure to (2) can activate T suppressor cells in addition to T helper, T effector, and B cells. T suppressor cells can prevent the activity of T helper cells, and, since T helper cells are involved with the activation of B cells and (3) , both (4) and cell-mediated responses can be inhibited. In addition, T suppressor cells can directly prevent B cell and T effector cell activity. Thus T suppressor cell activity is a mechanism for (5) the specific immune system. An example of other control mechanisms of the specific immune system is (6) , in which there is no response to an antigen.

1. _____

2. _____

3. _____

4. _____

5. _____

6. _____

☞ Tolerance may occur when there is a strong T suppressor cell response, when antigen-binding receptors on the surface of T or B cells are blocked or altered, or when lymphocytes capable of acting against self-antigens are eliminated.

Antibody-Mediated Immunity

" *Exposure of the body to an antigen can lead to activation of B cells and to the production of antibodies.* **"**
Antibody-mediated immunity is most effective against extracellular microorganisms.

A. Match these terms with the correct statement or definition:

Constant region   Immunoglobulins
Gamma globulins   Variable region

_____ 1. Other terms used for antibodies.

_____

_____ 2. Part of the antibody that combines with the antigenic determinant of the antigen; determines the specificity of the antibody.

_____ 3. Part of the antibody that is responsible for the ability to activate complement, thus promoting inflammation.

_____ 4. Part of the antibody that normally binds to mast cells or basophils; involved in inflammation.

_____ 5. Part of the antibody that macrophages bind to during phagocytosis.

☞ Antibodies can directly affect antigens by (a) binding to the antigenic determinant of an antigen and interfering with its function, or (b) combining with the antigenic determinant on two different antigens, rendering the antigens ineffective and making them more susceptible to phagocytosis.

462

B. Match these terms with the
   correct parts of the diagram
   labeled in Figure 22-3:

Complement binding site
Constant region
Heavy chain

Light chain
Variable region (antigen-binding
   site)

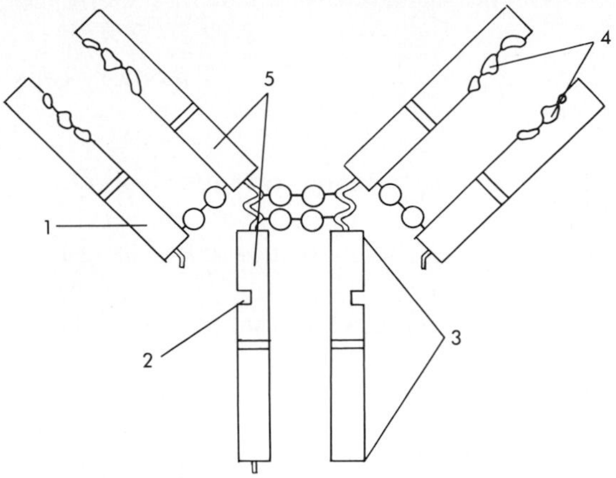

1. _____

2. _____

3. _____

4. _____

5. _____

Figure 22-3

C. Match these terms with the
   correct statement or definition:

Memory cells
Plasma cells

Primary response
Secondary, or memory response

_____ 1. Cell division, cell differentiation, and antibody production from the first
exposure of a B cell to an antigen.

_____ 2. Derived from B cells, these enlarged lymphocytes produce antibodies.

_____ 3. Cells that divide and produce plasma and memory cells when exposed
to a previously encountered antigen.

_____ 4. Antibody-mediated response that is the fastest and produces the most
antibodies.

D. Match these terms with the
   correct statement or definition:

Active artificial immunity
Active natural immunity
Immunization

Passive artificial immunity
Passive natural immunity

_____ 1. Natural exposure to an antigen that causes the body to mount an
immune system response against the antigen.

_____ 2. Deliberate introduction of an antigen into the body; the type of
immunity produced by vaccination.

_____ 3. Type of immunity produced by transfer of antibodies from mother to
child.

_____ 4. Type of immunity produced by antivenins, antisera, and antitoxins.

_____ 5. The general term that includes active and passive artificial immunity.

Cell-Mediated Immunity

"Cell-mediated immunity is a function of T cells and is most effective against intracellular microorganisms such as virus, fungi, intracellular bacteria and parasites."

Match these terms with the correct statement or definition:

Lymphokines   T memory cells
T effector cells

_____ 1. Cells that produce lymphokines and cause lysis of cells.

_____ 2. Cells that provide a secondary response and long-lasting immunity.

_____ 3. Glycoproteins that activate additional components of the immune system.

☞ One type of lymphocyte is called a null cell because it does not have the typical B or T cell surface molecules. Natural killer (NK) cells and killer (K) cells are subpopulations of null cells, which lyse tumor cells and virus-infected cells.

QUICK RECALL

1. List three basic functions of the lymphatic system.

2. List three factors that compress lymph vessels and move lymph toward the circulatory system.

3. List the functions of the lymph nodes, spleen, and thymus.

4. Name the two ways that complement is activated.

5. List the steps that occur when a cell is protected against a viral infection by interferon.

6. List the two cell types responsible for most of the phagocytosis in the body.

7. Name the function that basophils and mast cells have in common. Name the cell type that counteracts this function.

8. Name the cell type that produces antibodies and the cell type that produces most lymphokines.

9. List the four cell types involved in the activation and regulation of B cells and T cells.

10. Name the two cell types responsible for the secondary response.

11. Give the two basic ways that antibody-mediated immunity (antibodies) act against an antigen.

12. Give the two basic ways that cell-mediated immunity acts against an antigen.

MASTERY LEARNING ACTIVITY

Place the letter corresponding to the correct answer in the space provided.

_____ 1. The lymphatic system
   a. removes excess fluid from tissues.
   b. absorbs fats from the digestive tract.
   c. defends the body against microorganisms and other foreign substances.
   d. all of the above

_____ 2. Lymph capillaries
   a. have a basement membrane.
   b. are less permeable than blood capillaries.
   c. prevent backflow of lymph into the tissues.
   d. all of the above

_____ 3. Lymph is moved through lymph vessels due to
   a. contraction of surrounding skeletal muscle.
   b. contraction of smooth muscle within the lymph vessel wall.
   c. pressure changes in the thorax during respiration.
   d. all of the above

_____ 4. Lymph vessels
   a. do not have valves.
   b. empty into lymph nodes.
   c. from the right upper limb join the thoracic duct.
   d. all of the above

_____ 5. The tonsils
   a. consist primarily of diffuse lymphatic tissue.
   b. consists of three groups.
   c. are located in the nasal cavity.
   d. increase in size in adults.

_____ 6. Lymph nodes
   a. filter lymph.
   b. produce lymphocytes.
   c. contain a network of reticular fibers.
   d. all of the above

_____ 7. The spleen
   a. contains dense accumulation of lymphocytes called red pulp.
   b. has red pulp tissue surrounding arteries.
   c. destroys worn out red blood cells.
   d. is surrounded by trabeculae.

_____ 8. The thymus
   a. increases in size in adults.
   b. produces lymphocytes that move to other lymph tissue.
   c. responds to foreign substances in the blood.
   d. all of the above

_____ 9. A group of chemicals that are activated by a series of reactions, in which each component of the series activates the next component and the activated chemicals promote inflammation and phagocytosis?
   a. histamine
   b. antibodies
   c. complement
   d. kinins
   e. lysozyme

_____ 10. A substance that is produced by cells in response to infection by viruses and that can enter other cells and prevent viral replication in the other cells?
   a. interferon
   b. complement
   c. fibrinogen
   d. antibodies
   e. histamine

_____ 11. Neutrophils
   a. enlarge to become macrophages.
   b. account for most of the dead cells in pus.
   c. are usually the last cell type to enter infected tissues.
   d. are usually located in lymph and blood sinuses.

_____ 12. Which of the following cells is most important in the release of histamine, which promotes inflammation?
   a. monocyte
   b. macrophage
   c. eosinophil
   d. mast cell

_____ 13. During the inflammatory response
   a. histamine and other chemical mediators are released.
   b. chemotaxis of phagocytes occur.
   c. fibrin enters tissue from the blood.
   d. blood vessels vasodilate.
   e. all of the above

_____ 14. Antigens
   a. are foreign substances introduced into the body.
   b. are molecules produced by the body.
   c. stimulate a specific immune system response.
   d. all of the above

_____ 15. B cells
   a. are processed in the thymus.
   b. originate in red bone marrow.
   c. once released into the blood stay in the blood.
   d. all of the above

_____ 16. The activation of B cells
   a. involves the accumulation of antigen by macrophages.
   b. involves the production of interleukin by B cells.
   c. occurs when members of a B cell clone come together.
   d. begins when a foreign substance binds to antigenic determinants on the B cell.

_____ 17. T suppressor cells
   a. destroy B cell clones.
   b. inhibit the activity of T helper cells.
   c. produce antibodies.
   d. prevent the accumulation of antigen by macrophages.

_____ 18. Variable amino acid sequences on the arms of the antibody molecule
   a. make the antibody specific for a given antigen.
   b. enable the antibody to activate complement.
   c. enable the antibody to attach to basophils and mast cells.
   d. all of the above

_____ 19. Antibodies
   a. prevent antigens from binding together.
   b. promote phagocytosis.
   c. inhibit inflammation.
   d. block complement activation.

_____ 20. The secondary antibody response
   a. is slower than the primary response.
   b. produces less antibodies than the primary response.
   c. prevents the appearance of disease symptoms.
   d. a and b

_____ 21. The type of lymphocyte that is responsible for the secondary antibody response?
   a. memory cell
   b. B cell
   c. T cell
   d. T helper cell

_____ 22. Antibody-mediated immunity
   a. works best against intracellular antigens.
   b. is more important than cell-mediated immunity in graft rejection.
   c. cannot be transferred from one person to another person.
   d. is responsible for immediate hypersensitivity reactions.

_____ 23. The activation of T cells can result in the
   a. lysis of virus infected cells.
   b. production of lymphokines.
   c. production of T memory cells.
   d. all of the above

_____ 24. Lymphokines
   a. inhibit T helper cells.
   b. activate macrophages.
   c. block interferon production.
   d. isolate the antigen from neutrophils.

_____ 25. Null cells
   a. are a special type of B cell.
   b. produce antibodies.
   c. lyse tumor cells and virus-infected cells.
   d. all of the above

FINAL CHALLENGES

Use a separate sheet of paper to complete this section.

1. The central nervous system and bone marrow do not have lymph vessels. Why doesn't edema occur in these tissues?

2. Filarial worms enter the body and become lodged inside lymph vessels. What effect would this have on the lower limbs?

3. During radical breast surgery malignant lymph nodes in the axilla are removed, and their lymph vessels are tied off to prevent metastasis (spread) of the cancer. Predict the consequences of tying off the lymph vessels.

4. Ivy Mann developed a poison ivy rash after a camping trip. His doctor prescribed a cortisol ointment to relieve the inflammation. A few weeks later Ivy scraped his elbow, which became inflamed. Since he had some of the cortisol oinment left over, he applied it to the scrape. Explain why the ointment was or was not a good idea for the poison ivy and for the scrape.

5. Explain why booster shots are given.

6. A young lady has just had her ears pierced. To her dismay she finds that, when she wears inexpensive (but tasteful) jewelry, by the end of the day there is an inflammatory (allergic) reaction to the metal in the jewelry. Is this due to antibodies or lymphokines?

7. A child appears to be healthy until about 9 months of age. Then he develops severe bacterial infections one after another. Fortunately the infections are successfully treated with antibiotics. When infected with the measles and other viral diseases the child recovered without unusual difficulty. Explain the different immune system response to these infections. Why did it take so long for this disorder to become apparent (hint: IgG).

8. A baby was born with severe combined immunodeficiency disease (SCID). In an attempt to save the patient's life a bone marrow transplant was performed. Explain why the bone marrow transplant might help the patient. Unfortunately there was a graft rejection, and the patient died. Explain what happened.

9. A patient is taking a drug to prevent graft rejection. Would you expect the patient to be more or less likely to develop cancer? Explain.

10. Sometimes antilymphocyte sera is given to prevent rejection of an organ transplant. The sera consists of antibodies that can bind to the patient's lymphocytes. Explain why the sera prevents graft rejection and why it could produce harmful side effects.

ANSWERS TO CHAPTER 22

1. germinal
2. trabecula
3. reticulocytes; reticuloendothelial
4. antigen; antibody; antiviral
5. antigen

6. epitope
7. epitope
8. lymphokine
9. cytotoxic
10. motility

CONTENT LEARNING ACTIVITY

Introduction

1. Lymph
2. Lacteal

3. Chyle
4. Lymphocyte

Lymphatic System

1. Lymph capillaries
2. Lymph vessels
3. Thoracic duct

4. Cisterna chyli
5. Right lymphatic duct

Lymph Organs

A. 1. Diffuse lymphatic tissue
   2. Lymph nodules
   3. Peyer's patches
   4. Palatine tonsils
   5. Pharyngeal tonsils

B. 1. Trabeculae
   2. Reticular fibers
   3. Lymph tissue
   4. Lymph sinus
   5. Cortex
   6. Medullary cords
   7. Afferent lymph vessel
   8. Germinal center

C. 1. Capsule
   2. Germinal center
   3. Afferent lymph vessel
   4. Lymph sinus
   5. Efferent lymph vessel
   6. Trabeculae

D. 1. Capsule
   2. White pulp
   3. Red pulp
   4. Venous sinuses

E. 1. Trabecula
   2. Capsule
   3. Red pulp
   4. White pulp
   5. Periarterial sheath
   6. Lymph nodule

F. 1. Lobule
   2. Cortex
   3. Medulla
   4. Blood-thymic barrier

G. 1. Lymph node
   2. Spleen
   3. Thymus
   4. Thymus

Nonspecific Immune System

1. Mechanical mechanisms
2. Chemicals
3. Inflammation
4. Cells

Chemicals

A. 1. Lysozyme
    2. Mucus
    3. Histamine

B. 1. Complement
    2. Complement cascade
    3. Chemical pathway
    4. Alternate pathway
    5. Properdin

C. 1. Viral
    2. Nucleic acids
    3. Does not
    4. Does not
    5. Antiviral
    6. Specific

Cells

A. 1. Chemotactic factors
    2. Chemotaxis
    3. Phagocytosis

B. 1. Neutrophils
    2. Macrophages
    3. Macrophages
    4. Macrophages

C. 1. Basophils
    2. Mast cells
    3. Eosinophils

Inflammatory Response

A. 1. Chemical mediators
    2. Chemotactic
    3. Vascular permeability
    4. Fibrin
    5. Complement and kinins

B. 1. Local inflammation
    2. Systemic inflammation
    3. Neutrophils
    4. Pyrogens
    5. Vascular permeability

Specific Immune System

A. 1. Antigen
    2. Hapten
    3. Foreign antigen
    4. Self antigen

B. 1. B cells
    2. T effector cells
    3. T helper cells
    4. Antibody-mediated immunity

Origin and Development of Lymphocytes

1. Stem cells
2. Red bone marrow
3. Thymus
4. Clones

Activation of Lymphocytes

1. Antigenic determinants
2. Antigen-binding receptors
3. Macrophages and dendritic cells
4. T helper cells
5. Interleukin II

Regulation of Lymphocyte Activity

1. T suppressor cells
2. Antigens
3. T effector cells
4. Antibody-mediated
5. Dampening
6. Tolerance

Antibody-Mediated Immunity

A. 1. Gamma globulins; immunoglobulins
    2. Variable region
    3. Constant region
    4. Constant region
    5. Constant region

B. 1. Light chain
    2. Complement binding site
    3. Constant region
    4. Variable region
    5. Heavy chain

C. 1. Primary response
    2. Plasma cells
    3. Memory cells
    4. Secondary, or memory response

D. 1. Active natural immunity
    2. Active artificial immunity
    3. Passive natural immunity
    4. Passive artificial immunity
    5. Immunization

Cell-Mediated Immunity

1. T effector cells
2. T memory cells

3. Lymphokines

1. Maintains fluid balance, absorbs fat and other substances from the intestines, and defends against microorganisms and other foreign substances
2. Skeletal muscle contraction, contraction of lymph vessel smooth muscle, and thoracic pressure changes
3. Lymph nodes: filter lymph and remove substances by phagocytosis, stimulate and release lymphocytes.
Spleen: foreign substances stimulate lymphocytes in white pulp, foreign substances are phagocytized in the red pulp.
Thymus: processes T cells, and produces and releases T cells.
4. Alternate pathway: by spontaneous complement protein activation and properdin; classical pathway: by antibody activation
5. Interferon is produced by a virally infected cell, moves to other cells, and protects them from viral infection.
6. Neutrophils and macrophages
7. Basophils and mast cells release inflammatory chemicals. Eosinophils release enzymes that reduce inflammation.
8. B cells produce antibodies, T cells produce most lymphokines.
9. Macrophages, dendritic cells, T helper cells, and T suppressor cells
10. B and T memory cells
11. Direct effects: inactivate antigen or bind antigens together. Indirect effects (activates other mechanisms), act as opsonins, activate complement, and increase inflammation by attaching to mast cells or basophils and causing release of inflammatory chemicals.
12. T effector cells lyse cells and produce lymphokines.

MASTERY LEARNING ACTIVITY

1. D. The lymphatic system is involved in tissue fluid balance, fat absorption, and defense against microorganisms.

2. C. Lymph capillaries are simple squamous epithelium but unlike most other epithelial tissue do not have a basement membrane. In addition, the epithelial cells are loosely attached to each other. Consequently, lymph capillaries are far more permeable than blood capillaries. The lymph capillaries prevent backflow of lymph into the tissue because the epithelial cells overlap each other and function as one-way valves.

3. D. Compression of the lymph vessel by surrounding skeletal muscle, smooth muscle of the lymph vessel wall, and thoracic pressure changes all cause movement of lymph.

4. B. Most lymph vessels empty into a lymph node. Lymph vessels have valves that ensure one-way flow of lymph. Lymph vessels from the right upper limb, right thorax, and right side of the head and neck join the right lymphatic duct, which empties into the right subclavian vein. The lymph vessels from the rest of the body join the thoracic duct, which empties into the left subclavian vein.

5. B. There are three groups of tonsils: the palatine, pharyngeal, and lingual tonsils. The tonsils consists of large groups of lymph nodules, and they decrease in size in adults and may disappear. The tonsils are found in the oropharynx and nasopharynx (back of throat).

6. D. Lymph nodes contain a network of reticular fibers. Within lymph sinuses the reticular fiber network and phagocytic cells filter the lymph to remove microorganisms and other foreign substances. Within the lymphatic tissue of the lymph node, lymphocytes are produced in germinal centers.

7. C. The spleen detects and responds to foreign substances in the blood, destroys worn out red blood cells, and acts as a reservoir for red blood cells. The spleen is surrounded by a capsule and subdivided by trabeculae. White pulp, dense accumulations of lymphocytes, are found surrounding arteries. Red pulp is associated with venous sinuses and consists of a network of reticular fibers filled with blood cells and macrophages.

8. B. The thymus produces lymphocytes that move to other lymph tissues. The blood-thymic barrier prevents the thymus from responding to foreign substances in the blood. With increasing age in the adult, the thymus decreases in size.

9. C. The series of reactions is the complement cascade.

10. A. Interferon does not protect the cell producing the interferon, but does enter other cells and prevent viral replication.

11. B. Neutrophils are phagocytic cells that usually die after a single phagocytic event. They are usually the first cell type to enter infected tissues, and they account for most of the dead cells in pus. Monocytes enter tissues to become macrophages that can be found in lymph and blood sinuses and beneath the free surfaces of the body (e.g., skin, mucous membranes, serous membranes). Macrophages appear in infected tissue after neutrophils and are responsible for the cleanup of neutrophils and other cellular debris.

12. D. Mast cells and basophils release histamine, which promotes inflammation. Eosinophils release enzymes that break down histamine and inhibit inflammation.

13. E. Damage to tissue causes the release of histamine and other chemical mediators that cause vasodilation, attract phagocytes (chemotaxis), and increase vascular permeability (allowing the entry of fibrin, which walls off infected areas).

14. D. Antigens are large-molecular-weight molecules that stimulate a specific immune system response. They can be foreign substances (foreign antigens) or molecules produced by the body (self antigens).

15. B. B cells originate and are processed in red bone marrow. T cells originate in red bone marrow and are processed in the thymus. Once B and T cells are released into the blood, they circulate between the blood and lymph tissue.

16. A. One way of activating B (or T cells) begins when a macrophage (or dendritic cell) accumulates antigen and combines with a T helper cell and a B cell. Another way occurs when the macrophage produces interleukin I, which stimulates a T helper cell to produce interleukin II. The interleukin II activates the B cell. Antigens have specific regions, called antigenic determinants, which bind to antigen-binding receptors on the surface of lymphocytes.

17. B. The activity of B and T cells is regulated by T suppressor cells, which inhibit T helper cells. Since T helper cells are involved in the activation of B and T cells, the suppression of T helper cells results in less B and T cell activity.

18. A. The variable region, a sequence of amino acids on the antibody molecule, makes the antibody specific for a given antigen. The constant region is involved with the activation of complement and enables the antibody to attach to various cells.

19. B. Antibodies promote phagocytosis, activate the complement cascade, stimulate inflammation, and bind antigens together.

20. C. The secondary response is more rapid and produces more antibodies than the primary response. Thus the secondary response effectively destroys the antigen and prevents the appearance of disease symptoms.

21. A. During the primary response memory cells are formed. They are responsible for the secondary response.

22. D. Antibody-mediated immunity is responsible for immediate hypersensitivity reactions, and it works best against extracellular antigens. It can be transferred to another person through antibodies in plasma. Cell-mediated immunity is more important in graft rejection and control of intracellular antigens.

23. D. When activated T cells produce T effector cells that can directly lyse target cells and can produce lymphokines. Also formed are T memory cells, which are responsible for a secondary response (rapid, strong) to the antigen.

24. B. Lymphokines promote phagocytosis and inflammation. They attract macrophages, neutrophils, and other cells and convert ordinary macrophages into more effective phagocytic cells. One type of interferon is a lymphokine.

25. C. Null cells are distinct from either B cells or T cells. They lyse tumor cells and virus-infected cells.

 ## FINAL CHALLENGES

1. Since the brain and bone marrow are encased by bone, the bone prevents the tissues from swelling.

2. Blockage of the lymph vessels would produce edema, and the lower limbs can eventually swell to several times their normal size. This condition is called elephantiasis.

3. Tying off the lymph vessels would prevent lymph drainage. Edema in the axillary region and in the upper limb would be expected.

4. The ointment was a good idea for the poison ivy, which caused a delayed hypersensitivity reaction, i.e., too much inflammation. For the scrape it is a bad idea, since a normal amount of inflammation is beneficial and helps to fight infection in the scrape.

5. The first shot produces a primary response. The booster shot produces a secondary response with a higher production of antibodies. This provides greater, longer-lasting immunity.

6. Since antibodies and lymphokines both produce inflammation, the fact that the metal in the jewelry resulted in inflammation is not enough information to answer the question. However, the fact that it took most of the day (many hours) to develop the reaction would indicate a delayed hypersensitivity reaction and therefore lymphokines.

7. The child's antibody-mediated system is not functioning properly, whereas the child's cell-mediated system is functioning properly. This explains the susceptibility to extracellular bacterial infections and the resistance to intracellular viral infections. It took so long to become apparent because IgG from the mother crossed the placenta and provided the infant with protection. Once these antibodies were degraded, the infant began to get sick.

8. Bone marrow is the source of lymphocytes. If a bone transplant were successful the baby would have started producing lymphocytes and would have had a functioning immune system. In this case there was a graft versus host rejection in which the transplanted lymphocytes in the red marrow mounted an immune system attack against the baby's tissues, resulting in death.

9. The drug is probably suppressing cell-mediated immunity, which is responsible for rejecting grafts and is involved in tumor control. One might expect a higher incidence of cancer in such a patient, and there is evidence that this is true for some cancers. However, it is not true for many cancers, which would indicate that other components of the immune system are involved in tumor control, for example, null cells.

10. When the antibodies bind to the lymphocytes, mechanisms are activated that destroy the lymphocytes, e.g., opsonization or activation of complement. Without the lymphocytes an effective immune system response to the graft does not occur. The side effect could be an increased susceptibility to infections.

Respiratory System

FOCUS: The respiratory system consists of the nasal cavity, which warms and filters air; the pharynx, which connects the nasal cavity to the larynx; the larynx, which contains the vocal cords; a tubular air-conducting system consisting of the trachea, bronchi, and bronchioles; and the site of gas exchange with the blood, the alveoli. The diaphragm and thoracic wall muscles change the volume of the thoracic cavity, producing pressure gradients responsible for the movement of air into and out of the lungs. The elastic recoil of lung tissue and water surface tension make the lungs collapse, a tendency that is opposed by intrapleural pressure and surfactant. The partial pressure of a gas is the portion of the total pressure exerted by a gas in a mixture of gases. A gas moves from areas of high to areas of low partial pressure. Movement of a gas through the respiratory membrane is influenced by the partial pressure of the gas, the diffusion coefficient of the gas, and the thickness and surface area of the respiratory membrane. Once in the blood, oxygen is mostly transported bound to hemoglobin. The oxygen-hemoglobin dissociation curve shows how hemoglobin is saturated with oxygen in the lungs and how hemoglobin releases oxygen in the tissues. Most carbon dioxide is transported in the blood as bicarbonate ions. Neural control of respiration is achieved by the inspiratory, expiratory, pneumotaxic, and apneustic centers in the brain, and by the Hering-Breur reflex. The most important regulators of resting respiration are blood carbon dioxide and pH levels, although low blood oxygen levels can increase respiration. Respiration during exercise is mostly determined by the cerebral motor cortex and by feedback from proprioceptors.

WORD PARTS

Give an example of a new vocabulary word that contains each word part.

WORD PART	MEANING	EXAMPLE
aryten-	a pitcher or ladle	1._____
crico-	a ring	2._____
thyro-	a shield	3._____
pleur-	the side	4._____

WORD PART	MEANING	EXAMPLE
pneumo-	lungs	5. _____
tax-	arrangement	6. _____
-spire-	breathe	7. _____
dia-	across	8. _____
-phragm	fence; partition	9. _____
glob-	a ball	10. _____

CONTENT LEARNING ACTIVITY

Nose and Nasal Cavity

"*Air enters the external nares and passes through the nasal cavity into the pharynx.***"**

A. Match these terms with the correct statement or definition:

Conchae   Meatus
External nares (nostrils)   Soft palate
Hard palate   Vestibule
Internal nares (choanae)

_____ 1. Posterior openings from the nasal cavity into the pharynx.

_____ 2. Anterior portion of the nasal cavity, just inside the external nares.

_____ 3. Muscle covered by mucosa, which forms the posterior part of the floor of the nasal cavity.

_____ 4. Three bony ridges on the lateral wall of each nasal cavity.

_____ 5. Passageways between the conchae.

☞ The openings for the paranasal sinuses are found in the superior and median meatus, and the opening of the nasolacrimal duct is found in the inferior meatus.

B. Match these terms with the correct statement or definition:

Cilia   Mucous membrane
Mucus   Nasal hair

_____ 1. Structures located in the vestibule that trap large dust particles.

_____ 2. Substance secreted by goblet cells that traps debris in the incoming air.

_____ 3. Structures on the surface of the mucous membrane that move mucus posteriorly to the pharynx.

_____ 4. Structure responsible for the addition of moisture to inspired air.

Pharynx

66_The pharynx is the common opening of both the digestive and respiratory systems._**99**

A. Match these terms with the correct statement or definition:

Fauces
Laryngopharynx

Nasopharynx
Oropharynx

_____ 1. Superior portion of the pharynx that extends from the internal nares to the uvula.

_____ 2. Portion of the pharynx that extends from the uvula to the epiglottis.

_____ 3. Portion of the pharynx that contains the openings of the auditory tubes.

_____ 4. Portion of the pharynx that contains two sets of tonsils.

_____ 5. Opening of the oral cavity into the oropharynx.

B. Match these terms with the correct parts labeled on the diagram in Figure 23-1:

Conchae
Epiglottis
External naris (nostril)
Hard palate
Internal naris (choana)
Laryngopharynx
Larynx

Meatus
Nasopharynx
Oropharynx
Soft palate
Uvula
Vestibule

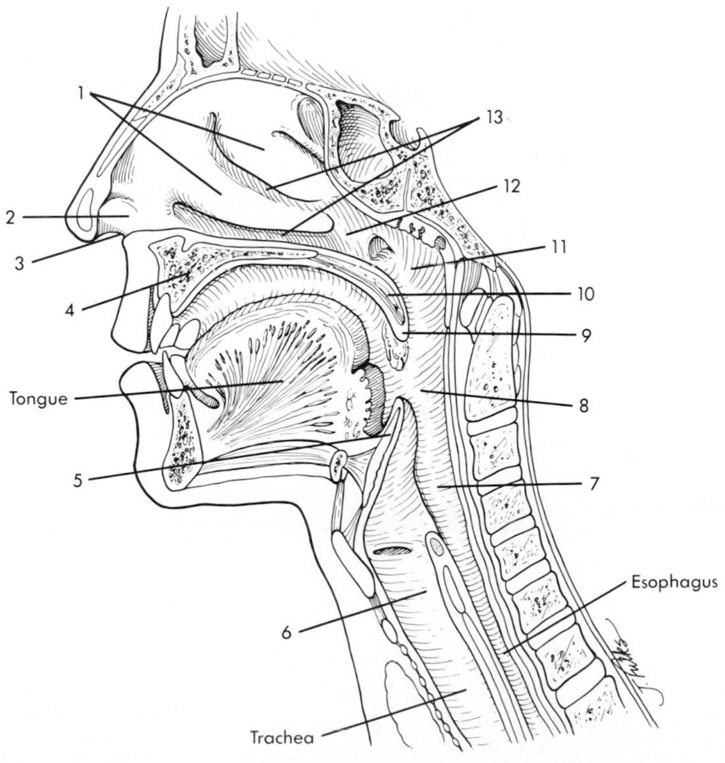

1._____

2._____

3._____

4._____

5._____

6._____

7._____

8._____

9._____

10._____

11._____

12._____

13._____

Figure 23-1

477

Larynx

66 *The larynx consists of an outer covering of nine cartilages that are connected to each other by muscles and ligaments.* **99**

A. Match these terms with the correct statement or definition:

Cricoid cartilage   Thyroid cartilage
Epiglottis   True vocal cords
Laryngitis   Vestibular folds

_____ 1. Largest and most superior of the cartilages in the larynx; Adam's apple.

_____ 2. Elastic cartilage that covers the opening into the larynx during swallowing.

_____ 3. Superior pair of ligaments which extend from the arytenoid cartilages to the thyroid cartilage; close to prevent the movement of air, food, or liquids through the larynx.

_____ 4. Inferior pair of ligaments, which extend from the arytenoid cartilages to the thyroid cartilage; involved in sound production.

_____ 5. Inflammation of the mucosal epithelium of the vocal cords.

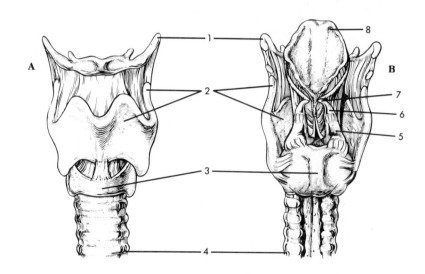

Figure 23-2, A and B

B. Match these terms with the correct parts labeled on the diagram in Figure 23-2:

Arytenoid cartilage
Cricoid cartilage
Corniculate cartilage
Cuneiform cartilage
Epiglottis
Hyoid bone
Thyroid cartilage
Tracheal cartilage

1. _____

2. _____

3. _____

4. _____

5. _____

6. _____

7. _____

8. _____

Trachea, Bronchi and Lungs

❝*The trachea and bronchi serve as passageways for air between the larynx and lungs; the lungs*❞
are the principal organs of respiration.

A. Match these terms with the
correct statement or definition:

Alveolar sac
Alveoli
Asthma attack
Cartilage plates
Pseudostratified ciliated columnar
epithelium

C-shaped cartilage rings
Hilum
Lobes
Lobules
Smooth muscle
Simple squamous epithelium

_____ 1. Sections of the lung separated by connective tissue but not visible as
surface fissures.

_____ 2. Structures that reinforce the walls of the trachea and bronchi.

_____ 3. Support structures present in the walls of secondary and tertiary
bronchi.

_____ 4. Cell type lining the trachea and bronchi.

_____ 5. Result of forceful contraction of smooth muscle in the bronchiole walls.

_____ 6. Air sacs in the lungs.

_____ 7. Two or more alveoli that share a common opening.

B. Using the following diagram, arrange the structures in the order a
molecule of oxygen would pass through them to enter the blood.

Alveolar duct
Alveoli
Primary bronchi
Respiratory bronchiole
Terminal bronchiole
Trachea

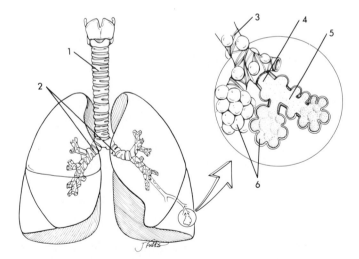

1._____

2._____

3._____

4._____

5._____

6._____

Figure 23-3

Pleura

" *The serous membranes lining the pleural cavity and covering the lungs are called pleura.* **"**

Match these terms with the
correct statement or definition:

Parietal pleura   Pleural fluid
Pleural cavity   Visceral pleura

_____ 1. Cavity surrounding each lung.

_____ 2. Serous membrane that lines the pleural cavity; the membrane in contact
with the thoracic wall.

_____ 3. Lubricant for the lungs that helps hold the pleural membranes together.

Blood Supply

" *There are two major blood flow routes to the lung.* **"**

Match these terms with the
correct statement or definition:

Bronchial arteries   Pulmonary arteries
Bronchial veins   Pulmonary veins

_____ 1. Blood vessels that carry deoxygenated blood to the lungs.

_____ 2. Blood vessels that branch from the thoracic aorta and carry oxygenated
blood to the bronchi.

_____ 3. Blood vessels that carry oxygenated blood from the lungs to the heart.

_____ 4. Blood vessels that carry deoxygenated blood from the bronchi to the
azygos system.

Muscles of Respiration and the Thoracic Wall

" *The diaphragm and thoracic wall form the boundaries of the thoracic cavity.* **"**

Match these terms with the
correct statement or definition:

Costal cartilages   Internal intercostal muscles
Diaphragm   External intercostal muscles

_____ 1. Large dome of skeletal muscles that, when contracted, expands the
superior-inferior dimension of the thoracic cavity.

_____ 2. Structures that allow lateral rib movement and lateral expansion of the
thoracic cavity.

_____ 3. Muscles of inspiration that, when contracted, elevate the ribs.

☞ During quiet breathing, expiration occurs when the muscles of inspiration relax and
a passive decrease in thoracic volume occurs; during labored breathing, the muscles
of expiration can contract, producing a more rapid and greater decrease in
thoracic volume.

Pressure Differences and Air Flow

66*The flow of air into the lungs requires a pressure gradient from the outside of the body to the alveoli.*99

A. Match these terms with the statements below:

Decreases
Increases

_____ 1. Effect of inspiration on thoracic volume.

_____ 2. Effect of increased thoracic volume on lung volume.

_____ 3. Effect of expansion of the lungs on alveolar volume.

_____ 4. Effect on intrapulmonary pressure if alveolar volume increases.

_____ 5. Effect on air movement into the alveoli if intrapulmonary pressure decreases below atmospheric pressure.

☞ Ventilation is the process of moving air into and out of the lungs.

Factors That Prevent the Lungs from Collapsing

66*Two factors tend to make the lungs collapse, and two factors keep the lungs from collapsing.*99

Match these terms with the correct statement or definition:

Elastic recoil
Intrapleural pressure
Intrapulmonary

Surface tension of alveolar fluid
Surfactant

_____ 1. Two factors that cause the lungs to tend to collapse.

_____

_____ 2. Two factors that keep the lungs from collapsing.

_____

_____ 3. Mixture of lipoprotein molecules produced by the secretory cells of the alveolar epithelium.

_____ 4. Pressure within the pleural cavity.

☞ Pneumothorax is the introduction of air into the pleural cavity, which causes an equalization of intrapleural and atmospheric pressures and results in collapse of the lung.

Compliance of the Lungs and Thorax

❝Compliance is a measure of the expansibility of the lungs and thorax.**❞**

Match these terms with the
correct statement:

Decrease
Increase

_____ 1. Effect on the ease of lung expansion when compliance decreases.

_____ 2. Effect on compliance when the collapsing force of the lungs increases,
as in respiratory distress syndrome.

_____ 3. Effect of emphysema on compliance.

Pulmonary Volumes and Capacities

❝Factors such as sex, age, body size and physical conditioning cause variations in respiratory**❞**
volumes and capacities from one individual to another.

A. Match these terms with the
correct definition and the
following diagram:

Expiratory reserve volume
Functional residual capacity
Inspiratory capacity
Inspiratory reserve volume
Residual volume
Tidal volume
Total lung capacity
Vital capacity

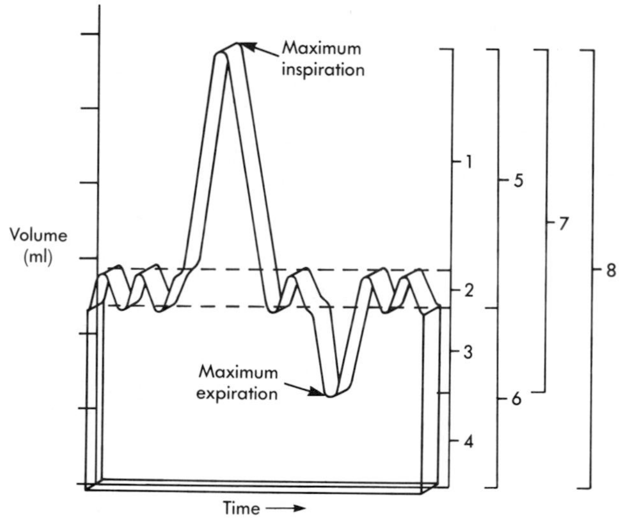

Figure 23-4

_____ 1. The volume of air that can be forcefully inspired after inspiration of the
normal tidal volume.

_____ 2. The volume of air inspired or expired in a normal inspiration or
expiration.

_____ 3. The volume of air that can be forcefully expired after expiration of the
normal tidal volume.

_____ 4. The volume of air still in the respiratory passages after the most forceful
expiration.

_____ 5. The tidal volume plus the inspiratory reserve volume.

_____ 6. The expiratory reserve volume plus the residual volume.

_____ 7. The sum of the expiratory reserve, tidal and inspiratory reserve volumes.

_____ 8. The sum of the inspiratory reserve, expiratory reserve, tidal and
residual volumes.

The rate at which lung volume changes during direct measurement of the vital capacity is called the forced expiratory vital capacity.

Minute Respiratory Volume and Alveolar Ventilation Rate

66 *Minute respiratory volume and alveolar ventilation rate are measurements of gas movement* 99 *in the respiratory system.*

Match these terms with the correct statement or definition:

Alveolar ventilation rate
Anatomical dead air space

Minute respiratory volume
Physiological dead air space

_____ 1. Total amount of air moved in and out of the respiratory system each minute; tidal volume times respiratory rate.

_____ 2. Nasal cavity, pharynx, larynx, trachea, bronchi, bronchioles, and terminal bronchioles; gas exchange does not take place.

_____ 4. Anatomical dead air space plus the volume of any nonfunctional alveoli.

_____ 5. Volume of air that is available for gas exchange each minute.

Partial Pressure

66 *It is traditional to designate the partial pressure of individual gases in a mixture as* 99 *P_{N_2}, P_{O_2}, or P_{CO_2}, for example.*

Match these terms with the correct statement or definition:

Dalton's Law
Partial pressure

Vapor pressure

_____ 1. In a mixture of gases the portion of the total pressure due to each type of gas is determined by the percentage of the total volume represented by each gas type.

_____ 2. Pressure exerted by each type of gas in a mixture.

_____ 3. Partial pressure of water molecules in the gaseous form.

Partial Pressure of Respiratory Gases

66 *The composition of alveolar and expired air is not identical to the composition of dry atmospheric air.* 99

Using the terms provided, complete the following statements:

Greater
Smaller

Compared to atmospheric air, alveolar air contains a _(1)_ partial pressure of oxygen, _(2)_ partial pressure of carbon dioxide, and _(3)_ vapor pressure.

1._____

2. _____

3. _____

Diffusion of Gases Through Liquids

66 *When a gas comes into contact with a liquid such as water, there is a tendency for the gas to dissolve in the liquid.* 99

A. Match these terms with the correct statement or definition:

Henry's law
Solubility coefficient

_____

1. Concentration of a dissolved gas is its partial pressure multiplied by its solubility coefficient.

_____

2. Measure of how easily a gas dissolves in a liquid.

B. Match these terms with the correct statements:

Decreases
Increases

_____

1. Effect on diffusion rate of a gas if the partial pressure gradient is increased.

_____

2. Effect on diffusion rate of a gas if the diffusion coefficient is decreased.

_____

3. Effect on diffusion rate of a gas if the thickness of the membrane is increased.

_____

4. Effect on diffusion rate of a gas if the surface area of the membrane is increased.

Diffusion of Gases Through the Respiratory Membrane

66 *The respiratory membranes of the lungs are in the respiratory bronchioles, alveolar ducts, and alveoli.* 99

Using the diagram in Figure 23-5, arrange each of the structures in the correct order from inside the alveoli to the blood:

Alveolar epithelium
Alveolar epithelial basement membrane
Capillary endothelium
Capillary endothelium basement membrane
Thin fluid layer
Thin interstitial space

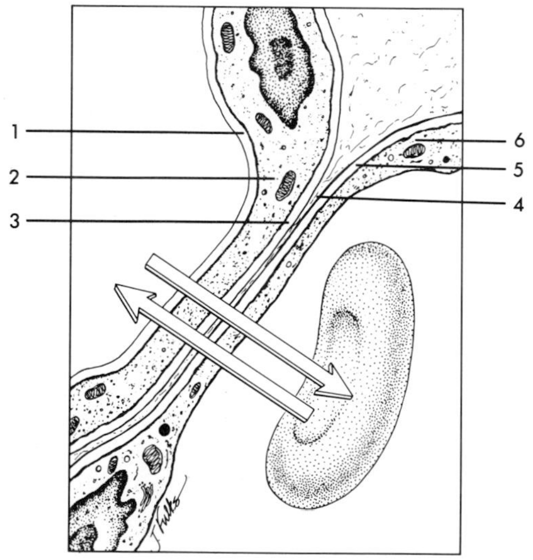

Figure 23-5

1._____   3. _____   5. _____

2. _____   4. _____   6. _____

Relationship Between Ventilation and Capillary Blood Flow

66 *There are two ways that the optimum relationship between ventilation and blood flow can be affected.* **99**

Match these problems with the
correct statement or definition:

Inadequate cardiac output
Shunted blood

_____

1. In this situation ventilation may exceed the ability of the blood to pick up oxygen.

_____

2. Blood that is not completely oxygenated; this occurs when ventilation is not great enough to oxygenate the blood in the alveolar capillaries.

Oxygen and Carbon Dioxide Diffusion Gradients

66 *To diffuse efficiently, carbon dioxide partial pressure gradients do not have to be as great* **99** *as the partial pressure gradients for oxygen.*

Match these partial pressures with
the correct location on the
following diagram:

104 mm Hg
95 mm Hg

45 mm Hg
40 mm Hg

1. _____

2. _____

3. _____

4. _____

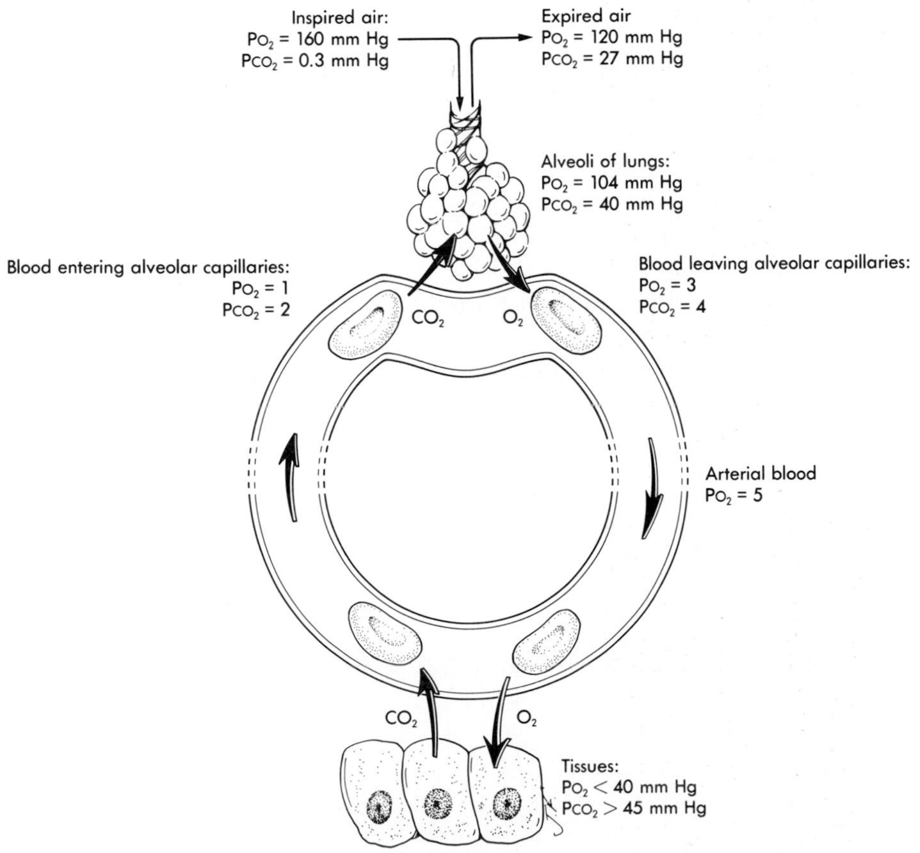

Inspired air:
P_{O_2} = 160 mm Hg
P_{CO_2} = 0.3 mm Hg

Expired air
P_{O_2} = 120 mm Hg
P_{CO_2} = 27 mm Hg

Alveoli of lungs:
P_{O_2} = 104 mm Hg
P_{CO_2} = 40 mm Hg

Blood entering alveolar capillaries:
P_{O_2} = 1
P_{CO_2} = 2

CO_2 O_2

Blood leaving alveolar capillaries:
P_{O_2} = 3
P_{CO_2} = 4

Arterial blood
P_{O_2} = 5

CO_2 O_2

Tissues:
P_{O_2} < 40 mm Hg
P_{CO_2} > 45 mm Hg

Figure 23-6

Hemoglobin and Oxygen Transport

"About 97% of the oxygen transported in the blood from the lungs to the tissues is transported in combination**"** with the hemoglobin in the red blood cells and the remaining 3% is dissolved in the water portion of plasma.

Match these terms with the correct statement or definition:

Bohr effect
Carbonic anhydrase
Decreases
Increases

Oxygen-hemoglobin dissociation curve
Reverse Bohr effect
Saturated hemoglobin

_____ 1. Hemoglobin molecule with an oxygen molecule bound to each of its four heme groups.

_____ 2. Shift to the right of the oxygen-hemoglobin dissociation curve when the affinity of hemoglobin for oxygen decreases; normally occurs in the tissues.

_____ 3. Shift to the left of the oxygen-hemoglobin dissociation curve when the affinity of hemoglobin for oxygen increases; normally occurs in the lungs.

_____ 4. Effect on the amount of oxygen bound to hemoglobin when pH decreases (P_{CO_2} increases).

_____ 5. Effect on the amount of oxygen bound to hemoglobin when temperature increases.

Transport of Carbon Dioxide

"Carbon dioxide is transported in the blood in three major ways.**"**

A. Match these terms with the correct statement or definition:

Bicarbonate ions
Carbamino compounds (including carbaminohemoglobin)

Dissolved in plasma

_____ 1. 72% of carbon dioxide is transported in this form.

_____ 2. 20% of carbon dioxide is transported in this form.

_____ 3. 8% of carbon dioxide is transported in this form.

☞ Carbon dioxide binds to the globin portion of hemoglobin.

B. Using the terms provided, complete the following statements:

Bicarbonate ions
Buffer
Carbonic acid
Carbonic anhydrase

Chloride ions
Chloride shift
Hemoglobin

Most of the carbon dioxide inside the red blood cell reacts with water to form (1); the reaction is catalyzed by (2) inside the red blood cell. The carbonic acid then dissociates to form (3) and hydrogen ions. Because more bicarbonate ions are inside the cells than outside, the bicarbonate ions readily diffuse out of the red blood cells into the plasma. In response to this movement of negatively charged ions out of the red blood cells, (4) move from the plasma into the red blood cells. The exchange of chloride ions for the bicarbonate ions across the membranes of the red blood cells is called the (5). The hydrogen ions formed by the dissociation of carbonic acid bind to (6) in the red blood cells. This binding prevents hydrogen ions from leaving the cells; thus hemoglobin acts as a (7) to prevent a decrease in blood pH.

1. _____

2. _____

3. _____

4. _____

5. _____

6. _____

7. _____

Nervous Control of Rhythmic Ventilation

"Nerve impulses responsible for controlling the respiratory muscles originate within**"** neurons of the medulla oblongata.

Match these terms with the correct statement or definition:

Apneustic center
Expiratory center
Hering-Breuer reflex

Inspiratory center
Pneumotaxic center

_____ 1. Spontaneously active; establishes the basic rhythm of respiration.

_____ 2. Remains inactive during quiet respiration, but becomes active when the rate and depth of ventilation are increased.

_____ 3. Has an inhibitory effect on the inspiratory and apneustic centers.

_____ 4. Stimulates the inspiratory center.

_____ 5. Occurs when action potentials from stretch receptors in the bronchi and bronchioles inhibit the inspiratory center.

Chemical Control of Respiration

"The respiratory system maintains the concentrations of oxygen and carbon dioxide**"** and the pH of the body fluids within a normal range of values.

A. Match these terms with the correct statement or definition:

Carotid and aortic bodies
Chemosensitive area

Decrease
Increase

_____ 1. Area of the medulla that is sensitive to changes in carbon dioxide and pH of the blood.

_____ 2. Small vascular sensory organs that monitor the partial pressure of oxygen.

_____ 3. Effect of increased blood carbon dioxide on breathing rate and depth.

_____ 4. Effect of decreased blood pH on breathing rate and depth.

_____ 5. Effect of greatly decreased blood oxygen level on breathing rate and depth.

Small changes in blood carbon dioxide and blood pH levels produce much larger changes in respiratory rate than do small changes in blood oxygen levels.

The Effect of Exercise on Respiratory Movements

66 *There are factors other than carbon dioxide, pH, and oxygen that influence the respiratory center during exercise.* 99

Match these terms with the correct statements:

Decrease
Increase

_____ 1. Effect of action potentials that come through collateral fibers from motor pathways on breathing rate during exercise.

_____ 2. Effect of stimulation of proprioceptors on respiratory rate during exercise.

_____ 3. Effect of stimulation of touch, thermal, and pain receptors in the skin on respiratory rate.

Voluntary hyperventilation can increase, or voluntary apnea can decrease the rate and depth of respiratory movements, resulting in changes in blood carbon dioxide and oxygen levels.

QUICK RECALL

1. Name three things that happen to inspired air on its way through the nasal cavity.

2. Trace the path of inspired air from the trachea to the alveoli by naming the structures the air passes through.

3. Describe the relationship between the volume and the pressure of a gas.

4. List two factors that tend to cause the lungs to collapse and two factors that prevent the lungs from collapsing.

5. List the four pulmonary volumes and define pulmonary capacity.

6. Give the formula for calculating alveolar ventilation rate.

7. List the six layers of a respiratory membrane through which gases must diffuse.

8. Name four factors that influence the rate of gas diffusion across the respiratory membrane.

9. List two ways oxygen is transported in the blood, and indicate their relative importance.

10. List three ways that carbon dioxide is transported in the blood, and indicate their relative importance.

11. List the events that result in a decrease in blood pH when blood carbon dioxide levels increase.

12. List three chemical factors that influence respiration, the location in the body where the levels of this chemical are monitored, and give the effect of increasing or decreasing their concentration.

MASTERY LEARNING ACTIVITY

Place the letter corresponding to the correct answer in the space provided.

_____ 1. The nasal cavity
   a. is mostly lined with pseudostratified ciliated epithelium.
   b. has a vestibule, which contains the olfactory epithelium.
   c. is connected to the pharynx by the external nares.
   d. has passageways called conchae.

_____ 2. The nasopharynx
   a. has openings from the paranasal sinuses.
   b. contains the pharyngeal tonsils.
   c. opens into the oral cavity via the fauces.
   d. extends to the tip of the epiglottis.

_____ 3. The larynx
   a. connects the oropharynx to the trachea.
   b. has one unpaired cartilage and eight paired cartilages.
   c. contains the vocal cords.
   d. all of the above

_____ 4. The trachea possess
   a. skeletal muscle.
   b. pleural fluid glands.
   c. C-shaped rings of cartilage.
   d. walls only one cell in thickness.

_____ 5. Gas exchange between the air in the lungs and the blood takes place in the
   a. alveoli.
   b. bronchi.
   c. terminal bronchioles.
   d. trachea.

_____ 6. During an asthma attack, the patient has difficulty breathing because of constriction of the
   a. trachea.
   b. bronchi.
   c. terminal bronchioles.
   d. alveoli.
   e. respiratory membrane.

_____ 7. The parietal pleura
   a. covers the surface of the lung.
   b. covers the inner surface of the pleural cavities.
   c. is the connective tissue membrane that divides the thoracic cavity into the right and left pleural cavities.
   d. covers the inner surface of the alveoli.
   e. is the membrane across which gas exchange occurs.

_____ 8. Given the following muscles:
   1. diaphragm
   2. external intercostals
   3. internal intercostals
   4. rectus abdominis

   Which of the muscles are involved with increasing the volume of the thorax?
   a. 1, 2
   b. 1, 3
   c. 2, 3
   d. 2, 4
   e. 3, 4

_____ 9. During the process of inspiration, which of the following pressures decrease when compared to the resting condition?
   a. intrapulmonary
   b. intrapleural
   c. atmospheric
   d. a and b
   e. all of the above

_____ 10. During the process of expiration the pressure within the alveoli must be
   a. greater than the pressure in the pleural spaces.
   b. greater than atmospheric pressure.
   c. less than atmospheric pressure.
   d. a and b

11. Contraction of the bronchiolar smooth muscle has which of the following effects?
    a. decreased resistance to gas flow through the respiratory tree
    b. a smaller pressure gradient is required to get the same rate of gas flow when compared to the normal bronchioles
    c. increased resistance to gas flow through the respiratory tree
    d. does not effect gas flow through the respiratory tree
    e. a and b

12. The lung does not normally collapse because of
    a. surfactant.
    b. intrapleural pressure.
    c. elastic recoil.
    d. a and b
    e. all of the above

13. Immediately after the creation of an opening through the thorax into the pleural cavity,
    a. air would flow into the pleural cavity through the hole.
    b. air would flow out of the pleural cavity through the hole.
    c. air would neither flow out nor in through the hole.
    d. the lung would be forced to protrude through the hole.
    e. b and d

14. Compliance of the lungs and thorax is
    a. the volume by which the lungs and thorax change for each unit change of intrapulmonary pressure.
    b. increased in emphysema.
    c. decreased due to a lack of surfactant.
    d. all of the above

15. A patient expires normally; then, using forced ventilation, he blows as much air as possible into a spirometer. This would measure the
    a. inspiratory reserve.
    b. expiratory reserve.
    c. residual volume.
    d. tidal volume.
    e. vital capacity.

16. Given the following lung volumes:
    1. tidal volume= 500 ml
    2. residual volume = 1000 ml
    3. inspiratory reserve= 2500 ml
    4. expiratory reserve= 1000 ml
    5. dead air space= 1000 ml

    The vital capacity would be
    a. 3000 ml
    b. 3500 ml
    c. 4000 ml
    d. 5000 ml
    e. 6000 ml

17. The alveolar ventilation rate is
    a. tidal volume times respiratory rate.
    b. the minute respiratory volume plus the dead air space.
    c. the amount of air available for gas exchange in the lungs.
    d. all of the above

18. If the total pressure of a gas were 760 mm Hg and its composition 20% oxygen, 0.04% carbon dioxide, 75% nitrogen, and 5% water vapor, the partial pressure of oxygen would be
    a. 15.2 mm Hg
    b. 20 mm Hg
    c. 148 mm Hg
    d. 152 mm Hg
    e. 740 mm Hg

19. Which of the following layers must gases cross to pass from the alveolus to the blood within the alveolar capillaries?
    a. endothelium
    b. basement membrane
    c. simple squamous epithelium
    d. a and b
    e. all of the above

20. Which of the following would increase the rate of gas exchange across the respiratory membrane?
    a. increase of the thickness of the respiratory membrane
    b. decrease of the surface area of the respiratory membrane
    c. increase of the partial pressure differences of gases across the respiratory membrane
    d. all of the above

491

_____ 21. Which gas diffuses most rapidly across the respiratory membrane?
a. carbon dioxide
b. oxygen

_____ 22. In which of the following sequences does P_{O_2} progressively decrease?
a. arterial blood, alveolar air, body tissues
b. body tissues, arterial blood, alveolar air
c. body tissues, alveolar air, arterial blood
d. alveolar air, arterial blood, body tissues

_____ 23. Oxygen is mostly transported in the blood
a. dissolved in plasma.
b. bound to blood proteins.
c. within bicarbonate ions.
d. bound to the heme portion of hemoglobin.

_____ 24. If the alveolar partial pressure of oxygen decreased to 77 mm Hg,
a. significantly less oxygen would be bound to hemoglobin.
b. the subject would rapidly die of asphyxiation.
c. approximately 50% of the hemoglobin would be saturated with oxygen.
d. nearly all of the hemoglobin would be saturated with oxygen.

_____ 25. The oxygen-hemoglobin dissociation curve is most adaptive if
a. it shifts to the right in the alveolar capillaries and to the left in the tissue capillaries.
b. it shifts to the left in the alveolar capillaries and to the right in the tissue capillaries.
c. it doesn't shift.

_____ 26. During exercise, the temperature of skeletal muscle tissue rises, pH of interstitial fluids falls, and carbon dioxide levels rise. This affects the oxygen-dissociation curve by shifting it to the _____ and is the _____.
a. left, Bohr effect
b. right, Bohr effect
c. left, reverse Bohr effect
d. right, reverse Bohr effect

_____ 27. Carbon dioxide is mostly transported in the blood
a. dissolved in plasma.
b. bound to blood proteins.
c. within bicarbonate ions.
d. bound to the globin portion of hemoglobin.

_____ 28. The partial pressure of carbon dioxide in the venous blood is
a. greater than in the tissue spaces.
b. less than in the tissue spaces.
c. less than in the alveoli.
d. less than in arterial blood.

_____ 29. The chloride shift
a. occurs primarily in alveolar capillaries.
b. occurs when chloride ions replace bicarbonate ions within red blood cells.
c. results in an increase in blood pH.
d. all of the above

_____ 30. If one could directly stimulate neurons within the inspiratory center, the frequency of action potentials within which of the following nerves would immediately increase?
a. intercostal nerves
b. phrenic nerves
c. vagus nerves
d. glossopharyngeal nerves
e. a and b

_____ 31. The pneumotaxic center
a. stimulates the inspiratory center.
b. inhibits the apneustic center.
c. is located in the medulla oblongata.
d. all of the above

_____ 32. The Hering-Breuer reflex
a. decreases inspiratory volume.
b. increases inspiratory volume.
c. occurs in response to changes in carbon dioxide levels in the blood.
d. a and b
e. b and c

_____ 33. The chemosensitive area
a. stimulates the respiratory center when blood carbon dioxide levels increase.
b. stimulates the respiratory center when blood pH increases.
c. is located in the pons.
d. all of the above

34. Blood oxygen levels
   a. are more important than carbon dioxide in the regulation of respiration.
   b. need to change only slightly to cause a change in respiration.
   c. are detected by sensory receptors in the cartoid and aortic bodies.
   d. all of the above

35. During exercise respiration rate and depth increases primarily due to
   a. increased blood carbon dioxide levels.
   b. decreased blood oxygen levels.
   c. decreased blood pH.
   d. input to the respiratory center from the cerebral motor cortex and from proprioceptors.

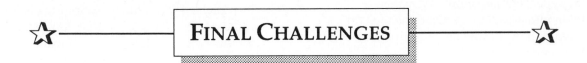

FINAL CHALLENGES

Use a separate sheet of paper to complete this section.

1. Marty Blowhard used a spirometer with the following results:
   a. After a normal inspiration, a normal expiration was 500 ml.
   b. Following a normal expiration, he was able to expel an additional 1000 ml.
   c. Taking as deep a breath as possible then forcefully exhaling all the air possible, yielded an output of 4500 ml.
   On the basis of these measurements, what is Marty's inspiratory reserve?

2. A patient has pneumonia, and fluid builds within the alveoli. Explain why this results in an increased rate of respiration that can be returned to normal with oxygen therapy.

3. A patient has severe emphysema that has extensively damaged the alveoli and reduced the surface area of the respiratory membrane. Although the patient is receiving oxygen therapy, he still has a tremendous urge to take a breath, i.e., he does not feel as if he is getting enough air. Why does this occur?

4. A patient has a bronchial tumor that greatly restricts airflow into his left lung. If the oxygen-hemoglobin dissociation curve of this patients's left lung were compared to the oxygen-hemoglobin dissociation curve of his right lung (which is normal), would it show a Bohr or a reverse Bohr effect? Explain why this would be advantageous or disadvantageous.

5. Consider the oxygen-hemoglobin dissociation curve graphed below:

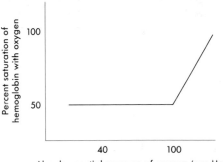

If an animal has such an oxygen-hemoglobin dissociation curve, would you expect to find it living at sea level or at high altitudes? Explain.

6. President Reagan was shot with a 22-caliber pistol in 1981. The bullet entered the thoracic cavity on his left side and lodged in his mediastinum. Explain why his respiratory depth increased before he arrived at the hospital (hint: the bullet would caused a pnuemothorax).

7. Patients with diabetes mellitus may occasionally take too little insulin. The result is the rapid metabolism of lipids and the accumulation of acidic by-products of lipid metabolism in the circulatory system. What effect would this have on respiration? Why is this change beneficial?

8. A hysterical patient is hyperventilating. The doctor makes the patient breathe into a paper bag. Since you are an especially astute physiology student you say to the doctor, "When the patient was hyperventilating he was blowing off carbon dioxide, and blood carbon dioxide levels decreased. When he breathed into the paper bag, carbon dioxided was trapped, and, as a result blood carbon dioxide levels increased. When blood carbon dioxide levels increase, the 'urge' to breathe should also increase. Therefore breathing into a paper bag should make the patient hyperventilate more. Why then is breathing into a paper bag recommended for hyperventilation?" How do you think the doctor would respond?

9. The effects of changes of arterial oxygen concentration on ventilation rates were determined under two different circumstances:
   1. Arterial carbon dioxide and pH levels were held at a constant value while arterial oxygen concentration was varied.
   2. Arterial carbon dioxide and pH levels were allowed to change while arterial oxygen concentration was varied.
   The results of these two experiments are graphed below:

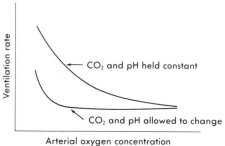

Arterial oxygen concentration

These data show that, when blood carbon dioxide and pH levels are allowed to flucuate as normally occurs in the body (2), arterial oxygen concentration exerts very little effect on ventilation rate until very low blood oxygen levels are reached. On the other hand, when blood carbon dioxide and pH levels are held at a constant level (1), arterial oxygen concentration can significantly affect ventilation rate. Propose an explanation.

10. Shown below is the blood pH of a runner following the start of a race:

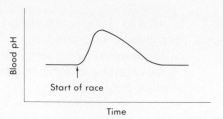

Two physiology students were puzzled over these results. They knew that ventilation rate would increase with exercise. Further, their instructor had told them that decreased pH was responsible for stimulating the respiratory centers and causing an increase in respiration rate. Propose an explanation that would account for the increased pH values following the start of the race (hint: don't say the instructor is wrong!).

ANSWERS TO CHAPTER 23

1. arytenoid
2. cricoid
3. thyroid
4. pleural, pleurisy
5. pneumotaxic, pneumothorax

6. pneumotaxic
7. inspiration, expiration
8. diaphragm
9. diaphragm
10. hemoglobin

CONTENT LEARNING ACTIVITY

Nose and Nasal Cavity

A. 1. Internal nares (choanae)
   2. Vestibule
   3. Soft palate
   4. Conchae
   5. Meatus

B. 1. Nasal hair
   2. Mucus
   3. Cilia
   4. Mucous membrane

Pharynx

A. 1. Nasopharynx
   2. Oropharynx
   3. Nasopharynx
   4. Oropharynx
   5. Fauces

B. 1. Conchae
   2. Vestibule
   3. External naris (nostril)
   4. Hard palate

5. Epiglottis
6. Larynx
7. Laryngopharynx
8. Oropharynx
9. Uvula
10. Soft palate
11. Nasopharynx
12. Internal naris (choana)
13. Meatus

Larynx

A. 1. Thyroid cartilage
   2. Epiglottis
   3. Vestibular folds
   4. True vocal cords
   5. Laryngitis

B. 1. Hyoid bone
   2. Thyroid cartilage
   3. Cricoid cartilage
   4. Tracheal cartilage
   5. Arytenoid cartilage
   6. Corniculate cartilage
   7. Cuniculate cartilage
   8. Epiglottis

Trachea, Bronchi and Lungs

A. 1. Lobules
   2. C-shaped cartilage rings
   3. Cartilage plates
   4. Pseudostratified ciliated columnar epithelium
   5. Asthma attack
   6. Alveoli
   7. Alveolar sac

B. 1. Trachea
   2. Primary bronchi
   3. Terminal bronchiole
   4. Respiratory bronchiole
   5. Alveolar duct
   6. Alveoli

Pleura

1. Pleural cavity
2. Parietal pleura

3. Pleural fluid

Blood Supply

1. Pulmonary arteries
2. Bronchial arteries

3. Pulmonary veins
4. Bronchial veins

Muscles of Respiration and Thoracic Wall

1. Diaphragm
2. Costal cartilages

3. External intercostal muscles

Pressure Differences and Air Flow

1. Increases
2. Increases
3. Increases

4. Decreases
5. Increases

Factors That Prevent the Lungs from Collapsing

1. Elastic recoil; surface tension of alveolar fluid
2. Surfactant; intrapleural pressure

3. Surfactant
4. Intrapleural pressure

Compliance of the Lungs and Thorax

1. Decrease
2. Decrease

3. Increase

Pulmonary Volumes and Capacities

1. Inspiratory reserve volume
2. Tidal volume
3. Expiratory reserve volume
4. Residual volume

5. Inspiratory capacity
6. Functional residual capacity
7. Vital capacity
8. Total lung capacity

Minute Respiratory Volume and Alveolar Ventilation Rate

1. Minute respiratory volume
2. Anatomical dead air space
3. Physiological dead air space
4. Alveolar ventilation rate

Partial Pressure

1. Dalton's law
2. Partial pressure
3. Vapor pressure

Partial Pressure of Respiratory Gases

1. Smaller
2. Greater
3. Greater

Diffusion of Gases Through Liquids

A. 1. Henry's law
2. Solubility coefficient

B. 1. Increases
2. Decreases
3. Decreases
4. Increases

Diffusion of Gases Through the Respiratory Membrane

1. Thin fluid layer
2. Alveolar epithelium
3. Alveolar epithelium basement membrane
4. Thin interstitial space
5. Capillary endothelium basement membrane
6. Capillary endothelium

Relationship Between Ventilation and Capillary Blood Flow

1. Inadequate cardiac output
2. Shunted blood

Oxygen and Carbon Dioxide Diffusion Gradients

1. 40 mm Hg
2. 45 mm Hg
3. 104 mm Hg
4. 40 mm Hg

Hemoglobin and Oxygen Transport

1. Saturated hemoglobin
2. Bohr effect
3. Reverse Bohr effect
4. Decreases
5. Decreases

Transport of Carbon Dioxide

A. 1. Bicarbonate ions
   2. Carbamino compounds (including carbaminohemoglobin)
   3. Dissolved in plasma

B. 1. Carbonic acid
   2. Carbonic anhydrase
   3. Bicarbonate ions
   4. Chloride ions
   5. Chloride shift
   6. Hemoglobin
   7. Buffer

Nervous Control of Rhythmic Ventilation

1. Inspiratory center
2. Expiratory center
3. Pneumotaxic center

4. Apneustic center
5. Hering-Breuer reflex

Chemical Control of Respiration

1. Chemosensitive area
2. Carotid and aortic bodies
3. Increase

4. Increase
5. Increase

The Effect of Exercise on Respiratory Movements

1. Increase
2. Increase

3. Increase

QUICK RECALL

1. Filtered, warmed, and moistened
2. Trachea, primary bronchi, secondary bronchi, tertiary bronchi, bronchioles, terminal bronchioles, respiratory bronchioles, alveolar duct, alveoli
3. Increasing volume decreases pressure, and decreasing volume increases pressure.
4. Recoil of elastic fibers and water surface tension cause the lungs to collapse; surfactant and intrapleural pressure prevent the collapse of the lungs.
5. The pulmonary volumes are tidal volume, inspiratory reserve volume, expiratory reserve volume, and residual volume. Pulmonary capacities are the sum of two or more pulmonary volumes.
6. Alveolar ventilation rate = respiratory rate times the difference between the tidal volume and the dead air space. AVR = RR(TV - DAS)
7. Thin layer of fluid, alveolar epithelium, alveolar epithelium basement membrane, interstitial space, basement membrane of capillary endothelium, and capillary endothelium
8. Thickness of membrane, diffusion coefficient, surface area of membrane, and partial pressure gradient
9. Hemoglobin 97%, dissolved in plasma 3%
10. Bicarbonate ions 72%, blood proteins 20%, and dissolved in plasma 8%
11. Carbon dioxide and water combine to form carbonic acid, which dissociates into hydrogen ions and bicarbonate ions.
12. Carbon dioxide and pH: chemosensitive area of medulla; stimulated with increasing carbon dioxide and decreasing pH
    Oxygen: carotid and aortic bodies; stimulated with greatly decreased oxygen

1. A. The pseudostratified ciliated epithelium of the nasal cavity catches debris in the air and moves it to the pharynx (through the internal nares). The vestibule is the most anterior portion of the nasal cavity and is lined with stratified squamous epithelium. The olfactory epithelium is located in the most superior part of the nasal cavity. The conchae are bony ridges that form passageways called meatus.

2. B. The paryngeal tonsils (adenoids) are located in the nasopharynx. The auditory tubes open into the nasopharynx (the paranasal sinuses open into the nasal cavity). The oropharynx opens into the oral cavity through the fauces and extends to the tip of the epiglottus.

3. C. The larynx contains the vocal cords. There are three unpaired cartilages (thyroid cartilage, cricoid cartilage, and epiglottis) and six paired cartilages (arytenoid, corniculate, and cuneiform cartilages). The larynx is located between the laryngopharynx and the trachea.

4. C. The trachea is held open by C-shaped cartilage rings. The ends of the rings are held together by smooth muscle, which can alter the diameter of the trachea by contracting.

5. A. Gas exchange takes place mainly in the alveoli and to a lesser extent in the respiratory bronchioles.

6. C. The terminal bronchioles have smooth muscle and no cartilage in their walls. Contraction of the smooth muscle, which can occur during an asthma attack, can impede air flow.

7. B. The parietal pleura covers the inner surface of the right and left pleural cavities. It is continuous at the hilum with the visceral pleura, which covers the surface of the lung. The space between the parietal and visceral pleura is a potential space and contains a small amount of fluid to reduce friction.

8. A. The diaphragm and the external intercostals increase thoracic volume, whereas the internal intercostals and rectus abdominis decrease thoracic volume.

9. D. During inspiration the intrapleural and intra pulmonary volumes increase. The result is that the pressure in those areas decrease ($P=RT/V$). Thus a pressure gradient is established between the atmospheric pressure and the pressure in the intrapulmonic space, and air flows into the lungs (flow = pressure difference/resistance).

10. B. The relationship that describes pressure changes within the alveoli is $P = nRT/V$. During expiration the volume of the thorax and the lungs decreases. The pressure within the alveoli therefore increases during expiration. Gas then flows out of the lungs because the pressure increases above atmospheric pressure.

11. C. Flow is directly proportional to the pressure gradient and inversely proportional to the resistance. Contraction of the bronchiolar smooth muscle increases the resistance to gas flow and therefore, for a given pressure gradient, decreases gas flow.

12. D. Surfactant, which reduces the surface tension of the film of water that lines the alveoli, and intraplural pressure prevent the lungs from collapsing. Elastic recoil caused by elastic fibers in the alveoli causes the lungs to collapse.

13. A. Since the pressure within the pleural cavity is less than atmospheric pressure, a hole as described would allow air to flow into the pleural cavity. This condition is called pneumothorax and the result is collapse of the lung.

14. D. Compliance is a measure of the expansibility of the lungs. The lungs are more expansible in emphysema due to the destruction of elastic tissue. Compliance is decreased when the lungs are less expansible, as when surfactant levels are too low.

15. B. The expiratory reserve is the amount of air that can be forcefully expired after expiration of normal tidal volume.

16. C. Vital capacity is the sum of the inspiratory reserve plus the tidal volume plus the expiratory reserve.

17. C. The alveolar ventilation rate is the respiratory rate times the tidal volume minus the dead air space. The minute respiratory volume is the tidal volume times the respiratory rate, i.e., the amount of air moved into and out of the respiratory system each minute.

18. D. The partial pressure of a gas can be determined by multiplying the percent composition of the gas by the total pressure. In this case, 760 mm Hg x 0.20 = 152 mm Hg.

19. E. Gases within the alveolus must cross the respiratory membrane: the fluid that lines the alveolus, the epithelial wall of the alveolus, a basement membrane, a thin interstitial space, a basement membrane, and the endothelium of the alveolar capillary. The alveolar wall and the endothelium are simple squamous epithelium.

20. C. The greater the difference in partial pressure of a gas across the respiratory membrane, the greater the rate of gas exchange. Increasing the thickness or decreasing the surface area of the respiratory membrane decreases the rate of gas exchange.

21. A. Carbon dioxide has a diffusion coefficient that is 20 times greater than that of oxygen.

22. D. The partial pressure of O_2 decreases as it moves from the alveoli to the arterial blood and from the arterial blood to the body tissues. These partial pressure gradients are responsible for the diffusion of oxygen.

23. D. About 97% of the oxygen transported in blood is bound to the heme portion of hemoglobin inside red blood cells. About 3% is transported dissolved in plasma.

24. D. Review the oxygen-hemoglobin dissociation curve. The partial pressure of oxygen must decrease below about 70 mm Hg before the hemoglobin becomes significantly less than 100% saturated with oxygen.

25. B. A shift of the oxygen-hemoglobin dissociation curve to the left (reverse Bohr effect) results in an increased tendency of hemoglobin to become saturated with oxygen at a lower Po_2. A shift to the right (Bohr effect) results in a decreased tendency of oxygen to become saturated with oxygen at a lower Po_2 or increases the likelihood that hemoglobin will release its oxygen. These tendencies are adaptive because they result in increased oxygen bound to hemoglobin in the lung and an increased tendency for hemoglobin to release oxygen bound to it in the tissue capillaries. Therefore a greater amount of oxygen is made available to cells.

26. B. A shift of the oxygen dissociation curve to the right is the Bohr effect, and a shift to the left is a reverse Bohr effect.

27. C. About 72% of the carbon dioxide in blood is transported in the form of bicarbonate ions, 20% in combination with blood proteins (mostly the globin portion of hemoglobin), and 8% is dissolved in the plasma.

28. B. The partial pressure of carbon dioxide is normally greatest in the tissue spaces, less in the venous blood, and still less in the alveoli. Carbon dioxide moves by diffusion from the tissue spaces into the capillaries and finally from the alveolar capillaries into the alveoli. Diffusion occurs from areas of high to low partial pressure of carbon dixoide.

29. B. The chloride shift occurs when bicarbonate ions diffuse out of red blood cells and chloride ions diffuse into red blood cells. This normally takes place within tissues (i.e. not alveolar capillaries where the reverse chloride shift occurs). At the same time, the movement of hydrogen ions out of the red blood cell can decrease blood pH.

30. E. Nerves that innervate muscles of inspiration would be stimulated if the inspiratory center were stimulated. Of those listed, the intercostal and phrenic nerves would carry a greater frequency of action potentials to stimulate the external intercostal muscles and diaphragm, respectively.

500

31. B. The pneumotaxic center inhibits the apneustic and inspiratory centers. The pneumotaxic center is located in the pons.

32. A. The Hering-Breuer reflex decreases inspiratory volume. It occurs in response to stretch of the lungs during inspiration.

33. A. The chemosensitive area stimulates the respiratory center (resulting in increased rate and depth of respiration) when blood carbon dioxide levels increase or blood pH decreases. The chemosensitive area is located in the medulla oblongata.

34. C. Blood oxygen levels are detected by sensory receptors in the carotid and aortic bodies. However, the arterial partial pressure of oxygen must decrease about 50% before a large stimulatory effect on respiration results. Carbon dioxide (and pH) are normally much more important than oxygen in regulating respiration.

35. D. Normally there is little change in blood carbon dioxide, oxygen, and pH levels during exercise; i.e., homeostasis is maintained. Increased respiration is due to impulses from the cerebral motor cortex and proprioceptors.

 FINAL CHALLENGES

1. The inspiratory reserve is 3000 ml. It is equal to the vital capacity minus the sum of the tidal volume and the expiratory reserve, i.e., 4500 - (500 + 1000) = 3000.

2. The build-up of fluid increases the thickness of the respiratory membrane. This reduces the exchange of gases between the alveolar air and the blood. However, since carbon dioxide diffuses across the respiratory membrane 20 times more rapidly than oxygen, only the levels of oxygen are significantly affected. The decreased oxygen levels stimulate the increased respiration. With oxygen therapy, blood oxygen levels are restored due to the increased partial pressure gradient for oxygen that results when the concentration of oxygen in alveolar air increases.

3. The lungs have been sufficiently damaged that there is inadequate diffusion of oxygen and carbon dioxide. The oxygen therapy only addresses the oxygen problem. With a buildup of carbon dioxide in the blood, the patient feels like he needs to breath more, even though he is getting enough oxygen.

4. In a patient with such a bronchial tumor there would be a buildup of carbon dioxide in the left lung due to inadequate ventilation. Consequently, there would be a Bohr effect (shift of the oxygen-hemoglobin dissociation curve to the right). This would be disadvantageous because the blood would pick up less oxygen in the lungs than normal.

5. The normal oxygen-hemoglobin dissociation curve indicates that the alveolar partial pressure of oxygen can decrease to about 70 mm Hg before hemoglobin becomes significantly less than 100% saturated with oxygen. The curve presented indicates that even a slight decrease in alveolar partial pressure of oxygen would result in far less than 100% of the hemoglobin being saturated with oxygen. Since the total atmospheric pressure decreases with altitude, the partial pressure of oxygen within the alveoli also decreases with altitude. Therefore the likelihood of surviving at high altitudes is not good and one would expect to find the animal at sea level.

6. The bullet caused a pneumothorax, and the left lung collapsed. Since the lung was collapsed, the stretch receptors in his left lung were not stimulated during inspiration. This interruption of the Hering-Breuer reflex on the left side caused his respiratory depth to increase. Because the other lung wasn't damaged, he probably was not suffering from lack of oxygen or increased carbon dioxide levels.

7. The acidic by-products of lipid metabolism cause a decrease in blood pH. The decreased pH stimulates the chemosensitive area of the medulla, resulting in an increased rate and depth of respiration. Consequently, carbon dioxide is lost from the blood at a faster rate, and pH increases. The changes in respiration help to correct the acid increase produced by the lipid by-products.

8. The cause of hyperventilation is due to stimulation of the respiratory centers by higher areas of the brain (due to the hysteria). The results are decreased blood carbon dioxide levels, inhibition of the vasomotor center, decreased peripheral resistance, decreased blood pressure, inadequate delivery of blood to the brain, and dizziness or a faint feeling (see Chapter 21). The dizziness could further contribute to the hysteria. By breathing into the paper bag blood carbon dioxide levels would increase and help to correct the feeling of dizziness. Concentrating on breathing might also take the patient's mind off the emotional stimulus that caused the hyperventilation. It is not likely that raising blood carbon dioxide levels from below normal to normal is going to significantly affect respiration rate until the influence of higher brain centers returns to normal.

9. This phenomena is known as the braking effect of carbon dioxide and pH on the regulation of respiration. As oxygen decreases, through the chemoreceptor reflexes respiration rate will increase. However, this results in a "blow off" of carbon dioxide and an increase in blood pH. Lowered blood carbon dioxide levels and pH both inhibit the respiratory center and thus oppose an increase in respiration rate. Therefore, when carbon dioxide and pH are allowed to fluctuate, oxygen levels are much less potent in affecting changes in respiration rate.

10. At the start of the race respiration rate initially increased more than was necessary to meet the requirements of exercise. Thus, hyperventilation occurred and this removed carbon dioxide from the blood and pH increased. The increased pH quickly inhibits the respiratory center and brings respiration rates back down to the appropriate level, and pH levels decrease. The increase in respiration rate when exercise is initiated seems to be controlled by cerebral motor areas and by proprioceptors.

Digestive System

FOCUS: The digestive system is specialized to ingest food, propel the food through the gastrointestinal tract, and digest and absorb the food. In the oral cavity the teeth mechanically break up the food and salivary amylase from the salivary glands begins the digestion of carbohydrates. The bolus of food formed is swallowed (partially a voluntary and partially a reflex activity) and passes through the esophagus to the stomach. The stomach stores the food, begins protein digestion (pepsin), and converts the food into chyme. The chyme enters the small intestine, where bile from the liver and enzymes from the pancreas and the lining of the small intestine complete the digestive process. Most of the products of digestion are absorbed in the small intestine. In the large intestine water and vitamins produced by intestinal bacteria are absorbed, and feces is formed. The feces is eliminated by means of the defecation reflex. Regulation of the digestive tract is controlled through the nervous system (local reflexes and central nervous system reflexes) and hormones.

WORD PARTS

Give an example of a new vocabulary word that contains each word part.

WORD PART	MEANING	EXAMPLE
sulc-	a furrow or groove	1. _____
bucc-	the cheek	2. _____
gingiv-	the gums	3. _____
api-	the tip	4. _____
viscos-	sticky	5. _____
cata-	downward	6. _____
lys-	loosening	7. _____

WORD PART	MEANING	EXAMPLE
mastic-	chew	8. _____
parot-	beside the ear	9. _____
hepato-	liver	10. _____

CONTENT LEARNING ACTIVITY

General Structures

❝*The intestinal tube consists of four layers or tunics.*❞

A. Match these terms with the correct statement or definition:

Mucosa
Muscularis
Serosa (adventitia)
Submucosa

_____ 1. Connective tissue tunic that forms the outermost layer of the digestive tract.

_____ 2. Tunic that consists of two layers of smooth muscle, located between the submucosa and serosa tunics.

_____ 3. Thick connective tissue tunic between the mucosa and the muscularis tunics.

B. Match these parts of the tunics with the correct statement or definition:

Intramural plexus
Lamina propria
Mucous epithelium
Muscularis mucosa
Myenteric plexus
Submucosal plexus

_____ 1. Layer of mucosa composed of squamous or columnar epithelial cells.

_____ 2. Layer of mucosa composed of loose, irregular connective tissue.

_____ 3. Layer of mucosa composed of a thin, smooth muscle layer.

_____ 4. Nerve plexus consisting of nerve fibers and parasympathetic cell bodies that is located between smooth muscle layers.

_____ 5. Collective name for both the submucosal and myenteric plexuses.

Oral Cavity

" *The first section of the digestive tract is the oral cavity.* **"**

A. **M**atch these terms with the correct statement or definition:

Buccal fat pad   Lingual tonsil
Buccinator muscle   Mastication
Extrinsic muscles   Speech
Fauces   Terminal sulcus
Frenulum   Vestibule
Intrinsic muscles

_____ 1. Posterior boundary of the oral cavity; the opening into the pharynx.

_____ 2. Space between the lips or cheeks and the alveolar processes.

_____ 3. Structure that rounds out the profile of the side of the face.

_____ 4. Two important functions of the lips, cheeks, and tongue.

_____

_____ 5. Thin fold of tissue that attaches the tongue to the floor of the mouth.

_____ 6. Muscles attached to the tongue; responsible for moving the tongue (e.g. protrusion and retraction).

_____ 7. Groove that divides the tongue into two portions.

_____ 8. Large aggregation of lymphatic tissue on the posterior one third of the tongue.

B. **M**atch these numbers with the correct description:

1   3
2

_____ 1. Number of incisors in each quadrant of the mouth.

_____ 2. Number of canines in each quadrant of the mouth.

_____ 3. Number of premolars in each quadrant of the mouth.

_____ 4. Number of molars in each quadrant of the mouth.

C. **M**atch these terms with the correct statement or definition:

Alveoli   Periodontal membrane
Gingiva (gums)   Primary teeth
Periodontal ligaments   Secondary teeth

_____ 1. Deciduous teeth; also called milk teeth.

_____ 2. Sockets containing the teeth.

_____ 3. Dense fibrous connective tissue and striated squamous epithelium that covers the alveolar ridges.

_____ 4. Connective tissue that holds the teeth in the alveoli.

_____ 5. Membrane that lines the alveolar walls.

D. Match these terms with the correct statement or definition:

Apical foramen   Neck
Cementum   Pulp
Crown   Pulp cavity
Dentin   Root
Enamel   Root canal

_____ 1. Cutting or chewing surface of the tooth, with one or more cusps (points).

_____ 2. Blood vessels, nerves, and connective tissue in the center of the tooth.

_____ 3. Pulp cavity within the root of the tooth.

_____ 4. Hole at the point of each root through which nerves and blood vessels enter and exit.

_____ 5. Living, cellular, calcified tissue that surrounds the pulp cavity of the tooth.

_____ 6. Extremely hard, nonliving, acellular substance that covers the dentin and protects the tooth from abrasion and acids produced by bacteria.

_____ 7. Substance that covers the root, and helps anchor the tooth in the jaw.

E. Match these terms with correct parts of the diagram labeled in Figure 24-1:

Apical foramen
Cementum
Crown
Dentin
Enamel
Neck
Periodontal ligaments
Pulp cavity
Root
Root canal

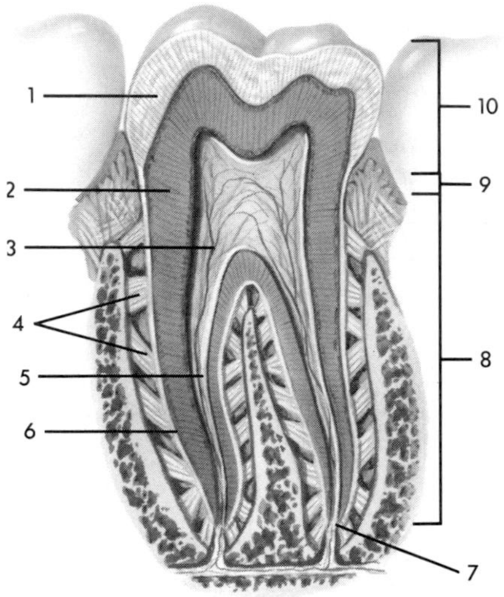

Figure 24-1

1. _____

2. _____

3. _____

4. _____

5. _____

6. _____

7. _____

8. _____

9. _____

10. _____

F. Match these terms with the correct statement or definition:

Hard palate   Soft palate
Muscles of mastication
Palatine tonsils   Uvula

_____ 1. Responsible for movement of the mandible.

_____ 2. Anterior bony structure that separates the nasal and oral cavities.

_____ 3. Posterior, skeletal muscle and connective tissue structure that separates the nasal and oral cavities.

_____ 4. Projection from the posterior edge of the soft palate.

_____ 5. Collection of lymphoid tissue, located in the lateral wall of the fauces.

☞ The palate consists of the hard palate and the soft palate. The palate prevents food from entering the nasal cavity.

G. Match these terms with the correct statement or definition:

Parotid glands   Submandibular glands
Sublingual glands

_____ 1. Largest salivary glands, located just anterior to the ear.

_____ 2. Salivary glands that become infected with mumps.

_____ 3. Salivary glands located along the inferior border of the posterior half of the mandible.

_____ 4. Salivary glands located immediately below the mucous membrane of the floor of the mouth.

☞ Numerous small glands in the lining of the oral cavity and tongue produce secretions that help to keep the oral cavity moist.

Pharynx and Esophagus

"_In the digestive tract, the pharynx communicates anteriorly with the oral cavity and posteriorly with the esophagus._**"**

Match these terms with the correct statement or definition:

Esophagus   Nasopharynx
Laryngopharynx   Oropharynx

_____ 1. Portions of the pharynx that transmit food.

_____ 2. Portion of the pharynx superior to the oropharynx.

_____ 3. Portion of the digestive tract that extends between the pharynx and stomach.

☞ An upper esophageal sphincter and lower esophageal sphincter regulate movement of materials into and out of the esophagus.

Stomach

“The stomach is an enlarged segment of the digestive tract in the left superior margin of the abdomen.”

A. Match these terms with the correct statement or definition:

Muscular layer   Rugae
Pyloric sphincter

_____ 1. Relatively thick ring of smooth muscle that surrounds the opening between the stomach and the small intestine.

_____ 2. Part of the stomach wall that consists of longitudinal, circular, and oblique layers of smooth muscle.

_____ 3. Large folds formed from the mucosal and submucosal layers when the stomach is empty.

B. Match these terms with correct parts of the diagram labeled in in Figure 24-2:

Body   Pyloric opening
Cardiac region   Pyloric sphincter
Fundus   Rugae
Gastroesophageal (cardiac)
    opening

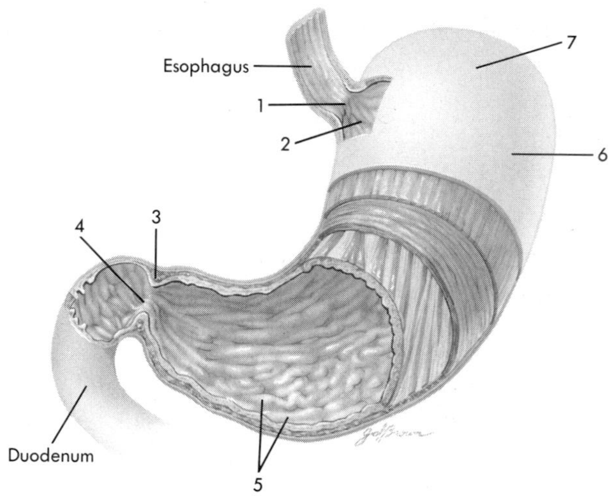

1. _____

2. _____

3. _____

4. _____

5. _____

6. _____

7. _____

Figure 24-2

C. Match these terms with the correct statement or definition:

Chief (zymogenic) cells   Mucous neck cells
Endocrine cells   Parietal (oxyntic) cells
Gastric glands   Surface mucous cells
Gastric pits

_____ 1. Tubelike openings in the mucosal surface of the stomach.

_____ 2. Glands in the stomach that open into the gastric pits.

_____ 3. Epithelial cells on the surface of the mucosa and lining the gastric pits that produce mucus.

_____ 4. Epithelial cells in the gastric glands that produce hydrochloric acid and intrinsic factor.

_____ 5. Epithelial cells in the gastric glands that produce pepsinogen.

Small Intestine

"The small intestine consists of three portions: the duodenum, the jejunum, and the ileum."

A. Match these terms with the correct statement or definition:

Greater duodenal papilla   Lesser duodenal papilla
Hepatopancreatic ampullar
 sphincter

_____ 1. Large mound where the hepatopancreatic ampulla empties into the duodenum.

_____ 2. Small mound where the accessory pancreatic duct opens into the duodenum.

_____ 3. Smooth muscle that usually keeps the hepatopancreatic ampulla closed.

B. Match these terms with the correct statement or definition:

Absorptive cells   Lacteals
Duodenal glands   Microvilli
Endocrine cells   Plicae circulares (circular folds)
Goblet cells   Villi
Intestinal glands

_____ 1. Perpendicular folds formed from the mucosa and submucosa that run perpendicular to the long axis of the duodenum.

_____ 2. Tiny, fingerlike projections of the mucosa of the duodenum.

_____ 3. Cytoplasmic extensions of villi.

_____ 4. Lymph capillary found in a villus.

_____ 5. Simple columnar epithelial cells in the duodenum that are specialized to produce digestive enzymes and absorb food.

_____ 6. Simple columnar epithelial cells in the duodenum that produce a protective mucus.

_____ 7. Tubular invaginations of the mucosa at the base of the villi.

_____ 8. Mucous glands located in the submucosa of the duodenum.

☞ The plicae circulares, villi, and microvilli function to increase surface area in the small intestine.

C. Match these terms with the correct statement or definition:

Ileocecal sphincter   Jejunum
Ileocecal valve   Peyer's patches
Ileum

_____ 1. Two parts of the small intestine that are the major sites of nutrient absorption.

_____

_____ 2. Ring of smooth muscle located at the junction of the ileum and the large intestine.

_____ 3. One-way valve at the junction of the ileum and the large intestine.

_____ 4. Aggregations of lymph nodules in the ileum.

Liver and Gallbladder

66 *The liver is the largest internal organ of the body.* **99**

A. Match these terms with the correct statement or definition:

Bare area   Cystic duct
Caudate and quadrate   Gallbladder
Common bile duct   Left and right
Common hepatic duct   Porta

_____ 1. Two minor lobes of the liver.

_____ 2. Location on the inferior surface of the liver where various vessels, ducts, and nerves enter and exit the liver.

_____ 3. Small sac on the inferior surface of the liver.

_____ 4. Duct formed from the junction of the right and left hepatic ducts.

_____ 5. Duct from the gallbladder.

_____ 6. Duct formed from the junction of the common hepatic duct and the cystic duct.

☞ The wall of the gallbladder has rugae, which allow gallbladder expansion, and smooth muscle, which enables the gallbladder to contract.

B. Match these terms with the correct statement or definition:

Endothelial and phagocytic cells   Hepatocytes
Hepatic cords   Lobules
Hepatic sinusoids   Portal triads

_____ 1. Portions of the liver divided by connective tissue septa.

_____ 2. Corner of a liver lobule where three vessels are commonly located.

_____ 3. Functional cells of the liver; produce bile.

_____ 4. Structures located between the central vein and the septa of each lobule; consist of hepatocytes.

_____ 5. Blood channels that separate the hepatic cords.

_____ 6. Cells that line the liver sinusoids.

C. Using the terms provided, complete the following statements:

1. _____

Bile canaliculus   Hepatic portal vein
Central vein   Hepatic veins
Hepatic ducts   Hepatocytes

2. _____

3. _____

In the liver, blood from the _(1)_ and the hepatic artery flows into
the sinusoids and becomes mixed. The mixed blood then flows to
the _(2)_ , where it exits the lobule and then exits the liver through
the _(3)_ . Bile, which is produced by the hepatocytes, flows
through the _(4)_ toward the triad and exits the liver through
the _(5)_ .

4. _____

5. _____

Pancreas

66 *The pancreas is a complex organ composed of both endocrine and exocrine tissues that perform several functions.* **99**

A. Using the terms provided, complete the following statements:

1. _____

Acini   Main pancreatic duct
Head   Pancreatic islets (Islets of
Intercalated ducts   Langerhans)
Interlobular ducts   Tail
Intralobular ducts

2. _____

3. _____

4. _____

The pancreas consists of a _(1)_ located within the curvature of
the duodenum, a body, and a _(2)_ , which extends to the spleen.
The exocrine portion of the pancreas consists of _(3)_ , which
produce digestive enzymes. Clusters of acini are connected by
small _(4)_ to _(5)_ , which leave the lobules to join _(6)_ between
the lobules. The interlobular ducts attach to the _(7)_ , which
joins the common hepatic duct at the hepatopancreatic ampulla.
Insulin and glucagon are produced by cells within the _(8)_ .

5. _____

6. _____

7. _____

8. _____

B. Match these terms with the   Common bile duct   Hepatic duct
correct parts of the diagram   Common hepatic duct   Hepatopancreatic duct
labeled in Figure 24-3:   Cystic duct   Pancreatic duct

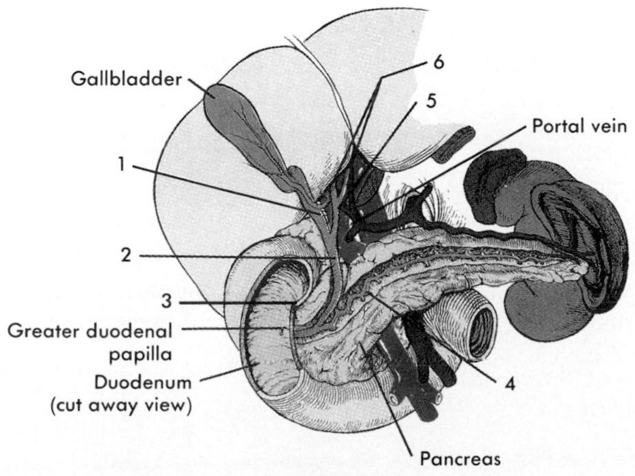

1. _____

2. _____

3. _____

4. _____

5. _____

6. _____

Figure 24-3

511

Large Intestine

"The large intestine extends from the ileocecal junction to the inferior end of the rectum."

Match these terms with the
correct statement or definition:

Anal canal		Haustra
Anus		Rectum
Ascending colon		Sigmoid colon
Cecum		Taenia coli
Crypts		Transverse colon
Descending colon		Vermiform appendix
Epiploic appendages		

_____ 1. Proximal end of the large intestine that extends inferiorly past the ileocecal junction as a blind sac.

_____ 2. Small blind tube that is attached to the cecum.

_____ 3. Portion of the colon that extends from the right colic flexure to the left colic flexure.

_____ 4. Portion of the colon that forms an S-shaped tube that ends at the rectum.

_____ 5. Straight tubular glands located in the epithelium of the large intestine.

_____ 6. Three bands of longitudinal smooth muscle that run the length of the colon.

_____ 7. Pouches formed in the colon when the taenia coli contract.

_____ 8. Small, fat-filled connective tissue pouches attached to the outer surface of the colon.

_____ 9. Straight, muscular tube that begins at the termination of the sigmoid colon and ends at the anal canal.

_____ 10. Last 2 to 3 cm of the digestive tube.

☞ Smooth muscle forms the internal anal sphincter at the superior end of the anal canal, and skeletal muscle forms the external anal sphincter at the inferior end of the anal canal.

Peritoneum

"The body walls and organs of the abdominal cavity are lined with serous membranes."

Match these terms with the
correct statement or definition:

Coronary ligament		Omental bursa
Falciform ligament		Parietal peritoneum
Greater omentum		Retroperitoneal
Lesser omentum		Transverse and Sigmoid
Mesenteries		mesocolon
Mesentery proper		Visceral peritoneum

_____   1. Serous membrane that covers the abdominal organs.

_____   2. General term for connective tissue sheets that hold abdominal organs in place.

_____   3. Abdominal organs that lie against the abdominal wall and have no mesenteries.

_____   4. Mesentery connecting the lesser curvature of the stomach to the liver and diaphragm.

_____   5. Cavity or pocket formed in the greater omentum as it extends inferiorly.

_____   6. Mesentery that attaches the liver to the anterior abdominal wall and divides the liver into right and left halves.

_____   7. Mesenteries of the small intestine.

_____   8. Mesenteries of the colon.

Functions of the Digestive System

66_Each segment of the digestive tract is specialized to assist in moving its contents from the oral end to the anal end._**99**

Using the terms provided, complete the following statements:

1. _____

Effector organ   Local reflex
Intramural plexus   Receptors

2. _____

3. _____

The processes of secretion, movement, and absorption are regulated by elaborate nervous and hormonal mechanisms. However, the digestive tract does have a unique regulatory mechanism, called the _(1)_ , which does not involve the spinal cord or brain. Stimuli (e.g., distension of the digestive tract) activate _(2)_ within the wall of the digestive tract, and action potentials are generated in the neurons of the _(3)_ . The action potentials travel up or down the intramural plexus and produce a response in a(n) _(4)_ (e.g. in smooth muscle or a gland).

4. _____

Oral Cavity

66_Food is masticated in the mouth, as secretions are added to the food._**99**

A. Match these terms with the   Amylase   Mucin
   correct statement or definition: Chewing (mastication) reflex Salivary amylase

_____   1. Reflex integrated in the medulla oblongata that controls chewing.

_____   2. Digestive enzyme in the serous portion of saliva.

_____   3. Proteoglycan secreted by the submandibular and sublingual glands that adds lubrication to the saliva.

B. Using the terms provided, complete the following statements:

Antibacterial   Premolars and molars
Incisors and canines   Starch
Parasympathetic

1. _____

2. _____

3. _____

4. _____

5. _____

Food taken into the mouth is chewed, or masticated, by the teeth. The most anterior teeth, the _(1)_ primarily cut and tear food, whereas the _(2)_ primarily crush and grind the food. The amylase in saliva starts the digestive process by breaking the covalent bonds between glucose molecules in _(3)_ . In addition, saliva prevents bacterial infection in the mouth by washing the oral cavity, and it contains substances (e.g., lysozyme) with weak _(4)_ action. Salivary gland secretion is stimulated mainly by the _(5)_ nervous system, but also by the sympathetic nervous system. Tactile stimulation, certain tastes, and higher brain centers also affect the activity of the salivary glands.

Deglutition

66 *Deglutition, or swallowing, can be divided into three different phases.* 99

A. Match these phases of deglutition with the correct statement or definition:

Esophageal phase
Pharyngeal phase
Voluntary phase

_____

1. Phase of swallowing that involves forcing a bolus of food into the oropharynx.

_____

2. Phase of swallowing that involves closing the nasopharynx, forcing food through the pharynx, and occluding the glottis.

_____

3. Phase of swallowing that is responsible for moving food from the pharynx to the stomach.

B. Match these terms with the correct statement or definition:

Epiglottis   Pharyngeal constrictor muscles
Peristaltic waves   Swallowing center

_____

1. Area in the medulla oblongata that controls swallowing.

_____

2. Muscles that contract and force food through the pharynx.

_____

3. Part of the larynx that covers the opening into the larynx.

_____

4. Muscular contractions of the esophagus.

☞ The presence of food in the esophagus stimulates the intramural plexus, which initiates the peristaltic waves.

514

Stomach

❝_The stomach functions primarily as a storage and mixing chamber for ingested food._**❞**

A. Match these terms with the correct statement or definition:

Gastrin
Hydrochloric acid
Intrinsic factor

Mucus
Pepsin
Pepsinogen

_____ 1. Viscous, alkaline substance that covers the surface of epithelial cells.

_____ 2. Glycoprotein secreted by parietal cells that binds with vitamin B12 and makes it more readily absorbed in the ileum.

_____ 3. Stomach enzyme that catalyzes the cleavage of some covalent bonds in proteins.

_____ 4. Inactive form of pepsin, secreted by chief cells.

_____ 5. Hormone that increases gastric secretion and increases stomach emptying.

B. Match these phases of stomach secretion with the correct statement:

Cephalic phase
Gastric phase
Intestinal phase

_____ 1. Phase of stomach secretion that responds to taste, smell, and the sensations of chewing and swallowing.

_____ 2. Phase of gastric secretion that is stimulated by food in the stomach.

_____ 3. Phase of gastric secretion that is stimulated by the entrance of acidic chyme into the duodenum.

C. Using the terms provided, complete the following statements:

Decrease

Increase

Several mechanisms regulate gastric secretions. Through the medulla the smell, taste, or thought of food can _(1)_ stimulation of parietal, chief cells and endocrine cells. The endocrine cells secrete gastrin, which travels in the blood to parietal cells and causes a(n) _(2)_ in hydrochloric acid secretion. In the stomach, distension and the presence of amino acids activates CNS and local myenteric reflexes, that _(3)_ gastric secretions. In the duodenum, distension and increased acidic chyme can activate the gastroenteric reflex and cause a _(4)_ in gastric secretions. Increased acidity in the duodenum also stimulates the secretion of secretin, which acts to _(5)_ gastric acid secretion, whereas fatty acids in the duodenum stimulate the secretion of cholecystokinin and gastric inhibitory peptide, which function to _(6)_ gastric secretions.

1. _____

2. _____

3. _____

4. _____

5. _____

6. _____

 Motility in the stomach is regulated by many of the hormonal and neural mechansims that stimulate stomach secretions. For instance, distension of the stomach stimulates local reflexes, central nervous system reflexes, and the release of gastrin, all of which increase stomach motility and relaxation of the pyloric sphincter.

515

D. Match these terms with the correct statement or definition:

Chyme
Mixing waves

Peristaltic waves
Pyloric pump

_____ 1. Semifluid material formed from ingested food mixed with stomach gland secretions.

_____ 2. Strong waves of contraction that force chyme toward the pyloric sphincter.

_____ 3. Movement of chyme through the partially closed pylorus by the force of peristaltic contraction.

Small Intestine

66 *The small intestine is the major area of digestion and absorption in the digestive tract.* 99

A. Match these terms with the correct description:

Disaccharidases
Mucus

Nucleases
Peptidases

_____ 1. Secreted by duodenal glands and goblet cells.

_____ 2. Enzymes on the intestinal microvilli that break down disaccharides to monosaccharides.

_____ 3. Enzymes on the intestinal microvilli that break peptide bonds between amino acid chains.

_____ 4. Enzymes on the intestinal microvilli that break down nucleic acids.

☞ Duodenal gland secretion is stimulated by vagal stimulation, secretin, and chemical or tactile irritation of the duodenal mucosa. Goblet cells produce mucus in response to tactile and chemical stimulation of the mucosa.

B. Match these terms with the correct statement or definition:

Decreases
Increases

Peristaltic contractions
Segmental contractions

_____ 1. Propagated only for short distances and mix the intestinal contents.

_____ 2. Effect of small intestine distension on intestinal smooth muscle contraction.

_____ 3. Effect of low pH, amino acids and peptides on small intestine contraction.

_____ 4. Effect of parasympathetic stimulation on small intestine contraction.

_____ 5. Effect of cecal distension on constriction of the ileocecal sphincter.

Liver and Gallbladder

66 *The liver produces bile, whereas the gallbladder stores and concentrates bile.* **99**

A. Match these terms with the correct statement:

Inhibits
Stimulates

_____ 1. Effect of secretin on bile secretion.

_____ 2. Effect of vagal stimulation or increased blood flow through the liver on bile secretion.

_____ 3. Effect of cholecystokinin on contraction of the gall bladder.

_____ 4. Effect of vagal stimulation on contraction of the gall bladder.

_____ 5. Effect of cholecystokinin on the hepatopancreatic ampullar sphincter.

☞ Most bile salts are reabsorbed in the ileum and are carried back to the liver in the blood, where they stimulate further bile secretion.

B. Using the terms provided, complete the following statements:

Bile pigments	Interconversion
Bile salts	Phospholipids
Blood proteins	Store
Detoxify	Transformed
Glycogen	Urea

Although bile does not contain digestive enzymes, it does have _(1)_ which emulsify fats. Bile also contains excretory products such as _(2)_ and cholesterol. Hepatocytes can also _(3)_ sugar, fat, vitamins A, D, E, and K, copper and iron. The liver is an important regulator of blood sugar levels because hepatocytes can remove sugar from the blood and store it as _(4)_, or they can break down glycogen into sugar that is released into the blood. Another function that hepatocytes perform is _(5)_ of nutrients, in which the proportion of nutrients is controlled by changing one type of nutrient into another (e.g., amino acids into glucose). Substances not readily usable by cells are _(6)_ within the liver. Ingested fats, for example, are combined with choline and phosphorus in the liver to produce _(7)_, and vitamin D is converted to its active form by hepatocytes. The hepatocytes also _(8)_ many harmful substances by altering their structure, such as converting ammonia to _(9)_. The liver can also produce its own unique new compounds, including albumins and other _(10)_.

1. _____

2. _____

3. _____

4. _____

5. _____

6. _____

7. _____

8. _____

9. _____

10. _____

Pancreas

The exocrine secretions of the pancreas are called pancreatic juice.

A. Match these components of pancreatic juice with the correct statement or definition:

Aqueous component
Enzymatic component

_____ 1. Portion of pancreatic juice that contains sodium ions, potassium ions, and bicarbonate ions.

_____ 2. Portion of pancreatic juice that is important for the digestion of food.

_____ 3. Portion of pancreatic juice produced by the epithelial cells of the small pancreatic ducts.

_____ 4. Portion of pancreatic juice produced by the acinar cells of the pancreas.

_____ 5. Portion of pancreatic juice that has an increased secretion rate due to secretin.

_____ 6. Portion of pancreatic juice that has an increased secretion rate due to cholecystokinin.

_____ 7. Portion of pancreatic juice that has an increased secretion rate due to parasympathetic stimulation.

B. Match these terms with the correct statement or definition:

Amylase   Lipase
Chymotrypsin   Trypsin
Enterokinase

_____ 1. Major proteolytic enzymes in pancreatic juice.

_____

_____ 2. Proteolytic enzyme that cleaves trypsinogen to trypsin.

_____ 3. Enzyme that hydrolyzes polysaccharides.

_____ 4. Enzyme that digests lipids.

Large Intestine

While in the colon, chyme is converted to feces, where it is stored until eliminated by defecation.

A. Match these terms with the correct statement or definition:

Bicarbonate ions   Mucus
Flatus   Water, sodium and chloride ions
Microorganisms

_____ 1. Substances absorbed by the colon.

_____ 2. Substance secreted by goblet cells in the colon.

_____ 3. Secretions that neutralize the acid produced by colic bacteria.

_____ 4. Source of vitamin K synthesis, and 30% of the dry weight of feces.

_____ 5. Gases produced by bacterial action in the colon.

Mucus secretion by the colon wall is stimulated by local reflexes and parasympathetic stimulation.

B. Match these terms with the correct statement or definition:

Defecation reflex
Duodenocolic reflex

Gastrocolic reflex
Mass movements

_____ 1. Strong peristaltic contractions of large portions of the colon.

_____ 2. Strong peristaltic contractions of the colon initiated by the stomach.

_____ 3. Distension of the rectal wall initiates this reflex.

_____ 4. Results in reinforcement of peristaltic contractions in the lower colon and rectum and relaxation of the internal anal sphincter.

 Local reflexes cause weak contractions and relaxation of the internal sphincter. Parasympathetic reflexes cause stronger contractions and are normally responsible for most of the defecation reflex.

QUICK RECALL

1. Name the four layers or tunics of the digestive tract.

2. List four types of teeth found in humans, and state the function of each type.

3. List the three large pairs of multicellular salivary glands, and name the digestive enzyme found in saliva.

4. Name six sphincters that control movement of materials through the digestive tract.

5. Name the five types of epithelial cells in the stomach, and list their secretions.

6. List the three structural modifications that increase surface area in the small intestine.

7. List the three major types of cells found in the intestinal mucosa.

8. Name the three phases of swallowing.

9. List the types of contraction (movement) that occur in the stomach, small intestine, and large intestine.

10. List the three phases of gastric secretion.

11. Name the substance found in pancreatic juice that is responsible for each of the following activities: neutralizes acid, digests proteins, digests fats, digests carbohydrates.

12. List three functions for bile.

13. List four major functions of the liver in addition to the production of bile.

14. List three major functions of the colon.

15. In the following table, indicate if the control mechanism stimulates (S), inhibits (I), or has no effect (O) on the activity:

	GASTRIN	CHOLE-CYSTO-KININ	SECRETIN	PARA-SYMPA-THETIC
Stomach secretion	_____	_____	_____	_____
Bile secretion	_____	_____	_____	_____
Pancreas secretion	_____	_____	_____	_____
Contraction of gall bladder	_____	_____	_____	_____
Gastric motility	_____	_____	_____	_____

Place the letter corresponding to the correct answer in the space provided.

_____ 1. Which layer of the digestive tract is in direct contact with the contents (the food that is consumed)?
a. tunica mucosa
b. tunica muscularis
c. tunica physh
d. tunica serosa
e. tunica submucosa

_____ 2. The intramural plexus is found in the
a. tunica submucosa.
b. tunica muscularis.
c. tunica serosa.
d. a and b
e. all of the above

_____ 3. The tongue
a. holds food in place during mastication.
b. is involved in speech.
c. helps to form words during speech.
d. all of the above

_____ 4. Dentin
a. forms the surface of the crown of teeth.
b. holds the teeth to the periodontal membrane.
c. is found in the pulp cavity.
d. makes up most of the structure of teeth.
e. is harder than enamel.

_____ 5. The number of premolar, deciduous teeth is
a. 0.
b. 4.
c. 8.
d. 12.

_____ 6. Which of the following glands secrete saliva into the oral cavity?
a. submandibular glands.
b. sublingual glands.
c. parotid glands.
d. all of the above

_____ 7. The stomach
a. has large folds in the submucosa and mucosa called rugae.
b. has two layers of smooth muscle in the tunica muscularis.
c. opening into the small intestine is the cardiac opening.
d. all of the above

_____ 8. Which of the following stomach cell types is NOT correctly matched with its function?
a. surface mucous cells: produce mucus
b. parietal cells: produce hydrochloric acid
c. chief cells: produce intrinsic factor
d. endocrine cells: produce regulatory hormones

_____ 9. Which of the following structures function to increase the mucosal surface of the small intestine?
a. plicae circulares
b. villi
c. microvilli
d. the length of the small intestine
e. all of the above

_____ 10. Given the following parts of the small intestine:
1. duodenum
2. ileum
3. jejunum

Choose the arrangement that lists the parts in the order food would encounter them as the food passes through the small intestine.
a. 1, 2, 3
b. 1, 3, 2
c. 2, 1, 3
d. 2, 3, 1

_____ 11. The structure that releases digestive enzymes in the small intestine?
a. duodenal glands
b. goblet cells
c. endocrine cells
d. absorptive cells

_____ 12. The hepatic sinusoids
   a. receive blood from the hepatic artery.
   b. receive blood from the hepatic portal vein.
   c. empty into the central veins.
   d. all of the above

_____ 13. Given the following ducts:
   1. common bile duct
   2. common hepatic duct
   3. cystic duct
   4. hepatic ducts

   Choose the arrangement that lists the ducts in the order bile would pass through them when moving from the bile canaliculi of the liver to the small intestine.
   a. 3, 4, 2
   b. 3, 2, 1
   c. 4, 2, 1
   d. 4, 1, 2

_____ 14. Given the following structures:
   1. ascending colon
   2. descending colon
   3. sigmoid colon
   4. transverse colon

   Choose the arrangement that lists the structures in the order food would encounter them as food passes from the small intestine to the rectum.
   a. 1, 2, 3, 4
   b. 1, 4, 2, 3
   c. 2, 3, 1, 4
   d. 2, 4, 1, 3

_____ 15. Given the following structures:
   1. coronary ligament
   2. greater omentum
   3. lesser omentum
   4. transverse mesocolon

   Choose the arrangement that lists the structures in the order they would be encountered as one moves along the mesenteries from the diaphragm to the abdominal wall.
   a. 1, 2, 3, 4
   b. 1, 3, 2, 4
   c. 2, 3, 4, 1
   d. 3, 2, 4, 1

_____ 16. A local reflex
   a. involves the spinal cord.
   b. occurs in the intramural plexus.
   c. causes mesenteries to contract.
   d. is the relaxation of smooth muscle when it is suddenly stretched.

_____ 17. The portion of the digestive tract in which digestion begins is the
   a. oral cavity.
   b. esophagus.
   c. stomach.
   d. duodenum.
   e. jejunum.

_____ 18. During swallowing,
   a. movement of food is primarily due to gravity.
   b. the swallowing center in the medulla oblongata is activated.
   c. food is pushed into the oropharynx during the pharyngeal phase.
   d. the soft palate closes off the opening to the larynx.

_____ 19. HCl
   a. is an enzyme.
   b. creates the acid condition necessary for pepsin to work.
   c. is secreted by the small intestine.
   d. all of the above

_____ 20. Why doesn't the stomach digest itself?
   a. The stomach wall isn't composed of protein, and so there are no digestive enzymes to attack it.
   b. The digestive enzymes in the stomach are not efficient enough.
   c. The lining of the stomach is too tough to be attacked by digestive enzymes.
   d. The stomach wall is protected by copious amounts of mucus.

_____ 21. Which of the following hormones stimulate stomach secretions?
   a. cholecystokinin
   b. gastric inhibitory peptide
   c. gastrin
   d. secretin

_____ 22. Which of the following phases of stomach secretion is correctly matched?
a. cephalic phase: the largest volume of secretion is produced
b. gastric phase: gastrin secretion is inhibited by distension of the stomach
c. gastric phase: initiated by chewing, swallowing, or thinking of food
d. intestinal phase: stomach secretions are inhibited

_____ 23. Which of the following statements accurately reflect the role that the stomach plays?
a. It receives relatively large amounts of materials during short periods of time and slowly releases it over a longer time.
b. It is the primary organ of digestion.
c. It performs no function other than storage.
d. It is the primary site of fat digestion.

_____ 24. The function of peristaltic waves in the stomach is to
a. move chyme into the small intestine.
b. increase the secretion of HCl.
c. empty the haustra.
d. all of the above

_____ 25. Which of the following would occur if a person suffered from a severe case of hepatitis that impaired liver function?
a. Fat digestion may be hampered.
b. Byproducts of hemoglobin break down may accumulate in the blood.
c. Plasma proteins may decrease in concentration.
d. b and c
e. all of the above

_____ 26. The gallbladder
a. produces bile.
b. stores bile.
c. contracts and releases bile in response to secretin.
d. contracts and releases bile in response to sympathetic stimulation.

_____ 27. The aqueous component of pancreatic secretions
a. is secreted by the islets of Langerhans.
b. contains bicarbonate ions.
c. is released primarily in response to cholecystokinin.
d. all of the above

_____ 28. Which of the following statements is consistent with inflammation of the pancreas (pancreatitis)?
a. Abdominal pain is enhanced due to the escape of activated digestive enzymes from the pancreas into surrounding tissues.
b. Release of digestive enzymes into the duodenum is reduced, resulting in disturbed digestion, nausea, and vomiting.
c. Production of hepatic enzymes compensates for the reduction in pancreatic enzyme production.
d. Accumulation of pancreatic enzymes in the gallbladder results until the inflammation subsides.
e. a and b

_____ 29. Which of the following is a function of the large intestine?
a. storage of wastes
b. absorption of certain vitamins
c. absorption of water and salts
d. production of mucus
e. all of the above

_____ 30. Defecation
a. can be initiated by stretch of the rectum.
b. can occur as a result of mass movements.
c. involves local reflexes.
d. involves parasympathetic reflexes mediated by the spinal cord.
e. all of the above

Use a separate sheet of paper to complete this section.

1. You and your anatomy and physiology instructor are lost in the desert without water. Your instructor suggests that you place a pebble in your mouth. What would you do and why?

2. Suppose you were given a histological slide from the digestive tract with the following characteristics: (1) muscosal epithelial cells were columnar, (2) many goblet cells, (3) the surface was highly folded to form leaflike projections, and (4) each fold contained a capillary and a lymphatic vessel. What portion of the digestive tract did the slide come from?

3. Why might cutting the vagal nerve supply to the stomach help someone with a peptic ulcer? What side effects might be expected from such a procedure?

4. If a friend had a peptic ulcer, would you recommend a diet high in fats or high in proteins?

5. If a friend had a duodenal peptic ulcer, would you recommend two large meals or six small meals per day? Explain.

6. An experimenter performed two experiments to determine the factors that affect stomach secretion. In the first experiment he measured the amount of gastric juice secreted as a function of the concentration of fats entering the duodenum. The results are graphed below:

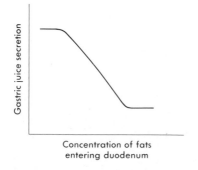

Concentration of fats
entering duodenum

In experiment two, animal A was given food that contained a large amount of fat and animal B was given food that contained a low amount of fat. Then, plasma was removed from animal A and injected into animal C; and, plasma from animal B was injected into animal D. Finally, the gastric juice secretion in animals C and D was determined with the following results:

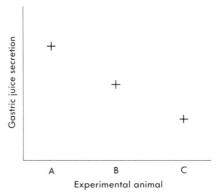

Experimental animal

On the basis of these data, what conclusions can be reached regarding the regulation of gastric juice secretion?

7. Upon autopsy, a 50-year-old male was found to have an abnormal liver. It was yellowish, enlarged, and appeared to be infiltrated with connective tissue. Before his death, this person probably suffered from which of the following symptoms: jaundice, edema, easy fatigue, and weight loss?

8. Explain why an enema results in defecation.

9. Many people have a bowel movement following a meal, especially breakfast. Why does this occur?

10. A patient has a spinal cord injury that has damaged the sacral region of the spinal cord. How will this affect her ability to defecate?

ANSWERS TO CHAPTER 24

WORD PARTS

1. sulcus
2. buccal, buccinator
3. gingiva
4. apical
5. viscosity

6. catalyze, catalysis, catalytic
7. catalyze, catalysis, catalytic
8. masticate
9. parotid
10. hepatic, hepatocytes, pancreatohepatic

CONTENT LEARNING ACTIVITY

General Structure

A. 1. Serosa (adventitia)
    2. Muscularis
    3. Submucosa

B. 1. Mucous epithelium
    2. Lamina propria
    3. Muscularis mucosa
    4. Myenteric plexus
    5. Intramural plexus

Oral Cavity

A. 1. Fauces
    2. Vestibule
    3. Buccal fat pad
    4. Mastication; speech
    5. Frenulum
    6. Extrinsic muscles
    7. Terminal sulcus
    8. Lingual tonsil

B. 1. 2
    2. 1
    3. 2
    4. 3

C. 1. Primary
    2. Alveoli
    3. Gingiva (gums)
    4. Periodontal ligament
    5. Periodontal membrane

D. 1. Crown
    2. Pulp
    3. Root canal
    4. Apical foramen
    5. Dentin
    6. Enamel
    7. Cementum

E. 1. Enamel
    2. Dentin
    3. Pulp cavity
    4. Periodontal ligament
    5. Root canal
    6. Cementum
    7. Apical foramen
    8. Root
    9. Neck
    10. Crown

F. 1. Muscles of mastication
    2. Hard palate
    3. Soft palate
    4. Uvula
    5. Palatine tonsils

G. 1. Parotid glands
    2. Parotid glands
    3. Submandibular glands
    4. Sublingual glands

Pharynx and Esophagus

1. Oropharynx; laryngopharynx
2. Nasopharynx

3. Esophagus

Stomach

A. 1. Pyloric sphincter
2. Muscular layer
3. Rugae

5. Rugae
6. Body
7. Fundus

B. 1. Gastroesophageal opening
2. Cardiac region
3. Pyloric sphincter
4. Pyloric opening

C. 1. Gastric pits
2. Gastric glands
3. Surface mucous cells
4. Parietal (oxyntic) cells
5. Chief (zymogenic) cells

Small Intestine

A. 1. Greater duodenal papilla
2. Lesser duodenal papilla
3. Hepatopancreatic ampulla

6. Goblet cells
7. Intestinal glands
8. Duodenal glands

B. 1. Plicae circulares
2. Villi
3. Microvilli
4. Lacteals
5. Absorptive cells

C. 1. Ileum; jejunum
2. Ileocecal sphincter
3. Ileocecal valve
4. Peyer's patches

Liver and Gallbladder

A. 1. Caudate and quadrate
2. Porta
3. Gall bladder
4. Common hepatic duct
5. Cystic duct
6. Common bile duct

4. Hepatic cords
5. Hepatic sinusoids
6. Endothelial and phagocytic cells

B. 1. Lobules
2. Portal triads
3. Hepatocytes

C. 1. Hepatic portal vein
2. Central vein
3. Hepatic veins
4. Bile canaliculi
5. Hepatic ducts

Pancreas

A.
1. Head
2. Tail
3. Acini
4. Intercalated ducts
5. Intralobular ducts
6. Interlobular ducts
7. Main pancreatic duct
8. Pancreatic islets (islets of Langerhans)

B.
1. Cystic duct
2. Common bile duct
3. Hepatopancreatic duct
4. Pancreatic duct
5. Common hepatic duct
6. Hepatic ducts

Large Intestine

1. Cecum
2. Vermiform appendix
3. Transverse colon
4. Sigmoid colon
5. Crypts
6. Taenia coli
7. Haustra
8. Epiploic appendages
9. Rectum
10. Anal canal

Peritoneum

1. Visceral peritoneum
2. Mesenteries
3. Retroperitoneal
4. Lesser omentum
5. Omental bursa
6. Falciform ligament
7. Mesentery proper
8. Transverse and sigmoid mesocolon

Functions of the Digestive System

1. Local reflex
2. Receptors
3. Intramural plexus
4. Effector organ

Oral Cavity

A.
1. Chewing (mastication) reflex
2. Amylase
3. Mucin

B.
1. Incisors and canines
2. Premolars and molars
3. Starch
4. Antibacterial
5. Parasympathetic

Deglutition

A.
1. Voluntary phase
2. Pharyngeal phase
3. Esophageal phase

B.
1. Swallowing center
2. Pharyngeal constrictor muscles
3. Epiglottis
4. Peristaltic waves

Stomach

A. 1. Mucus
    2. Intrinsic factor
    3. Pepsin
    4. Pepsinogen
    5. Gastrin

B. 1. Cephalic phase
    2. Gastric phase
    3. Intestinal phase

C. 1. Increase
    2. Increase
    3. Increase
    4. Decrease
    5. Decrease
    6. Decrease

D. 1. Chyme
    2. Peristaltic waves
    3. Pyloric pump

Small Intestine

A. 1. Mucus
    2. Disaccharidases
    3. Peptidases
    4. Nucleases

B. 1. Segmental contractions
    2. Increases
    3. Increases
    4. Increases
    5. Increases

Liver and Gallbladder

A. 1. Stimulates
    2. Stimulates
    3. Stimulates
    4. Stimulates
    5. Inhibits

B. 1. Bile salts
    2. Bile pigments
    3. Store

    4. Glycogen
    5. Interconversion
    6. Transformed
    7. Phospholipids
    8. Detoxify
    9. Urea
    10. Blood proteins

Pancreas

A. 1. Aqueous component
    2. Enzymatic component
    3. Aqueous component
    4. Enzymatic component
    5. Aqueous component
    6. Enzymatic component
    7. Enzymatic component

B. 1. Chymotrypsin; trypsin
    2. Enterokinase
    3. Amylase
    4. Lipase

Large Intestine

A. 1. Water, sodium, and chloride ions
   2. Mucus
   3. Bicarbonate ions
   4. Microorganisms
   5. Flatus

B. 1. Mass movements
   2. Gastrocolic reflex
   3. Defecation reflex
   4. Defecation reflex

QUICK RECALL

1. Mucosa, submucosa, muscularis, and serosa or adventitia.
2. Incisors and canines: cutting and tearing food; molars and premolars: crushing and grinding food.
3. Parotid, submandibular, and sublingual glands. Amylase is the enzyme in saliva.
4. Upper esophageal sphincter, lower esophageal sphincter, pyloric sphincter, ileocecal sphincter, internal anal sphincter, external anal sphincter.
5. Surface mucous cells: mucus; mucous neck cells: mucus; parietal cells: hydrochloric acid and intrinsic factor; chief cells: pepsinogen; endocrine cells: gastrin.
6. Plicae circulares, villi, and microvilli
7. Absorptive cells, goblet cells, and endocrine cells
8. Voluntary, pharyngeal, and esophageal
9. Stomach: mixing waves and peristaltic waves; small intestine: segmental contractions and peristaltic contractions; large intestine: segmental movements and mass movements.

10. Cephalic, gastric, and intestinal.
11. Bicarbonate ions neutralize acid; trypsin, chymotrypsin, and carboxypeptidase digest protein; lipase digests fats and amylase digests starch.
12. Neutralizes stomach acids, emulsifies fats, carries out excretory products
13. Regulate blood glucose, interconversion of nutrients, detoxification, transformation of substances, and formation of blood proteins
14. Reabsorb water and salts, secretion of mucus, absorption of vitamins produced by microorganisms, storage of feces
15. Stomach secretion: S,I,I,S.
    Bile secretion: O,O,S,S.
    Pancreas secretion: O,S,S,S.
    Contraction of gall bladder: O,S,O,S.
    Gastric motility: S,I,I,I.

1. A. From the inner lining of the digestive tract to the outer layers are the tunica mucosa, tunica submucosa, tunica muscularis, and tunica serosa.

2. D. The intramural plexus consists of the submucosa plexus (located in the tunica submucosa) and the myenteric plexus (found in the tunica muscularis).

3. D. The tongue moves food about and helps (with the lips and cheeks) to hold the food between the teeth during mastication. Movement of the tongue is also involved with speech. The tongue contains taste buds, which are involved with the sense of taste.

4. D. Most of the tooth is made up of dentin. Dentin is a coumpound of hydroxyapatite crystals that is generally less dense than bone. Enamel is very hard and covers the outer surface of the tooth (crown). The periodontal membrane secretes cementum, which helps to hold the tooth in its socket. The pulp cavity is found inside the tooth. It contains blood vessels, nerves, and pulp.

5. A. The typical number of teeth is as follows:

Tooth	Deciduous	Permanent
Incisors	8	8
Canines	4	4
Premolars	0	8
Molars	8	12
Total	20	32

6. D. There are three sets of salivary glands: the parotid, submandibular, and sublingual salivary glands. They secrete the saliva that enters the oral cavity.

7. A. The folds in the stomach, which allow the stomach to stretch, are called rugae. The tunica muscularis of the stomach has three layers of smooth muscle due to the presence of an additional oblique layer of smooth muscle. The pyloric opening is the opening between the small intestine and the stomach (the cardiac or gastroesophageal opening is the opening between the stomach and esophagus).

8. C. Chief cells produce pepsinogen. Intrinsic factor is produced by parietal cells.

9. E. All of the structures listed increase the mucosal surface area of the small intestine, except the serosa, which is not part of the mucosa.

10. B. Food passes from the stomach into the duodenum. From the duodenum food passes through the jejunum to the ileum, which empties into the large intestine.

11. D. The absorptive cells produce digestive enzymes that breakdown food. The digested food is then absorbed. Duodenal glands and goblet cells produce mucus, and endocrine cells release regulatory hormones.

12. D. Blood from the hepatic artery and hepatic portal vein mixes in the hepatic sinusoids and then flows to the central veins. The central veins join the hepatic veins, which exit the liver.

13. C. Bile would pass through the hepatic ducts, the common hepatic duct, and the common bile duct. Bile could pass through the cystic duct into the gallbladder and then back out into the common bile duct, but that route was not given.

14. B. The order would be ascending, transverse, descending, and sigmoid colon.

15. B. The coronary ligament connects the diaphragm and liver, the lesser omentum connects the liver and stomach, the greater momentum connects the stomach and the large intestine, and the transverse mesocolon connects the large intestine and abdominal wall.

16. B. A local reflex occurs in the intramural plexus and does not involve the spinal cord or brain. Stimulation of receptors within the digestive tract wall results in action potentials that are propagated by means of intramural plexus neurons to effector organs in the digestive tract wall such as smooth muscle and glands.

17. A. Both mechanical and chemical digestion begin in the oral cavity. Mastication begins the process of mechanial digestion, and salivary amylase begins the process of chemical digestion.

18. B. Swallowing involves a voluntary phase during which food is pushed into the oropharynx. In the pharyngeal and esophageal phases the swallowing center is activated, and muscular reflexes push the food toward the stomach (gravity assists but is not as importmant as the muscular contractions. The soft palate closes off the nasopharynx (the epiglottis closes off the larynx).

19. B. HCl is a strong acid secreted by the stomach. The pH in the stomach is reduced by HCl, which stops carbohydrate digestion and creates the acid conditions necessary for pepsin to work.

20. D. The stomach is protected from digestive enzymes due to the large amount of mucus that is secreted by the epithelial cells of the stomach lining.

21. C. Gastrin stimulates hydrochloric acid secretion by parietal cells. The other hormones inhibit gastric secretions.

22. D. The intestinal phase first stimulates, then inhibits, gastric secretions. The cephalic phase occurs as a result of chewing, swallowing, or thinking of food. The gastric phase produces the largest volume of secretions.

23. A. Food enters the stomach during a meal. That material is mixed with gastric secretions and is then slowly released, in a regulated fashion, into the duodenum. Although protein digestion begins in the stomach, most digestion take place in the duodenum.

24. A. Peristaltic waves move chyme from the stomach into the small intestine. The mixing waves of the stomach function to mix the secretions of the stomach and the food within it. Haustra are found in the large intestine, not the stomach.

25. E. The liver functions to produce many of the plasma proteins. It produces bile, which is important in fat emulsification in the small intestine, and eliminates pigments that are by-products of hemoglobin breakdown in the bile. Therefore all of the conditions listed would occur if the liver were inflamed and its functioning impaired.

26. B. The gallbladder stores and concentrates bile. The bile is produced by the liver. The gallbladder releases bile in response to cholecystokinin and parasympathetic stimulation.

27. B. The aqueous component contains bicarbonate ions that neutralize stomach acid. It is produced by the cells lining the smaller ducts of the pancreas and is released in response to secretin. The enzymatic component contains digestive enzymes, is produced by acinar cells, and is released in response to cholecystokinin. The islets of Langerhans are the endocrine portion of the pancreas, and they produce hormones (e.g., insulin and glucagon).

28. E. Pancreatitis, inflammation of the pancreas, is accompanied by the escape of digestive enzymes from pancreatic tissue into the surrounding tissues. The enzymes digest the tissues with which they come into contact and aggravate the inflammation, which results in severe abdominal pain. Also, the reduction in pancreatic enzymes delivered to the duodenum results in hampered digestion. The disrupted digestion and the abdominal pain produce nausea and vomiting. The liver does not produce digestive enzymes nor does the gallbladder function to store digestive enzymes.

29. E. The large intestine stores waste until they are eliminated. Water, salts, and vitamins (e.g., vitamin K produced by intestinal bacteria) are absorbed. Mucus protects the large intestine and provides adhesion of fecal matter.

30. E. Stretch of the rectum initiates local and spinal cord reflexes that result in defecation. Often the stretch is due to the movement of feces into the rectum as a result of a mass movement.

1. Tactile stimulation in the mouth increases saliva production, making the mouth feel less dry.

2. The slide is from the small intestine. The stomach, small intestine, and large intestine are lined by columnar epithelial cells (the esophagus and rectum are lined with moist, stratified squamous epithelium). Villi are the folds that contain a capillary and a lymphatic vessel and they are lined with columnar epithelial cells, many of which are goblet cells. The stomach is not lined with goblet cells, and the large intestine does not have villi.

3. Cutting the vagus nerves would eliminate the parasympathetic stimulation of the stomach, reducing stomach acid secretion. The reduction in hydrochloric acid production could help in the treatment of the peptic ulcer. However, after a few months acid secretion often increases, and an ulcer may develop. Since elimination of parasympathetic stimulation also decreases stomach movements, food might be retained in the stomach. In some cases the stomach never really empties.

4. A diet high in fats would be most logical, since fatty acids in the duodenum stimulate the secretion of gastric inhibitory polypeptide and cholecystokinin, both of which inhibit gastric secretions. On the other hand, proteins (amino acids and polypeptides) in the stomach stimulate the release of gastrin, which increases gastric secretions.

5. Large meals would not be recommended because distension of the stomach promotes acid secretion and rapid stomach emptying. This would result in large amounts of acidic chyme entering the duodenum and aggravating the duodenal peptic ulcer. With smaller meals there would be less acid production, slower movement of chyme into the duodenum, and an increased ability to neutralize the chyme once it enters the duodenum. Note, however, that although these arguments seem logical, there is at present no conclusive proof that spreading meals throughout the day is truly effective.

6. The first experiment shows that, as the fat content within the duodenum is increased, gastric acid secretion decreases. The second experiment suggests that fats entering the duodenum cause the release of a hormone into the blood that inhibits gastric secretion. Possible hormones could include gastric inhibitory polypeptide and cholecystokinin. Although the data do not indicate it, there is also evidence for a gastroenteric reflex that slows gastric secretion in response to large amounts of fat entering the small intestine.

7. Gross observation of the liver suggests that the man suffered from cirrhosis of the liver. He probably experienced symptoms associated with compromised liver function. Normally the liver releases bilirubin (from the breakdown of hemoglobin) into the bile. Jaundice would be expected because normal bile flow would be impaired and the bilirubin levels would increase in the blood. Edema would occur because the liver would not produce adequate plasma proteins. The man would tire easily because nutrients normally processed and stored in the liver would not be available. Weight loss might occur because reduced emulsification of fat due to lack of bile salts would hamper digestion.

8. The enema causes stretch of the rectum, which initiates local and central nervous system reflexes that cause defecation.

9. Following a meal the gastrocolic and duodenocolic reflex initiate mass movements in the colon. Feces moves into the rectum, and the resulting stretch activates the defecation reflex.

10. Damage to the sacral region would eliminate the central nervous system reflex involved in defecation. Also lost would be voluntary control of defecation. However, local reflexes would still be functional, and defecation would occur, although not as well as before the injury.

Metabolism And Nutrition

FOCUS: Vitamins and minerals are necessary for normal metabolism, and both must be ingested in the diet. Carbohydrate metabolism begins in the oral cavity and is completed in the small intestine, protein digestion is accomplished in the stomach and small intestine, and lipid digestion occurs in the small intestine. Glucose, amino acids, and ions are absorbed in the small intestine and are carried to the liver through the hepatic portal system. Digested lipids enter lacteals and pass through the thoracic duct to the left subclavian vein. Glucose is converted to pyruvic acid in glycolysis. In anaerobic respiration (without oxygen) the pyruvic acid becomes lactic acid with a net gain of 2 ATPs. In aerobic respiration (with oxygen) 36 ATPs are produced when the pyruvic acid is converted to acetyl CoA, which enters the citric acid cycle. Molecules such as NADH and $FADH_2$ produced in aerobic respiration are used to produce ATPs in the electron transport chain. Lipids can be broken down by beta-oxidation to yield acetyl-CoA molecules that can be used to produce ATPs (through the citric acid cycle) or ketones (ketogenesis). Amino acids can be used as a source of energy in oxidative deamination reactions. When there is too much blood glucose, it can be stored (glycogenesis), and, when there is too little blood glucose, it can be produced (glycogenolysis and gluconeogenesis). The energy produced from foods is used for basal metabolism, physical acitivity, and the assimilation of food. Heat produced by metabolism, in conjunction with heat exchanged with the environment by radiation, conduction, and convection, is responsible for body temperature.

WORD PARTS

Give an example of a new vocabulary word that contains each word part.

WORD PART	MEANING	EXAMPLE
glyco-	sweet	1. _____
genesis	origin	2. _____
metab-	change	3. _____
aerob-	the air	4. _____

WORD PART	MEANING	EXAMPLE
neo-	new	5. _____
chylo-	juice; chyle	6. _____
post-	after	7. _____
-duct-	lead	8. _____
con-	with	9. _____
-vect-	carried	10. _____

CONTENT LEARNING ACTIVITY

Introduction

❝*Metabolism is the total of all the chemical changes that occur in the body.*❞

Match these terms with the correct statement or definition:

Anabolism   Nutrition
Catabolism

_____ 1. Energy-releasing process by which large molecules are broken down into smaller molecules.

_____ 2. Process by which food (nutrients) is obtained and used by the body.

Nutrients

❝*Nutrients are chemicals taken into the body that are used to produce energy, provide building blocks for new molecules, or function in other chemical reactions.*❞

A. Match these terms with the correct statement or definition:

Fat-soluble vitamins   Water-soluble vitamins
Provitamins

_____ 1. Portions of vitamins that can be assembled or modified by the body into functional vitamins; carotene is an example.

_____ 2. B-complex vitamins and vitamin C.

_____ 3. Vitamins A, D, E, and K.

_____ 4. Vitamins that can be stored in the body.

 Vitamins are compounds that exist in minute quantities in food and are essential to normal metabolism.

B. Match these vitamins with the correct deficiency symptom:

A (retinol) D (cholecalciferol)
B_{12} (cobalamin) K (phylloquinone)
C (ascorbic acid)

_____ 1. Scurvy - defective bone formation.

_____ 2. Night blindness, retarded growth, and skin disorders.

_____ 3. Excessive bleeding due to retarded blood clotting.

_____ 4. Pernicious anemia and nervous system disorders.

C. Match these minerals with the correct function:

Calcium Phosphorus
Chlorine Potassium
Iodine Sodium
Iron

_____ 1. Bone and teeth formation, blood clotting, muscle activity, and nerve function.

_____ 2. Osmotic pressure regulation; nerve and muscle function.

_____ 3. Blood acid-base balance; hydrochloric acid production in the stomach.

_____ 4. Bone and teeth formation; important in ATP formation, component of nucleic acids.

_____ 5. Component of hemoglobin; ATP production in electron transport system.

_____ 6. Thyroid hormone production; maintenance of normal metabolic rate.

Digestion, Absorption, and Transport

66 _Digestion, absorption, and transport are components of nutrition._ 99

Match these terms with the correct statement or definition:

Absorption Lipids and lipid-soluble
Digestion substances
Ions and water-soluble substances Transport

_____ 1. Begins in the mouth.

_____ 2. Begins in the stomach; most occurs in the duodenum and jejunum, although some occurs in the ileum.

_____ 3. Transported through the hepatic portal system to the liver.

_____ 4. Transported through the lymphatic system to the left subclavian vein and then to the liver or adipose tissue.

Carbohydrates

66 *Ingested carbohydrates consist primarily of starches, glycogen, sucrose, and small amounts of lactose and fructose.* 99

Match these terms with the correct statement or definition:

Disaccharidase
Glucose
Insulin

Pancreatic amylase
Salivary amylase

_____ 1. Enzyme that digests starch and is secreted into the oral cavity.

_____ 2. Enzyme that is bound to the microvilli of the intestinal epithelium.

_____ 3. Sugar transported by the circulatory system to cells that need energy.

_____ 4. Hormone that greatly increases the rate of glucose transport into most types of cells.

☞ Glucose enters the cells by the process of facilitated transport.

Lipids

66 *Lipids are molecules that are insoluble or only slightly soluble in water.* 99

Using the terms provided, complete the following statements:

Bile salts
Chylomicrons
Emulsification
Lacteal

Lipase
Micelles
Triglycerides

1. _____

2. _____

3. _____

4. _____

5. _____

6. _____

7. _____

Lipids include triglycerides, phospholipids, steroids, and fat-soluble vitamins. (1) consist of one glycerol molecule and three fatty acids covalently bound together. The first step in lipid digestion is (2) , which is the transformation of large lipid droplets into much smaller droplets. This process is accomplished by (3) secreted by the liver. (4) secreted by the pancreas digests lipid molecules. Once lipids are digested in the intestine, bile salts aggregate around the small droplets to form (5) . When these structures come into contact with epithelial cells of the small intestine, their lipid contents pass through the cell membrane of the epithelial cell by the process of simple diffusion. Within the smooth endoplasmic reticulum of the intestinal epithelial cells, fatty acids combine with glycerol to form triglycerides. Proteins in the epithelial cells coat droplets of triglycerides, phospholipids, and cholesterol to form (6) , which leave the epithelial cell to enter a (7) . From there the chylomicrons are transported to the blood and are carried to adipose tissue or the liver.

Proteins

"Proteins are taken into the body from a variety of dietary sources."

Match these terms with the
correct statement or definition:

Pepsin   Trypsin
Peptidase

_____ 1. Enzyme in the stomach that catalyzes the cleavage of covalent bonds in
proteins.

_____ 2. Enzyme produced by the pancreas that continues the digestion of
proteins started in the stomach; produces small peptide chains.

_____ 3. Enzyme bound to the microvilli and found inside intestinal epithelial
cells; completes the breakdown of small peptide chains.

☞ Transport of amino acids into the cells of the body is stimulated by growth hormone
and insulin.

Water and Ions

"Water can move in either direction across the wall of the small intestine."

Match these terms with the
correct description below:

Active transport   Osmosis
Into circulation   Simple diffusion
Into lumen of intestine

_____ 1. Mechanism responsible for water movement across the wall of the small
intestine.

_____ 2. Mechanism that moves sodium, potassium, magnesium, calcium and
phosphate into the epithelial cells of the small intestine.

_____ 3. Passive movement of negative ions (e.g., chloride ions) as they follow
positive ions (e.g., sodium ions) into intestinal epithelial cells.

_____ 4. Direction of water movement when the chyme is very concentrated.

_____ 5. Direction of water movement as nutrients are absorbed.

Metabolism

"The cellular metabolic processes are often referred to as cellular metabolism."

Match these terms with the
correct statement or definition:

ATP   Oxidation-reduction
Oxidized   Reduced

_____ 1. Energy currency of the cell; used to drive cell activities.

_____ 2. Type of chemical reaction responsible for the transfer of energy from the
chemical bonds of nutrient molecules to ATPs.

_____ 3. Molecule that gains electrons, hydrogen ions, and energy.

Glycolysis

66 *Glycolysis is a series of chemical reactions that result in the breakdown of glucose to two pyruvic acid molecules.* 99

Match these terms with the
correct statement or definition:

ATP   One
Four   Phosphorylation
NAD^+ Two
NADH

_____ 1. Reduced form of nicotinamide adenine dinucleotide.

_____ 2. Process of attaching a phosphate group to a molecule.

_____ 3. Number of ATPs required to start glycolysis for one glucose molecule.

_____ 4. Net number of ATPs produced from one glucose molecule by glycolysis.

_____ 5. Number of NADH molecules produced from one glucose molecule by glycolysis.

_____ 6. Number of pyruvic acid molecules produced from one glucose molecule by glycolysis.

Anaerobic Respiration

66 *Anaerobic respiration is the breakdown of glucose in the absence of oxygen.* 99

Match these terms with the
correct statement or definition:

ATP   NADH
Cori cycle   Oxygen debt
Lactic acid

_____ 1. Net energy gain from the anaerobic respiration of one molecule of glucose is two of these molecules.

_____ 2. Formed by the reduction of pyruvic acid.

_____ 3. Two of these molecules are produced in glycolysis and used (oxidized) when pyruvic acid is reduced.

_____ 4. Lactic acid released from cells is transported to the liver; the lactic acid is converted to glucose which is transported back to cells.

_____ 5. Oxygen necessary for the synthesis of the ATP used to convert lactic acid to glucose.

Aerobic Respiration

" *Aerobic respiration is the breakdown of glucose in the presence of oxygen to produce carbon dioxide,* " *water, and ATPs.*

A. Match these terms with the correct statement or definition:

Acetyl CoA formation
Aerobic respiration
Citric acid (Kreb's) cycle

Electron transport chain
Glycolysis

_____ 1. First phase of aerobic respiration; produces two ATPs and two NADH per glucose molecule.

_____ 2. Second phase of aerobic respiration in which pyruvic acid is modified to form acetyl CoA; produces two NADH per glucose molecule.

_____ 3. Third phase of aerobic respiration, in which acetyl CoA is combined with oxaloacetic acid to form citric acid; citric acid is then converted by a series of reactions into oxaloacetic acid; produces six NADH, two FADH2, and two ATP per glucose molecule.

_____ 4. Produces 36 ATPs for each glucose molecule broken down.

_____ 5. Process produces three ATPs for every NADH oxidized, and two ATPs for every $FADH_2$ oxidized; occurs within the mitochondria.

_____ 6. Process uses oxygen as a final electron acceptor, producing water.

B. Match these terms with the correct parts of the diagram labeled in Figure 25-1:

Acetyl CoA
ADP
ATP
Aerobic respiration
Anaerobic respiration

Citric acid cycle
Glycolysis
H_2O
Lactic acid
NADH

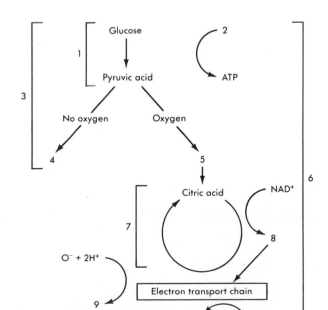

1. _____

2. _____

3. _____

4. _____

5. _____

6. _____

7. _____

8. _____

9. _____

10. _____

Figure 25-1

539

Lipids

❝ *Lipids are the body's main energy storage molecule.* **❞**

Match these terms with the
correct statement or definition:

Beta-oxidation   Ketogenesis
Free fatty acids   Triglycerides
Ketone bodies

_____   1. Primary storage form of lipids in adipose tissue.

_____   2. Fatty acids released into the blood from the breakdown of triglycerides;
used as an energy source by muscle and liver cells.

_____   3. Series of reactions in which two carbons are removed from the end of a
fatty acid chain to form acetyl CoA.

_____   4. Formation of ketone bodies from acetyl CoA.

_____   5. Acetoacetic acid, beta-hydroxybutyric acid, and acetone; used as an
energy source, especially by skeletal muscle.

Proteins

❝ *Once absorbed by the body, amino acids are quickly taken up by cells, especially in the liver.* **❞**

Using the terms provided, complete the following statements:

Citric acid (Kreb's) cycle   Oxidative deamination
Essential   Proteins
Keto acid   Stored
NADH   Transamination
Nonessential   Urea

Amino acids can be used to synthesize (1) or as a source of
energy, but, unlike carbohydrates and lipids, amino acids are not
 (2) in the body. The 20 amino acids in the human body can be
divided into two groups: (3) amino acids, which cannot be
synthesized by the body and must be obtained in the diet, and
 (4) amino acids, which can be produced in the body from other
molecules. The synthesis of a nonessential amino acid usually
begins with a (5), which is usually converted to an amino acid.
This process, called (6) involves the transfer of an amino group
from an amino acid to the keto acid. Amino acids can be used as
a source of energy in a(n) (7) reaction. In this reaction an amino
group is removed from an amino acid, leaving ammonia, a keto
acid, and (8) that can be used to produce ATP. Keto acids can
also enter the (9) or be converted into pyruvic acid or acetyl
CoA. The ammonia is converted to (10), which is eliminated by
the kidney.

1. _____

2. _____

3. _____

4. _____

5. _____

6. _____

7. _____

8. _____

9. _____

10. _____

Intraconversion of Nutrient Molecules

“_Many nutrient molecules may be converted to other nutrients as needed by the body._**”**

A. Using the terms provided, complete the following statements:

Gluconeogenesis   Glycogenolysis
Glycogenesis   Lipogenesis

If there is excess glucose, it can be used to form glycogen
through a process called _(1)_ . Once glycogen stores are filled,
glucose and amino acids are used to synthesize lipids, a process
called _(2)_ . When glucose is needed, glycogen can be broken
down into glucose-6-phosphate through a set of reactions called
_(3)_ . When liver glycogen levels are inadequate to supply
glucose, amino acids from proteins and glycerol from
triglycerides are used to produce glucose in a process called _(4)_ .

1._____

2._____

3._____

4._____

B. Match these terms with the
correct parts of the diagram
labeled in Figure 25-2:

Gluconeogenesis   Glycogenolysis
Glycogenesis   Glycolysis

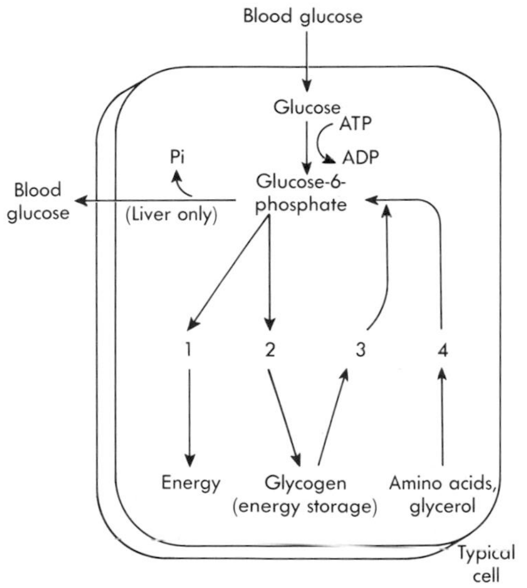

1. _____

2. _____

3. _____

4. _____

Figure 25-2

Metabolic States

“_There are two major metabolic states in the body; the absorptive state is the period immediately after a meal when_**”**
nutrients are being absorbed, and the postabsorptive state occurs after the absorptive state has concluded.

Match these terms with the
correct statement or definition:

Absorptive state
Postabsorptive state

_____   1. During this state most of the glucose entering the circulation is used by
cells; the remainder is converted into glycogen or fat.

_____   2. During this state blood glucose levels are maintained by converting
other molecules to glucose.

_____   3. During this state glycogen is used preferentially; then fats and ketones,
and then proteins.

Metabolic Rate

"The metabolic rate is the total amount of energy produced and used by the body per unit of time."

Match these terms with the correct statement or definition:

Basal metabolic rate   Increases
Calorie   Kilocalorie
Decreases

_____ 1. Amount of heat necessary to raise the temperature of 1 g of water from 14°,C to 15°,C.

_____ 2. Metabolic rate calculated in kilocalories per square meter of body surface per hour.

_____ 3. Effect of increased muscle tissue on BMR.

_____ 4. Effect of increasing age on BMR.

_____ 5. Effect of dieting or fasting on BMR.

_____ 6. Effect of physical activity on expenditure of energy.

_____ 7. Effect of specific dynamic activity on expenditure of energy.

👉 For every 3500 kcal above the necessary energy requirement, a pound of body fat can be gained; for every 3500 kcal below the necessary energy requirement, a pound of body fat can be lost.

Heat Production and Regulation

"We can regulate our body temperature rather than have our body temperature adjusted by the external environment."

Match these terms with the correct statement or definition:

Conduction   Free energy
Convection   Homeotherms
Decrease   Increase
Evaporation   Radiation

_____ 1. Total amount of energy that can be liberated by the complete catabolism of food.

_____ 2. Animals that can regulate their body temperature.

_____ 3. Exchange of heat between objects in direct contact with each other.

_____ 4. Exchange of heat between the body and the air.

_____ 5. Loss of water from the body, which carries heat with it.

_____ 6. Effect of vasodilation on skin temperature.

_____ 7. Effect of vasoconstriction on skin temperature.

👉 A negative feedback system accomplishes body temperature regulation. Specific body temperature is maintained by a set point in the hypothalamus.

542

1. List the seven major classes of nutrients.

2. List the breakdown products of carbohydrates, proteins, and fats.

3. List the locations in the digestive tract where carbohydrate digestion, lipid digestion, and protein digestion occur.

4. Name the routes by which water-soluble and lipid-soluble molecules leave the intestinal epithelial cells.

5. Give the total energy gain to the cell from one molecule of glucose from anaerobic respiration and aerobic respiration.

6. Name three energy-storing compounds produced in aerobic respiration.

7. List the endproducts formed by anaerobic and aerobic respiration.

8. Name the three phases of aerobic respiration. Assuming the electron transport chain is operating, give the number of ATPs produced by each phase.

9. Name the chemical reactions by which fatty acids and amino acids are used as a source of energy.

10. Name four processes that involve intraconversion of nutrient molecules.

11. List three ways that metabolic energy can be used.

Place the letter corresponding to the correct answer in the space provided.

_____ 1. Concerning vitamins;
  a. most can be synthesized by the body.
  b. they are normally broken down before they can be used by the body.
  c. A, D, E, and K are water-soluble vitamins.
  d. they function as coenzymes.

_____ 2. Minerals
  a. are inorganic nutrients.
  b. compose about 4% to 5% of total body weight.
  c. act as coenzymes, buffers, and osmotic regulators.
  d. all of the above

_____ 3. The breakdown products of carbohydrate digestion are
  a. monosaccharides.
  b. amino acids.
  c. monoglycerides and diglycerides.
  d. glycerol and fatty acids.

_____ 4. The enzyme responsible for the digestion of carbohydrates is produced by the
  a. salivary glands.
  b. pancreas.
  c. lining of the small intestine.
  d. a and b
  e. all of the above

_____ 5. Bile
  a. is an important enzyme for the digestion of fats.
  b. is made by the gallbladder.
  c. contains breakdown products from hemoglobin.
  d. emulsifies fats.
  e. c and d

_____ 6. Micelles are
  a. lipids surrounded by bile salts.
  b. produced by the pancreas.
  c. released into lacteals.
  d. all of the above

_____ 7. If the thoracic duct were tied off, which of the following classes of nutrients would NOT enter the circulatory system at their normal rate?
  a. amino acids
  b. glucose
  c. lipids
  d. fructose

_____ 8. Two enzymes involved in the digestion of proteins are
  a. pepsin and lipase.
  b. trypsin and hydrochloric acid.
  c. pancreatic amylase and bile.
  d. pepsin and trypsin.

_____ 9. Anaerobic respiration occurs in the _____ of oxygen and produces _____ energy for the cell than does aerobic respiration.
  a. absence, less
  b. absence, more
  c. presence, less
  d. presence, more

_____ 10. Which of the following reactions take place in both anaerobic and aerobic respiration?
  a. glycolysis
  b. Kreb's cycle
  c. electron-transport chain
  d. a and b
  e. all of the above

_____ 11. A molecule that moves electrons from the Krebs cycle to the electron-transport chain is
  a. tRNA.
  b. mRNA.
  c. ADP.
  d. NADH.

_____ 12. The production of ATPs by the electron-transport chain is accompanied by the production of
  a. alcohol.
  b. water.
  c. oxygen.
  d. lactic acid.
  e. glucose.

_____ 13. The carbon dioxide you breathe out comes from
   a. glycolysis.
   b. electron-transport chain.
   c. anaerobic respiration.
   d. the food you eat.

_____ 14. Lipids are
   a. stored primarily as triglycerides.
   b. synthesized by beta-oxidation.
   c. broken down by oxidative deamination.
   d. all of the above

_____ 15. Amino acids
   a. in the human body are classified as essential or nonessential.
   b. are synthesized in a transamination reaction.
   c. can be used as a source of energy.
   d. all of the above

_____ 16. Ammonia is
   a. a by-product of lipid metabolism.
   b. formed during ketogenesis.
   c. converted into urea in the liver.
   d. all of the above

_____ 17. The conversion of amino acids and glycerol into glucose is called
   a. gluconeogenesis.
   b. glycogenolysis.
   c. glycogenesis.
   d. ketogenesis.

_____ 18. Which of the following events takes place during the absorptive state?
   a. Glycogen is converted into glucose.
   b. Glucose is converted into fats.
   c. Ketones are produced.
   d. Proteins are converted into glucose.

_____ 19. The major use of energy by the body is used for
   a. basal metabolism.
   b. physical activity.
   c. specific dynamic activity.

_____ 20. The loss of heat due to the loss of water from the surface of the body is called
   a. radiation.
   b. evaporation.
   c. conduction.
   d. convection.

☆ ──────── FINAL CHALLENGES ──────── ☆

Use a separate sheet of paper to complete this section.

1. When you chew bread for a few minutes it tastes sweet. Explain how this happens.

2. Explain how drugs that bind to bile salts in the small intestine can cause a decrease in blood cholesterol levels.

3. Two ATPs are required to start beta-oxidation of a fatty acid chain. For each acetyl CoA formed (except for the last one), one NADH and one $FADH_2$ are also produced during beta-oxidation. Given this information, does the metabolism of a six-carbon fatty acid yield more or less energy than the aerobic metabolism of a six-carbon molecule of glucose?

4. When a person is trying to lose weight, a reduction in caloric input and exercise is recommended. Give three reasons why exercise would help to reduce weight.

5. In some diseases the infection results in a high fever. The patient is on the way to recovery when the crisis is over and body temperature begins to return to normal. If you were looking for symptoms in a patient who had just passed through the crisis state, would you look for a dry, pale skin or for a wet, flushed skin? Explain.

ANSWERS TO CHAPTER 25

1. glycogen, glycogenesis, glycogenolysis
2. gluconeogenesis, glycogenesis, ketogenesis
3. metabolism, metabolic
4. aerobic, anaerobic
5. gluconeogenesis

6. chylomicron
7. postabsorptive
8. conduction
9. convection, conduction
10. convection

CONTENT LEARNING ACTIVITY

Introduction

1. Catabolism

2. Nutrition

Nutrients

A. 1. Provitamins
   2. Water-soluble vitamins
   3. Fat-soluble vitamins
   4. Fat-soluble vitamins

B. 1. C (ascorbic acid)
   2. A (retinol)
   3. K (phylloquinone)
   4. B_{12} (cobalamin)

C. 1. Calcium
   2. Sodium
   3. Chlorine
   4. Phosphorus
   5. Iron
   6. Iodine

Digestion, Absorption, and Transport

1. Digestion
2. Absorption

3. Ions and other water-soluble substances
4. Lipids and lipid-soluble substances

Carbohydrates

1. Salivary amylase
2. Disaccharidase

3. Glucose
4. Insulin

Lipids

1. Triglycerides
2. Emulsification
3. Bile salts
4. Lipase

5. Micelles
6. Chylomicrons
7. Lacteal

Proteins

1. Pepsin
2. Trypsin

3. Peptidase

Water and Ions

1. Osmosis
2. Active transport
3. Into circulation

4. Into lumen of intestine
5. Into circulation

Metabolism

1. ATP
2. Oxidation-reduction

3. Reduced

Glycolysis

1. NADH
2. Phosphorylation
3. Two

4. Two
5. Two
6. Two

Anaerobic Respiration

1. ATP
2. Lactic acid
3. NADH
4. Cori cycle
5. Oxygen debt

Aerobic Respiration

A.
1. Glycolysis
2. Acetyl CoA formation
3. Citric acid (Kreb's) cycle
4. Aerobic respiration
5. Electron transport chain
6. Electron transport chain

4. Lactic acid
5. Acetyl CoA
6. Aerobic respiration
7. Citric acid cycle
8. NADH
9. H_2O
10. ATP

B.
1. Glycolysis
2. ADP
3. Anaerobic respiration

Lipids

1. Triglycerides
2. Free fatty acids
3. Beta-oxidation

4. Ketogenesis
5. Ketone bodies

Proteins

1. Proteins
2. Stored
3. Essential amino acids
4. Nonessential amino acids
5. Keto acids

6. Transamination
7. Oxidative deamination
8. NADH
9. Citric acid (Kreb's) cycle
10. Urea

Intraconversion of Nutrient Molecules

A. 1. Glycogenesis
2. Lipogenesis
3. Glycogenolysis
4. Gluconeogenesis

B. 1. Glycolysis
2. Glycogenesis
3. Glycogenolysis
4. Gluconeogenesis

Metabolic States

1. Absorptive state
2. Postabsorptive state

3. Postabsorptive state

Metabolic Rate

1. Calorie
2. Basal metabolic rate
3. Increases
4. Decreases

5. Decreases
6. Increases
7. Increases

Heat Production and Regulation

1. Free energy
2. Homeotherms
3. Conduction
4. Convection

5. Evaporation
6. Increase
7. Decrease

QUICK RECALL

1. Carbohydrates, lipids, proteins, vitamins, minerals, oxygen, and water
2. Carbohydrates: monosaccharides; proteins: amino acids; fats: fatty acids and glycerol
3. Carbohydrate digestion: mouth, small intestine; lipid digestion: small intestine; protein digestion: stomach, small intestine
4. Water-soluble molecules enter the hepatic portal system; lipid-soluble molecules enter the lacteals
5. Anaerobic respiration: two ATPs; aerobic respiration: 36 ATPs

6. NADH, $FADH_2$, and ATP
7. Anaerobic respiration: lactic acid; aerobic respiration: carbon dioxide and water
8. Glycolysis: six ATPs; pyruvic acid - acetyl CoA conversion: six ATPs; citric acid (Kreb's) cycle: 24 ATP
9. Fatty acids: beta-oxidation; amino acids: oxidative deamination
10. Glycogenesis, glycogenolysis, lipogenosis, and gluconeogenesis
11. Basal metabolic rate, muscular energy, and specific dynamic activity

MASTERY LEARNING ACTIVITY

1. D. Vitamins function as coenzymes, parts of coenzymes, or parts of enzymes. Most vitamins cannot be synthesized by the body, and they are not broken down before use. Vitamins A, D, E, and K are fat-soluble vitamins.

2. D. Minerals are inorganic nutrients that are necessary for normal metabolic functions.

3. A. Carbohydrates are composed of many monosaccharides linked together by covalent bonds. The complete breakdown of carbohydrates yields monosaccharides. The breakdown of proteins results in amino acids, and the breakdown of fats produces glycerol and fatty acids.

4. E. The salivary glands produce salivary amylase, the pancreas produces pancreatic amylase, and the lining of the small intestine produces a number of disaccharidases.

5. E. Bile is secreted by the liver, stored in the gall bladder, and enters the duodenum through the common bile duct. It contains bile pigments (breakdown products of hemoglobin) and bile salts that emulsify fats. Bile has no enzymatic activity.

6. A. Bile salts surround lipids within the small intestine to form micelles. The lipids diffuse from the micelles into intestinal cells, are coated with proteins, and are released into lacteals as chylomicrons.

7. C. Large amounts of lipids enter the lacteals, are carried to the thoracic duct, and finally to the venous circulation at the base of the left subclavian vein. If the thoracic duct were tied off, the movements of lipids into the general circulation would be hampered.

8. D. Pepsin (from the stomach) and trypsin (from the pancreas) are enzymes that digest proteins. Amylase digests starches and lipase digests lipids. Hydrochloric acid and bile have no enzymatic activity.

9. A. Anaerobic respiration, by definition, occurs in the absence of oxygen. It produces a net gain of two ATPs for every molecule of glucose that is degraded. Aerobic respiration, which takes place in the presence of oxygen, produces 36 ATPs.

10. A. Glycolysis is the breakdown of glucose to pyruvic acid. In aerobic respiration pyruvic acid is converted to acetyl CoA, which enters the Kreb's cycle.

11. D. NADH is the transport molecule. RNA is involved in protein synthesis, and ADP combines with a phosphate group to form ATP.

12. B. At the end of the electron transport chain, hydrogen combines with oxygen to produce water (often called metabolic water).

13. D. Remember that glucose (and other organic food molecules) is made up of carbon molecules bonded together. At the end of respiration, all the glucose is gone. Each carbon bond is broken, and the energy is transferred to ATPs. The carbon molecules combine with oxygen, and you breathe it out as a waste product, carbon dioxide. These reactions take place after glycolysis and in the Kreb's cycle. Therefore part of the food you eat is breathed out in carbon dioxide.

14. A. Lipids are stored primarily as triglycerides. The fatty acids of lipids are broken down by beta-oxidation into acetyl CoA molecules.

15. D. Amino acids are essential (must be ingested) or nonessential (synthesized by transamination reaction). Amino acids can be used as an energy source in an oxidative deamination reaction.

16. C. Ammonia is a by-product of oxidative deamination (the breakdown of amino acids for energy). Ammonia is toxic to cells and is converted into the less toxic urea in the liver. The urea is then eliminated by the kidneys.

17. A. Proteins (amino acids) and lipids (glycerol) can be used as a source of glucose. Glycogenesis is the conversion of glucose into glycogen, and glycogenolysis is the conversion of glycogen into glucose. Ketogenesis is the formation of ketones during lipid metabolism.

18. B. In the absorptive state glucose is used as an energy source or is stored (glycogen or fats). The other events listed take place in the postabsorptive state.

19. A. Basal metabolism accounts for 60% of energy expenditure, physical activity 30%, and assimilation of food (specific dynamic activity) 10%.

20. B. Evaporation is the loss of water (i.e., heat)

 FINAL CHALLENGES

1. Salivary amylase digests the starch in the bread, breaking the starch down to maltose and isomaltose. These sugars are responsible for the sweet taste.

2. The bile salts normally emulsify fats, which greatly increases the efficiency of the enzymes that digest fats. Bile salts (micelles) also transport fats to the intestinal wall, where the fats are absorbed. A drug that binds to bile salts and prevents these functions would prevent the absorption of fats such as cholesterol.

3. Beta-oxidation of the six-carbon fatty acid chain produces three acetyl CoA molecules. For all but the last acetyl CoA molecule formed, an NADH and $FADH_2$ are also produced. The electron-transport chain converts each NADH to three ATPs and each $FADH_2$ to two ATPs. Thus, for each acetyl CoA (except the last) there are five ATPs produced. For the six-carbon fatty acid therefore there are 10 ATPs produced from the NADHs and the $FADH_2$s. In addition, each acetyl CoA can enter the citric acid cycle to yield 12 ATPs. Since there are three acetyl CoA molecules, this yields 36 ATPs. Remembering that it takes two ATPs to start beta-oxidation, the total production of ATPs from the six-carbon fatty acid chain is 44 ATPs (10 + 36 - 2). Beta-oxidation of the fatty acid chain produces more ATPs than the metabolism of glucose, which yields 36 ATPs. In other words, fats can store more energy than carbohydrates.

4. First, exercise increases energy (kilocalorie) usage. Second, following exercise basal metabolic rate is elevated due to elevated body temperature and due to repayment of the oxygen debt. Third, exercise increases the proportion of muscle tissue to adipose tissue in the body. Since muscle tissue is metabolically more active than adipose tissue, basal metabolic rate increases. Fourth, during exercise epinephrine levels increase, resulting in increased blood sugar levels (see Chapter 18), which can depress the hunger center in the brain and reduce food consumption following exercise.

5. During fever production the body produces heat by shivering. The body also conserves heat by vasoconstriction of blood vessels in the skin (producing a pale skin) and by reduction in sweat loss (producing a dry skin). When the fever breaks, i.e., "the crisis is over," heat is lost from the body to lower body temperature to normal. This is accomplished by vasodilation of blood vessels in the skin (producing a flushed skin) and increased sweat loss (producing a wet skin).

Urinary System

FOCUS: The urinary system consists of the kidneys, ureters, urinary bladder, and urethra. The kidneys function to remove waste products from the blood, produce vitamin D, stimulate red blood cell production, and regulate blood volume, electrolyte levels, and pH. The ureters carry urine from the kidneys to the urinary bladder, where the urine is stored. Stretch of the urinary bladder initiates reflexes that cause the smooth muscles of the bladder to contract, and the urine passes to the outside of the body through the urethra. The functional unit of the kidney is the nephron, which consists of the glomerulus (a capillary) and a tubule. Materials enter the tubule from the glomerulus by passing through a filtration membrane, which prevents the entry of blood cells and large molecules. As the filtrate passes through the tubule, most of the useful materials are reabsorbed, leaving waste products in the urine. In some parts of the tubule hydrogen and potassium ions are secreted into the urine. The concentration of urine is regulated by aldosterone, which increases sodium reabsorption, and ADH, which increases water reabsorption. In addition, renin secreted by the kidneys results in increased angiotensin II production, which causes a decrease in urine volume.

WORD PARTS

Give an example of a new vocabulary word that contains each word part.

WORD PART	MEANING	EXAMPLE
juxta-	near	1. _____
rect-	straight	2. _____
vasa-	vessel	3. _____
corpusc-	little body	4. _____
papill-	nipple	5. _____
lamin-	layer	6. _____

WORD PART	MEANING	EXAMPLE
propri-	peculiar	7. _____
natri-	natrium (sodium)	8. _____
ure-	urine	9. _____
mictur-	urinate	10. _____

CONTENT LEARNING ACTIVITY

Introduction

❝The urinary system participates with other organs to maintain homeostasis in the body by regulating the❞ interstitial fluid composition within a narrow range of values.

Using the terms provided, complete the following statements:

Blood   Red blood cell
Ions   Toxic
pH   Vitamin D

The kidneys remove wastes, many of which are _(1)_, from the blood and play a major role in controlling _(2)_ volume, the concentration of _(3)_ in the blood, and the _(4)_ of the blood. The kidneys are also involved in the control of _(5)_ production and _(6)_ metabolism.

1. _____

2. _____

3. _____

4. _____

5. _____

6. _____

Kidneys

❝The kidneys are bean-shaped organs, each about the size of a tightly -clenched fist.❞

A. Match these terms with the correct statement or definition:

Hilum   Renal fat pad
Major calyces   Renal pelvis
Minor calyces   Renal sinus
Renal capsule   Ureter
Renal fascia

_____ 1. Fibrous connective tissue that surrounds each kidney.

_____ 2. Dense deposit of adipose tissue surrounding the renal capsule.

_____ 3. Thin layer of connective tissue that attaches the kidneys and surrounding adipose tissue to the abdominal wall.

_____ 4. Cavity filled with fat and connective tissue into which the hilum opens.

_____ 5. Enlarged urinary channel that is found in the center of the renal sinus.

_____ 6. Urinary channels that extend from the kidney tissue and open into the renal pelvis.

_____ 7. Tube that extends from the renal pelvis to the urinary bladder.

B. Match these terms with the correct statement or definition:

Cortex
Medulla
Medullary rays

Renal columns
Renal papilla
Renal pyramids

_____ 1. Outer portion of the kidney.

_____ 2. Cone-shaped structures found in the medulla of the kidney.

_____ 3. Extensions from the base of the renal pyramid that are located in the cortex.

_____ 4. Apex of the renal pyramid, found in the medulla and surrounded by a minor calyx.

_____ 5. Cortical tissue that extends between the pyramids to the renal sinus.

C. Match these terms with the correct parts of the diagram labeled in Figure 26-1:

Major calyx
Minor calyx
Renal capsule
Renal column

Renal papilla
Renal pelvis
Renal pyramid
Ureter

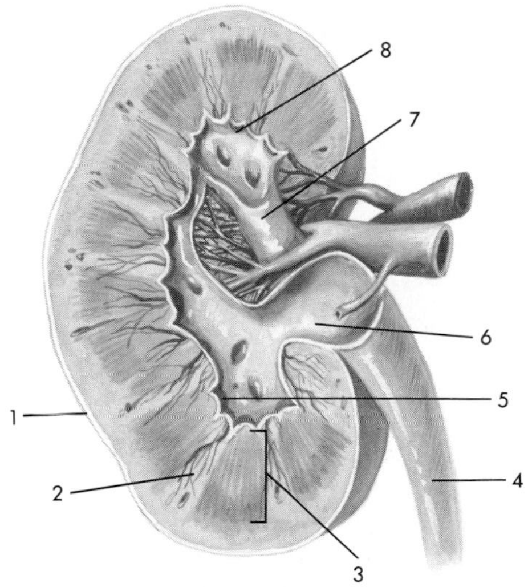

Figure 26-1

1. _____

2. _____

3. _____

4. _____

5. _____

6. _____

7. _____

8. _____

D. Match these terms with the correct statement or definition:

Filtration membrane   Nephron
Bowman's capsule   Podocytes
Glomerulus   Renal corpuscle

_____ 1. Histological and functional unit of the kidney.

_____ 2. Tubule portion of the renal corpuscle, composed of a parietal and visceral layer.

_____ 3. Capillary portion of the renal corpuscle.

_____ 4. Specialized cells found in the visceral layer of Bowman's capsule; gaps between their processes are the filtration slits.

_____ 5. Collective name for the capillary epithelium, basement membrane, and the podocytes.

E. Match these terms with the correct statement or definition:

Ascending limb   Juxtaglomerular apparatus
Collecting duct   Juxtaglomerular cells
Descending limb   Macula densa
Distal convoluted tubule   Proximal convoluted tubule

_____ 1. Smooth muscle cells that form a cuff around the afferent arteriole.

_____ 2. Specialized tubule cells found in the distal convoluted tubule, where it is adjacent to the afferent and efferent arterioles.

_____ 3. Part of the nephron between Bowman's capsule and the loop of Henle.

_____ 4. Portion of the loop of Henle connected to the distal convoluted tubule.

_____ 5. Duct to which the distal convoluted tubules of many nephrons join.

F. Match these terms with the correct parts of the diagram labeled in Figure 26-2:

Afferent arteriole   Loop of Henle
Collecting duct   Proximal convoluted tubule
Distal convoluted tubule   Renal corpuscle
Efferent arteriole

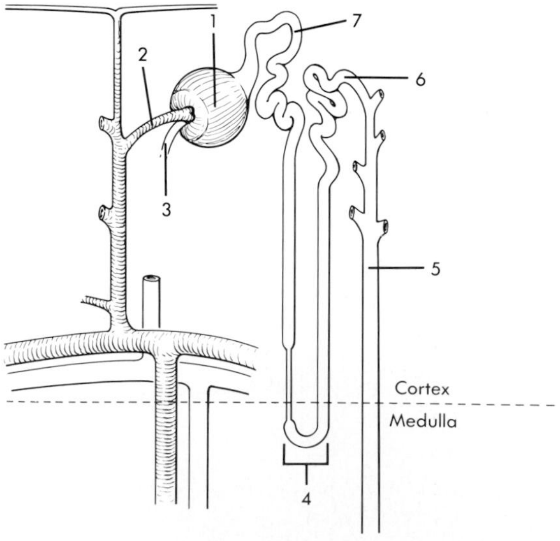

1. _____

2. _____

3. _____

4. _____

5. _____

6. _____

7. _____

Figure 26-2

G. Place the following vessels in the correct sequence that blood would pass through them, from the abdominal aorta to the interlobular veins.

Afferent arteriole
Arcuate artery
Efferent arteriole
Glomerulus
Interlobar artery

Interlobular artery
Peritubular capillaries
Renal artery
Segmental artery

1. Abdominal aorta _____

2. _____

3. _____

4. _____

5. _____

6. _____

7. _____

8. _____

9. _____

10. _____

11. Interlobular veins _____

☞ The vasa recta are specialized portions of the peritubular capillaries that dip into the medulla, along with the loops of Henle.

Ureters and Urinary Bladder

"The ureters extend from the renal pelvis to reach the urinary bladder, which functions to store urine.**"**

Match these terms with the correct statement or definition:

External urinary sphincter
Internal urinary sphincter
Smooth muscle

Transitional epithelium
Trigone

_____

_____

_____

_____

1. Triangular area of the bladder wall between the two ureters posteriorly and the urethra anteriorly.

2. Skeletal muscle that surrounds the urethra as it extends through the pelvic floor.

3. Thickness of the walls of the urinary bladder are mainly due to layers of these cells.

4. This internal lining permits changes in size of the urinary bladder and ureter.

Urine Production

"Because nephrons are the smallest structural components capable of producing urine,**"** they are called the function units of the kidney.

A. Match these terms with the correct statement or definition:

Filtration
Reabsorption

Secretion

_____

_____

_____

1. Movement of plasma across the filtration membrane due to a pressure gradient.

2. Movement of substances from the filtrate back into the blood.

3. Active transport of substances into the nephron.

555

Urine consists of constituents filtered and secreted into the nephron minus those substances that are reabsorbed.

B. Match these terms with the correct statement or definition:

Filtration fraction   Renal fraction
Glomerular filtration rate

_____ 1. Portion of the total cardiac output that passes through the kidneys; used to calculate renal blood flow rate.

_____ 2. Portion of the plasma volume that is filtered through the filtration membrane.

_____ 3. Amount of filtrate produced per minute.

C. Match these terms with the correct statement or definition:

Capsule pressure   Filtration pressure
Colloid osmotic pressure   Glomerular capillary pressure

_____ 1. Blood pressure within the glomerulus.

_____ 2. Pressure of filtrate already in Bowman's capsule; opposes the glomerular capillary pressure.

_____ 3. Pressure caused by unfiltered plasma proteins remaining within the glomerular capillary.

_____ 4. Net pressure gradient that forces fluid from the glomerular capillary through the filtration membrane into Bowman's capsule.

 The filtration barrier of the renal corpuscle is 100 to 1000 times more permeable than a typical capillary.

D. Using the terms provided, complete the following statements:

Bowman's capsule   Increased
Decreased   Peritubular capillaries

As the diameter of a vessel decreases, the resistance to flow through the vessel increases. Because the efferent arteriole has a small diameter, there is a(n) (1) resistance to blood flow, and blood pressure within the glomerulus is high. Consequently, filtration pressure is high, and filtrate moves into (2) . After the efferent arteriole, vessel diameters increase and blood pressure is (3) , therefore, fluid moves out of the interstitial spaces and into the (4) . Changing the diameter of the efferent arteriole can alter filtration pressure. For example, vasoconstriction of the efferent arteriole would result in (5) filtration pressure, and (6) production of urine.

1. _____

2. _____

3. _____

4. _____

5. _____

6. _____

E. Using the terms provided, complete the following statements:

Active transport   Osmosis
Antidiuretic hormone (ADH)   Out of
Chloride ions   Permeable
Cotransport   Sodium ions
Into

Proteins, amino acids, glucose, fructose, and sodium, potassium, calcium, bicarbonate and chloride ions all move from the lumen of the nephron to the interstitial spaces by the process of _(1)_ . Amino acids and glucose are transported across the membrane of the microvilli with sodium ions, through a process called _(2)_ . Because the proximal convoluted tubule is permeable to water, as solute particles are transported to the interstitial spaces, water follows by _(3)_ . As filtrate moves through the thin segment of the descending limb of the loop of Henle, water moves _(4)_ the nephron, and solutes diffuse _(5)_ the nephron. The cells of the ascending limb of the loop of Henle actively transport _(6)_ from the lumen of the tubule to the interstitial spaces. Because chloride ions have a negative charge, _(7)_ , which have a positive charge, move with them. Water does not follow them because the ascending limb of the loop of Henle is not _(8)_ to water. Sodium ions are actively transported across the wall of the distal convoluted tubule. The permeability of the distal convoluted tubule and the collecting duct is controlled by _(9)_ . When this substance increases, water moves by osmosis _(10)_ the distal convoluted tubule and collecting duct.

1. _____

2. _____

3. _____

4. _____

5. _____

6. _____

7. _____

8. _____

9. _____

10. _____

☞ Urea, urate ions, creatinine, sulfates, phosphates, and nitrates are reabsorbed, but at a much lower rate than water; therefore a greater percentage of them are eliminated in the urine.

F. Match these terms with the correct statements as they pertain to tubular secretion:

Active transport
Diffusion

_____   1. Method of movement of hydrogen and potassium ions into the distal convoluted tubules and the collecting ducts.

_____   2. Method of movement of ammonia into the lumen of the nephron.

Urine Concentration Mechanism

66 *The urine concentrating mechanism is regulated so that either dilute or concentrated urine is formed by the kidney.* 99

A. Match these terms with the correct statement or definition:

Loop of Henle
Vasa recta

_____ 1. Countercurrent system that supplies blood to the kidney medulla without disturbing the interstitial fluid concentration.

_____ 2. Countercurrent system that adds solutes to the interstitial fluid of the medulla, increasing the concentration of the interstitial fluid.

_____ 3. System that carries away extra water and solutes from the medulla.

☞ The formation of a concentrated urine depends on the production and maintenance of a high interstitial fluid concentration in the medulla. This is accomplished by the loop of Henle and the vasa recta.

B. Match these terms with the correct statement or definition:

Ascending limb of the loops of Henle
Descending limb of the loops of Henle

Distal convoluted tubules
Proximal convoluted tubules

_____ 1. 65% of the filtrate is reabsorbed at this location.

_____ 2. As the filtrate moves through these, water moves out and solutes diffuse into the nephron; osmolality increases up to about 1200 mOsm/L.

_____ 3. This part of the nephrons is not permeable to water, but chloride ions are transported out, sodium ions follow passively, and urea diffuses out.

_____ 4. In this part of the nephrons and in collecting ducts, if ADH is present, water diffuses out of the nephron and collecting ducts, resulting in a concentrated urine.

Regulation of Urine Concentration and Volume

66 *Regulation of urine production involves hormonal mechanisms, autoregulation, and sympathetic nervous system stimulation.* 99

A. Match these terms with the correct statement:

Decreases
Increases

_____ 1. Effect of aldosterone on sodium and chloride ion transport into the blood.

_____ 2. Effect of the hyposecretion of aldosterone on urine volume and on concentration of solutes in urine.

_____ 3. Effect of hypersecretion of aldosterone on potassium and hydrogen ion secretion.

_____ 4. Effect of increased concentrations of potassium ions in the plasma or interstitial fluid on aldosterone secretion.

_____ 5. Effect of angiotensin II on the secretion of aldosterone by adrenal cortex cells.

B. Match these terms with the correct location in the following statements as they pertain to the renin-angiotensin-aldosterone system of the kidney:

Decrease(s)
Increase(s)

If blood pressure in the afferent arteriole decreases, the rate of renin secretion by the juxtaglomerular apparatus _(1)_ . Renin converts angiotensinogen to angiotensin I, and other proteolytic enzymes convert angiotensin I to angiotensin II. Increased angiotensin II _(2)_ blood pressure in two ways. First, angiotensin II is a potent vasoconstrictor, causing a(n) _(3)_ in peripheral resistance, which causes an increase in blood pressure. Second, it _(4)_ the rate of aldosterone secretion, which leads to a(n) _(5)_ in water retention by the kidney, and a(n) _(6)_ in urine volume. An increase in sodium and chloride ions in the filtrate passing through the juxtaglomerular apparatus _(7)_ renin secretion. An increase in sodium and chloride ions can occur due to diet or a(n) _(8)_ in glomerular filtration rate.

1. _____

2. _____

3. _____

4. _____

5. _____

6. _____

7. _____

8. _____

C. Match these terms with the correct statement:

Decreases
Increases

_____ 1. Effect of increased blood osmolality on ADH secretion.

_____ 2. Effect of increased ADH secretion on blood osmolality.

_____ 3. Effect of increased blood pressure on ADH secretion.

_____ 4. Effect of decreased ADH secretion on blood pressure.

_____ 5. Effect of increased blood pressure on atrial natriuretic factor secretion.

_____ 6. Effect of atrial natriuretic factor on ADH secretion.

_____ 7. Effect of large increases in arterial blood pressure on urine production.

_____ 8. Effect of constriction of the afferent arteriole on renal blood flow and filtration pressure.

_____ 9. Effect of sympathetic stimulation on renal blood flow and filtrate formation.

 Autoregulation, which involves changes in the degree of constriction of the afferent and efferent arterioles, maintains a relatively constant filtration rate despite comparatively large changes in systemic blood pressure.

Clearance and Tubular Maximum

66 *Clearance and tubular maximum are concerned with removal of substances from the plasma and filtrate.* **99**

Match these terms with the correct statement or definition:

Plasma clearance   Tubular maximum
Tubular load

_____ 1. Total amount of a substance that filters through the filtration membrane into the nephrons each minute.

_____ 2. Maximum rate at which a substance can be actively reabsorbed.

☞ If the tubular load for a substance exceeds its tubular maximum, the excess amount of that substance will remain in the urine.

Urine Movement

66 *Urine is produced in the nephrons of the kidney, moves into the renal pelvis, then into the ureters,* **99** *next into the urinary bladder and out through the urethra.*

A. Match these terms with the correct statement or definition:

Hydrostatic pressure
Peristaltic contractions

_____ 1. Mechanism responsible for urine flow from the nephron into the renal pelvis.

_____ 2. Mechanism responsible for urine flow through the ureters into the urinary bladder.

B. Match these terms with the correct statement or definition:

External urinary sphincter   Micturition reflex
Higher brain centers

_____ 1. Reflex initiated by stretching of the bladder wall, which results in contraction of the bladder and inhibition of the urinary sphincters.

_____ 2. Part of the nervous system responsible for inhibition or stimulation of the micturition reflex.

_____ 3. This structure is kept tonically contracted by the higher brain centers.

1. List five major parts of a nephron.

2. Name the two parts of the juxtaglomerular apparatus.

3. List the three steps in urine formation.

4. Write the formula for filtration pressure.

5. Complete the following table by placing a + under the location where the condition exists.

CONDITION	Proximal Convoluted Tubule	Descending Limb	Ascending Limb	Distal Convoluted Tubule
Na$^+$ actively reabsorbed from tubule	_____	_____	_____	_____
Na$^+$ passively diffuses into tubule	_____	_____	_____	_____
Na$^+$ passively diffuses out of tubule	_____	_____	_____	_____
Water moves out of tubule by osmosis	_____	_____	_____	_____

6. Name two ions actively secreted into the distal convoluted tubule.

ANSWERS TO CHAPTER 26

1. juxtaglomerular, juxtamedullary
2. vasa recta
3. vasa recta
4. corpuscle
5. renal papilla

6. lamina propria
7. lamina propria
8. natriuretic
9. natriuretic, diuretic, ureter, urethra
10. micturition

CONTENT LEARNING ACTIVITY

Introduction

1. Toxic
2. Blood
3. Ions

4. pH
5. Red blood cells
6. Vitamin D

Kidneys

A. 1. Renal capsule
    2. Renal fat pad
    3. Renal fascia
    4. Renal sinus
    5. Renal pelvis
    6. Major calyces
    7. Ureter

B. 1. Cortex
    2. Renal pyramids
    3. Medullary rays
    4. Renal papilla
    5. Renal columns

C. 1. Renal capsule
    2. Renal column
    3. Renal pyramid
    4. Ureter
    5. Minor calyx
    6. Renal pelvis
    7. Major calyx
    8. Renal papilla

D. 1. Nephron
    2. Bowman's capsule
    3. Glomerulus
    4. Podocytes
    5. Filtration membrane

E. 1. Juxtaglomerular cells
    2. Macula densa
    3. Proximal convoluted tubule
    4. Ascending limb
    5. Collecting duct

F. 1. Glomerulus
    2. Afferent arteriole
    3. Efferent arteriole
    4. Loop of Henle
    5. Collecting duct
    6. Distal convoluted tubule
    7. Proximal convoluted tubule

G. 2. Renal artery
    3. Segmental artery
    4. Interlobar artery
    5. Arcuate artery
    6. Interlobular artery
    7. Afferent arteriole
    8. Glomerulus
    9. Efferent arteriole
    10. Peritubular capillaries

Ureters and Urinary Bladder

1. Trigone
2. External urinary sphincter

3. Smooth muscle
4. Transitional epithelium

Urine Production

A. 1. Filtration
   2. Reabsorption
   3. Secretion

B. 1. Renal fraction
   2. Filtration fraction
   3. Glomerular filtration rate

C. 1. Glomerular capillary pressure
   2. Glomerular capsule pressure
   3. Colloid osmotic pressure
   4. Filtration pressure

D. 1. Increased
   2. Bowman's capsule
   3. Decreased
   4. Peritubular capillaries
   5. Increased
   6. Increased

E. 1. Active transport
   2. Cotransport
   3. Osmosis
   4. Out of
   5. Into
   6. Chloride ions
   7. Sodium ions
   8. Permeable
   9. Antidiuretic hormone (ADH)
   10. Out of

F. 1. Active transport
   2. Diffusion

Urine Concentration Mechanism

A. 1. Vasa recta
   2. Loop of Henle
   3. Vasa recta

B. 1. Proximal convoluted tubule
   2. Descending limb of the loops of Henle
   3. Ascending limb of the loops of Henle
   4. Distal convoluted tubule

Regulation of Urine Concentration and Volume

A. 1. Increases
   2. Increases
   3. Increases
   4. Increases
   5. Increases

B. 1. Increases
   2. Increases
   3. Increase
   4. Increases
   5. Increase
   6. Decrease
   7. Decreases
   8. Increase

C. 1. Increases
   2. Decreases
   3. Decreases
   4. Decreases
   5. Increases
   6. Decreases
   7. Increases
   8. Decreases
   9. Decreases

Clearance and Tubular Maximum

1. Tubular load

2. Tubular maximum

Urine Movement

A. 1. Hydrostatic pressure
   2. Peristaltic contractions

B. 1. Micturition reflex
   2. Higher brain centers
   3. External urinary sphincter

QUICK RECALL

1. Glomerulus, Bowman's capsule, proximal convoluted tubule, loop of Henle, distal convoluted tubule
2. Macula densa and juxtaglomerular cells of the afferent arterioles
3. Filtration, reabsorption, and secretion
4. Filtration pressure = glomerular capillary pressure minus glomerular capsule pressure minus colloid osmotic pressure.
5.

	PCT	DL	AL	DCT
Na^+ actively reabsorbed	+	-	-	+
Na^+ passively diffuses in	-	+	-	-
Na^+ passively diffuses out	-	-	+	-
Water moves out by osmosis	+	+	-	+

6. Potassium ions and hydrogen ions
7.

	PCT	TLH	DCT
100 mOsm	-	-	+
300 mOsm	+	-	-
1200 mOsm	-	+	-

8.

	PCT	DL	AL	DCT	CD
65%	+	-	-	-	-
15%	-	+	-	-	-
10%	-	-	-	+	-
9% (if ADH present)	-	-	-	-	+

9. Aldosterone: increased sodium ion reabsorption, resulting in decreased urine concentration and volume; renin: causes angiotensin II production, which stimulates aldosterone production, resulting in decreased urine concentration and volume; ADH: decreased urine volume; atrial natriuretic factor: inhibits ADH production, resulting in increased urine volume.
10. Stretch of bladder, reflex stimulated, bladder contracts, and urinary sphincters inhibited

1. E. The kidneys perform many important functions that you should be familiar with.

2. B. The glomeruli are found primarily in the cortex (outer layer) of the kidney. The renal pyramids are found in the medulla (inner layer) of the kidney. The renal fat pad surrounds the kidney. The hilum is an indentation through which arteries enter the kidney and veins and the ureter leave.

3. C. The vasa recta and the loops of Henle pass from the cortex into the medulla and then back into the cortical area. The collecting ducts pass through the medulla as they project toward the pelvis. Thus the vasa recta, the loops of Henle, and the collecting ducts are found primarily in the renal medulla and make up the greatest portion of it.

4. C. The collecting tubules converge at the renal papilla and the urine enters the minor calyx. The minor calyx empties into a major calyx, which connects to the renal pelvis. The ureter drains the renal pelvis.

5. E. The glomerulus is a knot of coiled capillaries from which materials enter the tubule portion of nephrons by filtration. The vasa recta is a capillary system that supplies the loop of Henle and the collecting ducts.

6. A. The interlobar arteries branch from the segmental arteries and pass through the renal columns. The arcuate arteries branch from the interlobar arteries, and arch over the base of the renal pyramids. The interlobular arteries branch from the arcuate arteries and extend into the cortex. The afferent arterioles are derived from the interlobular arteries.

7. D. The urinary bladder is located in the pelvic cavity. It is connected to the kidneys by the ureters and to the outside of the body by the urethra. It is a hollow organ consisting of smooth muscle and lined by transitional epithelium.

8. E. Filtration occurs when materials move from the glomerulus into Bowman's capsule. Reabsorption (e.g., sodium, glucose) and secretion (e.g., potassium and hydrogen ions) occur throughout the tubule system.

9. A. The glomerular filtration rate is the amount of plasma (filtrate) that enters Bowman's capsule per minute.

10. B. The glucose would pass through the fenestra of the capillary endothelium, the basement membranes of the capillary endothelium and podocytes, and the filtration slit formed by the pedicles of the podocytes.

11. B. The pressure gradient that develops across the glomerular membrane is responsible for glomerular filtration. Glomerular capillary pressure in the glomerular capillaries (40 mm Hg) is opposed by the capsule pressure within Bowman's capsule (10 mm Hg) and the colloid osmotic pressure of the plasma (30 mm Hg). Therefore the net filtration pressure is 0 mm Hg (40 - [30 +10]).

12. B. Recall that filtration pressure = glomerular capillary pressure - (glomerular capsule pressure +colloid osmotic pressure). Constriction of the afferent arterioles reduces glomerular capillary pressure (blood pressure) in the glomeruli, thus reducing filtration pressure. Elevated blood pressure or vasodilation of the afferent arteriole raises the glomerular capillary pressure in the glomeruli, causing an increase in filtration pressure. An increase in filtration pressure also results when plasma protein concentrations decrease because this decreases the colloid osmotoic pressure.

13. A. Normally all of the glucose, proteins, amino acids, and vitamins that enter Bowman's capsule are reabsorbed by active processes in the proximal convoluted tubule.

14. A. About 80% of the volume of the filtrate is reabsorbed in the proximal convoluted tubule. Reabsorption in the proximal convoluted tubule is neither regulated by aldosterone nor ADH. It is obligatory reabsorption due to such factors as the active transport of sodium, amino acids, and glucose from the proximal convoluted tubule.

15. C. This is an important fact to know if you are to understand renal physiology. Sodium (or chloride) ions move out of the tubules by active transport, and water passively follows by osmosis because of the resulting osmotic gradient.

16. D. Potassium is secreted into the distal convoluted and collecting tubules by active transport. Potassium is removed from the proximal convoluted tubule by active transport.

17. E. To produce a hyperosmotic solution in the interstitial spaces of the medulla, solutes (chloride and sodium) must be added. This process must necessarily be by active transport, since movement is against the concentration gradient. Chloride is actively transported, and sodium follows the chloride by diffusion. If the ascending limb were permeable to water, water would follow the chloride and sodium, resulting in dilution and destroying the hyperosmotic environment. Hence the necessity of the impermeability of the ascending limb to water.

18. D. Filtrate in the proximal convoluted tubule is isosmotic with the plasma in the glomerulus. As the filtrate move down to the bottom of the loop of Henle, it becomes hyperosmotic due to the counter current mechanism in Henle's loop. By the time the distal convoluted tubule is reached, the solution is hypoosmotic to blood.

19. E. As aldosterone concentrations increase, potassium secretion and sodium reabsorption increase. As sodium moves from the urine back into the blood, chloride and water passively follow. This increases the extracellular fluid volume and decreases urine volume.

20. C. Renin is secreted by the juxtaglomerulus apparatus, ADH and oxytocin by the neurohypophysis, and aldosterone by the adrenal cortex.

21. D. ADH governs the permeability of the distal convoluted tubules and collecting ducts and does not directly affect any of the other choices. When ADH levels increase, permeability increases.

22. C. When blood osmolality decreases, this inhibits the osmoreceptors in the hypothalamus, and there is less ADH secretion. As a consequence, the permeability of the collecting tubules decreases, and less water is reabsorbed, producing a less concentrated urine.

23. A. Despite large changes in blood pressure, there is usually a small change in glomerular capillary pressure and, therefore only a slight increase in glomerular filtration rate. A large change in glomerular capillary pressure is prevented by constriction of the afferent arterioles and is called autoregulation.

24. B. The tubular load is the amount of a substance that enters the renal capsule. Tubular maximum is the fastest rate at which a substance is reabsorbed from the nephron. Plasma clearance is the volume of plasma from which a substance is completely removed in a minute.

25. C. The transection would eliminate nervous impulses to and from the brain; thus volunatry control and the sensation of the need to urinate would be lost. The spinal cord center for urination would still be functional, so reflex emptying of the bladder would still occur.

1. Renal blood flow would be 1008 ml of blood per minute (5600 x 0.18), renal plasma flow would be 554 ml of plasma per minute (1008 x 0.55), glomerular filtration rate would be 105 ml of filtrate per minute (554 x 0.19), and urine production would be 0.84 ml of urine per minute (105 x 0.008) or 1.2 L of urine per day (0.84 x 1.44).

2. Arteriosclerosis would reduce blood pressure in the afferent arterioles and increase renin secretion by the juxtaglomerular apparatus. Renin converts angiotensinogen to angiotensin I, which is converted into angiotensin II. Angiotensin II increases aldosterone secretion. The decreased urine production that results from the increased angiotensin II and the aldosterone causes an increased blood volume and blood pressure. Through baroreceptors the increased blood pressure inhibits ADH secretion.

3. Urine production should increase for several reasons. First, the increased fluid intake would increase blood volume, blood pressure, and glomerular capillary pressure. Vasodilation of the afferent arterioles by the caffeine would also increase glomerular capillary pressure. As a result of the increased glomerular capillary pressure, the filtration pressure would increase, producing an increased glomerular filtration rate and therefore a greater volume of urine. Second, inhibition of ADH secretion would decrease the permeability of the collecting ducts to water, so a larger amount of filtrate would pass through as urine.

4. Excess aldosterone would lead to increased sodium and water reabsorption. Due to the water increase, blood volume would increase, causing a rise in blood pressure. Eventually, however, the increase in blood volume is opposed by mechanisms that regulate blood volume/pressure. The increased sodium in the extracellular fluid stimulates the thirst center, resulting in excessive drinking followed by excessive urine production. Diarrhea would be a symptom associated with hyposecretion of aldosterone. When aldosterone is absent or in low quantities, sodium reabsorption from the intestines is very poor, and the intestinal contents become hyperosmotic, water is retained, and diarrhea results.

5. There are several ways that a diuretic could increase water loss from the kidneys.
   A. It could block sodium reabsorption or stimulate sodium excretion. Since water follows sodium, producing a higher than normal amount of sodium in the tubules would result in a larger urine volume. The drug could act directly on the sodium pump mechanism, block the effects of aldosterone on the tubule, or even block aldosterone production.
   B. It could inhibit ADH production or interfere with the effects of ADH on the collecting ducts.
   C. It could add a solute (glycerol or mannitol) that is filtered but not reabsorbed. The solute then osmotically obligates large amounts of tubular water. In the descending limb of the loop of Henle, the water moves into the interstitial fluid, overloading the countercurrent mechanism. The osmolality of the interstitial fluid is reduced, due to the influx of water; and, as a result, less water will be reabsorbed from the collecting ducts.

6. If only water were lost, blood osmolality would increase and stimulate ADH secretion. The ADH would increase the permeability of the collecting ducts to water, increasing water reabsorption and reducing the effectiveness of the diuretic. If sodium is also lost, blood osmolality does not increase, and the ADH response does not occur.

7. A low-salt diet would tend to reduce the osmolality of the blood. Consequently, ADH secretion would increase, producing a dilute urine and thus eliminating water. This in turn would reduce blood volume and blood pressure.

8. As the loops of Henle become longer, the countercurrent mechanism becomes more and more effective, raising the concentration of the interstitial fluid. The maximum concentration for urine is determined by the concentration of the interstitial fluid at the ends of the loops of Henle. Thus the longer the loops of Henle, the greater the concentration of urine it is possible to produce.

9. Here is one possible approach: use an experimental animal that has had both kidneys removed and is being maintained on the dialysis machine. It is then possible to increase the quantity of salt without increasing the water. Even if salt concentrations rise as much as 20% above normal, no hypertension develops. On the other hand, if the volume is increased with out a change in total body salt, hypertension does develop. These results are obtained when the kidneys are completely nonfunctional. If the kidneys are even partially functional, as fluid volume increases so does salt, because the kidneys will not retain water without salt. For this reason clinicians think in terms of salt instead of water retention, even though water retention results in hypertension.

Water, Electrolytes, And Acid-Base Balance

FOCUS: Intracellular fluid (i.e., the fluid inside cells) is different from the extracellular fluid (i.e., the fluid outside cells). Sodium ions are responsible for most of the osmotic pressure of extracellular fluid. Consequently, mechanisms that regulate blood volume or blood osmolality such as ADH levels are important for the regulation of blood sodium ion concentrations. Potassium ion concentrations affect resting membrane potentials and are regulated by aldosterone. Calcium ion concentrations, which are regulated by parathyroid hormone and vitamin D, affect the electrical activity of cells. Water levels are regulated by a balance between water intake (thirst) and water output (evaporation, feces, and urine). Blood pH is normally 7.4. A decrease in blood pH below normal is called acidosis, and an increase above normal is termed alkalosis. Blood pH is controlled by buffers, chemicals that resist a change in pH; by the respiratory system, which changes pH by changing blood carbon dioxide levels; and by the kidneys, which can produce an acidic or alkaline urine.

WORD PARTS

Give an example of a new vocabulary word that contains each word part.

WORD PART	MEANING	EXAMPLE
electro-	electricity	1. _____
-lyte	that which may be loosed	2. _____
baro-	pressure	3. _____
osmo-	pushing	4. _____
diure-	urinate	5. _____

WORD PART	MEANING	EXAMPLE
anti-	against	6. _____
para-	beside; near	7. _____
lumen-	light; opening	8. _____
permea-	pass through	9. _____
im-	not	10. _____

CONTENT LEARNING ACTIVITY

Body Fluids

"_There are two major fluid compartments in the body._**"**

Match these terms with the
correct statement or definition:

Extracellular fluid
Intracellular fluid

_____

1. Accounts for about 40% of the total body weight and includes the small amount of fluid in trillions of cells.

_____

2. Accounts for about 20% of the total body weight and includes lymph, cerebrospinal fluid, plasma, synovial fluid and interstitial fluid.

☞ The osmotic concentration of most extracellular fluid compartments is approximately equal.

Regulation of Intracellular Fluid Concentration

"_The composition of intracellular fluid is substantially different from that of extracellular fluid._**"**

Using the terms provided, complete the following statements:

Decreases   Osmosis
Electrolytes Proteins
Increases

1. _____

2. _____

3. _____

Cell membranes are differentially permeable, being relatively impermeable to _(1)_ ; whereas the movement of _(2)_ is determined by transport processes, as well as electrical charge differences. Water movement across the cell membrane by _(3)_ is influenced by the composition of the extracellular fluid. During conditions of dehydration the concentration of solutes in the extracellular fluid _(4)_ , resulting in the movement of water from the intracellular space into the extracellular space. When water intake increases following a period of dehydration, the concentration of solutes in the extracellular fluid _(5)_ , resulting in the movement of water into the cells.

4. _____

5. _____

Regulation of Ion Concentrations

66*The regulation of water and electrolytes involves the coordinated participation of several organ systems.***99**

A. Match these terms with the correct statement or definition:

Aldosterone   Sodium ions
Kidneys   Sweat

_____

1. Dominant extracellular cation; over 90% of the osmotic pressure is due to these ions and the negative ions associated with them.

_____

2. Major route by which sodium ions are secreted.

_____

3. Hormone that increases reabsorption of sodium ions from the distal convoluted tubule and collecting duct.

☞ The most abundant anion in the extracellular fluid is chloride; it is regulated by the same mechanisms that regulate sodium ions.

B. Match these terms with the correct location on the following diagram by placing a D in the blank for those substances that are decreased and an I for those that are increased.

Decreased
Increased

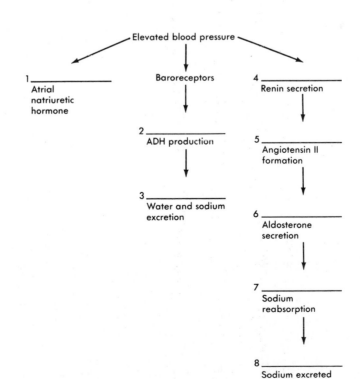

Figure 27-1

575

C. Using the terms provided, complete the following statements:

Decrease   Hyperpolarization
Depolarization   Increase
Distal convoluted tubules   Proximal convoluted tubule

The extracellular concentration of potassium ions must be
maintained within a narrow range of concentrations. An
increase in the extracellular potassium ion concentration leads to
 (1) of the resting membrane potential, and a decrease in extracellular potassium ion concentration leads to (2) of the resting
membrane potential. Potassium ions are actively reabsorbed in
the (3) and are actively secreted in the (4) . This secretion is
primarily responsible for controlling the extracellular
concentration of potassium ions. The hormone aldosterone
plays a major role in regulating extracellular potassium ions by
causing a(n) (5) in the rate of potassium ion secretion.
Aldosterone secretion is regulated by blood potassium
levels; in response to a(n) (6) in blood potassium levels,
aldosterone secretion is increased.

1. _____

2. _____

3. _____

4. _____

5. _____

6. _____

D. Match these terms with the   Decrease(s)
   correct statement or definition: Increase(s)

_____   1. Effect of reduced extracellular calcium concentration on spontaneous
                             action potential generation.

_____   2. Effect of parathyroid hormone on extracellular calcium ion levels.

_____   3. Effect of decreased calcium levels on the secretion of parathyroid
                             hormone.

_____   4. Effect of parathyroid hormone on calcium reabsorption from the
                             kidneys.

_____   5. Effect of parathyroid hormone on production of active vitamin D.

_____   6. Effect of vitamin D on calcium absorption in the gastrointestinal tract.

_____   7. Effect of calcitonin on extracellular calcium concentration.

_____   8. Effect of elevated extracellular calcium levels on secretion of calcitonin.

E. Match these terms with the correct location on the following diagram by placing a D in the blank for those substances that are decreased and an I for those substances that are increased.

Decreased
Increased

Figure 27-2

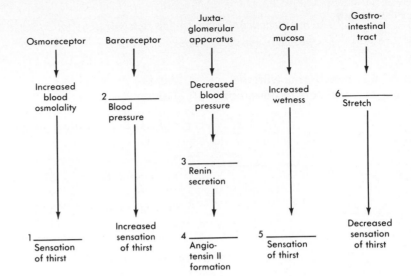

Osmoreceptor   Baroreceptor   Juxta-glomerular apparatus   Oral mucosa   Gastro-intestinal tract

↓ Increased blood osmolality

2 _____ Blood pressure

↓ Decreased blood pressure

↓ Increased wetness

6 _____ Stretch

3 _____ Renin secretion

1 _____ Sensation of thirst

Increased sensation of thirst

4 _____ Angio-tensin II formation

5 _____ Sensation of thirst

Decreased sensation of thirst

 Water loss from the body occurs through three major routes, including passage in urine, through evaporation, and in the feces.

F. Using the terms provided, complete the following statements:

Body temperature   Kidneys
Decrease   Plasma volume
Feces   Solutes
Increase   Sensible perspiration
Insensible perspiration

The water lost through simple evaporation through the skin is called _(1)_ . Sweat, or _(2)_ , is secreted by the sweat glands and, in contrast to insensible perspiration, contains _(3)_ . Evaporation of water is a major mechanism by which _(4)_ is regulated. The loss of a large volume of sweat causes a(n) _(5)_ in plasma volume and a(n) _(6)_ in hematocrit. The greater amount of water is lost through evaporation and urine, while a smaller volume is lost in the _(7)_ . The _(8)_ are the primary organs that regulate the composition and volume of body fluids by controlling the volume of water and the concentration of solutes excreted in the form of urine.

1. _____

2. _____

3. _____

4. _____

5. _____

6. _____

7. _____

8. _____

G. Using the terms provided, complete the following statements:

Decrease(s)   Increase(s)

Increased extracellular fluid osmolality _(1)_ ADH secretion from the posterior pituitary and _(2)_ aldosterone release from the adrenal cortex. Decreased blood pressure also _(3)_ ADH secretion and renin secretion from the kidneys. These hormonal change _(4)_ water and sodium reabsorption and _(5)_ urine volume and the concentration of solutes in the urine. Conversely, a(n) _(6)_ in extracellular osmolality or a(n) _(7)_ in blood pressure cause hormonal changes that result in the formation of a large volume of dilute urine.

1. _____

2. _____

3. _____

4. _____

5. _____

6. _____

7. _____

Regulation of Acid-Base Balance

"*The maintenance of hydrogen ion concentration within a narrow range of values is essential*"
for normal metabolic reactions.

A. Match these terms with the
   correct statement or definition:

   Acids   Neutral
   Bases   pH scale

_____   1. Substances that release hydrogen ions into a solution.

_____   2. Measurment of the acidity of a solution; numbers above 7 are basic,
                                those below 7 are acidic.

_____   3. Substance with a pH of 7.

☞ Strong acids completely dissociate in a solution so that all the hydrogen ions are released
   into the solution; weak acids release hydrogen ions but do not dissociate completely.

B. Using the terms provided, complete the following statements: 1. _____

   Acidosis   Hyperexcitability   2. _____
   Alkalosis   Increased
   Bicarbonate ions   Lactic acid   3. _____
   Decreased   Urine
   Depression   4. _____

   When the pH value of the body fluids is below 7.35, the condition   5. _____
   is referred to as _(1)_ ; and, when the pH is above 7.45, the
   condition is called _(2)_ . The major effect of acidosis is _(3)_ of the   6. _____
   central nervous system, whereas a major effect of alkalosis is (4)
   of the nervous system. Respiratory acidosis may occur with _(5)_   7. _____
   elimination of carbon dioxide from the body fluids through the
   respiratory system. Metabolic acidosis may occur through loss of   8. _____
   _(6)_ through diarrhea or vomiting; ingestion of acidic drugs;
   untreated diabetes mellitus; or _(7)_ buildup from severe exercise, 9. _____
   heart failure, or shock. Respiratory alkalosis may occur with _(8)_
   elimination of carbon dioxide from the body fluids through the
   respiratory system. Metabolic alkalosis may occur when loss of
   large amounts of acidic stomach contents occurs, when alkaline
   substances are ingested, or when there is a higher-than-normal
   loss of hydrogen ions in the _(9)_ .

C. Match these terms with the
   correct statement or definition:

   Bicarbonate buffer system   Protein buffer system
   Phosphate buffer system

_____   1. Provides three-fourths of the buffer system of the body; includes
                                hemoglobin and histone.

_____   2. Plays an exceptionally important role in controlling the pH of
                                extracellular fluid; involves carbonic acid.

_____   3. An important intracellular buffer system, but concentrations of this
                                system are low compared to the other two buffer systems.

Buffers resist changes in pH by chemically binding hydrogen ions when they are added to a solution or by releasing hydrogen ions when their concentration in a solution begins to fall.

D. Match these terms with the correct statement or definition:

Respiratory system
Urinary system

_____ 1. System that responds most rapidly to pH change.

_____ 2. System with the greatest regulatory capacity for acid-base balance, but with a slower response to pH change.

E. Using the terms provided, complete the following statements:

Carbonic acid-bicarbonate   Decreases
Carbonic anhydrase   Increases

The ability of the respiratory system to regulate acid-base balance depends on the _(1)_ buffer system. The reaction between carbon dioxide and water is catalyzed by _(2)_ , which is found in fairly high concentration within red blood cells. As carbon dioxide levels increase, pH of the body fluids _(3)_ , neurons in the medullary respiratory center of the brain are stimulated, and the rate and depth of ventilation _(4)_ . Carbon dioxide elimination then _(5)_ , and the concentration of carbon dioxide in the body fluids _(6)_ . This causes hydrogen ions to combine with bicarbonate ions to form carbonic acid, which then dissociates to form carbon dioxide and water, and pH _(7)_ to its normal range.

1. _____

2. _____

3. _____

4. _____

5. _____

6. _____

7. _____

F. Using the terms provided, complete the following statements:

Bicarbonate ions   Hydrogen ions
Carbonic acid   Increase
Decrease   Sodium ions

The kidneys can secrete _(1)_ into the urine from cells in the distal portion of the nephron. Within these cells carbon dioxide and water combine to form _(2)_ , which dissociates into hydrogen and bicarbonate ions. The hydrogen ions are then secreted into the filtrate by an active transport pump that exchanges _(3)_ for hydrogen ions. The _(4)_ then pass into the extracellular fluid and combine with hydrogen ions. The loss of hydrogen ions into the filtrate and the combination of bicarbonate ions with hydrogen ions in the extracellular fluid cause the body fluid pH to _(5)_ .

1. _____

2. _____

3. _____

4. _____

5. _____

G. Using the terms provided, complete the following statements:

Bicarbonate ion   Carbonic acid
Carbon dioxide   Hydrogen ions

Bicarbonate ions are not lost in the urine when the body pH is
elevated. This occurs because bicarbonate ions in the filtrate
combine with _(1)_ to form carbonic acid, which dissociates into
carbon dioxide and water. The _(2)_ diffuses into tubule cells and
combines with water to form _(3)_ , which dissociates into
hydrogen ions and bicarbonate ions. The _(4)_ are secreted into
the filtrate, and the _(5)_ reenter the extracellular fluid. If body
pH is elevated, not enough _(6)_ are secreted into the filtrate to
combine with bicarbonate ions, and they pass out of the body in
the urine. Excretion of excess bicarbonate ions results in a
decrease in extracellular fluid pH.

1. _____

2. _____

3. _____

4. _____

5. _____

6. _____

 Buffers (e.g., bicarbonate ions, phosphate ions, ammonia, and lactic acid) in the urine
combine with secreted hydrogen ions to increase the pH of the urine. The increased
pH allows additional secretion of hydrogen ions.

1. List three hormones that influence sodium ion levels, and state if they cause an increase or decrease in blood sodium levels.

2. Name the major hormone that regulates potassium ion concentration, and describe its effect.

3. List three compounds and their influence on extracellular calcium levels.

4. List three ways the sensation of thirst is increased and two ways it is decreased.

5. Name two stimuli that increase ADH, aldosterone, and renin secretion; state the effect of these substances on urine production and concentration.

6. Name three important buffer systems in the body.

7. State the effect on blood pH when respiration rate increases above normal and decreases below normal.

8. State the effect on blood pH when the acidity of the urine increases or decreases.

9. List two mechanisms that influence the number of hydrogen ions secreted into the distal part of the nephron.

Place the letter corresponding to the correct answer in the space provided.

_____ 1. Extracellular fluid
    a. is much like intracellular fluid in composition.
    b. includes interstitial fluid.
    c. osmotic concentration tends to be very different in the different fluid compartments of the body.
    d. all of the above

_____ 2. Which of the following would result in an increased blood sodium concentration?
    a. decrease in ADH secretion
    b. decrease in aldosterone secretion
    c. increase in atrial natriuretic hormone
    d. decrease in renin secretion

_____ 3. A decrease in extracellular potassium
    a. produces hypopolarization of cell membranes.
    b. results when aldosterone levels increase.
    c. occurs when tissues are damaged (e.g. in burn patients).
    d. all of the above

_____ 4. Calcium ion concentration in the blood decreases when
    a. vitamin D levels are lower than normal.
    b. calcitonin secretion decreases.
    c. parathyroid hormone secretion increases.
    d. all of the above

_____ 5. The sensation of thirst increases when
    a. the levels of angiotensin II increases.
    b. the osmolality of the blood decreases.
    c. blood pressure increases.
    d. renin secretion decreases.

_____ 6. Insensible perspiration
    a. is lost through sweat glands.
    b. results in heat loss from the body.
    c. increases when ADH secretion increases.
    d. results in the loss of solutes such as sodium and chloride.

_____ 7. An acid
    a. solution has a pH that is greater than 7.
    b. is a substance that releases hydrogen ions into a solution.
    c. is considered weak if it completely dissociates in water.
    d. all of the above

_____ 8. Buffers
    a. release hydrogen ions when pH increases.
    b. resist changes in the pH of a solution.
    c. include the proteins in blood.
    d. all of the above

_____ 9. Which of the systems regulating blood pH is the fastest acting?
    a. buffer
    b. respiratory
    c. kidney

_____ 10. An increase in blood carbon dioxide
    levels is followed by a (an) _____ in
    hydrogen ions and a (an) _____ in
    blood pH.
    a. decrease, decrease
    b. decrease, increase
    c. increase, increase
    d. increase, decrease

_____ 11. High levels of bicarbonate ions in the
    urine indicate
    a. a low level of hydrogen ion secretion
        into the urine.
    b. the kidneys are causing blood pH to
        decrease.
    c. that urine pH was increasing.
    d. all of the above

_____ 12. High levels of ammonium ions in the
    urine indicate
    a. a high level of hydrogen ion
        secretion into the urine.
    b. that the kidneys are causing blood
        pH to decrease.
    c. that urine pH is too alkaline.
    d. all of the above

_____ 13. Blood plasma pH is normally
    a. slightly acidic.
    b. strongly acidic.
    c. slightly alkaline.
    d. strongly alkaline.
    e. neutral.

_____ 14. Respiratory alkalosis is due to _____
    and can be compensated for by
    production of an _____ urine.
    a. hypoventilation, alkaline
    b. hypoventilation, acidic
    c. hyperventilation, alkaline
    d. hyperventilation, acidic

_____ 15. Acidosis
    a. increases neuron excitability.
    b. can produce coma by affecting the
        peripheral nervous system.
    c. may produce convulsions through
        the central nervous system.
    d. may lead to coma.

Use a separate sheet of paper to complete this section.

1. A researcher knows that under normal circumstances the level of sodium in the plasma is very precisely regulated, rarely rising or falling more than 1% from day to day. To determine the effect of the ADH-thirst mechanism versus aldosterone on control of sodium levels, the following two experiments were performed:
   A. Experiment 1: after the aldosterone system was blocked, sodium plasma levels were measured as daily sodium intake was increased.
   B. Experiment 2: after the ADH-thirst mechanism was blocked, sodium plasma levels were measured as daily sodium intake was increased. The results of these two experiments are graphed below:

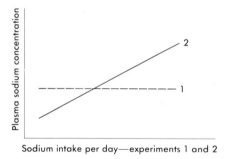

Sodium intake per day—experiments 1 and 2

On the basis of these two experiments, is aldosterone more important than the ADH-thirst mechanism in the regulation of plasma sodium levels? Explain the results of these experiments.

2. A patient has suffered lesions in the hypothalamus and has lost the ability to produce ADH. Despite this loss, normal or near normal levels of sodium are maintained in the blood plasma. Can you conclude that ADH is unimportant for the regulation of plasma sodium levels? Explain.

3. Would the pH in skeletal muscle be higher, lower, or the same as the pH of lung tissue?

4. John Uptight has a gastric ulcer. One day he consumes 10 packages of antacid tablets (mainly sodium bicarbonate). What effect would their consumption have on blood pH, urine pH, and respiration rate?

5. Many animals use panting as a mechanism to keep cool. Panting, of course, increases the respiratory rate dramatically. Despite this increase, serious respiratory alkalosis does not develop. Another student suggests to you that respiratory alkalosis is prevented by the production of an alkaline urine. Explain why you would agree or disagree with this explanation.

6. Chlorothiazide produces diuresis by inhibiting sodium loss from the ascending limb of the loop of Henle. Thus large amounts of sodium enter the distal convoluted tubule. Much of this sodium cannot be reabsorbed in the distal convoluted tubule, resulting in the retention of water in the tubule and the diuretic effect. However, reabsorption of sodium in the distal convoluted tubule is more rapid than normal because of the higher concentration of sodium present. This rapid reabsorption leaves an excess of buffer ions (e.g., phosphate) in the tubule. These ions are then able to combine with hydrogen ions. What effect does chlorothiazide have on blood pH, urine pH, and respiration rate?

ANSWERS TO CHAPTER 27

WORD PARTS

1. electrolyte
2. electrolyte
3. baroreceptor
4. osmoreceptor, osmolality
5. diuretic, antidiuretic

6. antidiuretic
7. parathyroid
8. lumen
9. permeable, impermeable
10. impermeable

CONTENT LEARNING ACTIVITY

Body Fluids

1. Intracellular fluid

2. Extracellular fluid

Regulation of Intracellular Fluid Concentration

1. Proteins
2. Electrolytes
3. Osmosis

4. Increases
5. Decreases

Regulation of Ion Concentrations

A. 1. Sodium ions
   2. Kidneys
   3. Aldosterone

B. 1. Increaseed
   2. Decreased
   3. Increased
   4. Decreased
   5. Decreased
   6. Decreased
   7. Decreased
   8. Increased

C. 1. Depolarization
   2. Hyperpolarization
   3. Proximal convoluted tubule
   4. Distal convoluted tubule
   5. Increase
   6. Increase

D. 1. Increases
   2. Increases
   3. Increases

4. Increases
5. Increases
6. Increases
7. Decreases
8. Increases

E. 1. Increased
   2. Decreased
   3. Increased
   4. Increased
   5. Decreased
   6. Increased

F. 1. Insensible perspiration
   2. Sensible perspiration
   3. Solutes
   4. Body temperature
   5. Decrease
   6. Increase
   7. Feces
   8. Kidneys

G. 1. Increases
   2. Increases
   3. Increases
   4. Increase

   5. Decrease
   6. Decreases
   7. Increase

Regulation of Acid-Base Balance

A. 1. Acids
   2. pH scale
   3. Neutral

B. 1. Acidosis
   2. Alkalosis
   3. Depression
   4. Hyperexcitability
   5. Decreased
   6. Bicarbonate ions
   7. Lactic acid
   8. Increased
   9. Urine

C. 1. Protein buffer system
   2. Bicarbonate buffer system
   3. Phosphate buffer system

D. 1. Respiratory system
   2. Urinary system

E. 1. Carbonic acid-bicarbonate
   2. Carbonic anhydrase
   3. Decreases
   4. Increases
   5. Increases
   6. Decreases
   7. Increases

F. 1. Hydrogen ions
   2. Carbonic acid
   3. Sodium ions
   4. Bicarbonate ions
   5. Increase

G. 1. Hydrogen ions
   2. Carbon dioxide
   3. Carbonic acid
   4. Hydrogen ions
   5. Bicarbonate ions
   6. Hydrogen ions

QUICK RECALL

1. Aldosterone: increases blood sodium; ADH: increases blood sodium; atrial natriuretic hormone: decreases blood sodium
2. Aldosterone increases potassium secretion in the distal convoluted tubule
3. Parathyroid hormone: increases extracellular calcium levels; calcitonin: decreases extracellular calcium level; vitamin D: increases calcium uptake in intestine
4. Increased osmolality of body fluid, reduction in plasma volume, and decrease in blood pressure cause increased thirst. Wetting of oral mucosa and stretching gastrointestinal tract decrease thirst.
5. Urine, evaporation, and feces
6. Decreased blood pressure and increased tissue osmolality cause an increase in ADH, aldosterone, and renin production, producing less urine that is more highly concentrated.
7. Carbonic acid-bicarbonate, protein, and phosphate buffer systems
8. $CO_2 + H_2O \rightleftharpoons H_2CO_3 \rightleftharpoons HCO_3^- + H^+$
9. Above normal: increase in blood pH; below normal: decrease in blood pH
10. If the acidity of the urine increases, the blood pH increases; if the acidity of the urine decreases, the blood pH decreases.
11. Aldosterone and body fluid pH

1. B. Extracellular fluid is all of the fluid outside of cells and includes interstitial fluid. Most extracellular fluids have approximately the same osmotic concentration. However, extracellular fluid composition is substantially different from that of intracellular fluid.

2. A. Decreased ADH secretion would result in increased water loss in the urine. As blood volume decreased, sodium concentration would increase. The other changes listed would decrease blood sodium concentration.

3. B. An increase in aldosterone results in more secretion of potassium ions into the urine. Increased extracellular potassium produces hyperpolarization and occurs when cells are damaged and release their potassium.

4. A. Vitamin D is required for normal absorption of calcium from the small intestine. A decrease in calcitonin or an increase in parathyroid hormones causes an increase in blood calcium levels.

5. A. Angiotensin II stimulates thirst by acting on the brain. A decrease in blood pressure stimulates renin secretion, which leads to the production of angiotensin II. An increased blood osmolality also stimulates thirst.

6. B. Water lost by insensible perspiration results in heat loss. Insensible perspiration is water loss from the respiratory passages and skin (but not sweat glands). There is no loss of solutes.

7. B. Acids release hydrogen ions into a solution, have a pH that is less than 7, and are considered strong acids if they completely dissociate (i.e., release all their hydrogen ions).

8. D. Buffers resist changes in pH by releasing or binding hydrogen ions. Blood proteins are responsible for about three fourths of the buffering ability of the blood.

9. A. The buffer system reacts within fractions of a second. The respiratory system takes several minutes, and the kidney system requires several hours or even days to respond.

10. D. When blood carbon dioxide levels increase, the carbon dioxide combines with water to produce carbonic acid. The carbonic acid dissociates to form hydrogen ions and bicarbonate ions. The increase of hydrogen ions lowers the pH; i.e., increases the acidity of the blood.

11. D. High levels of bicarbonate ions occur in the urine when hydrogen ion secretion is low. Consequently, blood pH decreases and urine pH increases.

12. A. Ammonia acts as a buffer in the urine by combining with hydrogen ions to form ammonium ions. High levels of ammonium ions would indiate a high level of hydrogen ions in the urine and therefore an acidic urine. Excretion of large amounts of hydrogen ions causes blood pH to increase.

13. C. Blood pH is normally about 7.4 or slightly alkaline.

14. C. Hyperventilation reduces plasma carbon dioxide levels resulting in alkalosis. The kidneys compensate by producing an alkaline urine. This conserves hydrogen ions and restores pH to normal levels.

15. D. One of the effects of acidosis is a depression of the central nervous system, which can lead to coma and death. Alkalosis causes excitation of the nervous system. Usually the peripheral nervous sytem is affected first, producing tetany due to repetitive firing of the neurons. Tetany of the respiratory muscles can cause death. Overexcitation of the central nervous system leads to extreme nervousness and convulsions.

1. When the ADH-thirst mechanism is blocked but the aldosterone system is still operating (experiment 2), the levels of plasma sodium increase when daily sodium intake increases. This indicates that aldosterone plays a very minor role in regulation of plasma sodium levels. Indeed, even when the aldosterone system is blocked (experiment 1), plasma sodium levels stay relatively constant, despite increasing sodium intake. These data suggest that the ADH-thirst mechanism is more important than aldosterone in regulating plasma sodium levels. It is consistent with the data to assume that the ADH-thirst mechanism maintains sodium levels as follows: if sodium levels rise, the thirst center is stimulated, fluids are ingested, and sodium concentration decreases. If sodium levels fall, ADH secretion is inhibited, a hypoosmotic urine is produced, water is lost, and sodium concentration increases. It seems paradoxical that aldosterone, which controls sodium reabsorption, is not effective in regulating plasma sodium levels, whereas ADH, which regulates water reabsorption is effective. The ineffectiveness of aldosterone results from the following effect: when aldosterone increases, sodium reabsorption increases, water follows sodium, and extracellular fluid volume increases, producing a rise in arterial blood pressure. The increased blood pressure increases the glomerular filtration rate and the rapid movement of filtrate through the tubule compensates for the increased reabsorption of sodium caused by the aldosterone.

2. The conclusion might seem reasonable at first. However, careful thought may lead to the opposite conclusion. All that has been lost is the ability to produce ADH. If plasma osmolality increases, the thirst mechanism is activated and the patient drinks more fluids, thus restoring plasma osmolality to normal levels. Conversely, if plasma osmolality decreases, the patient reduces fluid intake and plasma osmolality increases. Thus the thirst mechanism plays an important role in regulating sodium levels.

Because there is no ADH present, the patient will produce a dilute urine, and the daily urine production may increase to 5 to 15 L/day. This, of course, requires a corresponding increase of fluid intake. This condition is called diabetes insipidis. The importance of the ADH, coupled with the thirst mechanism, can be demonstrated when there is inadequate intake of water (such as during unconsciousness). Then, plasma sodium increase tremendously, because without ADH the kidneys are unable to conserve water.

3. The skeletal muscle would have a slightly higher carbon dioxide content than lung tissue. Consequently, the pH of skeletal muscle would be slightly lower than the pH of lung tissue.

4. The overdose of the alkaline antacid tablets raises blood pH, producing metabolic alkalosis. By secreting fewer hydrogen ions a more alkaline urine is formed, and hydrogen ion concentration in the blood increases, compensating for the alkalosis. Respiration rate also decreases to compensate. Reduced respiration would increase plasma CO_2 levels which would reduce the pH through the production of carbonic acid.

5. Although production of an alkaline urine helps to compensate for respiratory alkalosis, it would take the kidneys hours or even days to exert their effect. Instead, panting animals prevent respiratory alkalosis by decreasing tidal volume. As a result, with each breath very little "new" air enters the alveoli. This prevents the loss of excessive amounts of carbon dioxide and the development of respiratory alkalosis.

6. The hydrogen ions combine with the buffer and are excreted as an acidic urine. The loss of hydrogen ions produces alkalosis. The subsequent increase in pH inhibits the respiratory center and lowers respiration rate.

Reproductive System

FOCUS: The male reproductive system consists of the testes, where spermatozoa and testosterone are produced; the scrotum and cremaster muscles, which help to maintain the testes at a temperature suitable for spermatozoa development; a series of ducts that carry the spermatozoa to the outside of the body; glands that provide secretions that provide energy molecules to the spermatozoa and neutralize the acidic environment of the the male urethra and female reproductive tract; and the penis, which, when erect, functions in the transfer of spermatozoa to the female vagina. The functions of the male reproductive system are under hormonal and neural control. The hypothalamus releases GnRH, which stimulates the adenhypophysis to secrete FSH and LH. The FSH stimulates spermatogenesis and the LH stimulates testosterone production. Testosterone is responsible for the development of male sexual characteristics. The nervous system controls the process of erection, emission, ejaculation, and orgasm. The female reproductive system consists of the ovaries, where the secondary oocyte is formed and hormones such as estrogen and progesterone are produced; the uterine tubes, where the secondary oocyte is normally fertilized and which carry the developing organism to the uterus; the uterus, where implantation followed by embryonic and fetal development occurs; the vagina, which receives the penis during intercourse and serves as a passageway to the outside of the body for the fetus; the external genitalia, which form a cover for the openings to the vagina and urinary tract; and the breast, where milk is produced for the infant. The ovaries undergo a cyclic production of oocytes and hormones. The hormones act on the uterine lining to prepare the uterus for implantation. During pregnancy hormones produced by the placenta (HCG, estrogen, progesterone) prevent a spontaneous abortion. Parturition involves changes in estrogen, progesterone, and oxytocin levels. Lactation results from increased prolactin, which stimulates milk production, and increased oxytocin, which stimulates milk letdown. Menopause occurs when the ovaries stop producing oocytes and hormones.

WORD PARTS

Give an example of a new vocabulary that contains each word part.

WORD PART	MEANING	EXAMPLE
semin-	semen	1. _____
-fer	to bear	2. _____

WORD PART	MEANING	EXAMPLE
andro-	male	3. _____
-gen	origin	4. _____
ejacul-	to throw out	5. _____
-metr-	the uterus	6. _____
myo-	muscle	7. _____
lacto-	milk	8. _____
gyn-	female	9. _____
mast-	breast	10. _____

CONTENT LEARNING ACTIVITY

Scrotum

❝*The scrotum contains the testes and is divided into two internal compartments by a connective tissue septum.***❞**

Match these terms with the
correct statement or definition:

Cremaster muscles   Raphe
Dartos muscle

_____ 1. Irregular ridge on the midline of the scrotum.

_____ 2. Layer of cutaneous muscle surrounding the scrotum; this muscle
   contracts, causing the skin of the scrotum to become firm and wrinkled.

_____ 3. Extensions of abdominal muscles into the scrotum that contract and
   pull the testes closer to the body.

 Both the dartos and cremaster muscles operate to keep the testes at the proper temperature
for spermatogenesis to occur.

Perineum

"The area between the thighs which is bounded by the pubis anteriorly, the coccyx posteriorly, and ischial tuberosities laterally, is called the perineum."

Match these terms with the correct statement or definition:

Anal triangle
Urogenital triangle

_____ 1. Anterior portion of the perineum, which contains the external genitalia.

_____ 2. Posterior portion of the perineum, which contains the anal opening.

Testes

"The testes are small, ovoid organs within the scrotum."

A. Match these terms with the correct statement or definition:

Efferent ductules   Seminiferous tubules
Interstitial cells (cells of Leydig)   Septa
Rete testis   Tunica albuginea

_____ 1. Thick white capsule surrounding the testis.

_____ 2. Connective tissue of the tunica albuginea that divides the testis into lobules.

_____ 3. Structures in which sperm develop.

_____ 4. Endocrine cells located in the connective tissue stroma around the seminiferous tubules.

_____ 5. Tubules that connect the rete testis and the epididymis.

B. Match these terms with the correct statement or definition:

Inguinal canal   Tunica vaginalis
Process vaginalis

_____ 1. Passageway from the abdominal cavity to the scrotum, through which the testes descend.

_____ 2. Outpocketing of peritoneum that precedes the testis as it moves to the scrotum.

_____ 3. Small, closed sac that covers most of the testis.

591

C. Match these terms with the correct statement or definition:

Acrosome
Germ cells
Primary spermatocytes
Secondary spermatocytes
Sertoli cells
Spermatids
Spermatogenesis
Spermatogonia
Spermatozoa

_____ 1. Production of spermatozoa.

_____ 2. Large cells that nourish the germ cells, probably produce hormones, and form the blood-testis barrier.

_____ 3. Most peripheral germ cells; they divide by mitosis.

_____ 4. Germ cells produced from spermatogonia that are passing through the first meiotic division.

_____ 5. Germ cells produced from primary spermatocytes by the first meiotic division.

_____ 6. Germ cells produced from secondary spermatocytes by the second meiotic division.

_____ 7. Cap containing enzymes necessary for penetrating the egg, found on the head of the sperm.

Ducts

❝ _After their production in the seminiferous tubules, spermatozoa leave the testes through the_ ❞ _efferent ductules and pass through a series of ducts to reach the exterior of the body._

Match these terms with the correct statement or definition:

Ductus (vas) deferens
Ductus epididymis
Ejaculatory duct
Epididymis
Membranous urethra
Prostatic urethra
Spermatic cord
Spongy urethra

_____ 1. Comma-shaped structure on the posterior testis in which maturation of spermatozoa occurs.

_____ 2. Single convoluted ductule located primarily in the body of the epididymis.

_____ 3. Duct running from the tail of the epididymis to the ejaculatory duct; the end enlarges to form the ampulla.

_____ 4. Duct formed when the duct of the seminal vesicle joins the ductus deferens.

_____ 5. The ductus deferens, the blood vessels and nerves that supply the testis and the cremaster muscle compose this structure.

_____ 6. Portion of the urethra that passes through the prostate gland into which the ejaculatory ducts empty.

_____ 7. Shortest portion of the urethra; extends from the prostate through the urogenital diaphragm.

Penis

66 *The penis is the male organ of copulation and functions in the transfer of spermatozoa from the male to the female.* **99**

Match these terms with the
correct statement or definition:

Bulb of the penis
Corpora cavernosa
Corpus spongiosum
Crus of the penis

Glans penis
Prepuce (foreskin)
Root of the penis

_____ 1. Two erectile columns that form the dorsum and sides of the penis.

_____ 2. Smallest erectile column that occupies the ventral portion of the penis;
the spongy urethra passes through it.

_____ 3. Cap, formed from the corpus spongiosum, over the distal end of the
penis.

_____ 4. Expansion of the corpus spongiosum at the base of the penis.

_____ 5. Expansion of the corpora cavernosa at the base of the penis.

_____ 6. Collective name for the crus of the penis and the bulb of the penis;
attaches the penis to the coxae.

_____ 7. Loose fold of skin that covers the glans penis; removed by circumcision.

Glands

66 *Several glands secrete substances into the reproductive system.* **99**

A. Match these terms with the
correct statement or definition:

Bulbourethral glands
Bulbourethral secretions
Ejaculation
Emission
Prostate gland

Prostate secretions
Semen
Seminal vesicle secretions
Seminal vesicles

_____ 1. Sac-shaped glands located near the ampullae of the ductus deferens.

_____ 2. Gland the size and shape of a walnut located dorsal to the pubic
symphysis at the base of the bladder; surrounds part of the urethra and
the ejaculatory ducts.

_____ 3. Pair of small mucous glands near the membranous portion of the
urethra.

_____ 4. Discharge of semen into the prostatic urethra; stimulated by
sympathetic impulses.

_____ 5. Mucous secretion that lubricates the urethra, neutralizes the contents of
the spongy urethra, provides lubrication during intercourse, and
reduces the acidity of the vagina.

_____ 6. Thick, mucoid secretions with a relatively low pH and containing
fructose, fibrinogen, and prostaglandins.

_____ 7. Thin, milky, alkaline secretions that neutralize the urethra and vagina
and contain clotting factors that cause fibrinogen to aggregate.

B. Match these terms with the
correct parts of the diagram
labeled in Figure 28-1:

Bulbourethral gland
Ductus deferens
Ejaculatory duct
Epididymis
External urethral orifice
Membranous urethra
Penis
Prostate
Prostatic urethra
Scrotum
Seminal vesicle
Spongy urethra
Testis

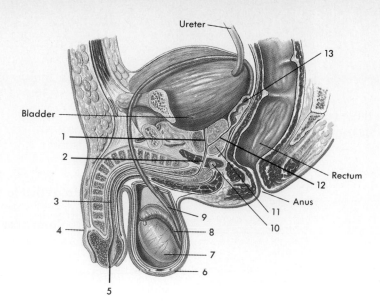

Figure 28-1

1. _____

2. _____

3. _____

4. _____

5. _____

6. _____

7. _____

8. _____

9. _____

10. _____

11. _____

12. _____

13. _____

Regulation of Hormone Secretion

66 *Hormonal mechanisms that influence the male reproductive system involve the hypothalamus,* 99
the pituitary gland, and the testes.

Match these terms with the
correct statement or definition:

Androgen
Follicle stimulating hormone (FSH)
Gonadotropins
Gonadotropin-releasing hormone
   (GnRH)

Inhibin
Luteinizing hormone (LH)
Testosterone

_____ 1. Small peptide hormone released from the hypothalamus that stimulates
cells in the adenohypophysis.

_____ 2. General term for hormones that affect the gonads; secreted by the
adenohypophysis.

_____ 3. Hormone that binds to the cells of Leydig in the testes and stimulates
testosterone synthesis and secretion.

_____ 4. Hormone that binds to the Sertoli cells in the seminiferous tubules of
the testes, and promotes spermatogenesis.

_____ 5. Major male hormone secreted by the testes.

_____ 6. General name for male hormones that stimulate the development of male secondary sexual characteristics.

_____ 7. Released from the testes, this hormone inhibits FSH secretion from the adenohypophysis.

Puberty

66 *The reproductive system becomes functional at puberty, which normally begins when a boy is 12 to 14 years old.* **99**

Using the terms provided, complete the following statements:

HCG   Spermatogenesis
GnRH   Testosterone
LH and FSH

 (1) , which is secreted by the placenta, stimulates the synthesis and secretion of testosterone before birth. After birth only small amounts of testosterone are secreted until puberty, when the hypothalamus becomes less sensitive to the inhibitory effect of (2) and the rate of (3) secretion increases, leading to increased (4) release from the adenohypohpysis. Elevated FSH levels promote (5) , and elevated LH levels cause the interstitial cells to secrete (6) .

1. _____

2. _____

3. _____

4. _____

5. _____

6. _____

Effects of Testosterone

66 *Testosterone is by far the major androgen in males.* **99**

Using the terms provided, complete the following statements:

Body fluids   Protein synthesis
Coarser   Rapid bone growth
Hypertrophy   Sebaceous glands
Melanin   Testes

Testosterone causes the enlargement and differentiation of the male reproductive system, it is necessary for spermatogenesis, and it is required for the descent of the (1) during fetal development. Testosterone stimulates hair growth in several regions and causes the texture of the hair to become (2) . Testosterone increases the quantity of (3) in the skin, causing it to become darker, and increases the rate of secretion of (4) , frequently resulting in acne. Testosterone also causes (5) of the larynx, deepening the voice. Testosterone stimulates red blood cell production, stimulates metabolism, and causes retention of sodium in the body, resulting in an increase of (6) . Testosterone promotes (7) , resulting in an increased skeletal muscle mass, and (8) , resulting in an increase in height.

1. _____

2. _____

3. _____

4. _____

5. _____

6. _____

7. _____

8. _____

Male Sexual Behavior and the Male Sex Act

66*Testosterone is required to initiate and maintain normal male sexual behavior.***99**

Match these terms with the correct statement or defintion:

Ejaculation   Impotence
Emission   Orgasm
Erection   Resolution

_____ 1. Pleasurable climax sensation associated with ejaculation.

_____ 2. Period after ejaculation when the penis becomes flaccid, and the male is unable to achieve erection and a second ejaculation.

_____ 3. Inability to accomplish the male sex act; due to psychic or physical factors.

_____ 4. Occurs when parasympathetic or sympathetic impulses cause the dilation of the arteries that supply blood to the erectile tissues in the penis.

_____ 5. Stimulated by sympathetic centers in the spinal cord as the level of sexual tension increases; secretions of the prostate and seminal vesicles are released.

_____ 6. Rhythmic contractions that force semen out of the urethra, triggered by efferent somatic impulses.

☞ Psychic stimuli (e.g., sight, sound, odor, or thoughts) have a major effect on sexual reflexes.

Ovaries

66*The ovaries are small organs located on each side of the uterus.***99**

A. Match these terms with the correct statement or definition:

Cortex   Ovarian follicle
Medulla   Ovarian ligament
Mesovarium   Suspensory ligament
Oocyte   Tunica albuginea
Ovarian (germinal) epithelium

_____ 1. Peritoneal fold that attaches the ovaries to the broad ligament.

_____ 2. Ligament that extends from the mesovarium to the body wall.

_____ 3. Peritoneum covering the surface of the ovary.

_____ 4. Layer of dense fibrous connective tissue that surrounds the ovary.

_____ 5. Looser inner portion of the ovary.

_____ 6. Small vesicles, each of which contains an oocyte, distributed throughout the cortex of the ovary.

_____ 7. Egg cell.

B. Match these terms with the correct statement or definition:

Oogenesis   Primary oocyte
Oogonia   Secondary oocyte
Polar body

_____ 1. Production of a secondary oocyte within the ovaries.

_____ 2. Cells from which oocytes develop.

_____ 3. Oogonia that have started meiosis but stopped at prophase I.

_____ 4. Two structures produced from a primary oocyte by the first meiotic division.

_____

C. Match these terms with the correct statement or definition:

Antrum   Primary follicle
Corona radiata   Secondary follicle
Cumulus mass   Theca
Graafian (vesicular) follicle   Zona pellucida

_____ 1. Primary oocyte with the surrounding granulosa cells.

_____ 2. Follicle that is not yet mature, but contains an antrum, cumulus mass, and corona radiata.

_____ 3. Mass of follicular cells that surrounds the oocyte in a secondary follicle.

_____ 4. Innermost cells of the cumulus mass.

_____ 5. Fully mature follicle.

_____ 6. Layers of cells molded around the secondary and graafian follicle to form a capsule.

_____ 7. Layer of clear, viscous fluid that is deposited around a primary oocyte.

D. Using the terms provided, complete the following statements:

Corpus albicans   Polar body
Corpus luteum   Progesterone and estrogen
Fertilization   Secondary oocyte
Ovulation   Zygote

The release of the secondary oocyte from the follicle is called _(1)_. During ovulation the development of the _(2)_ has stopped at metaphase II. Continuation of the second meiotic division is triggered by _(3)_, the entry of the sperm into the secondary oocyte. If fertilization does not occur, meiosis is not completed; if fertilization occurs, meiosis II is completed, and a second _(4)_ is formed. The fertilized oocyte is now called a _(5)_. After ovulation the follicle is transformed into a glandular structure called the _(6)_. The granulosa cells and the theca interna enlarge and begin to secrete _(7)_. If pregnancy occurs the corpus luteum enlarges and remains throughout pregnancy. If pregnancy does not occur, the connective tissue cells in the corpus luteum become enlarged and clear and give the whole structure a whitish color; therefore it is called the _(8)_.

1. _____

2. _____

3. _____

4. _____

5. _____

6. _____

7. _____

8. _____

597

Uterine Tubes

❝The two uterine tubes are located along the superior margin of the broad ligament.❞

Match these terms with the
correct statement or definition:

Ampulla   Muscular layer
Fimbriae   Mucosa
Infundibulum   Ostium
Isthmus   Serosa
Mesosalpinx   Uterine (intramural) part

_____ 1. Portion of the broad ligament most directly associated with the uterine tube.

_____ 2. Funnel-shaped end of the uterine tube.

_____ 3. Opening in the funnel-shaped end of the uterine tube.

_____ 4. Long thin processes that surround the opening into the uterine tube.

_____ 5. Portion of the uterine tube closest to the infundibulum; the widest and longest part of the tube.

_____ 6. Narrow, thin-walled portion of the uterine tube closest to the uterus.

_____ 7. Middle layer of the uterine tube wall.

_____ 8. Inner layer of the uterine tube wall.

☞ Fertilization usually occurs in the ampulla of the uterine tube.

Uterus

❝The uterus is the size and shape of a medium-sized pear.❞

A. Match these terms with the
correct statement or definition:

Body   Isthmus
Broad ligament   Ostium
Cervical canal   Round ligaments
Cervix   Uterine cavity
Fundus   Uterosacral ligaments

_____ 1. Larger, rounded portion of the uterus, directed superiorly.

_____ 2. Narrower portion of the uterus, directed inferiorly.

_____ 3. Slight constriction that marks the junction between the body of the uterus and the cervix.

_____ 4. Cavity inside the cervix of the uterus.

_____ 5. Opening of the cervix into the vagina.

_____ 6. Major ligament that extends from the uterus through the inguinal canal to the external genitalia.

_____ 7. Major ligament that spreads on both sides of the uterus and to which the ovaries and uterine tubes are attached.

B. Match these terms with the correct statement or definition:

Basal layer   Myometrium (muscular coat)
Endometrium (mucous membrane)   Perimetrium (serous coat)
Functional layer

_____ 1. The outer coat of the uterus; peritoneum.

_____ 2. The innermost layer of the uterus.

_____ 3. The thin, deep layer of the endometrium that is continuous with the myometrium.

_____ 4. The thicker, superficial layer of the endometrium; undergoes the greatest change during the menstrual cycle.

C. Match these terms with the correct parts of the diagram labeled in Figure 28-2:

Body of uterus
Cervix
Fundus of uterus
Ovarian ligament
Ovary
Round ligament
Uterine tube
Vagina

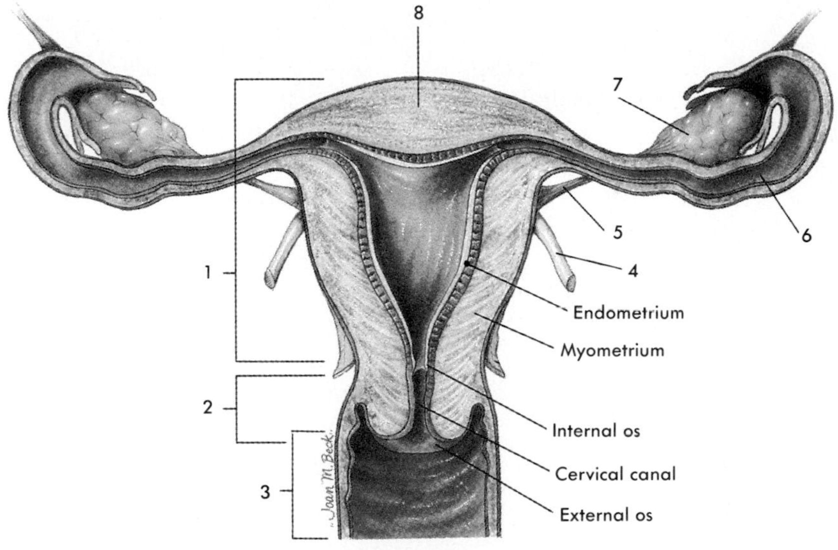

Figure 28-2

1. _____   5. _____   9. _____

2. _____   6. _____

3. _____   7. _____

4. _____   8. _____

Vagina

"_The vagina is the female organ of copulation and functions to receive the penis during intercourse,_ **"**
and allows menstrual flow and childbirth.

Match these terms with the correct statement or definition:

Columns	Hymen
Fornix	Rugae

_____ 1. Longitudinal ridges that extend the length of the vaginal walls.

_____ 2. Transverse ridges in the vagina.

_____ 3. Superior, domed portion of the vagina.

_____ 4. Thin mucous membrane that may cover the external vaginal orifice.

External Genitalia

"_The external genitalia, also referred to as the vulva or pudendum, consist of the vestibule_ **"**
and its surrounding structures.

A. Match these terms with the correct statement or definition:

Bulb of the vestibule	Lesser vestibular glands
Clitoris	Mons pubis
Greater vestibular glands	Prepuce
Labia majora	Pudendal cleft
Labia minora	Vestibule

_____ 1. Space into which the vagina and urethra open.

_____ 2. Thin, longitudinal skin folds bordering the vestibule.

_____ 3. Small erectile structure that corresponds to the corpora cavernosa in the male; located in the anterior margin of the vestibule; functions to intensify sexual tension.

_____ 4. Fold of skin over the clitoris, formed anteriorly by the union of the two labia minora.

_____ 5. Erectile tissue that corresponds to the corpus spongiosum in the male; functions to narrow the vaginal orifice.

_____ 6. Ducts of these glands open on either side of the vestibule between the vaginal opening and the labia minora.

_____ 7. These small mucous glands are located near the clitoris and urethral opening.

_____ 8. Mound located over the pubic symphysis where the labia majora unite anteriorly.

B. Match these terms with the correct parts of the diagram labeled in Figure 28-3:

Clitoris
Labia majora
Labia minora
Mons pubis
Prepuce
Urethra
Vagina
Vestibule

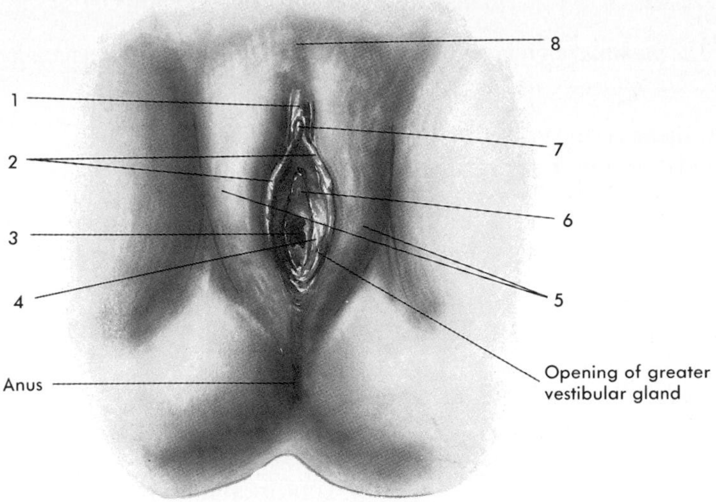

Figure 28-3

Anus

Opening of greater vestibular gland

1. _____   4. _____   7. _____

2. _____   5. _____   8. _____

3. _____   6. _____

Perineum

❝*The area between the thighs which is bounded by the pubis anteriorly, the coccyx posteriorly,*❞ *and the ischial tuberosities laterally, is called the perineum.*

Match these terms with the correct statement or definition:

Anal triangle   Episiotomy
Clinical perineum   Urogenital triangle

_____ 1. Anterior portion of the perineum that contains the external genitalia.

_____ 2. Region between the vagina and the anus.

_____ 3. Incision in the clinical perineum to prevent tearing during childbirth.

Mammary glands

"The mammary glands are the organs of milk production and are located within the mammae, or breasts.**"**

A. Match these terms with the correct statement or definition:

Alveoli		Lobes
Areola		Lobules
Areolar glands		Mammary (Cooper's) ligaments
Gynecomastia		Nipple
Lactiferous duct		

_____ 1. Circular, pigmented area surrounding the nipple.

_____ 2. Rudimentary mammary glands located just below the surface of the areola.

_____ 3. Condition in which male breasts become enlarged.

_____ 4. Glandular compartments of the breast, each of which possesses a single lactiferous duct.

_____ 5. Smaller subcompartments of the breast that contain the milk-producing glands.

_____ 6. Duct opening to the nipple that carries milk from one lobe; contains a sinus that accumulates milk.

_____ 7. Secretory sacs in the milk-producing breast.

_____ 8. Ligaments that support and hold the breast in place.

B. Match these terms with the correct parts of the diagram labeled in Figure 28-4:

Alveoli
Areola
Lactiferous duct
Lactiferous sinus
Lobe
Lobule
Nipple

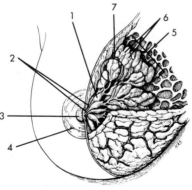

1. _____

2. _____

3. _____

4. _____

5. _____

6. _____

7. _____

Figure 28-4

Puberty

66 *Puberty in the female is marked by the first episode of menstrual bleeding, which is called menarche.* 99

Using the terms provided, complete the following statements:

Estrogen and progesterone   GnRH
FSH and LH

Changes associated with puberty are due primarily to elevated levels of _(1)_ secreted by the ovaries. These hormones are secreted in response to an increasing and cyclic pattern of _(2)_ secretion by the adenohypophysis. These hormones in turn are secreted in response to an increasing _(3)_ secretion by the hypothalamus. The cyclic surge of _(4)_ results in ovulation, and the monthly changes in secretion of _(5)_, produce changes in the uterus, which characterize the menstrual cycle.

1. _____

2. _____

3. _____

4. _____

5. _____

The Menstrual Cycle

66 *The term menstrual cycle technically refers to the series of changes that occur in sexually mature,* 99 *nonpregnant females and culminate in menses.*

A. Match these terms with the correct statement or definition:

Follicular (proliferative) stage   Menses
Luteal (secretory) stage

_____   1. Period of mild hemorrhage during which the uterine epithelium is sloughed and expelled from the uterus.

_____   2. Time between the ending of menses and ovulation.

_____   3. Time between ovulation and the beginning of menses.

☞ Typically the menstrual cycle is 28 days long, with ovulation occurring about day 14.

B. Using the terms provided, complete the following statements:

Adenohypophysis
FSH
FSH surge
HCG
LH surge

Luteal
Ovarian cycle
Ovulation
Progesterone

The _(1)_ specifically refers to the series of events that occur in a regular fashion in the ovaries of sexually mature, nonpregnant women. FSH and LH are released from the _(2)_ in large amounts just before ovulation. As a result, a number of follicles begin to mature during each menstrual cycle, under the influence of _(3)_, but normally only one is ovulated. An increase in _(4)_ blood levels causes the primary oocyte to complete the first meiotic division and triggers several events that result in _(5)_. After ovulation, the granulosa cells enlarge and increase in number to become _(6)_ cells, which secrete large amounts of _(7)_ and some estrogen. If fertilization does not occur, the cells of the corpus luteum atrophy. If fertilization of the ovulated oocyte takes place, _(8)_, an LH-like hormone is secreted by the developing embryo and keeps the corpus luteum from degenerating.

1. _____
2. _____
3. _____
4. _____
5. _____
6. _____
7. _____
8. _____

C. Using the terms provided, complete the following statements:

Decrease(s)
Increase(s)

Negative feedback
Positive feedback

The theca interna cells of the developing follicle secrete estrogen, which _(1)_ GnRH secretion from the hypothalamus, which in turn _(2)_ FSH and LH secretion from the adenohypophysis. This leads to a _(3)_ system, in which FSH and LH _(4)_ estrogen secretion from the ovary, which _(5)_ GnRH secretion and FSH and LH production. Near the time of ovulation, estrogen secretion _(6)_, and progesterone secretion _(7)_. Both estrogen and progesterone secretion _(8)_ after the corpus luteum is formed, and this has a _(9)_ effect on GnRH release from the hypothalamus. As a result, LH and FSH secretion by the adenohypophysis _(10)_, the corpus luteum degenerates, and the levels of estrogen and progesterone _(11)_.

1. _____
2. _____
3. _____
4. _____
5. _____
6. _____
7. _____
8. _____
9. _____
10. _____
11. _____

D. Match these terms with the correct statement or definition:

Spiral arteries
Spiral tubular glands
Uterine cycle

_____ 1. Changes that occur primarily in the endometrium of the uterus during the menstrual cycle.

_____ 2. Glands formed when columnar epithelial in the endometrium are thrown into folds; secrete a glycogen-rich fluid.

_____ 3. Arteries that project between spiral glands to supply nutrients to the endometrial cells.

E. Match these terms with the correct statement:

Decrease(s)
Increase(s)

_____ 1. Effect of estrogen on the thickness of the endometrium; affects cell division.

_____ 2. Effect of estrogen on the sensitivity of the endometrial cells to progesterone.

_____ 3. Effect of progesterone on the thickness of the endometrium and the myometrium; affects cell size.

_____ 4. Effect of progesterone on the secretory ability of endometrial cells.

_____ 5. Effect of progesterone on smooth muscle contraction.

_____ 6. Effect of declining progesterone levels on blood supply to the endometrium.

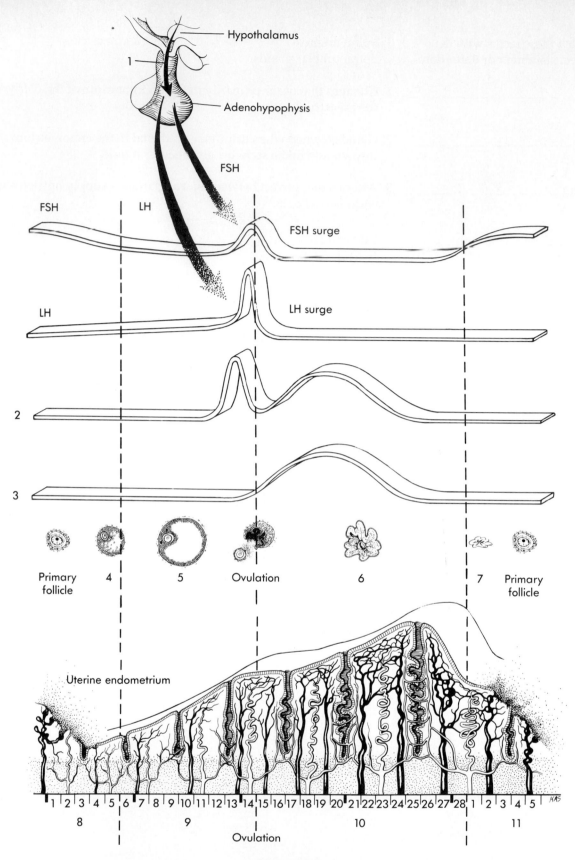

Figure 28-5

F. Match these terms with the correct parts of the diagram labeled in Figure 28-5:

Corpus albicans
Corpus luteum
Estrogen
GnRH
Graafian (vesicular) follicle

Menses
Progesterone
Proliferative phase
Secondary follicle
Secretory phase

1. _____

2. _____

3. _____

4. _____

5. _____

6. _____

7. _____

8. _____

9. _____

10. _____

11. _____

Female Sexual Behavior and the Female Sex Act

66 *Sexual drive in females, like sexual drive in males, is dependent upon hormones.* 99

Using the terms provided, complete the following statements:

Clitoris
Fertilization
Orgasm
Psychic factors

Resolution
Sacral
Vagina

Androgens and possibly estrogens affect cells and influence sexual behavior; however, (1) also play a role in sexual behavior. The (2) region of the spinal cord is the area that integrates sexual reflexes, which are modulated by cerebral influences. During sexual excitement, parasympathetic stimulation causes erectile tissue in the (3) and around the vaginal opening to become engorged with blood. Secretions from the (4) provide lubrication for the movement of the penis. Tactile stimulation during intercourse, as well as psychological stimuli, normally trigger a(n) (5), the female climax. After the sexual act, there is a period of (6), characterized by an overall sense of satisfaction and relaxation. Although orgasm is a pleasurable component of sexual intercourse, it is not required for (7) to occur.

1. _____

2. _____

3. _____

4. _____

5. _____

6. _____

7. _____

Female Fertility and Pregnancy

66 *Spermatozoa must travel from the vagina to the ampulla of the uterine tube for fertilization to occur.* **99**

Match these terms with the
correct statement or definition:

Capacitation   Placenta
HCG   Swimming
Oxytocin and prostaglandins   Trophoblast

_____ 1. One of the forces that propels the sperm.

_____ 2. Two hormones that stimulate smooth muscle contraction, moving the
sperm toward the ampulla.

_____ 3. Enables the spermatozoa to release acrosomal enzymes that allow
penetration of the oocyte.

_____ 4. Outer layer of the developing embryonic mass responsible for
implantation.

_____ 5. Hormone secreted by the trophoblast that causes the corpus luteum to
remain functional.

_____ 6. Organ responsible for secreting estrogen and progesterone during most
of pregnancy.

Parturition

66 *Parturition refers to the process by which the baby is born.* **99**

A. Match these terms with the
correct statement or definition:

First stage of labor   Third stage of labor
Second stage of labor

_____ 1. Delivery of the placenta.

_____ 2. Stage of labor from the onset of regular uterine contractions until
maximum cervical dilation occurs.

_____ 3. Stage of labor from maximal cervical dilation until delivery is complete.

B. Match these terms with
the correct statement:

Decrease(s)
Increase(s)

_____ 1. Effect of stress on fetal ACTH production.

_____ 2. Effect of ACTH on glucocorticoid production in the fetus.

_____ 3. Effect of fetal glucocorticoids on maternal progesterone production.

_____ 4. Effect of fetal glucocorticoids on maternal estrogen production.

_____ 5. Effect of declining progesterone and increasing estrogen levels on
smooth muscle contraction.

_____ 6. Effect of prostaglandins on uterine contractions.

_____ 7. Effect of oxytocin on uterine contractions.

_____ 8. Effect of stretching the uterus on oxytocin production.

_____ 9. Effect of progesterone on oxytocin release.

☞ Uterine contractions and oxytocin form a positive-feedback mechanism that ends when delivery occurs and the uterus is no longer stretched.

Lactation

❝*Lactation is the production of milk by the breasts (mammary glands).*❞

Match these terms with the correct statement or definition:

Colostrum	Prolactin inhibiting factor (PIF)
Estrogen	Prolactin releasing factor (PRF)
Oxytocin	Progesterone
Prolactin	

_____ 1. Hormone primarily responsible for breast growth during pregnancy.

_____ 2. Hormone responsible for development of the breast's secretory alveoli.

_____ 3. Hormone responsible for milk production.

_____ 4. Substance secreted by the hypothalamus in response to mechanical stimulation of the breast.

_____ 5. Substance produced by the breasts for the first few days after parturition that contains little fat and less lactose than milk.

_____ 6. Hormone that causes cells around the alveoli to contract; responsible for milk letdown.

_____ 7. Substance secreted by the neurohypophysis in response to mechanical stimulation of the breasts.

Menopause

❝*The cessation of menstrual cycles is called menopause.***❞**

Using the terms provided, complete the following statements:

Estrogen and progesterone   LH and FSH
Female climacteric   Ovary
Hot flashes

The whole time period from the onset of irregular menstrual cycles to their complete cessation is called the _(1)_ . The major cause of menopause is age-related changes in the _(2)_ . Follicles become less sensitive to stimulation by _(3)_ , and fewer mature follicles and corpora lutea are produced. Gradual morphological changes occur in the female in response to reduced amounts of _(4)_ produced by the ovaries. Symptoms include _(5)_ , irritability, fatigue, anxiety, and occasionally severe emotional disturbances.

1. _____

2. _____

3. _____

4. _____

5. _____

QUICK RECALL

1. List the stages of development of spermatozoa.

2. List the following structures in the correct order from the site of spermatozoa production to the exterior of the body: ductus deferens, efferent ductules, ejaculatory duct, epididymis, membranous urethra, prostatic urethra, rete testis, seminiferous tubules, spongy urethra.

3. List three glands involved in reproduction in the male, and describe the secretions they produce.

4. For the male, list the hormones involved with the reproductive system that are secreted by the hypothalamus, the adenohypophysis, and the testes.

5. List six effects that testosterone has in the male.

6. Arrange the following stages in the development of follicle cells in the correct order: corpus albicans, corpus luteum, primary follicle, secondary follicle, vesicular (graafian) follicle.

7. List the stages of development of the oocyte, and state at what stage fertilization normally takes place.

8. Starting with the site of milk production, name the structures a drop of milk would pass through on the way to the outside of the woman's body.

9. Name the three phases of the menstrual cycle.

10. List the effect the following hormones have on the ovarian cycle: GnRH, FSH, and LH.

11. List the effects of estrogen and progesterone on the uterine cycle.

12. List the positive feedback mechanism that occurs during parturition.

13. List the effects of estrogen, progesterone, prolactin, and oxytocin on the female breast.

Place the letter corresponding to the correct answer in the space provided.

_____ 1. If an adult male jumped into a swimming pool of cold water, which of the following muscles would be expected to contract?

    a. cremaster muscle
    b. dartos tunic
    c. gubernaculum
    d. a and b
    e. all of the above

_____ 2. Early in development the testes
    a. are found inside the peritoneal cavity.
    b. move through the inguinal canal.
    c. produce a membrane that becomes the scrotum.
    d. all of the above

_____ 3. Testosterone is produced in the
    a. interstitial cells of Leydig.
    b. seminiferous tubules of the testes.
    c. anterior lobe of the pituitary gland.
    d. Sertoli cells.

_____ 4. Given the following structures:
    1. ductus deferens
    2. efferent ductule
    3. epididymis
    4. ejaculatory duct
    5. rete testis

Choose the arrangement that lists the structures in the order spermatozoa would pass through them from the seminiferous tubules to the urethra.
    a. 2, 3, 5, 4, 1
    b. 2, 5, 3, 4, 1
    c. 3, 2, 4, 1, 5
    d. 3, 4, 2, 1, 5
    e. 5, 2, 3, 1, 4

_____ 5. The site of spermatogenesis in the male is the
    a. ductus deferens.
    b. seminiferous tubules.
    c. epididymis.
    d. tubuli recti.
    e. rete testis.

_____ 6. The site of final maturation and storage of spermatoza before their ejaculation is the
    a. seminal vesicles.
    b. seminiferous tubules.
    c. glans penis.
    d. epididymis.
    e. sperm bank.

_____ 7. Concerning the penis,
    a. the membranous urethra passes through the corpora cavernosa.
    b. the glans penis is formed by the corpus spongiosum.
    c. the penis contains four columns of erectile tissue.
    d. the prepuce is attached to the crus of the penis.

_____ 8. Given the following glands:
    1. prostate gland
    2. bulbourethral gland
    3. seminal vesicle

Choose the arrangement that is in the order the glands would contribute their secretions to semen during its formation.
    a. 1, 2, 3
    b. 2, 1, 3
    c. 2, 3, 1
    d. 3, 1, 2
    e. 3, 2, 1

_____ 9. Enlargement of what gland could interfere with the reproductive ability of the male?
    a. prostate gland
    b. paraurethral gland
    c. seminal vesicle
    d. bulbourethral gland

_____ 10. Which of the following glands is correctly matched with the function of gland secretion?
    a. bulbourethral gland - neutralizes acidic contents of the urethra
    b. seminal vesicles - contains large amounts of fructose that nourishes the spermatozoa

c. prostate gland - contains clotting factors that causes coagulation of the semen

d. all of the above

_____ 11. LH in the male stimulates
   a. development of the seminiferous tubules.
   b. spermatogenesis.
   c. testosterone production.
   d. a and b
   e. all of the above

_____ 12. Before puberty,
   a. FSH levels are higher than after puberty.
   b. LH levels are higher than after puberty.
   c. GnRH release is inhibited by testosterone.
   d. all of the above

_____ 13. Which of the following cause a decrease in GnRH release?
   a. decreased inhibin
   b. increased testosterone
   c. decreased FSH
   d. decreased LH

_____ 14. Testosterone
   a. stimulates hair growth.
   b. decreases red blood cell count.
   c. prevents closure of the epiphyseal plate.
   d. all of the above

_____ 15. Which of the following is consistent with erection?
   a. parasympathetic stimulation
   b. vasodilation of arterioles
   c. engorgement of sinusoids with blood
   d. occlusion of veins
   e. all of the above

_____ 16. A polar body
   a. is normally formed before fertilization.
   b. is normally formed after fertilization.
   c. is a sun bathing eskimo.
   d. normally receives most of the cytoplasm.
   e. a and b

_____ 17. After ovulation the graafian follicle collapses, taking on a yellowish appearance to become the
   a. atretic follicle.
   b. corpus luteum.
   c. corpus albicans.
   d. tunica albuginea.

_____ 18. The infundibulum
   a. is the opening of the uterine tube into the uterus.
   b. has long, thin projections called the ostium.
   c. is connected to the isthmus of the uterine tube.
   d. is covered with a ciliated mucous membrane.

_____ 19. The layer of the uterus that undergoes the greatest change during the menstrual cycle is the
   a. serosa.
   b. hymen.
   c. endometrium.
   d. myometrium.

_____ 20. The vagina
   a. consists of skeletal muscle.
   b. has ridges called rugae.
   c. is lined with simple squamous epithelium.
   d. all of the above

_____ 21. During intercourse, which of the following structures fill with blood and cause the vaginal orifice to narrow?
   a. bulb of the vestibule
   b. clitoris
   c. mons pubis
   d. labia major

_____ 22. Given the following vestibular-perineal structures:
   1. vaginal orifice
   2. clitoris
   3. urethral opening
   4. anus

   Choose the arrange that lists the structures in their proper order from the anterior to the posterior aspect.
   a. 2, 3, 1, 4
   b. 2, 4, 3, 1
   c. 3, 1, 2, 4
   d. 3, 1, 4, 2
   e. 4, 2, 3, 1

23. Given the following structures:
    1. cervix
    2. pelvic cavity
    3. uterine cavity
    4. uterine tubes

    Assume a couple has just consummated the sex act and the sperm of the male has been deposited in the woman's vagina. Trace the pathway of the sperm through the female reproductive tract to the ovary.
    a. 1, 3, 2, 4
    b. 1, 3, 4, 2
    c. 3, 1, 2, 4
    d. 3, 1, 4, 2
    e. 4, 2, 1, 3

24. Concerning the breasts,
    a. lactiferous ducts open on the areola.
    b. each lactiferous duct supplies an alveoli.
    c. they are attached to the pectoralis major muscles by mammary ligaments.
    d. even before puberty, the female breast is quite different from the male breast.

25. The major secretory product of the graafian follicle is
    a. estrogen.
    b. progesterone.
    c. LH.
    d. FSH.
    e. relaxin.

26. An increase in LH initiates the process of ovulation approximately at day 14 in the average adult female. Shortly after the increase in LH, the rate at which it is secreted declines to very low levels. Which of the following events may play a role in decreasing the plasma LH levels following its rapid increase in the blood?
    a. corpus luteum function
    b. increasing progesterone concentration in the plasma
    c. very low estrogen and progesterone secretion
    d. elevated FSH secretion
    e. a and b

27. Which of the following processes or phases in the monthly reproductive cycle of the human female occur at the same time?
    a. maximal LH secretion and menstruation
    b. early follicular development and the secretory phase in the uterus
    c. regression of the corpus luteum and an increase in ovarian progesterone secretion
    d. ovulation and menstruation
    e. proliferation stage of uterus and increased estrogen production

28. During the secretory phase of the menstrual cycle, one would normally expect
    a. the highest levels of estrogen that occur during the menstrual cycle.
    b. a mature graafian follicle to be present in the ovary.
    c. an increase in the thickness of the endometrium.
    d. a and b
    e. all of the above

29. The cause of menses in the menstrual cycle appears to be
    a. increased progesterone secretion from the ovary, which produces blood clotting.
    b. increased estrogen secretion from the ovary, which stimulates the muscles of the uterus to contract.
    c. decreased progesterone and estrogen secretion by the ovary.
    d. decreased production of oxytocin, causing the muscles of the uterus to relax.

30. A woman with a typical 28-day menstrual cycle is most likely to become pregnant as a result of coitus on days
    a. 1 - 3.
    b. 5 - 8.
    c. 12 - 15.
    d. 22 - 24.
    e. 24 - 28.

31. After fertilization the successful development of a mature, full-term fetus depends on:
    a. the release of human chorionic gonadotropin (HCG) by the developing placenta.
    b. production of estrogen and progesterone by the placental tissues.
    c. the maintenance of the corpus luteum for 9 months.

d. a and b
e. all of the above

_____ 32. Onset of labor may be a result of which of the following?
    a. increased estrogen secretion by the placenta
    b. increased glucocorticoid secretion by the fetus
    c. increased secretion of oxytocin
    d. all of the above

_____ 33. The hormone involved in milk production is
    a. oxytocin.
    b. prolactin.
    c. estrogen.
    d. progesterone.
    e. ACTH.

_____ 34. Which of the following most appropriately predicts the consequence of removing the sensory neurons from the areola of a lactating rat (or human)?
    a. Blood levels of oxytocin decrease.
    b. Blood levels of prolactin decrease.
    c. Milk production and secretion decrease.
    d. all of the above

_____ 35. Menopause
    a. happens whenever a woman pauses to think about a man.
    b. occurs when a woman stops a man from making a pass.
    c. develops when follicles become less responsive to FSH.
    d. results from high estrogen levels in 40-50 year old women.
    e. c and d

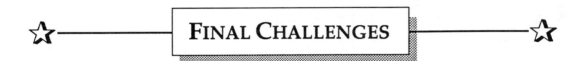

FINAL CHALLENGES

Use a separate sheet of paper to complete this section.

1. Two teenage girls wanted to make a douche (a solution used to remove spermatozoa from the vagina following intercourse). One girl suggested using a vinegar solution, and the other girl wanted to use a solution of baking soda. Although neither solution promises to be very effective, which solution would be most likely to succeed in preventing a pregnancy?

2. In response to an injection of a large amount of testosterone in an adult male, what would happen to testosterone production in the testes? Explain.

3. Suppose a 9 year-old boy had an interstitial cell tumor that resulted in very high levels of testosterone production. How would this affect his development?

4. Suppose that a man had a complete transection of his spinal cord at level L3. Would it be possible for him to achieve an erection through psychic stimulation or through stimulation of the genitals? Explain.

5. Birth control pills for women contain estrogen, progesterone, or a combination of both these hormones. Explain how these hormones can prevent pregnancy.

6. Birth control pills for women that consist of estrogen or progesterone are only taken for 21 days. The woman stops taking the birth control pill or takes a placebo pill for 7 days. Then she resumes taking the birth control pill. Why does she do this?

7. Sexually transmitted diseases such as gonorrhea can sometimes cause peritonitis in females. However, in males peritonitis from this cause does not develop. Explain.

8. Predict the consequences of removing the corpus luteum during the first trimester (first 3 months) of pregnancy in humans. What happens if the corpus luteum is removed during the second trimester (months 4 to 6)?

ANSWERS TO CHAPTER 28

WORD PARTS

1. seminiferous
2. seminiferous, lactiferous
3. androgen
4. androgen
5. ejaculate

6. endometrium, myometrium, perimetrium
7. myometrium
8. lactiferous, lactation, prolactin
9. gynecomastia
10. gynecomastia

CONTENT LEARNING ACTIVITY

Scrotum

1. Raphe
2. Dartos muscle

3. Cremaster muscles

Perineum

1. Urogenital triangle

2. Anal triangle

Testes

A. 1. Tunica albuginea
   2. Septa
   3. Seminiferous tubules
   4. Interstitial cells (cells of Leydig)
   5. Efferent ductules

B. 1. Inguinal canal
   2. Process vaginalis
   3. Tunica vaginalis

C. 1. Spermatogenesis
   2. Sertoli cells
   3. Spermatogonia
   4. Primary spermatocytes
   5. Secondary spermatocytes
   6. Spermatids
   7. Acrosome

Ducts

1. Epididymis
2. Ductus epididymis
3. Ductus (vas) deferens
4. Ejaculatory duct

5. Spermatic cord
6. Prostatic urethra
7. Membranous urethra

Penis

1. Corpora cavernosa
2. Corpus spongiosum
3. Glans penis
4. Bulb of the penis
5. Crus of the penis
6. Root of the penis
7. Prepuce (foreskin)

Glands

A.
1. Seminal vesicles
2. Prostate gland
3. Bulbourethral glands
4. Emission
5. Bulbourethral secretions
6. Seminal vesicle secretions
7. Prostate secretions

B.
1. Prostatic urethra
2. Membranous urethra
3. Spongy urethra

4. Penis
5. External urethral orifice
6. Scrotum
7. Testis
8. Epididymis
9. Ductus deferens
10. Bulbourethral gland
11. Prostate
12. Ejaculatory duct
13. Seminal vesicle

Regulation of Hormone Secretion

1. Gonadotropin-releasing hormone (GnRH)
2. Gonadotropins
3. LH
4. FSH
5. Testosterone
6. Androgen
7. Inhibin

Puberty

1. HCG
2. Testosterone
3. GnRH
4. LH and FSH
5. Spermatogenesis
6. Testosterone

Effects of Testosterone

1. Testes
2. Coarser
3. Melanin
4. Sebaceous glands
5. Hypertrophy
6. Body fluids
7. Protein synthesis
8. Rapid bone growth

Male Sexual Behavior and the Male Sex Act

1. Orgasm
2. Resolution
3. Impotence
4. Erection
5. Emission
6. Ejaculation

Ovaries

A. 1. Mesovarium
    2. Suspensory ligament
    3. Ovarian (germinal) epithelium
    4. Tunica albuginea
    5. Medulla
    6. Ovarian follicle
    7. Oocyte

B. 1. Oogenesis
    2. Oogonia
    3. Primary oocyte
    4. Secondary oocyte; polar body

C. 1. Primary follicle
    2. Secondary follicle
    3. Cumulus mass

4. Corona radiata
5. Graafian (vesicular) follicle
6. Theca
7. Zona pellucida

D. 1. Ovulation
    2. Secondary oocyte
    3. Fertilization
    4. Polar body
    5. Zygote
    6. Corpus luteum
    7. Progesterone and estrogen
    8. Corpus albicans

Uterine Tubes

1. Mesosalpinx
2. Infundibulum
3. Ostium
4. Fimbriae

5. Ampulla
6. Isthmus
7. Muscular layer
8. Mucosa

Uterus

A. 1. Fundus
    2. Cervix
    3. Isthmus
    4. Cervical canal
    5. Ostium
    6. Round ligaments
    7. Broad ligament

B. 1. Perimetrium (serous coat)
    2. Endometrium (mucous membrane)
    3. Basal layer
    4. Functional layer

C. 1. Body of uterus
    2. Cervix
    3. Vagina
    4. Round ligament
    5. Ovarian ligament
    6. Uterine tube
    7. Ovary
    8. Fundus of uterus

Vagina

1. Columns
2. Rugae
3. Fornix
4. Hymen

External Genitalia

A. 1. Vestibule
   2. Labia minora
   3. Clitoris
   4. Prepuce
   5. Bulb of the vestibule
   6. Greater vestibular glands
   7. Lesser vestibular glands
   8. Mons pubis

B. 1. Prepuce
   2. Labia minora
   3. Vagina
   4. Vestibule
   5. Labia majora
   6. Urethra
   7. Clitoris
   8. Mons pubis

Perineum

1. Urogenital triangle
2. Clinical perineum

3. Episiotomy

Mammary Glands

A. 1. Areola
   2. Areolar glands
   3. Gynecomastia
   4. Lobes
   5. Lobules
   6. Lactiferous duct
   7. Alveoli
   8. Mammary (Cooper's) ligaments

B. 1. Lactiferous sinus
   2. Lactiferous duct
   3. Nipple
   4. Areola
   5. Lobe
   6. Alveoli
   7. Lobule

Puberty

1. Estrogen and progesterone
2. FSH and LH
3. GnRH

4. FSH and LH
5. Estrogen and progesterone

The Menstrual Cycle

A. 1. Menses
   2. Follicular (proliferative) stage
   3. Luteal (secretory) stage

B. 1. Ovarian cycle
   2. Adenohypophysis
   3. FSH
   4. LH
   5. Ovulation
   6. Luteal
   7. Progesterone
   8. HCG

C. 1. Increases
   2. Increases
   3. Positive feedback
   4. Increase
   5. Increases
   6. Decreases
   7. Increases
   8. Increase
   9. Negative feedback
   10. Decrease
   11. Decrease

D. 1. Uterine cycle
   2. Spiral tubular glands
   3. Spiral arteries

E. 1. Increases
2. Increases
3. Increases
4. Increases
5. Decreases
6. Decrease

F. 1. GnRH
2. Estrogen

3. Progesterone
4. Secondary follicle
5. Graafian (vesicular) follicle
6. Corpus luteum
7. Corpus albicans
8. Menses
9. Proliferative phase
10. Secretory phase
11. Menses

Female Sexual Behavior and the Female Sex Act

1. Psychic factors
2. Sacral
3. Clitoris
4. Vagina

5. Orgasm
6. Resolution
7. Fertilization

Female Fertility and Pregnancy

1. Swimming
2. Oxytocin and prostaglandins
3. Capacitation

4. Trophoblast
5. HCG
6. Placenta

Parturition

A. 1. Third stage of labor
2. First stage of labor
3. Second stage of labor

B. 1. Increases
2. Increases

3. Decrease
4. Increase
5. Increase
6. Increase
7. Increases
8. Increases
9. Decreases

Lactation

1. Estrogen
2. Progesterone
3. Prolactin
4. Prolactin-releasing factor (PRF)
5. Colostrum

6. Oxytocin
7. Oxytocin

Menopause

1. Female climacteric
2. Ovary
3. FSH and LH

4. Estrogen and Progesterone
5. Hot flashes

1. Spermatogonia, primary spermatocyte, secondary spermatocyte, spermatid, spermatozoon
2. Seminiferous tubules, rete testis, efferent ductules, epididymis, ductus deferens, ejaculatory duct, prostatic urethra, membranous urethra, spongy urethra
3. Seminal vesicles: thick, mucoid secretions containing nutrients, fibrinogen, and prostaglandins; prostate gland: thin, milky, alkaline secretions containing clotting factors; bulbourethral glands: alkaline mucous secretions.
4. Hypothalamus: GnRH; adenohypophysis: FSH and LH; testes: testosterone.
5. Enlargement and differentiation of the reproductive system, descent of testes, spermatogenesis, hair growth, increased skin pigmentation, increased sebaceous secretions, increased muscle mass, increased body fluids, increased skeletal growth, laryngeal hypertrophy, increased metabolism, and increased blood cell count
6. Primary follicle, secondary follicle, Graafian (vesicular) follicle, corpus luteum, corpus albicans
7. Oogonia, primary oocyte, secondary oocyte; fertilization occurs in the secondary oocyte.
8. Alveoli, lobule, lobe, lactiferous sinus, lactiferous duct
9. Menses, proliferative (follicular) phase, secretory (luteal) phase
10. GnRH: stimulates release of FSH and LH from the adenohypophysis; FSH: stimulates follicle development; LH: triggers ovulation, development of corpus luteum.
11. Estrogen: thickening of endometrium; progesterone: thickening, glandularity, and vascularity of endometrium stimulated.
12. Stretch of the uterus stimulates oxytocin secretion, which stimulates uterine contraction, which increases uterine stretch.
13. Estrogen: development of duct system and fat deposition; progesterone: development of secretory alveoli; prolactin: milk production; oxytocin: milk letdown.

MASTERY LEARNING ACTIVITY

1. D. The cremaster muscle, which is a continuation of the internal abdominal oblique muscle, extends through the spermatic cord to the testes. The tunica dartos muscle layer is found beneath the skin of the scrotal sac. When temperatures are low, these muscles contract, bringing the testes closer to the body and thus a warmer environment. Conversely, if temperatures are too high, they relax, and the testes descend to a cooler climate.

2. B. The testes develop as retroperitoneal organs in the abdominopelvic cavity and pass through the inguinal canal into the scrotum.

3. A. The interstitial cells produce testosterone. The seminiferous tubules produce spermatozoa. Sertoli cells form a blood-testes barrier, produce hormones, and nourish the developing spermatozoa.

4. E. The spermatozoa would pass through the rete testis, efferent ductule, epididymis, ductus deferens, and ejaculatory duct.

5. B. Spermatozoa are produced in the seminferous tubules.

6. D. Spermatozoa undergo final maturation and are stored in the epididymis. During ejaculation smooth muscles within the epididymis contract and move sperm into the ductus deferens. A small amount of sperm are also stored in the ampulla of the ductus deferens.

7. B. There are three columns of erectile tissue in the penis: the corpus spongiosum and two corpora cavernosa. The spongy urethra passes through the corpus spongiosum. The distal end of the corpus spongiosum is the glans penis, which is covered by the prepuce. The proximal ends of the corpora cavernosa form the crura of the penis, which attach the penis to the coxae.

8. D. The seminal vesicles empty into the ejaculatory duct, the prostate gland into the prostatic urethra, and the bulbourethral glands into the membranous urethra.

9. A. The prostate glands surrounds the prostatic urethra and part of the ejaculatory duct. Enlargement of the gland can prevent passage of semen.

10. D. The bulbourethral glands neutralize the urethra before ejaculation. The seminal vesicles provide fructose and fibrinogen. The prostatic secretions neutralize semen and contain clotting factors that act on fibrinogen to cause coagulation of the semen.

11. C. LH stimulates the production of testosterone. FSH stimulates the development of the seminiferous tubules and spermatogenesis.

12. C. GnRH release is inhibited by small amounts of testosterone. Consequently, FSH and LH levels are low.

13. B. Testosterone has a negative feedback effect that decreases GnRH release from the hypothalamus.

14. A. Testosterone stimulates such factors as hair growth, red blood cell production, and increased metabolism. It also stimulates bone growth and closes off the epiphyseal plate.

15. E. When the penis is flaccid, the arterioles that supply it are partially contracted. During sexual excitation parasympathetic impulses from the sacral region of the spinal cord travel to the arterioles, causing relaxation and vasodilation. This allows more blood to enter the sinuosoids in the penis, causing them to enlarge. As the erectile tissue of the penis expands, it compresses the veins that drain the penis. The combination of increased blood flow into the penis and reduced blood flow out produce the erection.

16. E. A polar body is formed by the first meiotic division before fertilization. After fertilization the second meiotic division occurs to produce a second polar body. The polar bodies contain little cytoplasm.

17. B. A follicle goes through several stages: primary follicle, secondary follicle, graafian follicle, corpus luteum, and corpus albicans.

18. D. The infundibulum is the expanded end of the uterine tube next to the ovary. It has long projections, the fimbriae, which are covered by a ciliated mucous membrane. When an oocyte is released from the ovary, it is swept through the opening of the infundibulum, the ostium, into the uterine tube. The infundibulum is attached to the ampulla of the uterine tube.

19. C. The functional layer of the endometrium undergoes the greatest change during the menstrual cycle.

20. B. The vaginal walls have longitudinal ridges (columns) and transverse ridges (rugae) that allow the vagina to expand. The vagina consists of smooth muscle, which can stretch to allow expansion of the vagina, and is lined with moist, stratified squamous epithelium.

21. A. The bulb of the vestibule, which corresponds to the corpus spongiosum of the male penis, causes the vaginal orifice to narrow. The clitoris, which corresponds to the corpora cavernosa of the male penis, also fills with blood during inter-course. Although the clitoris expands in diameter, it does not affect the size of the vaginal orifice.

22. A. The order is clitoris, urethral opening, vaginal orifice, and anus.

23. B. The spermatozoa would pass through the cervical canal into the uterine cavity. From the uterus they would enter the uterine tube and pass into the pelvic cavity.

24. C. The mammary ligaments support the breasts. The lactiferous ducts supply the lobes of the breast and open onto the nipple. Before puberty male and female breasts are similar in structure.

25. A. The graafian follicle secretes primarily estrogen. The corpus luteum produces primarily progesterone. Relaxin is produced by the corpus luteum during pregnancy. LH and FSH are both produced in the anterior pituitary.

26. E. The sudden increase in plasma LH initiates ovulation. Following ovulation the ovulated follicle is rapidly converted to a functional corpus luteum and begins to produce large amounts of progesterone. The development of the corpus luteum and the progesterone that it secretes appears to play a role in decreasing the LH levels following the LH surge.

27. E. Increased estrogen levels produced by the developing follicle stimulate the endometrium to proliferate.

28. C. During the secretory phase of the menstrual cycle, the corpus luteum produces mainly progesterone and some estrogen (but less than the graafian follicle produces during the proliferation stage). The estrogen increases cell division, and the progesterone stimulates secretory activity, resulting in increased thickness of the endometrium.

29. C. During the last few days of the menstrual cycle, progesterone and estrogen production by the ovary drops. This results in vasospasms of vessels in the uterus and the cutting off of blood supply to the endometrium. The outer portion of the endometrium dies, and uterine contractions expel the menses.

30. C. Ovulation occurs about day 14 of the typical menstrual cycle, and the ovum remains viable for about 24 hours. Spermatozoa remain viable in the female reproductive tract for up to 72 hours. Therefore deposition of spermatozoa on days 12-15 could result in fertilization.

31. D. HCG stimulates the corpus luteum to produce estrogen and progesterone for the first 3 months of pregnancy. The estrogen and progesterone are necessary for the maintenance of the endometrium. After 3 months, HCG levels decline, and the corpus luteum becomes a corpus albicans. Meanwhile, the placenta produces estrogen and progesterone.

32. D. The initiation of labor involves several mechanisms. Production of fetal glucocorticoids causes the placenta to increase estrogen secretion and decreases progesterone secretion. This results in increased uterine excitability. Stretch of the cervix leads to the production of oxytocin, which in turn stimulates uterine contraction, leading to more stretch, and so on.

33. B. Prolactin stimulates milk production. Oxytocin is responsible for ejection of milk into the ducts of the breast. Estrogen stimulates duct development, and progesterone causes development of the secretory alveoli.

34. D. Sensory stimuli initiated by suckling pups is critical in the female rat (also in the human). It is important in initiating oxytocin secretion and maintaining prolactin secretion. The loss of sensory information due to a lack of suckling or due to severed neurons results in reduced oxytocin and prolactin secretion from the anterior pituitary and therefore eventually stops the secretion and production of milk.

35. C. A decrease in the number of follicles and a decrease in the sensitivity of the remaining follicles to FSH and LH are responsible for menopause.

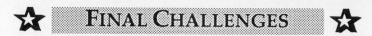

1. Spermatozoa are destroyed by the acidic environment of the vagina; one function of the secretions in semen is to neutralize acids. Consequently, the acidic vinegar solution would be better than the alkaline baking soda solution.

2. Increased testosterone, by a negative feedback mechanism, inhibits the production of gonadotropin-releasing hormone (GnRH) from the hypothalamus. Decreased GnRH results in decreased LH release from the adenohypophysis, which causes a reduction in the production of testosterone by the testes.

3. The testosterone would cause early and pronounced development of his sexual organs. He would also have rapid growth of muscle and bone, but early closure of the epiphyseal plate would result in a shorter height than he would have normally achieved.

4. Erection is mediated through parasympathetic reflex centers in the sacral region of the spinal cord. Injury at level L3 would not interrupt the reflex arc from the genitals, so stimulation of the genitals should cause an erection. Erection can also be achieved by sympathetic stimulation through centers at levels T12 to L1. Psychic stimulation can activate the sympathetic centers, which are superior to the spinal cord injury, but not the parasympathetic centers, which are inferior to the injury. It is possible that psychic stimulation could also cause an erection.

5. Estrogen and progesterone have a negative feed back effect on the hypothalamus, inhibiting GnRH secretion. Consequently, FSH and LH secretion is inhibited in the adenohypophysis. This decreases follicle development and inhibits ovulation.

6. By not taking the birth control pill estrogen and/or progesterone levels fall, resulting in menstruation.

7. In females the microorgansims can travel from the vagina to the uterus, to the uterine tubes, to the pelvic cavity. Infection of the peritoneum lining the pelvic cavity results in peritonitis. In males the microorganisms can move up the urethra to the bladder or into the ejaculatory duct to the ductus deferens. There is no direct connection into the pelvic cavity in the male, so peritonitis does not develop.

8. Progesterone and estrogen prepare the uterine endometrium for implantation and maintain the endometrium throughout pregnancy. In addition, progesterone inhibits uterine contractions. During the first trimester of pregnancy the corpus luteum is the primary source of progesterone. After the first trimester the corpus luteum becomes less important, and the placenta produces enough progesterone to maintain pregnancy. Therefore, during the first trimester if the corpus luteum is removed, the blood levels of progesterone decrease, the endometrium degenerates, and an abortion results. Removal of the corpus luteum in the second or third trimester does not interrupt pregnancy.

Development, Growth, And Aging

FOCUS: The lifespan of an individual can be divided into life stages. During the germinal period the fertilized oocyte develops into an embryo with the three primary germ layers, which give rise to the organ systems. In the embryonic stage the major organ systems develop, and in the fetal stage growth and maturation of the organ systems takes place. After birth, during the neonate stage, the fetal circulation pattern changes, and the digestive system becomes functional. In the infant muscular-nervous coordination increases, enabling the infant to control body movements. During childhood there is further growth, and in adolescence the child becomes a sexually functional adult. Aging results in a decrease in nervous and muscular cells, loss of elasticity in tissues, decreased ability to mount an immune system response, and a general loss of the ability to maintain homeostasis. Eventually some challenge to homeostasis cannot be overcome, and death results.

WORD PARTS

Give an example of a new vocabulary word that contains each word part.

WORD PART	MEANING	EXAMPLE
poly-	many	1. _____
-genic	genes	2. _____
nata-	birth	3. _____
neo-	new	4. _____
blast-	a bud	5. _____
-cyst-	bladder	6. _____

-cele	hollow	7. _____
pluri-	more, several	8. _____
noto-	the back	9. _____
-chord	a string	10. _____

CONTENT LEARNING ACTIVITY

Prenatal Development

❝_The nine months before birth comprise a critical part of an individual's existence._**❞**

Match these terms with the
correct statement or definition:

Clinical age   Fetal period
Embryonic period   Postovulatory age
Germinal period   Prenatal period

_____ 1. Period from conception to birth.

_____ 2. Period of development during which primitive germ layers are formed.

_____ 3. Period of development during which the major organ systems are formed.

_____ 4. Period of development in which the organ systems grow and become more mature.

_____ 5. The calculation of developmental age that uses the last menstrual period (LMP) as the starting point.

Fertilization

❝_Normally one sperm penetrates the oocyte cell membrane and enters the cytoplasm in the process of fertilization._**❞**

Match these terms with the
correct statement or definition:

Female pronucleus   Second meiotic division
Male pronucleus   Zygote

_____ 1. Division triggered by entrance of the sperm into the oocyte; a second polar is formed.

_____ 2. Haploid nucleus of the oocyte, after the second meiotic division.

_____ 3. Enlarged head of the sperm cell, containing the haploid number of chromosomes.

_____ 4. Result of the process of fertilization.

☞ The fusion of the male and female pronuclei restores the diploid number of chromosomes.

Early Cell Division: Morula and Blastocyst

66 *Through the process of cell division, the zygote produces an embryonic cell mass.* **99**

Match these terms with the
correct statement or definition:

Blastocele   Morula
Blastocyst   Pluripotent
Inner cell mass   Trophoblast

_____ 1. Ability of the embryonic cell mass, up to the 8-cell stage, to develop into any cell in the body.

_____ 2. Dividing embryonic cell mass, once it has 12 or more cells.

_____ 3. Embryonic cell mass, once a cavity begins to form inside.

_____ 4. Single layer of cells that surrounds most of the blastocele and develops into the placenta and the membranes surrounding the embryo.

_____ 5. Thickened area, several cell layers thick, at one end of the blastocyst, that will develop into the embryo.

Implantation of the Blastocyst and Development of the Placenta

66 *About 7 days after fertilization, the blastocyst attaches itself to the uterine wall.* **99**

Match these terms with the
correct statement or definition:

Cytotrophoblast   Syncytiotrophoblast
Lacunae

_____ 1. Nondividing, multinucleate cells that invade maternal tissues.

_____ 2. Cavities produced when the syncitiotrophoblast surrounds and digests away the walls of maternal bloodvessels.

_____ 3. Dividing population of trophoblast cells that protrudes into the lacunae and is followed by embryonic blood vessels; later disappears, leaving only a thin layer separating the maternal and fetal blood supplies.

Formation of Germ Layers

"All the tissues of the adult can be traced to three germ layers.**"**

Match these terms with the correct statement or definition:

Amniotic cavity   Mesoderm
Ectoderm   Notochord
Embryonic disk   Primitive streak
Endoderm   Yolk sac

_____ 1. New cavity formed inside the inner cell mass after implantation occurs.

_____ 2. Flat disk of tissue composed of two cell layers.

_____ 3. Cell layer of the embryonic disk that is adjacent to the amniotic cavity.

_____ 4. Third cavity produced by the endoderm that forms inside the blastocele.

_____ 5. Thickened line, formed by cells of the ectoderm, migrating to the center and caudal end of the embryonic disk.

_____ 6. Third germ layer, formed from ectodermal cells that migrate through the primitive streak and emerge between the ectoderm and endoderm.

Neural Tube and Neural Crest Formation

"The neural tube and neural crest are formed from ectoderm.**"**

Match these terms with the correct statement or definition:

Neural crest cells   Neural plate
Neural crests   Neural tube

_____ 1. Thickened layer of cells produced when the notochord stimulates overlying ectoderm cells.

_____ 2. Lateral edges of the neural plate that rise and move toward each other.

_____ 3. Structure formed when the neural crests meet at the midline and fuse; becomes the brain and spinal cord.

_____ 4. Cells that become part of the peripheral nervous system, melanocytes, or varied structures in the head.

Somite, Gut, and Body Cavity Formation

"_During the early embryonic period, many structures are being formed._**"**

Match these terms with the
correct statement or definition:

Branchial arches   Oropharyngeal membrane
Celom   Pharyngeal pouches
Cloacal membrane   Somites
Evaginations   Somitomeres

_____ 1. Distinct segments formed from mesoderm adjacent to the neural tube.

_____ 2. First few somites in the head region that never become clearly divided.

_____ 3. Membrane formed in the hindgut, where it is in close contact with
                            ectoderm; this later becomes the urethra and the anus.

_____ 4. Outpocketings from the GI tract; these develop into structures such as
                            the anterior pituitary, lungs, liver, pancreas, and the urinary bladder.

_____ 5. Solid bars of mesenchyme that develop along the head.

_____ 6. Pockets of the pharynx that extend between the branchial arches; give
                            rise to the tonsils, thymus, and parathyroid glands.

_____ 7. Collective name for the body cavities, subdivided to form the
                            pericardial, pleural, and peritoneal cavities.

Limb Bud and Face Development

"_The arms and legs first appear as limb buds, but the face develops by fusion of five embryonic structures._**"**

Match these terms with the
correct statement or definition:

Apical ectodermal ridge   Nasal placodes
Frontonasal process   Primary palate
Mandibular processes   Secondary palate
Maxillary processes

_____ 1. Specialized thickening of ectoderm that develops on the lateral margin
                            of each limb bud.

_____ 2. Process that forms the forehead, nose, and midportion of the upper jaw
                            and lip.

_____ 3. Processes that form the lateral portion of the upper jaw and lip.

_____ 4. Structures that develop at the lateral margin of the frontonasal process
                            and become the nose and center of the upper jaw and lip.

_____ 5. Upper jaw and lip, formed by fusion of the nasal placodes and the
                            maxillary processes.

_____ 6. Roof of the mouth; failure to fuse is a cleft palate.

Development of the Organ Systems

"The major organ systems appear and begin to develop during the embryonic period.**"**

A. Match these terms with the correct structure that develops from them:

Mesoderm
Neural crest cells

Neural tube
Neural tube cavity

_____ 1. Dermis of the skin.

_____ 2. Melanocytes and sensory receptors in the skin.

_____ 3. Appendicular skeleton.

_____ 4. Ventricles of the brain and the central canal of the spinal cord.

B. Match these terms with the correct statement or definition:

Adenohypophysis
Myoblasts
Neurohypophysis

Olfactory bulb and nerve
Optic stalk and optic vesicle

_____ 1. Multinucleated cells that develop into skeletal muscle fibers.

_____ 2. These structures develop as an evagination from the telencephalon.

_____ 3. These structures develop as an evagination from the diencephalon.

_____ 4. Gland that is formed from an evagination of the floor of the diencephalon.

C. Match these terms with the correct statement or definition:

Adrenal cortex
Adrenal medulla
Pancreas

Parathyroid glands
Thyroid gland

_____ 1. Originates as an evagination from the floor of the pharynx in the region of the developing tongue.

_____ 2. Gland part that originates from mesoderm.

_____ 3. Gland part that arises from neural crest cells and consists of specialized postganglionic neurons of the sympathetic division.

_____ 4. Originates as two evaginations from the duodenum; they later fuse.

D. Match these terms with the correct statement or definition:

Blood islands
Bulbis cordis

Foramen ovale
Sinus venosus

_____ 1. The blood vessels are formed from these structures, which are located on the surface of the yolk sac and inside the embryo.

_____ 2. Site of blood entering the embryonic heart; part of this structure later becomes the SA node.

_____ 3. Site of blood exiting the embryonic heart; this structure is later incorporated into the ventricles.

_____ 4. Opening in the interatrial septum that allows blood to flow from the right to the left atrium.

The lungs develop as a single, midline evagination from the foregut in the region of the future esophagus.

E. Match these terms with the correct statement or definition:

Allantois   Metanephros
Cloaca   Pronephros
Mesonephros   Ureter

_____ 1. Consists of a duct and simple tubules connected to the celomic cavity; it is probably never functional in the embryo.

_____ 2. Consists of tubules that open into the mesonephric duct; the other end forms a glomerulus.

_____ 3. Common junction of the digestive, urinary, and genital systems.

_____ 4. Blind tube that extends into the umbilical cord; develops into the urinary bladder.

_____ 5. Distal end of the ureter enlarges and branches to form the duct system of this structure.

F. Match these terms with the correct statement or definition:

Gonads   Paramesonephric ducts
Mesonephric ducts   Primoridal germ cells

_____ 1. Cells that form on the surface of the yolk sac, migrate into the embryo, and enter the gonadal ridge.

_____ 2. Ducts that form just lateral to the mesonephric ducts; if no testosterone is present, these ducts become the internal female reproductive structures.

_____ 3. Under the influence of testosterone, will develop into the epididymis, ductus deferens, seminal vesicles, and prostate gland.

G. Match these terms with the structure that they later become:

Genital tubercle   Urogenital folds
Labioscrotal swellings

_____ 1. Penis

_____

_____ 2. Scrotum or labia majora

_____ 3. Clitoris

_____ 4. Labia minora

 If the closure of the urogenital folds is not complete when the penis is formed, a defect known as hypospadias results.

Growth of the Fetus

The embryo becomes a fetus about 60 days after fertilization.

Using the terms provided, complete the following statements:

Embryonic   Subcutaneous fat
Growing   Vernix
Lanugo

Most morphological changes occur in the _(1)_ phase of development, whereas the fetal period is primarily a _(2)_ phase. Fine, soft hair called _(3)_ covers the fetus, and a waxy coat of sloughed epithelial cells called _(4)_ protects the fetus from the somewhat toxic nature of the amniotic fluid. _(5)_ accumulates in the late fetus and provides a nutrient reserve, helps insulate the infant, and assists the infant in sucking by supporting the cheeks.

1. _____

2. _____

3. _____

4. _____

5. _____

Genetics

Genetics is the study of heredity, i.e., those characteristics inherited by children from their parents.

Match these terms with the correct statement or definition:

Autosomes   Gene
Congenital   Regulatory genes
DNA   Structural genes

_____ 1. Functional unit of heredity.

_____ 2. Molecule responsible for heredity.

_____ 3. DNA sequences that code for specific amino acid sequences in proteins such as enzymes, hormones, and structural proteins.

_____ 4. DNA sequences that are involved in controlling which structural genes are transcribed.

_____ 5. Disorder that is present at birth.

_____ 6. All chromosomes except those that determine sex.

☞ Each normal human somatic cell contains 23 pairs of chromosomes, or 46 total chromosomes.

Chromosomes

The DNA of each cell is packed into chromosomes.

Match these terms with the correct statement or definition:

Alleles   Homozygous
Dominant   Locus
Genotype   Phenotype
Heterozygous   Recessive

_____ 1. Specific location on a chromosome.

_____ 2. Genes occupying the same locus on homologous chromosomes.

_____ 3. Condition in which a person has two copies of the same gene; identical alleles for a trait specified at one gene locus.

_____ 4. Trait that covers up the opposite trait (its allele).

_____ 5. Trait that can be seen in an individual.

_____ 6. Actual genetic condition of an individual.

 Consideration of inheritance patterns is based on probability predictions.

The Newborn

66 *The newborn infant, or neonate, experiences several dramatic changes at the time of birth.* 99

A. Match these terms with the correct statement or definition:

Ductus arteriosus   Foramen ovale
Ductus venosus   Umbilical vein

_____ 1. Opening between the right and left atria that closes at birth to become the fossa ovalis.

_____ 2. Opening that is between the pulmonary trunk and the aorta; it closes 1 or 2 days after birth and becomes the ligamentum arteriosum.

_____ 3. Vessel that flows through the umbilical cord; it becomes nonfunctional at birth, and is called the round ligament of the liver.

_____ 4. Vessel that receives blood from the umbilical vein and passes through the liver; at birth it degenerates into the ligamentum venosum.

B. Using the terms provided, complete the following statements:

Amylase   Meconium
Bilirubin   Stomach pH
Lactose

Swallowed amniotic fluid, cells sloughed from the mucosal lining, mucus, and bile from the liver pass from the GI tract as a greenish discharge called _(1)_ . The pH of the stomach is nearly neutral at birth, but gastric acid secretion occurs, causing _(2)_ to decrease. The newborn digestive system is capable of digesting _(3)_ from the time of birth, but _(4)_ secretion by the salivary glands remains low until after the first year. The neonatal liver also lacks adequate amounts of the enzyme required to excrete _(5)_ , which may lead to jaundice.

1. _____

2. _____

3. _____

4. _____

5. _____

 The newborn infant may be evaluated for its physiological condition by the APGAR score; this acronym stands for appearance, pulse, grimace, activity, and respiratory effort.

Life Stages

"There are considered to be 8 stages of life from fertilization to death."

Match these life stages with
the correct description:

Adolescent	Fetus
Adult	Germinal
Child	Infant
Embryo	Neonate

_____ 1. Period from fertilization to 14 days.

_____ 2. Period from 14 to 60 days after fertilization.

_____ 3. Period from 60 days after fertilization to birth.

_____ 4. Period from birth to 1 month after birth.

_____ 5. Period from 1 or 2 years after birth to puberty.

_____ 6. Period from puberty to 20 years.

Aging

"The process of aging begins at fertilization."

Using the terms provided, complete the following statements:

Atherosclerosis	Embolus
Arteriosclerosis	Filtration
Autoimmunity	Genetic
Collagen	Heart
Cytologic aging	Thrombosis
Decrease	

As the individual ages, more and more cross-links are formed
between _(1)_ molecules, rendering the tissues more rigid and
less plastic. Death or damage to a nondividing cell produces
irreversible damage; as a result, the number of muscle cells and
neurons _(2)_ with age. The _(3)_ loses elastic recoil ability and
muscular contractility, causing a decline in cardiac output.
Reduced cardiac function may also result in decreased blood
flow to the kidneys, causing a decrease in _(4)_ . _(5)_ is the
deposit of lipid in the intima of large and medium-sized arteries.
These deposits become fibrotic and calcified, causing _(6)_ ,
which interfere with normal blood flow and may lead to a _(7)_
(a clot or plaque formed inside a vessel). A piece of plaque
(called a[n] _[8]_) can break loose and lodge in smaller arteries,
causing myocardial infarctions or strokes. _(9)_ , or cellular wear
and tear, is another factor that contributes to aging. Losing the
ability to respond to a foreign antigen or _(10)_ (responding to
one's own antigens) may be part of the aging process. Progeria
may indicate there is a _(11)_ component to aging.

1. _____

2. _____

3. _____

4. _____

5. _____

6. _____

7. _____

8. _____

9. _____

10. _____

11. _____

 Cerebral death is irreparable brain damage clinically manifested by the absence of
response to stimulation, the absence of respiration, and an isoelectric
electroencephalogram for at least 30 minutes.

1. List the following structures in the order in which they form during development: blastocyst, embryonic disk, mesoderm, morula, primitive streak, zygote.

2. Name the structures derived from the inner cell mass and the trophoblast.

3. Name an organ system that develops entirely from ectoderm.

4. Name two organ systems that develop entirely from mesoderm.

5. List the three kidneys that form during the development of the urinary system.

6. Name the two kinds of chromosomes, and give the number of each in humans.

7. List three circulatory changes that occur at birth.

8. List the eight life stages.

9. List four factors that may influence aging.

MASTERY LEARNING ACTIVITY

Place the letter corresponding to the correct answer in the space provided.

_____ 1. The major development of organs takes
place in the
a. organ period.
b. fetal period.
c. germinal period.
d. embryonic period.

_____ 2. Given the following structures:
1. blastocyst
2. morula
3. zygote

Choose the arrangement that lists the
structures in the order in which they are
formed during development.
a. 1, 2, 3
b. 1, 3, 2
c. 2, 3, 1
d. 3, 1, 2
e. 3, 2, 1

_____ 3. The embryo develops from the
a. inner cell mass.
b. trophoblast.
c. blastocele.
d. yolk sac.

_____ 4. The placenta
a. develops from the trophoblast.
b. allows maternal blood to mix with
embryonic blood.
c. invades the lacunae of the embryo.
d. all of the above

_____ 5. The embryonic disk
a. forms between the amniotic cavity
and the yolk sac.
b. contains the primitive streak.
c. becomes a three-layered structure.
d. all of the above

_____ 6. The brain develops from
a. endoderm.
b. ectoderm.
c. mesoderm.

_____ 7. Most of the skeletal system develops
from
a. endoderm.
b. ectoderm.
c. mesoderm.

_____ 8. Given the following structures:
1. neural crest
2. neural plate
3. neural tube

Choose the arrangement that lists the
structures in the order in which they
form during development.
a. 1, 2, 3
b. 1, 3, 2
c. 2, 1, 3
d. 2, 3, 1
e. 3, 2, 1

_____ 9. The somites give rise to the
a. circulatory system.
b. skeletal muscle.
c. skin.
d. kidneys.

_____ 10. The pericardial cavity forms from
a. evaginations of the early GI tract.
b. the neural tube.
c. the celom.
d. the brachial arches.

_____ 11. The parts of the limbs develop
a. in a proximal-to-distal sequence.
b. in a distal-to-proximal sequence.
c. at approximately the same time.

_____ 12. Concerning development of the face,
a. the face develops by the fusion of
five embryonic structures.
b. the maxillary processes normally
meet at the midline to form the lip.
c. the primary palate forms the roof of
the mouth.
d. cleft palates normally occur to one
side of the midline.

_____ 13. Concerning the development of the heart,
   a. the heart develops from a single tube.
   b. the sinus venosum becomes the SA node.
   c. the foramen ovale lets blood flow from the right to left atrium.
   d. all of the above

_____ 14. Given the following structures:
   1. mesonephros
   2. metanephros
   3. pronephros

   Choose the arrangement that lists the structures in the order in which they form during development.
   a. 1, 2, 3
   b. 1, 3, 2
   c. 2, 3, 1
   d. 3, 1, 2
   e. 3, 2, 1

_____ 15. A study of the early embryo indicates that the penis of the male develops from the same embryonic structure as which of the following female structures?
   a. labia majora
   b. labia minora
   c. uterus
   d. clitoris
   e. vagina

_____ 16. Which hormone causes differentiation of sex organs in the developing male fetus?
   a. FSH
   b. LH
   c. testosterone
   d. estrogen

_____ 17. Which of the following is correctly matched with its definition?
   a. autosome - an X or Y chromosome
   b. phenotype - the genetic makeup of an individual
   c. allele - genes occupying the same locus on homologous chromosomes
   d. heterozygous - having two identical genes for a trait
   e. recessive - a trait expressed when the genes are heterozygous

_____ 18. Following birth
   a. the ductus arteriosus closes.
   b. meconium is discharged from the GI tract.
   c. stomach pH decreases.
   d. all of the above

_____ 19. Which of the following life stages is correctly matched with the time that the stage occurs?
   a. neonate - birth to 1 month after birth
   b. infant - 1 month to 6 months
   c. child - 6 months to 5 years
   d. puberty - 10 to 12 years

_____ 20. Which of the following occurs as one gets older?
   a. Neurons replicate to replace lost neurons.
   b. Skeletal muscle cells replicate to replace lost muscle cells.
   c. Cross-links between collagen molecules increase.
   d. The immune system becomes less sensitive to the body's own antigens.
   e. all of the above

Use a separate sheet of paper to complete this section.

1. If a woman contracts rubella (German measles) while pregnant, the baby may be born with a congenital disorder such as cataracts, deafness, neurologic disturbances, or cardiac malformation. If the mother is infected in the first month of pregnancy versus the third month, what effect would this time difference have on the likelihood that the baby will have a congenital disorder? Explain.

2. A surgical procedure can remove the embryonic yolk sac from early mammalian embryos. How would this affect the development of the reproductive system?

3. Predict the consequences of removing the gonads in an early embryo (8 weeks old) that is a genetic female.

4. The ability to taste PTC (phenyl thiocarbamide) is an inherited trait. If two people, each of whom can taste PTC, have a child that cannot, is the trait for PTC dominant or recessive?

5. A woman who does not have hemophilia marries a man who has the disorder. Determine the genotype of both parents if half of their children have hemophilia.

ANSWERS TO CHAPTER 29

1. polygenic
2. polygenic
3. prenatal, neonatal
4. neonatal
5. blastocyst, blastocele, cytotrophoblast, syncitiotrophoblast, myoblast

6. blastocyst
7. blastocele, celom
8. pluripotent
9. notochord
10. notochord

CONTENT LEARNING ACTIVITY

Prenatal Development

1. Prenatal period
2. Germinal period
3. Embryonic period

4. Fetal period
5. Clinical age

Fertilization

1. Second meiotic division
2. Female pronucleus

3. Male pronucleus
4. Zygote

Early Cell Division: Morula and Blastocyst

1. Pluripotent
2. Morula
3. Blastocyst

4. Trophoblast
5. Inner cell mass

Implantation of the Blastocyst and Development of the Placenta

1. Syncitiotrophoblast
2. Lacunae

3. Cytotrophoblast

Formation of Germ Layers

1. Amniotic cavity
2. Embryonic disk
3. Ectoderm

4. Yolk sac
5. Primitive streak
6. Mesoderm

Neural Tube and Neural Crest Formation

1. Neural plate
2. Neural crests

3. Neural tube
4. Neural crest cells

Somite, Gut, and Body Cavity Formation

1. Somites
2. Somitomeres
3. Cloacal membrane
4. Evaginations

5. Branchial arches
6. Pharyngeal pouches
7. Celom

Limb Bud and Face Development

1. Apical ectodermal ridge
2. Frontonasal process
3. Maxillary processes

4. Nasal placodes
5. Primary palate
6. Secondary palate

Development of the Organ Systems

A. 1. Mesoderm
2. Neural crest cells
3. Mesoderm
4. Neural tube cavity

B. 1. Myoblasts
2. Olfactory bulb and nerve
3. Optic stalk and vesicle
4. Neurohypophysis

C. 1. Thyroid gland
2. Adrenal cortex
3. Adrenal medulla
4. Pancreas

D. 1. Blood islands
2. Sinus venosus
3. Bulbis cordis
4. Foramen ovale

E. 1. Pronephros
2. Mesonephros
3. Cloaca
4. Allantois
5. Metanephros

F. 1. Primordial germ cells
2. Paramesonephric ducts
3. Mesonephric ducts

G. 1. Genital tubercle; urogenital folds
2. Labioscrotal swellings
3. Genital tubercle
4. Urogenital folds

Growth of the Fetus

1. Embryonic
2. Growing
3. Lanugo

4. Vernix
5. Subcutaneous fat

Genetics

1. Gene
2. DNA
3. Structural genes

4. Regulatory genes
5. Congenital disorder
6. Autosomes

Chromosomes

1. Locus
2. Alleles
3. Homozygous

4. Dominant
5. Phenotype
6. Genotype

The Newborn

A. 1. Foramen ovale
    2. Ductus arteriosus
    3. Umbilical vein
    4. Ductus venosus

B. 1. Meconium
    2. pH
    3. Lactose
    4. Amylase
    5. Bilirubin

Life Stages

1. Germinal
2. Embryo
3. Fetus

4. Neonate
5. Child
6. Adolescent

Aging

1. Collagen
2. Decrease
3. Heart
4. Filtration
5. Atherosclerosis
6. Arteriosclerosis

7. Thrombosis
8. Embolus
9. Cytologic aging
10. Autoimmunity
11. Genetic

QUICK RECALL

1. Zygote, morula, blastocyst, embryonic disk, primitive streak, mesoderm
2. Inner cell mass: embryo; trophoblast: placenta and membranes surrounding the embryo.
3. Nervous system
4. Muscular system and circulatory system
5. Pronephros, mesonephros, and metanephros
6. Autosomes: 22 pairs (44); sex chromosomes one pair (2).

7. Ductus arteriosus closes, foramen ovale closes, and umbilical arteries and veins and ductus venosus close.
8. Germinal, embryo, fetus, neonate, infant, child, adolescent, adult
9. Cross-linking of collagen, loss of functional cells, atherosclerosis and arteriosclerosis, cytological aging, immune changes, and genetic components

1. D. Most organ development takes place in the embryonic period (second to eighth week). The germ layers develop in the germinal period (first 2 weeks) and growth and maturation occur in the fetal period (the last 7 months).

2. E. Fertilization of the secondary oocyte by a spermatozoa produces the single-celled zygote. After several divisions the morula is formed, which continues to develop and becomes the blastocyst.

3. A. The embryo develops from the inner cell mass, which is several layers of cells at one end of the blastocyst.

4. A. The placenta develops from the trophoblast. Trophoblast cells protrude into cavities (lacunae) formed within maternal blood vessels. However, there is no mixing of maternal and embryonic blood.

5. D. The embryonic disk is located in the inner cell mass between the amniotic cavity and yolk sac. It consists of an ectoderm and endoderm. The primitive streak forms in the embryonic disk, followed by development of the mesoderm. Thus the embryonic disk has three germ layers: ectoderm, mesoderm, and endoderm.

6. B. The brain, spinal cord, and nerves develop from ectoderm.

7. C. Most of the skeleton is derived from mesoderm. Same of the facial bones are derived from ectoderm.

8. C. The neural plate rises to form the neural crests, which come together to form the neural tube.

9. B. The somites develop into skeletal muscle, the vertebral column, and a portion of the skull.

10. C. The celom, an isolated cavity within the embryo, gives rise to the pericardial, pleural, and peritoneal cavities. Evaginations of the GI tract become, for example, the lungs, liver, anterior pituitary, and the urinary bladder.

11. A. The limb parts develop in a proximal-to-distal sequence.

12. A. The face develops from five embryonic structures. The maxillary processes fuse with the frontonasal process to form the lip and upper jaw. The secondary palate forms the roof of the mouth. A cleft palate normally occurs along the midline.

13. D. The heart develops from a single tube divided into the sinus venosus, an atrium, a ventricle, and the bulbus cordis. The sinus venosus initiates contraction of the tube heart and later becomes the SA node (pacemaker). The foramen ovale allows blood to bypass the lungs.

14. D. Three kidneys form during development in the following order: pronephros, mesonephros, and metanephros. The metanephros becomes the functional kidney, and the others degenerate.

15. D. Many of the sexual organs in males and females have the same embryonic origin but differentiate during development. The penis and clitoris form from the genital tubercle.

16. C. Testosterone stimulates the development of male sexual characteristics. Without testosterone female characteristics are formed.

17. C. Chromosomes come in pairs called homologues, and the location of the gene for a given trait on the chromosomes is called a locus. Alleles are genes for a given trait found at the same locus.

18. D. The ductus arteriosus closes to prevent blood flow from the pulmonary trunk to the aorta. It later becomes the ligamentum arteriosum. Meconium is a mixture of amniotic fluid, mucus, and bile. Once the alkaline amniotic fluid is removed and gastric secretions increase, stomach pH decreases.

19. A. The correct times for each stage are neonate (birth to 1 month), infant (1 month to 1 or 2 years), child (1 or 2 years to puberty), adolescent (puberty to 20 years), and adult (20 years to death). Puberty in females is about 11 to 13 years and in males about 12 to 15 years.

20. C. The cross-link between collagen molecules increases, making tissues less plastic. Neurons and skeletal muscle don't replicate after birth and the immune system becomes more sensitive to self-antigens resulting in autoimmune disease.

 FINAL CHALLENGES

1. During the time of formation of the organ systems (first 2 months of development), the individual is most susceptible to damage. About 50% develop congenital malformations if the infection occurs in the first month, 25% in the second month, and 10% if the infection occurs later in the pregnancy.

2. The yolk sac gives rise to the cells that eventually produce gametes (oocytes or spermatozoa). These cells migrate from the yolk sac to the gonadal ridge during development. If the yolk sac is destroyed before the cells have migrated, the animal will be sterile.

3. In the normal male the embryonic gonads secrete testosterone and Mullerian-inhibiting hormone. These hormones cause masculinization of the mesonephric duct system and the external genitalia. If the gonads are removed before 8 weeks of development, the embryo will develop morphologically as a female regardless of the its genotype.

4. The trait is dominant. The parents must be heterozygous (Tt), and the child homozygous-recessive (tt).

5. Hemophilia is an X-linked trait. Since the father has hemophilia he must be XhY. If the mother were XHXH, all their offspring would be normal. Since half the children have hemophilia, she must be XHXh.